Diffraction-Limited Imaging with Large and Moderate Telescopes

Swapan K. Saha
Indian Institute of Astrophysics
Bangalore, India

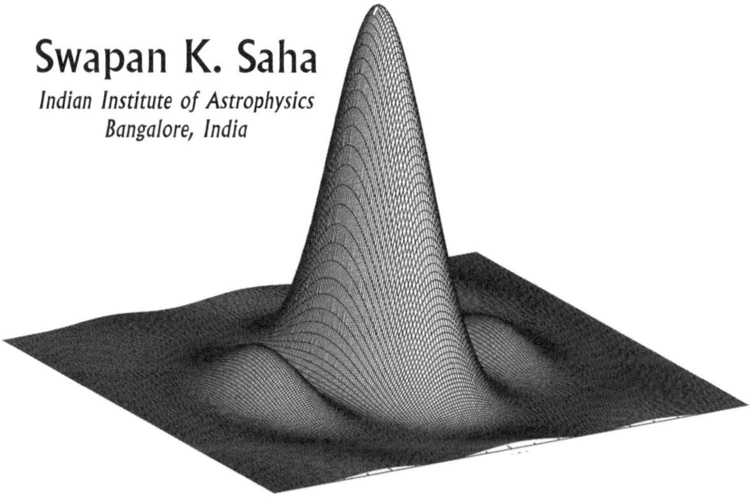

Diffraction-Limited Imaging with Large and Moderate Telescopes

World Scientific

NEW JERSEY • LONDON • SINGAPORE • BEIJING • SHANGHAI • HONG KONG • TAIPEI • CHENNAI

Published by

World Scientific Publishing Co. Pte. Ltd.

5 Toh Tuck Link, Singapore 596224

USA office: 27 Warren Street, Suite 401-402, Hackensack, NJ 07601

UK office: 57 Shelton Street, Covent Garden, London WC2H 9HE

British Library Cataloguing-in-Publication Data
A catalogue record for this book is available from the British Library.

DIFFRACTION-LIMITED IMAGING WITH LARGE AND MODERATE TELESCOPES

ISBN-13 978-981-270-777-2
ISBN-10 981-270-777-8

Printed in Singapore by World Scientific Printers (S) Pte Ltd

In memory of my wife,
KALYANI

Preface

Diffraction-limited image of an object is known as the image with a resolution limited by the size of the aperture of a telescope. Aberrations due to an instrumental defect together with the Earth's atmospheric turbulence set severe limits on angular resolution to $\sim 1''$ in optical wavelengths. Both the sharpness of astronomical images and the signal-to-noise (S/N) ratios (hence faintness of objects that can be studied) depend on angular resolution, the latter because noise comes from the sky as much as is in the resolution element. Hence reducing the beam width from, say, 1 arcsec to 0.5 arcsec reduces sky noise by a factor of four. Two physical phenomena limit the minimum resolvable angle at optical and infrared (IR) wavelengths – diameter of the collecting area and turbulence above the telescope, which introduces fluctuations in the index of refraction along the light beam. The cross-over between domination by aperture size ($\sim 1.22\lambda$/aperture diameter, in which λ is the wavelength of light) and domination by atmospheric turbulence ('seeing') occurs when the aperture becomes somewhat larger than the size of a characteristic turbulent element, that is known as atmospheric coherence length, r_0 (e.g. at 10- 30 cm diameter). Light reaching the entrance pupil of a telescope is coherent only within patches of diameters of order r_0. This limited coherence causes blurring of the image, blurring that is modeled by a convolution with the point-spread function (PSF), which prevents the telescope from reaching into deep space to unravel the secrets of the universe. The deployment of a space-bound telescope beyond the atmosphere circumvents the problem of atmosphere, but the size and cost of such a venture are its shortcomings.

This book has evolved from a series of talks given by the author to a group of senior graduate students about a decade ago, following which, a couple of large review articles were published. When Dr. K. K. Phua invited the author, for which he is indebted to, for writing a lecture note based on these articles, he took the opportunity to comply; a sequel of this note is also under preparation. This book is aimed to benefit graduate students, as well as researchers who intend to embark on a field dedicated to the high resolution techniques, and would serve as an interface between the astrophysicists and the physicists. Equipped with about two hundred illustrations and tens of footnotes, which make the book self-content, it addresses the basic principles of interferometric techniques in terms of both post-processing and on-line imaging that are applied in optical/IR astronomy using ground-based single aperture telescopes; several fundamental equations, Fourier optics in particular, are also highlighted in the appendices.

Owing to the diffraction phenomenon, the image of the point source (unresolved stars) cannot be smaller than a limit at the focal plane of the telescope. Such a phenomenon can be seen in water waves that spread out after they pass through a narrow aperture. It is present in the sound waves, as well as in the electro-magnetic spectrum starting from gamma rays to radio waves. The diffraction-limited resolution of a telescope refers to optical interference and resultant image formation. A basic understanding of interference phenomenon is of paramount importance to other branches of physics and engineering too. Chapters 1 through 3 of this book address the fundamentals of electromagnetic fields, wave optics, interference, and diffraction at length. In fact, a book of this kind calls for more emphasis on imaging phenomena and techniques, hence the fourth chapter discusses at length the imaging aspects of the same.

Turbulence and the concomitant development of thermal convection in the atmosphere distort the phase and amplitude of the incoming wavefront of the starlight; longer the path, more the degradation that the image suffers. Environment parameters, such as fluctuations in the refractive index of the atmosphere along the light beam, which, in turn, are due to density variations associated with thermal gradients, variation in the partial pressure of water vapour, and wind shear, produce atmospheric turbulence. Random microfluctuations of such an index cause the fluctuation of phase in the incoming random field and thereby, produce two dimensional interferences at the focus of the telescope. These degraded images are the product of dark and bright spots, known as speckles. The fifth chapter enumerates the origin, properties, and optical effects of turbulence in the Earth's atmosphere.

One of the most promising developments in the field of observational astronomy in visible waveband is the usage of speckle interferometry (Labeyrie, 1970) offering a new way of utilizing the large telescopes to obtain diffraction-limited spatial Fourier spectrum and image features of the object. Such a technique is entirely accomplished by *a posteriori* mathematical analysis of numerous images of the same field, each taken over a very short time interval. In recent years, a wide variety of applications of speckle patterns has been found in many areas. Though the statistical properties of the speckle pattern is complicated, a detailed analysis of this pattern is useful in information processing. Other related concerns, such as pupil plane interferometry, and hybrid methods (speckle interferometry with non-redundant pupils), have also contributed to a large extent. Chapter 6 enumerates the details of these post-detection diffraction-limited imaging techniques, as well as the relationship between image-plane techniques and pupil-plane interferometry.

Another development in the field of high angular resolution imaging is to mitigate the effects of the turbulence in real time, known as adaptive optics (AO) system. Though such a system is a late entry among the list of current technologies, it has given a new dimension to this field. In recent years, the technology and practice of such a system have become, if not in commonplace, at least well known in the defence and astronomical communities. Most of the astronomical observatories have their own AO programmes. Besides, there are other applications, namely vision research, engineering processing, and line-of-sight secure optical communications. The AO system is based on a hardware-oriented approach, which employs a combination of deformation of reflecting surfaces (i.e., flexible mirrors) and post-detection image restoration. A brief account of the development of such an innovative technique is presented in chapter 7.

The discovery of the corpuscular nature of light, beyond the explanation of the photo-electric effect, by Albert Einstein almost 100 years ago, in 1905, has revolutionized the way ultra-sensitive light detectors are conceived. Such a discovery has far reaching effects on the astrophysical studies, in general, and observational astronomy, in particular. The existence of a quantum limit in light detection has led to a quest, through the 20th century (and still going on), for the perfect detector which is asymptotically feasible. The advent of high quantum efficiency photon counting systems, vastly increases the sensitivity of high resolution imaging techniques. Such systems raise the hope of making diffraction-limited images of objects as faint as $\sim 15-16\ m_v$ (visual magnitude). Chapter 8 elucidates the development of various detectors that are being used for high resolution imaging.

It is well known that standard autocorrelation technique falls short of providing reconstruction of a true image. Therefore, the success of single aperture interferometry has encouraged astronomers to develop further image processing techniques. These techniques are indeed an art and for most part, are post-detection processes. A host of image reconstruction algorithms have been developed. The adaptive optics system also requires such algorithms since the real-time corrected images are often partial. The degree of compensation depends on the accuracy of the wavefront estimate, the spacing of the actuators in the mirror, and other related factors. The mathematical intricacies of the data processing techniques for both Fourier modulus and Fourier phase are analyzed in chapter 9. Various schemes of image restoration techniques are examined as well, with emphasis set on their comparisons.

Stellar physics is the study of physical makeup evolutionary history of stars, which is based on observational evidence gathered with telescopes collecting electromagnetic radiation. Single aperture high resolution techniques became an extremely active field scientifically with important contributions made to a wide range of interesting problems in astrophysics. A profound increase has been noticed in the contribution of such techniques to measure fundamental stellar parameters and to uncover details in the morphology of a range of celestial objects, including the Sun and planets. They have been used to obtain separation and position angle of close binary stars, to measure accurate diameter of a large number of giant stars, to determine shapes of asteroids, to resolve Pluto-Charon system, to map spatial distribution of circumstellar matter surrounding objects, to estimate sizes of expanding shells around supernovae, to reveal structures of active galactic nuclei (AGN) and of compact clusters of a few stars like R 136a complex, and to study gravitationally lensed QSO's. Further benefits have been witnessed from the application of adaptive optics systems of large telescopes, in spite of its limited capability of retrieving fully diffraction-limited images of these objects. The last two chapters (10 and 11) discuss the fundamentals of astronomy and applications of single aperture interferometry.

The author expresses his gratitude to many colleagues, fellow scientists, and graduate students at Indian Institute of Astrophysics and elsewhere, particularly to A. Labeyrie, J. C. Bhattacharyya, and M. K. Das Gupta (late) for their encouragement and to Luc Damé, A. K. Datta, L. N. Hazra, Sucharita Sanyal, Kallol Bhattacharyya, P. M. S. Namboodiri, N. K. Rao, G. C. Anupama, A. Satya Narayana, K. Sankar Subramanian, B. S. Nagabhushana, Bharat Yerra, K. E. Rangarajan, V. Raju, D. Som, and A. Vyas, for assistance as readers of draft chapters. He is indebted to S. C Som for careful editing of preliminary chapters. Thanks are also due to V. Chinnappan, A. Boccaletti, T. R. Bedding, S. Koutchmy, Y. Y. Balega, S. Morel, A. V. Raveendran, L. Close, M. Wittkowski, R. Osterbart, J. P. Lancelot, B. E. Reddy, P. Nisenson (late), R. Sridharan, K. Nagaraju and A. Subramaniam, for providing the images, figures etc., and granting permission for their reproduction. The services rendered by B. A. Varghese, P. Anbazhagan, V. K. Subramani, K. Sundara Raman, R. T. Gangadhara, D. Mohan, S. Giridhar, R. Srinivasan, L. Yeswanth, and S. Mishra are gratefully acknowledged.

Swapan K. Saha

Principal symbols

$\vec{E}$	Electric field vector
$\vec{B}$	Magnetic induction
$\vec{H}$	Magnetic vector
$\vec{D}$	Electric displacement vector
$\vec{J}$	Electric current density
$\vec{r}(=x,y,z)$	Position vector of a point in space
σ	Specific conductivity
μ	Permeability of the medium
ϵ	Permittivity or dielectric
q	Charge
$\vec{F}$	Force
$\vec{v}$	Velocity
$\vec{p}$	Momentum
$\vec{a}$	Acceleration
e	Electron charge
$S(\vec{r},t)$	Poynting vector
$V(\vec{r},t)$	Monochromatic optical wave
$\Re$ and $\Im$	Real and imaginary parts of the quantities in brackets
t	Time
κ	Wave number
ν	Frequency of the wave
A	Complex amplitude of the vibration
$U(\vec{r},t)$	Complex representation of the analytical signal
$I(\vec{x})$	Intensity of light
I_ν	Specific intensity
$\langle\ \rangle$	Ensemble average
$*$	Complex operator

λ	Wavelength
$\vec{x} = (x, y)$	Two-dimensional space vector
$P(\vec{x})$	Pupil transmission function
$\star$	Convolution operator
$\widehat{\ }$	Fourier transform operator
$\widehat{P}(\vec{u})$	Pupil transfer function
$S(\vec{x})$	Point spread function
$\widehat{S}(\vec{u})$	Optical transfer function
$\lvert\widehat{S}(\vec{u})\rvert^2$	Modulus transfer function
$\mathcal{R}$	Resolving power of an optical system
ω	Angular frequency
T	Period
$\vec{V}_j$	Monochromatic wave vector
j	= 1, 2, 3
$\mathcal{J}_{12}$	Interference term
$\Delta\varphi$	Optical path difference
λ_0	Wavelength in vacuum
c	Velocity of light
$\vec{\gamma}(\vec{r}_1, \vec{r}_2, \tau)$	Complex degree of (mutual) coherence
$\vec{\Gamma}(\vec{r}_1, \vec{r}_2, \tau)$	Mutual coherence
$\vec{\Gamma}(\vec{r}, \tau)$	Self coherence
τ_c	Temporal width or coherence time
$\Delta\nu$	Spectral width
l_c	Coherence length
$\vec{\gamma}(\vec{r}_1, \vec{r}_2, 0)$	Spatial coherence
$J(\vec{r}_1, \vec{r}_2)$	Mutual intensity function
$\mu(\vec{r}_1, \vec{r}_2)$	Complex coherence factor
$\mathcal{V}$	Contrast of the fringes
f	Focal length
v_a	Average velocity of a viscous fluid
l	Characteristic size of viscous fluid
R_e	Reynolds number
$n(\vec{r}, t)$	Refractive index of the atmosphere
$\langle\sigma\rangle$	Standard deviation
m_v	Apparent visual magnitude
M_v	Absolute visual magnitude
$L_\odot$	Solar luminosity
$L_\star$	Stellar luminosity

$M_\odot$	Solar mass
$M_\star$	Stellar mass
$R_\odot$	Solar radius
$R_\star$	Stellar radius
$\langle\sigma\rangle^2$	Variance
k_B	Boltzmann constant
g	Acceleration due to gravity
H	Scale height
n_0	Mean refractive index of air
P	Pressure
T	Temperature
ε	Energy dissipation
$\Phi_n(\vec{k})$	Power spectral density
k_0	Critical wave number
l_0	Inner scale length
k_{l_0}	Spatial frequency of inner scale
$\mathcal{C}_n^2$	Refractive index structure constant
$\mathcal{D}_n(\vec{r})$	Refractive index structure function
$\mathcal{B}_n(\vec{r})$	Covariance function
$\mathcal{D}_v(\vec{r})$	Velocity structure function
$\mathcal{C}_v^2$	Velocity structure constant
$\mathcal{D}_T(\vec{r})$	Temperature structure function
$\mathcal{C}_T^2$	Temperature structure constant
h	Height
$(\vec{x}, h)$	Co-ordinate
$\Psi_h(\vec{x})$	Complex amplitude at co-ordinate, $(\vec{x}, h)$
$\langle\psi_h(\vec{x})\rangle$	Average value of the phase at h
δh_j	Thickness of the turbulence layer
$\mathcal{D}_{\psi_j}(\vec{\xi})$	Phase structure function
$\mathcal{D}_n(\vec{\xi}, \zeta)$	Refractive index structure function
$\mathcal{B}_{h_j}(\vec{\xi})$	Covariance of the phase
$\mathcal{B}(\vec{\xi})$	Coherence function
γ	Distance from the zenith
r_0	Fried's parameter
$O(\vec{x})$	Object illumination
$\left\langle \widehat{S}(\vec{u}) \right\rangle$	Transfer function for long-exposure images
$\vec{u}$	Spatial frequency vector with magnitude u

$\widehat{I}(\vec{u})$	Image spectrum
$\widehat{O}(\vec{u})$	Object spectrum
$B(\vec{u})$	Atmosphere transfer function
$\mathcal{T}(\vec{u})$	Telescope transfer function
$F\#$	Aperture ratio
F	Flux density
$arg\vert\ \vert$	The phase of ' '
p_j	Sub-apertures
β_{123}	Closure phase
θ_i, θ_j	Error terms introduced by errors at the individual antennae
$\mathcal{A}\delta(\vec{x})$	Dirac impulse of a point source
$\otimes$	Correlation
$\widehat{\mathcal{N}}(\vec{u})$	Noise spectrum
$\left\langle \vert\widehat{I}(\vec{u})\vert^2 \right\rangle$	Image energy spectrum
θ_j	Apertures
UBV	Johnson photometric system
$B(T)$	Brightness distribution

List of acronyms

AAT	Anglo-Australian telescope
A/D	Analog-to-digital
AGB	Asymptotic giant branch
AGN	Active galactic nuclei
AMU	Atomic mass unit
AO	Adaptive optics
ASM	Adaptive secondary mirror
ATF	Atmosphere Transfer Function
BC	Babinet compensator
BDM	Bimorph deformable mirror
BID	Blind iterative deconvolution
BLR	Broad-line region
CCD	Charge Coupled Device
CFHT	Canada French Hawaii telescope
CHARA	Center for high angular resolution astronomy
CS	Curvature sensor
DM	Deformable mirror
EMCCD	Electron multiplying CCD
ESA	European space agency
ESO	European Southern Observatory
ESPI	Electronic speckle pattern interferometry
FOV	Field-of-view
DFT	Discrete Fourier Transform
FFT	Fast Fourier Transform
FT	Fourier Transform
FWHM	Full width at half maximum
Hz	Hertz

HF	High frequency
HR	Hertzsprung-Russell
HST	Hubble space telescope
ICCD	Intensified CCD
IDL	Interactive Data Language
IMF	Initial mass function
IR	Infrared
I2T	Interféromètre à deux Télescopes
KT	Knox-Thomson
kV	Kilovolt
laser	Light Amplification by Stimulated Emission of Radiation
LBOI	Long baseline optical interferometers
LBT	Large Binocular Telescope
LC	Liquid crystal
LF	Low frequency
LGS	Laser guide star
LHS	Left Hand Side
LSI	Lateral shear interferometer
L3CCD	Low light level CCD
maser	Microwave Amplification by Stimulated Emission of Radiation
MCAO	Multi-conjugate adaptive optics
MCP	Micro-channel plate
MEM	Maximum entropy method
MHz	Megahertz
MISTRAL	Myopic iterative step preserving algorithm
MMDM	Micro-machined deformable mirror
MMT	Multi mirror telescope
MOS	Metal-oxide semiconductor
MTF	Modulus Transfer Function
NGS	Natural guide star
NICMOS	Near Infrared Camera and Multi-Object Spectrograph
NLC	Nematic liquid crystal
NLR	Narrow-line region
NRM	Non-redundant aperture masking
NTT	New Technology Telescope
OPD	Optical Path Difference
OTF	Optical Transfer Function
PAPA	Precision analog photon address

PHD	Pulse height distribution
PMT	Photo-multiplier tube
PN	Planetary nebula
PSF	Point Spread Function
PTF	Pupil Transmission Function
PZT	Lead-zirconate-titanate
QE	Quantum efficiency
QSO	Quasi-stellar object
RA	Right Ascension
RHS	Right Hand Side
RMS	Root Mean Square
SAA	Shift-and-add
SDC	Static dielectric cell
SLC	Smectic liquid crystal
SH	Shack-Hartmann
SL	Shoemaker-Levy
SN	Supernova
S/N	Signal-to-noise
SOHO	Solar and heliospheric observatory
SUSI	Sydney University Stellar Interferometer
TC	Triple-correlation
TTF	Telescope Transfer Function
UV	Ultraviolet
VBO	Vainu Bappu Observatory
VBT	Vainu Bappu Telescope
VTT	Vacuum Tower Telescope
WFP	Wiener filter parameter
WFS	Wavefront sensor
YSO	Young stellar objects

Contents

Chapter 1

Introduction to electromagnetic theory

1.1 Introduction

Electromagnetism is a fundamental physical phenomena that is basic to many areas science and technology. This phenomenon is due to the interaction, called electromagnetic interaction, of electric and magnetic fields with the constituent particles of matter. This interaction is physically described in terms of electromagnetic fields, characterized by the electric field vector, $\vec{E}$ and the magnetic induction, $\vec{B}$. These field vectors are generally time-dependent as they are determined by the positions of the electric charges and their motions (currents) in a medium in which the electromagnetic field exists. The fields $\vec{E}$ and $\vec{B}$ are directly correlated by Ampère-Maxwell and Faraday-Henry laws that satisfy the requirements of special relativity. The time-dependent relations between the time-dependent vectors in these laws and Gauss' laws for electric and magnetic fields are given by Maxwell's equations that form the the basis of electromagnetic theory.

The electric charge and current distributions enter into these equations and are called the sources of the electromagnetic field, because if they are given Maxwell's equations may be solved for $\vec{E}$ and $\vec{B}$ under appropriate boundary conditions.

1.2 Maxwell's equations

In order to describe the effect of the electromagnetic field on matter, it is necessary to make use, apart from $\vec{E}$ and $\vec{B}$, of a set another three field vectors, viz., the magnetic vector, $\vec{H}$, the electric displacement vector, $\vec{D}$, and the electric current density, $\vec{J}$. The four Maxwell's equations may be written either in integral form or in differential form. In differential form, the Maxwell's equations are expressed as,

$$\nabla \times E(\vec{r},t) = -\frac{1}{c}\left[\frac{\partial B(\vec{r},t)}{\partial t}\right], \tag{1.1}$$

$$\nabla \times H(\vec{r},t) = \frac{1}{c}\left[4\pi J(\vec{r},t) + \frac{\partial D(\vec{r},t)}{\partial t}\right], \tag{1.2}$$

$$\nabla \cdot D(\vec{r},t) = 4\pi\rho(\vec{r},t) \qquad \text{and} \tag{1.3}$$

$$\nabla \cdot B(\vec{r},t) = 0. \tag{1.4}$$

In these equations, $c = 2.99,79 \times 10^8$meter(m)/second(s) is the velocity of light in free space, ρ the volume charge density, and Gaussian units are used for expressing the vector quantities, and ∇ represents a vector differential operator,

$$\nabla = \vec{i}\frac{\partial}{\partial x} + \vec{j}\frac{\partial}{\partial y} + \vec{k}\frac{\partial}{\partial z}.$$

The unit of the electric field intensity, $\vec{E}$, is expressed as volt (V) m^{-1}, and that for the magnetic flux density $|\vec{B}|$, tesla ($\mathrm{T} = \mathrm{Wb\,m}^{-2}$) in which $|\,|$ stands for the modulus. Equations (1.1 -1.4) represent Faraday-Henry law of induction, Ampére's law with the displacement current introduced by Maxwell, known as Ampére-Maxwell law, Gauss' electric and magnetic laws respectively. It is further assumed that the space and time derivatives of the field vectors are continuous at every point $(\vec{r},t)$ where the physical properties of the media are continuous. In order to describe the interaction of light with matter at thermal equilibrium, the Maxwell's equations are substituted by the additional equations,

$$\vec{J} = \sigma\vec{E}, \tag{1.5}$$

$$\vec{B} = \mu\vec{H}, \tag{1.6}$$

$$\vec{D} = \epsilon_0\vec{E}, \tag{1.7}$$

where σ is the specific conductivity, μ the permeability of the medium in which magnetic field acts, and $\epsilon_0 (= 8.8541 \times 10^{-12}$farads (F)/m) the permittivity or dielectric constant at vacuum.

Equations (1.5 - 1.7) describe the behavior of substances under the influence of the field. These relations are known as material equations. The electric and magnetic fields are also present in matter giving rise to the relations (in standard notation),

$$\vec{E}_m = \vec{E} + \frac{\vec{P}}{\epsilon_0}, \tag{1.8}$$

$$\vec{B}_m = \vec{B} + \mu_0 \vec{M}, \tag{1.9}$$

where $\vec{E}_m$ is the electric field corresponding to the dielectric displacement in *volts*(V) m^{-1}, $\vec{B}_m$ the magnetic field in the presence of medium, $\vec{P}$ the polarization susceptibility, $\vec{M}$ the magnetization, and $\mu_0 (= 4\pi k = 4\pi \times 10^{-7}$henrys (H)/m), the permeability in free space or in vacuum, and k the constant of proportionality.

In a medium of free space, by using the integral form of Gauss' electric law,

$$\int_{\mathrm{S}} \vec{E} \cdot \vec{n} d\mathrm{S} = 4\pi q, \tag{1.10}$$

and the relation between $\vec{E}$ and φ, i.e.,

$$E(\vec{r}) = -\nabla \varphi(\vec{r}), \tag{1.11}$$

the Poisson (S. D. Poisson, 1781-1840) partial differential equation for φ is obtained,

$$\nabla^2 \varphi = -4\pi \rho(\vec{r}), \tag{1.12}$$

in which the Lapacian operator, ∇^2, in Cartesian coordinates reads,

$$\nabla^2 = \frac{\partial^2}{\partial x^2} + \frac{\partial^2}{\partial y^2} + \frac{\partial^2}{\partial z^2}. \tag{1.13}$$

The equation (1.12) relates the electric potential $\varphi(\vec{r})$ with its electric charge $\rho(\vec{r})$. In regions of empty of charge, this equation turns out to be homogeneous, i.e.,

$$\nabla^2 \varphi = 0. \tag{1.14}$$

This expression is known as the Laplace (P. S. de Laplace, 1749-1827) equation.

1.2.1 *Charge continuity equation*

Maxwell added the second term of the right hand side (RHS) of equation (1.2), which led to the continuity equation. By taking divergence on both sides of the said equation (1.2),

$$\nabla \cdot (\nabla \times H(\vec{r}, t)) = \frac{4\pi}{c} \nabla \cdot J(\vec{r}, t) + \frac{1}{c} \nabla \cdot \frac{\partial D(\vec{r}, t)}{\partial t}. \tag{1.15}$$

Using the vector equation, $\nabla \cdot (\nabla \times \vec{A}) = 0$ for any vector field $\vec{A}$, the equation (1.15) translates into,

$$\nabla \cdot J(\vec{r}, t) = -\frac{1}{4\pi} \nabla \cdot \frac{\partial D(\vec{r}, t)}{\partial t}. \tag{1.16}$$

By substituting the equation (1.3) into equation (1.16), the following relationship emerges,

$$\nabla \cdot \frac{\partial D(\vec{r}, t)}{\partial t} = \frac{\partial \rho(\vec{r}, t)}{\partial t}. \tag{1.17}$$

The volume charge density, ρ and the current density, $J(\vec{r}, t)$ are the sources of the electromagnetic radiation[1]. The current density $\vec{J}$ associated with a charge density ρ moving with a velocity $\vec{v}$ is $\vec{J} = \rho \vec{v}$.

On replacing the value of $\nabla \cdot \partial \vec{D}/\partial t$ from the equation (1.16) in equation (1.17), one obtains,

$$\nabla \cdot J(\vec{r}, t) = -\frac{\partial \rho(\vec{r}, t)}{\partial t}. \tag{1.18}$$

Thus the equation of continuity is derived as,

$$\nabla \cdot \vec{J} + \frac{\partial \rho}{\partial t} = 0. \tag{1.19}$$

Equation (1.19) expresses the fact that the charge is conserved in the neighborhood of any point. By integrating this equation with the help of Gauss'

[1]Electromagnetic radiation is emitted or absorbed when an atom or a molecule moves from one energy level to another. It has a continuous energy spectrum, a graph, that depicts the intensity of light being emitted over a range of energies. This radiation may be arranged in a spectrum according to its frequency ranging from very high frequencies to the lowest frequencies. The highest frequencies, known as gamma rays whose frequencies range between 10^{19} to 10^{21} Hz ($\lambda \sim 10^{-11} - 10^{-13}$ m), are associated with cosmic sources. The other sources are being the gamma decay of radioactive materials and nuclear fission. The frequency range for X-ray falls between 10^{17} to 10^{19} Hz ($\lambda \sim 10^{-9} - 10^{-11}$ m), which is followed by ultraviolet with frequencies between 10^{15} to 10^{17} Hz ($\lambda \sim 10^{-7} - 10^{-9}$ m). The frequencies of visible light fall between 10^{14} and 10^{15} Hz ($\lambda \sim 10^{-6} - 10^{-7}$ m). The infrared frequencies are 10^{11} to 10^{14} Hz ($\lambda \sim 10^{-3} - 10^{6}$ m); heat radiation is the source for infrared frequencies. The lower frequencies such as radio waves having frequencies 10^4 to 10^{11} Hz ($\lambda \sim 10^4 - 10^{-3}$ m) and microwave (short high frequency radio waves with wavelength 1 mm-30 cm) are propagated by commutated direct-current sources. Only the optical and portions of the infrared and radio spectrum can be observed at the ground.

theorem,

$$\frac{d}{dt}\int_{\mathrm{V}} \rho d\mathrm{V} + \int_{\mathrm{S}} \vec{J} \cdot \vec{n} d\mathrm{S} = 0. \tag{1.20}$$

The charged particle is a small body with a charge density ρ and the total charge, $q = \int_{\mathrm{V}} \rho d\mathrm{V}$, contained within the domain can increase due to flow of electric current,

$$i = \int_{\mathrm{S}} \vec{J} \cdot \vec{n} d\mathrm{S}. \tag{1.21}$$

It is important to note that all the quantities that figure in the Maxwell's equations, as well as in the equation of continuity are evaluated in the rest frame of the observer and all surfaces and volumes are held fixed in that frame.

1.2.2 *Boundary conditions*

In free space, or vacuum, the vectors are $\vec{E}$ and $\vec{H}$, and the relations between the vectors $\vec{E}$, $\vec{B}$, $\vec{D}$, and $\vec{H}$ in a material are derived from the equations (1.6) and (1.7),

$$\begin{aligned} D(\vec{r}, t) &= \epsilon E(\vec{r}, t) = \epsilon_r \epsilon_0 E(\vec{r}, t), \\ H(\vec{r}, t) &= \frac{1}{\mu} B(\vec{r}, t) = \frac{1}{\mu_r \mu_0} B(\vec{r}, t), \end{aligned} \tag{1.22}$$

where ϵ is the permittivity of the medium in which the electric field acts, $\epsilon_r = \epsilon/\epsilon_0$, and $\mu_r = \mu/\mu_0$ the respective relative permittivity and permeability.

It is assumed that both ϵ and μ in equation (1.22) are independent of position ($\vec{r}$) and time (t), and that $\epsilon_r \geq 1, \mu_r \geq 1$. The field vectors can be determined in regions of space (Figure 1.1a) where both ϵ and μ are continuous functions of space from the set of Maxwell's equations, as well as from the material equations. From the Maxwell equation, $\nabla \cdot \vec{B} = 0$, one may write,

$$\int_{\mathrm{V}} \nabla \cdot \vec{B} d\mathrm{V} = 0. \tag{1.23}$$

Equation (1.23) implies the flux into the volume element is equal to the flux out of the volume. For a flat volume whose faces can be neglected, the integral form of Gauss' magnetic law may be written,

$$\oint_S \vec{B} \cdot \vec{n} dS = 0. \tag{1.24}$$

Similarly, the other Maxwell equation $\nabla \cdot \vec{D} = \rho$, may also be used. With boundary conditions at the interface between two different media, i.e., when the physical properties of the medium are discontinuous, the electromagnetic fields within a bounded region are given by,

$$\vec{n} \cdot (\vec{B}_2 - \vec{B}_1) = 0, \tag{1.25}$$

$$\vec{n} \cdot (\vec{D}_2 - \vec{D}_1) = \rho, \tag{1.26}$$

in which $\vec{n}$ is the unit vector normal (a line perpendicular to the surface) to the surface of discontinuity directed from medium 1 to medium 2.

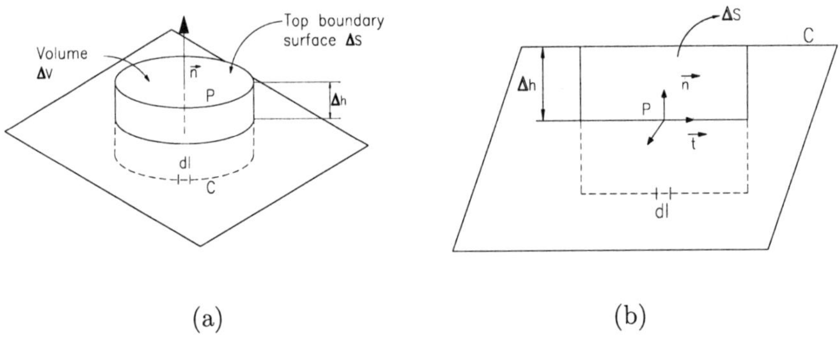

Fig. 1.1 Boundary conditions for (a) the normal components of the electromagnetic field, and (b) the tangential components of the said field.

Equations (1.25 and 1.26) may be written as,

$$B_{2n} - B_{1n} = 0, \tag{1.27}$$

$$D_{2n} - D_{1n} = \rho, \tag{1.28}$$

where $B_n = \vec{n} \cdot \vec{B}$ and the subscript n signifies the component normal to the boundary surface.

Equations (1.27) and (1.28) are the boundary conditions for the normal components of $\vec{B}$ and $\vec{D}$, respectively. The normal component of the magnetic induction is continuous, while the normal component of the electric displacement changes across the boundary as a result of surface charges. From the Ampére-Maxwell law, the condition for $\vec{H}$ can also be derived.

Choosing the integration path in a way that the unit vector is tangential to the interface between the media (Figure 1.1b). The integral form of the equation after applying Stokes formula yields,

$$\vec{t} \times (\vec{H}_2 - \vec{H}_1) = \frac{4\pi}{c}\vec{J}_s, \tag{1.29}$$

where $\vec{t}$ signifies the unit vector tangential to the interface between the media, and $\vec{J}_s$ the surface density of current tangential to the interface, locally perpendicular to both $\vec{t}$ and $\vec{n}$.

Similarly, for a static case a corresponding equation for the tangent component of electric field, $\oint_C \vec{E} \cdot d\vec{l} \equiv 0$, is written as,

$$\vec{t} \times (\vec{E}_2 - \vec{E}_1) = 0. \tag{1.30}$$

Equations (1.29 and 1.30) demonstrate respectively that the tangential components of the electric field vector are continuous across the boundary and the tangential component of the magnetic vector changes across the boundary as a result of a surface current density.

Since $\vec{B} = \mu\vec{H}$, from the equation (1.25), one obtains,

$$\mu_1(\vec{H}_1 \cdot \vec{n}) = \mu_2(\vec{H}_2 \cdot \vec{n}), \tag{1.31}$$

and for the normal component,

$$(\vec{H}_1)_n = \frac{\mu_2}{\mu_1}(\vec{H}_2)_n. \tag{1.32}$$

In the case of the equation of continuity for electric charge (equation 1.19), the boundary condition is given by,

$$\vec{n} \cdot (\vec{J}_2 - \vec{J}_1) + \nabla_s \cdot \vec{J}_s = -\frac{\partial \rho_s}{\partial t}. \tag{1.33}$$

This is the surface equation of continuity for electric charge; it is a statement of conservation of charge at a point on the surface.

1.3 Energy flux of electromagnetic field

When a point charge q moves with velocity, $\vec{v}$, in both electric and magnetic fields, $\vec{E}$ and $\vec{B}$, the total force exerted on charge, q, by the field is given by the Lorentz law,

$$\vec{F} = q\left(\vec{E} + \frac{\vec{v}}{c} \times \vec{B}\right). \tag{1.34}$$

The equation (1.34) describes the resultant force experienced by a particle of charge q moving with velocity $\vec{v}$, under the influence of both an electric field, $\vec{E}$ and a magnetic field $\vec{B}$. The total force at a point within the particle is the applied field together with the field due to charge in the particle itself (self field). In practical situation, the self force is negligible, therefore the total force on the particle is approximately the applied force. The expression (equation 1.34) referred as Lorentz force density, provides the connection between classical mechanics and electromagnetism.

The concepts such as energy, linear and angular momentum[2] may be associated with the electromagnetic field through the expression that is derived above. In classical mechanics, a particle of mass m, moving with velocity $\vec{v}$ at position $\vec{r}$ in an inertial reference frame, has linear momentum $\vec{p}$ (Goldstein, 1980, Haliday et al. 2001),

$$\vec{p} = m\frac{d\vec{r}}{dt} = m\vec{v}. \tag{1.35}$$

The total force applied to the particle, according to the Newton's second law, is given by,

$$\begin{aligned}\vec{F} &= \frac{d\vec{p}}{dt} = m\frac{d\vec{v}}{dt} \\ &= m\frac{d^2\vec{r}}{dt^2} = m\vec{a},\end{aligned} \tag{1.36}$$

in which, $\vec{a}$ indicates the acceleration (the rate of change of velocity) of the particle.

If the particle has charge e, the force on the particle of mass m due to electric field $\vec{E}$ is

$$\vec{F} = e\vec{E} = m\vec{a}. \tag{1.37}$$

The symbol e is used to designate the charge of a particle, say electron ($e = 1.6 \times 10^{-19}$coulomb (C)), instead of q. Since the force $\vec{F}$ on the particle is equal to the charge of a particle that is placed in a uniform electric field, i.e., $\vec{F} = e\vec{E}$. The force is in the same direction as the field if the charge is positive, and the force become opposite to the field if the charge is negative. If the particle is rest and the field is applied, the particle is accelerated uniformly in the direction of the field.

[2]Angular momentum is defined as the product of moment of inertia and angular velocity of a body revolving about an axis.

The work done by the applied force on the particle when it moves through the displacement $\Delta\vec{r}$ is defined as,

$$\Delta W = \vec{F} \cdot \Delta\vec{r}. \tag{1.38}$$

The rate at which the work is done is the power $\mathcal{P}$,

$$\begin{aligned}\mathcal{P} &= \lim_{\Delta t \to 0} \left(\frac{\Delta W}{\Delta t}\right) \\ &= \lim_{\Delta t \to 0} \left(\vec{F} \cdot \frac{\Delta\vec{r}}{\Delta t}\right) = \vec{F} \cdot \vec{v}.\end{aligned} \tag{1.39}$$

The energy in the case of of a continuous charge configuration $\rho(\vec{r})$ is expressed as,

$$W = \frac{1}{2} \iint \frac{\rho(\vec{r}')\rho(\vec{r})}{|\vec{r} - \vec{r}'|} d\mathrm{V} d\mathrm{V}' = \frac{1}{2} \int \varphi(\vec{r})\rho(\vec{r}) d\mathrm{V}, \tag{1.40}$$

where the potential of a charge distribution is,

$$\varphi(\vec{r}) = \int \frac{\rho(\vec{r}')}{|\vec{r} - \vec{r}'|} d\mathrm{V}'.$$

In this equation (1.73), the integration extends over the point $\vec{r} = \vec{r}'$, so that the said equation contains self energy parts which become infinitely large for point charges. The amount of electrostatic energy stored in an electric field in a region of space is expressed as,

$$\begin{aligned}W &= \frac{1}{2} \int \varphi(\vec{r})\rho(\vec{r}) d\mathrm{V} = \frac{1}{2} \int \frac{1}{4\pi} \left[\nabla \cdot \vec{E}(\vec{r})\right] \varphi(\vec{r}) d\mathrm{V} \\ &= -\frac{1}{8\pi} \int E(\vec{r}) \cdot \nabla\varphi(\vec{r}) d\mathrm{V} = \frac{1}{8\pi} \int E^2(\vec{r}) d\mathrm{V}.\end{aligned} \tag{1.41}$$

The integrand represents the energy density of the electric field, i.e.,

$$w_e = \frac{1}{8\pi} \vec{E}^2. \tag{1.42}$$

The power can be determined in terms of the kinetic energy (KE) of the particle, $\mathcal{K}$ by invoking equation (1.39),

$$\begin{aligned}\mathcal{P} &= \vec{F} \cdot \vec{v} = m \frac{d\vec{v}}{dt} \cdot \vec{v} \\ &= \frac{d}{dt} \left(\frac{1}{2} m |\vec{v}|^2\right) = \frac{d\mathcal{K}}{dt}.\end{aligned} \tag{1.43}$$

Thus, the rate at which work is done by the applied force - the power - is equal to the rate of increase in KE of the particle. The mechanical force of electromagnetic origin acting on the charge and current for a volume V of free space at rest containing charge density, ρ and current density, $\vec{J}$ is given by the Lorentz law,

$$\begin{aligned}\vec{F} &= \int_{\mathrm{V}} \left(\rho\vec{E} + \vec{J}\times\vec{B}\right) d\mathrm{V} \\ &= \int_{\mathrm{V}} \left(\rho\vec{E} + \rho\frac{\vec{v}}{c}\times\vec{B}\right) d\mathrm{V},\end{aligned} \tag{1.44}$$

where $\vec{J} = \rho\vec{v}$, and $\vec{v}$ is the velocity of the particle moving the current density within the particle.

The power $\mathcal{P}$ is deduced as,

$$\begin{aligned}\mathcal{P} &= \vec{v}\cdot\int_{\mathrm{V}} \left(\rho\vec{E} + \rho\frac{\vec{v}}{c}\times\vec{B}\right) d\mathrm{V} \\ &= \int_{\mathrm{V}} \left[\rho\vec{v}\cdot\vec{E} + \vec{v}\cdot\left(\rho\frac{\vec{v}}{c}\times\vec{B}\right)\right] d\mathrm{V}.\end{aligned} \tag{1.45}$$

Since the velocity is same at all points in the particle, $\vec{v}$ is moved under the integral sign. Because $\vec{v}\cdot\left(\vec{v}\times\vec{B}\right) = 0$, the magnetic field does no work on the charged particle. Thus the equation (1.45) is written as,

$$\mathcal{P} = \int_{\mathrm{V}} \vec{E}\cdot\vec{J}d\mathrm{V} = \frac{d\mathcal{K}}{dt}. \tag{1.46}$$

The equation (1.46) expresses the rate at which energy is exchanged between the electromagnetic field and the mechanical motion of the charged particle. When $\mathcal{P}$ is positive, the field supplies energy to the mechanical motion of the particle, and in the case of negative $\mathcal{P}$, the mechanical motion of the particle supplies energy to the field.

1.4 Conservation law of the electromagnetic field

The energy conservation law of the electromagnetic field was evolved by Poynting (John Henry Poynting, 1831-1879) in late Nineteenth century, from the Maxwell's equations (1.1 and 1.2), which results in

$$\vec{E}\cdot\left(\nabla\times\vec{H}\right) = \frac{4\pi}{c}\vec{E}\cdot\vec{J} + \frac{1}{c}\vec{E}\cdot\frac{\partial\vec{D}}{\partial t}, \tag{1.47}$$

$$\vec{H}\cdot\left(\nabla\times\vec{E}\right)=-\frac{1}{c}\vec{H}\cdot\frac{\partial\vec{B}}{\partial t}. \tag{1.48}$$

Equation (1.46) is applied to a general volume V. By subtracting equation (1.48) from equation (1.47), one gets,

$$\vec{E}\cdot\left(\nabla\times\vec{H}\right)-\vec{H}\cdot\left(\nabla\times\vec{E}\right)=\frac{4\pi}{c}\vec{E}\cdot\vec{J}+\frac{1}{c}\left(\vec{E}\cdot\frac{\partial\vec{D}}{\partial t}+\vec{H}\cdot\frac{\partial\vec{B}}{\partial t}\right). \tag{1.49}$$

The term $\vec{E}\cdot\vec{J}$ represents the work done by the field on the electric current density. By using the vector relation,

$$\vec{A}\cdot(\nabla\times\vec{B})-\vec{B}\cdot(\nabla\times\vec{A})=-\nabla\cdot(\vec{A}\times\vec{B}),$$

the left hand side (LHS) quantity of the equation (1.82) can be written as,

$$\vec{E}\cdot(\nabla\times\vec{H})-\vec{H}\cdot(\nabla\times\vec{E})=-\nabla\cdot(\vec{E}\times\vec{H}). \tag{1.50}$$

Therefore, the equation (1.82) turns out to be,

$$\frac{4\pi}{c}\vec{E}\cdot\vec{J}+\frac{1}{c}\left(\vec{E}\cdot\frac{\partial\vec{D}}{\partial t}+\vec{H}\cdot\frac{\partial\vec{B}}{\partial t}\right)+\nabla\cdot(\vec{E}\times\vec{H})=0. \tag{1.51}$$

Integrating equation (1.51) all through an arbitrary volume, and using Gauss' divergence theorem, $\int_{\rm V}\nabla\cdot\vec{A}d{\rm V}=\oint_{\rm S}\vec{n}\cdot\vec{A}d{\rm S}$, one finds

$$\int_{\rm V}\vec{E}\cdot\vec{J}d{\rm V}+\frac{1}{4\pi}\int_{\rm V}\left(\vec{E}\cdot\frac{\partial\vec{D}}{\partial t}+\vec{H}\cdot\frac{\partial\vec{B}}{\partial t}\right)d{\rm V}+\frac{c}{4\pi}\oint_{\rm S}(\vec{E}\times\vec{H})\cdot d\vec{S}=0. \tag{1.52}$$

The equation (1.52) represents the energy law of electromagnetic field. Let

$$S(\vec{r},t)=\frac{c}{4\pi}\left[E(\vec{r},t)\times H(\vec{r},t)\right], \tag{1.53}$$

then term, $\vec{S}$, is called the energy flux density of the electromagnetic field in the direction of propagation. It is known as the Poynting vector, or power surface density. The Poynting vector $\vec{S}$ has the units of energy per unit area per unit time (joule (J) $\rm m^{-2}s^{-1}$) or power per unit area watt (W)$\rm m^{-2}$. Its magnitude $|\vec{S}|$ is equal to the rate of flow per unit area element perpendicular to $\vec{S}$.

Thus far the expression obtained above is for the energy associated with the motion of a charged particle. In what follows, an expression for the energy that applies to the general volume distribution of charge, ρ and current $\vec{J}$ is derived. Let the equation (1.52) be written in the form,

$$\int_{\mathrm{V}} \vec{E} \cdot \vec{J} d\mathrm{V} + \frac{1}{4\pi} \int_{\mathrm{V}} \left(\vec{E} \cdot \frac{\partial \vec{D}}{\partial t} + \vec{H} \cdot \frac{\partial \vec{B}}{\partial t} \right) d\mathrm{V} + \oint_{\mathrm{S}} \vec{S} \cdot d\vec{S} = 0. \qquad (1.54)$$

This relation is known as Poynting theorem. The power carried away from a volume bounded by a surface S by the electromagnetic field is given by the term, $\oint_{\mathrm{S}} \vec{S} \cdot d\vec{S}$. This is equal to the rate at which electromagnetic energy is leaving volume by passing through its surfaces.

On using material equations $\vec{D} = \epsilon \vec{E}$ and $\vec{B} = \mu \vec{H}$, the second term of the Poynting theorem (equation 1.52) can be simplified. For the electric term, one gets,

$$\frac{1}{4\pi} \vec{E} \cdot \frac{\partial \vec{D}}{\partial t} = \frac{1}{4\pi} \vec{E} \cdot \frac{\partial}{\partial t} \left(\epsilon \vec{E} \right) = \frac{1}{8\pi} \frac{\partial}{\partial t} \left(\epsilon \vec{E}^2 \right) = \frac{1}{8\pi} \frac{\partial}{\partial t} \left(\vec{E} \cdot \vec{D} \right). \qquad (1.55)$$

Similarly, for the magnetic term one may derive as,

$$\frac{1}{4\pi} \vec{H} \cdot \frac{\partial \vec{B}}{\partial t} = \frac{1}{8\pi} \frac{\partial}{\partial t} \left(\mu \vec{H}^2 \right) = \frac{1}{8\pi} \frac{\partial}{\partial t} \left(\vec{H} \cdot \vec{B} \right). \qquad (1.56)$$

Thus, the second term of the equation (1.54) is rewritten as,

$$\int_{\mathrm{V}} \left(\vec{E} \cdot \frac{\partial \vec{D}}{\partial t} + \vec{H} \cdot \frac{\partial \vec{B}}{\partial t} \right) d\mathrm{V} = \frac{1}{8\pi} \frac{\partial}{\partial t} \int_{\mathrm{V}} \left(\vec{E} \cdot \vec{D} + \vec{H} \cdot \vec{B} \right) d\mathrm{V}. \qquad (1.57)$$

For an electrostatic field in a simple material, the energy stored in the electric field, as well as for a magnetostatic field in a simple material, the stored energy in the magnetic field are respectively given by,

$$w_e = \frac{1}{8\pi} \vec{E} \cdot \vec{D}; \qquad w_m = \frac{1}{8\pi} \vec{H} \cdot \vec{B}, \qquad (1.58)$$

where w_e and w_m are the electric and magnetic energy densities respectively.

From the expressions (equations 1.57, 1.58), the equation (1.51) is cast as,

$$\frac{4\pi}{c} \vec{E} \cdot \vec{J} + \nabla \cdot (\vec{E} \times \vec{H}) = \frac{\partial}{\partial t} (w_e + w_m). \qquad (1.59)$$

This expression (1.92) describes the transfer of energy during a decrease of the total energy density of the electromagnetic field in time. The Poynting theorem (equation 1.54) takes the form,

$$-\frac{dW}{dt} = \frac{d}{dt}\int_{\text{V}} (w_e + w_m)d\text{V} = \int_{\text{V}} \vec{E} \cdot \vec{J} d\text{V} + \oint_{\text{S}} \vec{S} \cdot d\vec{S}, \tag{1.60}$$

in which

$$W = \int_{\text{V}} (w_e + w_m)d\text{V}. \tag{1.61}$$

is total electric and magnetic energy.

The equation (1.60) represents the energy conservation law of electrodynamics. The term dW/dt is interpreted as the time rate of change of the total energy contained within the volume, V. Let the Lorentz law given by equation (1.34) be recalled, and assuming that all the charges e_k are displaced by $\delta\vec{x}_k$ (where $k = 1, 2, 3, \cdots$) in time δt, therefore the total work done is given by,

$$\begin{aligned}\delta A &= \sum_k e_k \left[\vec{E}_k + \frac{1}{c}\vec{v}_k \times \vec{B}\right] \cdot \delta\vec{x}_k \\ &= \sum_k e_k \vec{E}_k \cdot \delta\vec{x}_k = \sum_k e_k \vec{E}_k \cdot \vec{v}_k \delta t,\end{aligned} \tag{1.62}$$

with $\delta\vec{x}_k = \vec{v}_k \delta t$.

On introducing the total charge density ρ, one obtains,

$$\frac{\delta A}{\delta t} = \int_{\text{V}} \rho\vec{v} \cdot \vec{E} d\text{V}. \tag{1.63}$$

The current density, $\vec{J}$, is may be split into two parts,

$$\vec{J} = \vec{J}_c + \vec{J}_v, \tag{1.64}$$

where $\vec{J}_c = \sigma\vec{E}$ is the conduction current density, and $\vec{J}_v = \rho\vec{v}$ the convection current density.

Thus for an isothermal conductor, the energy is irreversibly transferred to a heat reservoir as Joule's heat (James Brescott Joule, 1818 - 1889), then one writes,

$$\mathcal{Q} = \int_{\text{V}} \vec{E} \cdot \vec{J}_c d\text{V} = \int_{\text{V}} \sigma \vec{E}^2 d\text{V}. \tag{1.65}$$

Here $\mathcal{Q}$ represents resistive dissipation of energy called Joule's heat in a conductor ($\sigma \neq 0$).

When the motion of the charge is instantaneously supplying energy to the electromagnetic field throughout the volume, the volume density of current due to the motion of the charge $\vec{J}_v$ is given by,

$$\frac{\delta A}{\delta t} = \int_{\mathrm{V}} \vec{E} \cdot \vec{J}_v d\mathrm{V}. \tag{1.66}$$

From the equations (1.63) and (1.64), one finds,

$$\int_{\mathrm{V}} \vec{J} \cdot \vec{E} d\mathrm{V} = \mathcal{Q} + \int_{\mathrm{V}} \vec{J}_v \cdot \vec{E} d\mathrm{V} = \mathcal{Q} + \frac{\delta A}{\delta t}. \tag{1.67}$$

Thus, equation (1.60) translates into,

$$\frac{dW}{dt} = -\mathcal{Q} - \frac{\delta A}{\delta t} - \oint_{\mathrm{S}} \vec{S} \cdot d\vec{S}. \tag{1.68}$$

where $\delta A/\delta t$ is the rate at which electromagnetic energy is being stored.

The interpretation of such a relation as a statement of conservation of energy within the volume, V, stands. Finally, in a nonconducting medium ($\sigma = 0$) where no mechanical work is done ($A = 0$), the energy law may be written in the hydrodynamical continuity equation for non-compressible fluids,

$$\frac{\partial w}{\partial t} + \nabla \cdot \vec{S} = 0, \tag{1.69}$$

with $w = w_e + w_m$.

The physical meaning of the equation (1.69) is that the decrease in the time rate of change of electromagnetic energy density within a volume is equal to the flow of energy out of the volume.

1.5 Electromagnetic wave equations

Consider the propagation of light in a medium, in which the charges or currents are absent, i.e., $\vec{J} = 0$ and $\rho = 0$, and therefore, the first two Maxwell's equations can be cast into the forms,

$$\nabla \times \vec{E} = -\frac{1}{c}\frac{\partial \vec{B}}{\partial t},$$

$$\nabla \times \vec{H} = \frac{1}{c}\frac{\partial \vec{D}}{\partial t}. \tag{1.70}$$

To proceed further, $\vec{B}$ is replaced with $\mu\vec{H}$ (equation 1.6) in the first equation (1.70), so that,

$$\nabla \times \vec{E} = -\frac{\mu}{c}\frac{\partial \vec{H}}{\partial t}, \tag{1.71}$$

or,

$$\frac{1}{\mu}\left(\nabla \times \vec{E}\right) = -\frac{1}{c}\frac{\partial \vec{H}}{\partial t}. \tag{1.72}$$

The curl of the equation (1.72) gives,

$$\nabla \times \left[\frac{1}{\mu}\left(\nabla \times \vec{E}\right)\right] = \nabla \times \left[-\frac{1}{c}\frac{\partial \vec{H}}{\partial t}\right] = -\frac{1}{c}\nabla \times \frac{\partial \vec{H}}{\partial t}. \tag{1.73}$$

Similarly, by replacing $\vec{D}$ with $\epsilon\vec{E}$ (equation 1.7) from the second equation (1.70), one writes,

$$\nabla \times \vec{H} = \frac{\epsilon}{c}\frac{\partial \vec{E}}{\partial t}. \tag{1.74}$$

Differentiating both sides of equation (1.74) with respect to time, and interchanging differentiation with respect to time and space, one gets,

$$\nabla \times \frac{\partial \vec{H}}{\partial t} = \frac{\epsilon}{c}\frac{\partial^2 \vec{E}}{\partial t^2}. \tag{1.75}$$

Substituting (1.75) in equation (1.73), the following relationship emerges,

$$\nabla \times \left[\frac{1}{\mu}\left(\nabla \times \vec{E}\right)\right] = -\frac{\epsilon}{c^2}\frac{\partial^2 \vec{E}}{\partial t^2}, \tag{1.76}$$

By using the vector triple product identity,

$$\nabla \times (\nabla \times \vec{A}) = \nabla(\nabla \cdot \vec{A}) - \nabla^2 \vec{A},$$

we may write,

$$\nabla \times \left[\frac{1}{\mu}\left(\nabla \times \vec{E}\right)\right] = \nabla\left(\frac{1}{\mu}\nabla \cdot \vec{E}\right) - \frac{1}{\mu}\nabla^2 \vec{E}. \tag{1.77}$$

When light propagates in vacuum, use of the Maxwell's equation $\nabla \cdot \vec{E} = 0$, in equation (1.77) yields,

$$\nabla \times \left[\frac{1}{\mu}\left(\nabla \times \vec{E}\right)\right] = -\frac{1}{\mu}\nabla^2 \vec{E}. \tag{1.78}$$

Invoking equation (1.76), this equation (1.78) takes the form,

$$\frac{1}{\mu}\nabla^2\vec{E} = \frac{\epsilon}{c^2}\frac{\partial^2\vec{E}}{\partial t^2}, \tag{1.79}$$

or, on rearranging this equation (1.79),

$$\left(\nabla^2 - \frac{\epsilon\mu}{c^2}\frac{\partial^2}{\partial t^2}\right)\vec{E} = 0. \tag{1.80}$$

Similarly, one derives for $\vec{H}$,

$$\left(\nabla^2 - \frac{\epsilon\mu}{c^2}\frac{\partial^2}{\partial t^2}\right)\vec{H} = 0. \tag{1.81}$$

The above expressions (equations 1.80-1.81) are known as the electromagnetic wave equations, which indicate that electromagnetic disturbances (waves) are propagated through the medium. This result gives rise to Maxwell's electromagnetic theory of light. The propagation velocity v of the waves obeying the wave equations is given by,

$$v = \frac{c}{\sqrt{\epsilon\mu}}, \tag{1.82}$$

therefore, one may express the wave equation (1.80) as,

$$\left(\nabla^2 - \frac{1}{v^2}\frac{\partial^2}{\partial t^2}\right)\vec{E} = 0. \tag{1.83}$$

For a scalar wave E propagating in the z-direction, the equation (1.83) is simplified to,

$$\frac{\partial^2 E}{\partial z^2} - \frac{1}{v^2}\frac{\partial^2 E}{\partial t^2} = 0. \tag{1.84}$$

The permittivity constant ϵ_0 and the permeability constant μ_0 in a vacuum are related to the speed of light c,

$$c = \frac{1}{\sqrt{\epsilon_0\mu_0}} = 2.99,79 \times 10^8\mathrm{m\,s^{-1}}. \tag{1.85}$$

1.5.1 *The Poynting vector and the Stokes parameter*

It is evident from Maxwell's equations that the electromagnetic radiation is transverse wave motion, where the electric and magnetic fields oscillate perpendicular to each other and also perpendicular to the direction of propagation denoted by $\vec{\kappa}$ (see Figure 1.2). These variations are described by the harmonic wave equations in the form,

$$E(\vec{r},t) = E_0(\vec{r},\omega)e^{i(\vec{\kappa}\cdot\vec{r}-\omega t)}, \tag{1.86}$$

$$B(\vec{r},t) = \vec{B}_0(\vec{r},\omega)e^{i(\vec{\kappa}\cdot\vec{r}-\omega t)}, \tag{1.87}$$

in which $E_0(\vec{r},\omega)$ and $B_0(\vec{r},\omega)$ are the amplitudes[3] of the electric and magnetic field vectors respectively, $\vec{r}(=x,y,z)$ the position vector, $\omega(=2\pi\nu)$ is the angular frequency, $\nu = 1/T$ represents the number of complete cycles of waves per unit time, called frequency, (the shorter the wavelength[4], the higher the frequency) and T the period[5] of motion, and

$$\vec{\kappa}\cdot\vec{r} = \kappa_x x + \kappa_y y + \kappa_z z, \tag{1.88}$$

represents planes in a space of constant phase (any portion of the wave cycle), and

$$\vec{\kappa} = \kappa_x\vec{i} + \kappa_y\vec{j} + \kappa_z\vec{k}. \tag{1.89}$$

The Cartesian components of the wave travel with the same propagation vector $\vec{\kappa}$ and frequency ω. The cosinusoidal fields are,

$$\begin{aligned} E(\vec{r},t) &= \Re\left[E_0(\vec{r},\omega)e^{i(\vec{\kappa}\cdot\vec{r}-\omega t)}\right] = E_0(\vec{r},\omega)\cos(\vec{\kappa}\cdot\vec{r}-\omega t), \\ B(\vec{r},t) &= \Re\left[\vec{B}_0(\vec{r},\omega)e^{i(\vec{\kappa}\cdot\vec{r}-\omega t)}\right] = B_0(\vec{r},\omega)\cos(\vec{\kappa}\cdot\vec{r}-\omega t). \end{aligned} \tag{1.90}$$

Assuming that $\vec{E}_0$ is constant, hence the divergence of the equation (1.86) becomes,

$$\begin{aligned} \nabla\cdot\vec{E} &= \vec{E}_0\cdot\nabla\left(e^{i[\vec{\kappa}\cdot\vec{r}-\omega t]}\right) \\ &= \vec{E}_0\cdot(i\vec{\kappa})e^{i[\vec{\kappa}\cdot\vec{r}-\omega t]} = (i\vec{\kappa})\cdot\vec{E}. \end{aligned} \tag{1.91}$$

[3]An amplitude of a wave defined as the maximum magnitude of the displacement from the equilibrium position during one wave cycle.

[4]Wavelength is defined as the least distance between two points in same phase in a periodic wave motion

[5]Period is defined by the shortest interval in time between two instants when parts of the wave profile that are oscillating in phase pass a fixed point and any portion of the wave cycle is called a phase. When two waves of equal wavelength travel together in the same direction they are said to be in phase if they are perfectly aligned in their cycle, and out of phase if they are out of step.

The curl of the electric field is derived as,

$$\begin{aligned}\nabla \times \vec{H} &= \frac{\epsilon}{c}\frac{\partial \vec{E}}{\partial t} = \frac{\epsilon}{c}\vec{E}_0\frac{\partial}{\partial t}\left(e^{i(\vec{\kappa}\cdot\vec{r} - \omega t)}\right) \\ &= -\frac{i\omega\epsilon}{c}\vec{E}_0 e^{i(\vec{\kappa}\cdot\vec{r} - \omega t)} = -\frac{i\omega\epsilon}{c}\vec{E}. \end{aligned} \tag{1.92}$$

Replacing ∇ to $i\vec{\kappa}$ and $\partial/\partial t$ to $-i\omega$, this equation (1.92) is recast as,

$$\vec{\kappa} \times \vec{H} = -\frac{\epsilon\omega}{c}\vec{E}. \tag{1.93}$$

Similarly, from the Maxwell's equation (1.1) one derives,

$$\vec{\kappa} \times \vec{E} = \frac{\omega\mu}{c}\vec{H}, \tag{1.94}$$

After rearranging equations (1.93, 1.94),

$$\vec{E} = -\frac{c}{\epsilon\omega}\vec{\kappa} \times \vec{H} = -\frac{1}{\omega}\sqrt{\frac{\mu}{\epsilon}}\vec{\kappa} \times \vec{H}, \tag{1.95}$$

$$\vec{H} = \frac{c}{\omega\mu}\vec{\kappa} \times \vec{E} = \frac{1}{\omega}\sqrt{\frac{\epsilon}{\mu}}\vec{\kappa} \times \vec{E}. \tag{1.96}$$

with $c = \sqrt{\epsilon\mu}$ and $i = \sqrt{-1}$.

In vacuum, ρ is assumed to be zero, therefore, the Maxwell equation for the electric field is written as, $\nabla \cdot \vec{E} = 0$. Hence from the equation (1.91), one finds,

$$\vec{\kappa} \cdot \vec{E} = 0. \tag{1.97}$$

Similarly, from the divergence of the magnetic field, i.e., $\nabla \cdot \vec{B} = 0$, one derives,

$$\vec{\kappa} \cdot \vec{B} = 0. \tag{1.98}$$

Scalar multiplication with $\vec{\kappa}$ provides us,

$$\vec{E} \cdot \vec{\kappa} = \vec{H} \cdot \vec{\kappa} = 0, \tag{1.99}$$

This shows that the electric and magnetic field vectors lie in planes normal to the direction of propagation. From the equation (1.99) one gets,

$$\sqrt{\mu}|\vec{H}| = \sqrt{\epsilon}|\vec{E}|. \tag{1.100}$$

The magnitude of a real vector $|\vec{E}|$ for a general time dependent electromagnetic field, $\vec{E}(\vec{r}, t)$ is represented by $\sqrt{\vec{E} \cdot \vec{E}}$. In Cartesian coordinates the quadratic term, $\vec{E} \cdot \vec{E}$, is written out as,

$$\vec{E} \cdot \vec{E} = E_x E_x + E_y E_y, \tag{1.101}$$

Thus, the Maxwell's theory leads to quadratic terms associated with the flow of energy, that is intensity (or irradiance), I, which is defined as the time average of the amount of energy carried by the wave across the unit area perpendicular to the direction of the energy flow in unit time, therefore, the time averaged intensity of the optical field. The unit of intensity is expressed as the joule per square meter per second, $(\mathrm{J\,m^{-2}s^{-1}})$, or watt per square meter, $(\mathrm{W\,m^{-2}})$.

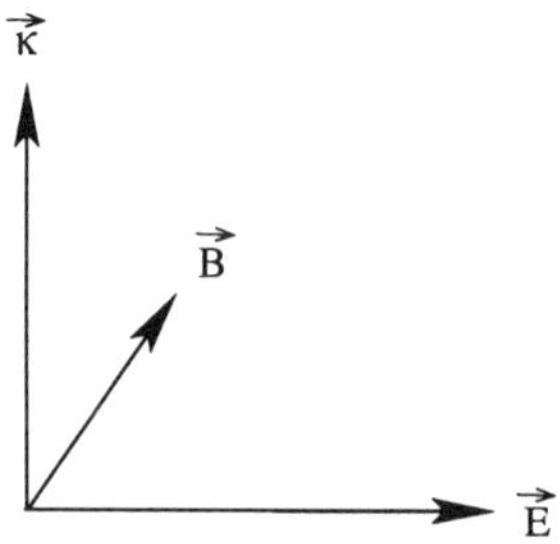

Fig. 1.2 The orthogonal triad of vectors.

It is observed from the equations (1.91-1.94) that in an electromagnetic wave, the field intensities $\vec{E}$, $\vec{H}$, and the unit vector in the propagation direction of the wave $\vec{\kappa}$ form a right handed orthogonal triad of vectors. To be precise, if an electromagnetic wave travels in the positive $x-$axis, the electric and magnetic fields would oscillate parallel to the $y-$ and $z-$axis respectively.

The energy crossing an element of area in unit time is perpendicular to the direction of propagation. In a cylinder with unit cross-sectional area, whose axis is parallel to $\vec{s}$, the amount of energy passing the base of the cylinder in unit time is equal to the energy that is contained in the portion of the cylinder of length v. Therefore, the energy flux is equal to vw, where

$$w = \frac{\epsilon}{4\pi}\left|\vec{E}\right|^2 = \frac{\mu}{4\pi}\left|\vec{H}\right|^2, \tag{1.102}$$

is the energy density. Hence the energy densities of both electric and magnetic fields are equal everywhere along an electromagnetic wave. The equation (1.102) is derived by considering the equations (1.58), and (1.100).

Thus, the Poynting vector is expressed as,

$$\begin{aligned}\vec{S} &= \frac{c}{4\pi}(\vec{E}\times\vec{H}) = \frac{c}{4\pi}\frac{\vec{\kappa}}{\omega}|\vec{E}||\vec{H}| \\ &= \frac{c}{4\pi}\sqrt{\frac{\epsilon}{\mu}}\frac{\vec{\kappa}}{\omega}|\vec{E}|^2 = \frac{c}{4\pi}\sqrt{\frac{\mu}{\epsilon}}\frac{\vec{\kappa}}{\omega}|\vec{H}|^2. \end{aligned} \tag{1.103}$$

Equation (1.103) relates that the electric and magnetic fields are perpendicular to each other in electromagnetic wave. By combining the two equations (1.102) and (1.103), one finds,

$$\vec{S} = \frac{c}{\sqrt{\epsilon\mu}}\frac{\vec{\kappa}}{\omega}w = \frac{\vec{\kappa}}{\omega}vw, \tag{1.104}$$

with $v = c/\sqrt{\epsilon\mu}$.

The Poynting vector represents the flow of energy, both with respect to its magnitude and direction of propagation. Expressing $\vec{E}$ and $\vec{H}$ in complex terms, then the time-averaged flux of energy is given by the real part of the Poynting vector,

$$\vec{S} = \frac{1}{2}\frac{c}{4\pi}(\vec{E}\times\vec{H}^*), \tag{1.105}$$

in which * represents for the complex conjugate of ‘ ’.

Thus one may write,

$$\vec{S} = \frac{c}{8\pi}\sqrt{\frac{\epsilon}{\mu}}\frac{\vec{\kappa}}{\omega}(\vec{E}\cdot\vec{E}^*). \tag{1.106}$$

In order to describe the strength of a wave, the amount of energy carried by the wave in unit time across unit area perpendicular to the direction of propagation is used. This quantity, known as intensity of the wave, according to the Maxwell's theory is given in equation (1.101). From the relationship that described in equation (1.103), one may derive the intensity as,

$$\begin{aligned} I = v\langle w\rangle &= \frac{c}{4\pi}\sqrt{\frac{\epsilon}{\mu}}\left\langle\vec{E}^2\right\rangle \\ &= \frac{c}{4\pi}\sqrt{\frac{\mu}{\epsilon}}\left\langle\vec{H}^2\right\rangle, \end{aligned} \tag{1.107}$$

where $\langle\,\rangle$ stands for the time average of the quantity.

The Poynting vector, $\vec{S}$, in terms of spherical coordinates is written as,

$$\vec{S} = \frac{c}{8\pi}\sqrt{\frac{\epsilon}{\mu}\frac{\vec{\kappa}}{\omega}}\left(E_\theta E_\theta^* + E_\phi E_\phi^*\right), \tag{1.108}$$

The quantity within the parentheses represents the total intensity of the wave field, known as the first Stokes parameter I. Thus the Poynting vector is directly proportional to the first Stokes parameter.

1.5.2 *Harmonic time dependence and the Fourier transform*

The Maxwell's equations for an electromagnetic field with time dependence are simplified by specifying a field with harmonic dependence (Smith, 1997). The harmonic time dependent electromagnetic fields are given by,

$$E(\vec{r}, t) = \Re\left[E_0(\vec{r}, \omega)e^{i\omega t}\right], \tag{1.109}$$

$$B(\vec{r}, t) = \Re\left[B_0(\vec{r}, \omega)e^{i\omega t}\right], \tag{1.110}$$

in which $\vec{E}_0$ is a complex vector with Cartesian rectangular components,

$$\begin{aligned} \vec{E}_{0x} &= a_1(\vec{r}, \omega)e^{i\psi_1(\vec{r}, \omega)}, \\ \vec{E}_{0y} &= a_2(\vec{r}, \omega)e^{i\psi_2(\vec{r}, \omega)}, \\ \vec{E}_{0z} &= a_3(\vec{r}, \omega)e^{i\psi_3(\vec{r}, \omega)}, \end{aligned} \tag{1.111}$$

where $a_j(\vec{r}, \omega)$ is the amplitude of the electric wave, $\vec{\kappa}$ the propagation vector, and $j = 1, 2, 3$.

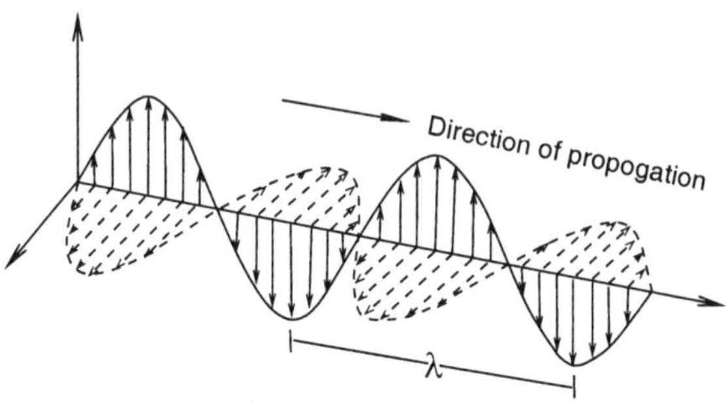

Fig. 1.3 Propagation of a plane electromagnetic wave; the solid and dashed lines represent respectively the electric and magnetic fields.

Figure (1.3) depicts the propagation of a plane electromagnetic wave. For a homogeneous plane wave, the amplitudes, $a_j(\vec{r}, \omega)$'s, are constant. Each component of the vector phasor $\vec{E}_0$ has a modulus a_j and argument ψ_j which depend on the position $\vec{r}$ and the parameter ω. The unit of this vector phasor $E_0(\vec{r}, \omega)$ for harmonic time dependence is Vm^{-1}. By differentiating the equation (1.109) with respect to the temporal variables, the Maxwell's equation (1.1) turns out to be,

$$\nabla \times E(\vec{r}, t) = \nabla \times \Re\left[E_0(\vec{r}, \omega)e^{i\omega t}\right]$$
$$= -\frac{\partial}{\partial t}\Re\left[E_0(\vec{r}, \omega)e^{i\omega t}\right] = \Re\left[-i\omega B_0(\vec{r}, \omega)e^{i\omega t}\right]. \quad (1.112)$$

By rearranging this equation (1.112),

$$\nabla \times E_0(\vec{r}, \omega)e^{i\omega t} = -i\omega B_0(\vec{r})e^{i\omega t}, \quad (1.113)$$

or,

$$\nabla \times E_0(\vec{r}, \omega) = -i\omega B_0(\vec{r}, \omega). \quad (1.114)$$

Similarly, the other Maxwell's equations may also be derived,

$$\nabla \times H_0(\vec{r}, \omega) = J_0(\vec{r}, \omega) + i\omega D_0(\vec{r}, \omega), \quad (1.115)$$
$$\nabla \cdot D_0(\vec{r}, \omega) = \rho(\vec{r}, \omega), \quad (1.116)$$
$$\nabla \cdot B_0(\vec{r}, \omega) = 0, \quad (1.117)$$
$$\nabla \cdot J_0(\vec{r}, \omega) = -i\omega\rho(\vec{r}, \omega). \quad (1.118)$$

These equations (1.115-1.118) are known as the Maxwell's equations for the frequency domain. The Maxwell's equations for the complex vector phasors, $E_0(\vec{r}, \omega)$, $B_0(\vec{r}, \omega)$, etc., are applied to electromagnetic systems in which the constitutive relations for all materials are time-invariant and linear. The Maxwell's equation with a cosinusoidal excitation are solved to obtain the vector phasors for the electromagnetic field $E(\vec{r}, t)$, $B(\vec{r}, t)$. For harmonic time dependence, $E(\vec{r}, t) = \Re[E_0(\vec{r}, \omega)e^{i\omega t}]$, the Hermitian magnitude of a complex vector is, $|\vec{E}_0| = [\vec{E}_0 \cdot \vec{E}_0^*]^{1/2}$. If the electromagnetic is harmonic in time, the instantaneous rate at which energy is exchanged between the field and the mechanical motion of the charge is the product of,

$$E(\vec{r}, t) \cdot J(\vec{r}, t) = \Re\left[E_0(\vec{r}, \omega)e^{i\omega t}\right] \cdot \Re\left[J_0(\vec{r}, \omega)e^{i\omega t}\right]. \quad (1.119)$$

The real terms are written as,

$$\Re\left[E_0(\vec{r},\omega)e^{i\omega t}\right] = \frac{1}{2}\left[E_0(\vec{r},\omega)e^{i\omega t} + E_0^*(\vec{r},\omega)e^{-i\omega t}\right],$$
$$\Re\left[J_0(\vec{r},\omega)e^{i\omega t}\right] = \frac{1}{2}\left[J_0(\vec{r},\omega)e^{i\omega t} + J_0^*(\vec{r},\omega)e^{-i\omega t}\right]. \quad (1.120)$$

The scalar product of these two terms provides,

$$\begin{aligned} E(\vec{r},t)\cdot J(\vec{r},t) &= \frac{1}{4}\left[E_0(\vec{r},\omega)\cdot J_0^*(\vec{r},\omega) + E_0^*(\vec{r},\omega)\cdot J_0(\vec{r},\omega)\right. \\ &\quad \left. + E_0(\vec{r},\omega)\cdot J_0(\vec{r},\omega)e^{2i\omega t} + E_0^*(\vec{r},\omega)\cdot J_0^*(\vec{r},\omega)e^{-2i\omega t}\right] \\ &= \frac{1}{2}\left\{\Re\left[E_0(\vec{r},\omega)\cdot J_0^*(\vec{r})\right] + \Re\left[E_0(\vec{r},\omega)\cdot J_0(\vec{r},\omega)e^{2i\omega t}\right]\right\}. \end{aligned} \quad (1.121)$$

Since the optical frequencies are very large, one can observe their time average[6] over a period of oscillation, $T = 2\pi/\omega$. Hence the time average of the product $\langle E(\vec{r},t)\cdot J(\vec{r},t)\rangle$ is expressed as,

$$\langle E(\vec{r},t)\cdot J(\vec{r},t)\rangle = \frac{1}{2}\Re\left(E_0(\vec{r},\omega)\cdot J_0^*(\vec{r},\omega)\right). \quad (1.122)$$

Similarly, the time average value of the Poynting vector product may also be derived as,

$$\begin{aligned} \langle S(\vec{r},t)\rangle &= \frac{1}{T}\int_0^T \frac{c}{4\pi}\left[E(\vec{r},t)\times H(\vec{r},t)\right]dt \\ &= \frac{c}{4\pi}\frac{1}{T}\int_0^T \frac{1}{4}\left[E_0(\vec{r},\omega)\times H_0^*(\vec{r},\omega) + E_0^*(\vec{r},\omega)\times H_0(\vec{r},\omega)\right. \\ &\quad \left. + E_0(\vec{r},\omega)\times H_0(\vec{r},\omega)e^{2i\omega t} + E_0^*(\vec{r},\omega)\times H_0^*(\vec{r},\omega)e^{-2i\omega t}\right]dt \\ &\simeq \frac{c}{16\pi}\left[E_0(\vec{r},\omega)\times H_0^*(\vec{r},\omega) + E_0^*(\vec{r},\omega)\times H_0(\vec{r},\omega)\right] \\ &= \frac{c}{8\pi}\Re\left[E_0(\vec{r},\omega)\times H_0^*(\vec{r},\omega)\right] = \Re[S_c(\vec{r},\omega)]. \end{aligned} \quad (1.123)$$

---

[6]The time average over a time that is large compared with the inverse frequency of the product of the two harmonic time-independent functions $\vec{a}$ and $\vec{b}$, of the same frequency is given by,

$$\left\langle \vec{a}(t)\cdot\vec{b}(t)\right\rangle = \frac{1}{T}\int_0^T \frac{1}{4}\left[\vec{a}e^{i\omega t} + \vec{a}^* e^{-i\omega t}\right]\cdot\left[\vec{b}e^{i\omega t} + \vec{b}^* e^{-i\omega t}\right]dt = \frac{1}{2}\Re \quad \vec{a}\cdot\vec{b}^* \quad .$$

Thus the complex Poynting vector is deduced as,

$$S_c(\vec{r},\omega) = \frac{1}{2}\left[E_0(\vec{r},\omega) \times H_0^*(\vec{r},\omega)\right]. \tag{1.124}$$

The real part of this Poynting vector is known as the time average of the Poynting vector. The law of conservation of energy takes a simple form. The complex Poynting theorem is given by,

$$\int_{\mathrm{V}} \left(\frac{1}{2}\vec{E}_0 \cdot \vec{J}_0^*\right) d\mathrm{V} - i\omega \int_{\mathrm{V}} \frac{1}{2}\left(\vec{E}_0 \cdot \vec{D}_0^* - \vec{H}_0^* \cdot \vec{B}_0\right) d\mathrm{V} + \oint_{\mathrm{S}} \vec{S}_c \cdot d\vec{S} = 0. \tag{1.125}$$

For non-conducting medium ($\sigma = 0$), where no mechanical work is done, the time average of equation (1.69) turns out to be,

$$\nabla \cdot \langle S(\vec{r},t)\rangle = 0. \tag{1.126}$$

By integrating this equation (1.127) over an arbitrary volume which contains no absorber or radiator of energy, one obtains after applying Gauss' theorem,

$$\oint_{\mathrm{S}} \langle S(\vec{r},t)\rangle \cdot \vec{n} d\vec{S} = 0, \tag{1.127}$$

in which $\vec{n}$ is the outward normal to the surface.

Thus the averaged total flux of energy through any closed surface is zero. The time average of the electric energy density is derived as,

$$\begin{aligned}\langle w_e\rangle &= \frac{1}{T}\int_0^T \frac{\epsilon}{8\pi}\vec{E}^2 dt \\ &= \frac{\epsilon}{8\pi}\frac{1}{T}\int_0^T \frac{1}{4}\left[\vec{E}_0^2 e^{2i\omega t} + \vec{E}_0 \cdot \vec{E}_0^* + \vec{E}_0{}^{*2} e^{-2i\omega t}\right] dt. \end{aligned} \tag{1.128}$$

Since T is assumed to be large, the integrals involving the exponentials are neglected. Therefore, one gets,

$$\langle w_e\rangle = \frac{\epsilon}{16\pi}\vec{E}_0 \cdot \vec{E}_0^* = \frac{\epsilon}{16\pi}|\vec{E}_0|^2. \tag{1.129}$$

Similarly, the time average of the magnetic energy density is also derived as,

$$\langle w_m\rangle = \frac{\mu}{16\pi}\vec{H}_0 \cdot \vec{H}_0^* = \frac{\mu}{16\pi}|\vec{H}_0|^2. \tag{1.130}$$

therefore, the difference in the time-average energies stored in the electric and magnetic fields with the volume,

$$\langle w_e \rangle - \langle w_m \rangle = \frac{\epsilon}{16\pi}|\vec{E}_0|^2 - \frac{\mu}{16\pi}|\vec{H}_0|^2 = 0. \tag{1.131}$$

Hence $\langle w_e \rangle = \langle w_m \rangle$. The total energy W is given by $W = 2\langle w_e \rangle = 2\langle w_m \rangle$.

A time dependent field is a linear superposition of fields that vary harmonically with time at different frequencies. This relationship is known as Fourier transform (FT; Bracewell, 1965). The free wave equation is a linear homogeneous differential equation, therefore any linear combination of its solution is a solution as well. From the Fourier equation (see Appendix B) associated with the harmonic function of frequency ω, the FT of an electric field is expressed as,

$$\widehat{E}_0(\vec{r}, \omega) = \int_{-\infty}^{\infty} E(\vec{r}, t) e^{-i\omega t} dt, \tag{1.132}$$

in which, the functions satisfy the relation, $f(t) \leftrightarrow \widehat{f}(\omega)$ and the notation, $\widehat{}$ stands for a Fourier transform (FT) of a particular physical quantity,

The Fourier transform of $\widehat{E}_0(\vec{r}, \omega)$ is a complex function of the variable ω and has the units of the electric field (Vm^{-1}) per unit frequency, i.e., $(V/m)/Hz$. By invoking the principle of Fourier transform of a temporal derivative of a function, $df(t)/dt \leftrightarrow i\omega\widehat{f}(\omega)$, the curl of the equation (1.133) is applied on Maxwell's equation, thus,

$$\begin{aligned} \nabla \times \widehat{E}_0(\vec{r}, \omega) &= \int_{-\infty}^{\infty} \nabla \times E(\vec{r}, t) e^{-i\omega t} dt \\ &= \int_{-\infty}^{\infty} \left[\frac{\partial}{\partial t} B(\vec{r}, t)\right] e^{-i\omega t} dt. \end{aligned} \tag{1.133}$$

The Fourier transform of the magnetic field is given by,

$$\widehat{B}_0(\vec{r}, \omega) = \int_{-\infty}^{\infty} B(\vec{r}, t) e^{-i\omega t} dt, \tag{1.134}$$

therefore, the equation (1.134) turns out to be,

$$\nabla \times \widehat{E}_0(\vec{r}, \omega) = -i\omega \widehat{B}_0(\vec{r}, \omega). \tag{1.135}$$

Similar operations may be applied with other Maxwell's equations in order to derive their Fourier transforms as well,

$$\nabla \times \widehat{H}_0(\vec{r}, \omega) = \widehat{J}_0(\vec{r}, \omega) + i\omega \widehat{D}_0(\vec{r}, \omega), \tag{1.136}$$

$$\nabla \cdot \widehat{D}_0(\vec{r}, \omega) = \widehat{\rho}(\vec{r}, \omega), \tag{1.137}$$

$$\nabla \cdot \widehat{B}_0(\vec{r}, \omega) = 0, \tag{1.138}$$

$$\nabla \cdot \widehat{J}_0(\vec{r}, \omega) = -i\omega\widehat{\rho}(\vec{r}, \omega). \tag{1.139}$$

These are the Maxwell's equations for the Fourier transform of the electromagnetic fields, $E_0(\vec{r}, \omega)$, $B_0(\vec{r}, \omega)$, etc.

Let the integral of the complex Poynting vector be examined,

$$\frac{2}{\pi}\int_0^\infty \vec{n} \cdot \Re[S_c(\vec{r}, \omega)]d\omega = \frac{2}{\pi}\int_0^\infty \vec{n} \cdot \left\{\frac{c}{8\pi}\Re\left[\widehat{E}_0(\vec{r}, \omega) \times \widehat{H}_0^*(\vec{r}, \omega)\right]\right\} d\omega. \tag{1.140}$$

By using Parseval's theorem (see appendix II), one finds,

$$\begin{aligned}\int_{-\infty}^\infty \vec{n} \cdot S(\vec{r}, t)dt &= \int_{-\infty}^\infty \vec{n} \cdot \frac{c}{4\pi}[E(\vec{r}, t) \times H(\vec{r}, t)]dt \\ &= \frac{2}{\pi}\int_0^\infty \vec{n} \cdot \left\{\frac{c}{8\pi}\Re\left[\widehat{E}_0(\vec{r}, \omega) \times \widehat{H}_0^*(\vec{r}, \omega)\right]\right\} d\omega \\ &= \frac{2}{\pi}\int_0^\infty \vec{n} \cdot \left\{\Re\left[\widehat{S}_c(\vec{r}, \omega)\right]\right\} d\omega. \end{aligned} \tag{1.141}$$

The LHS of equation (1.142) is the total electromagnetic energy passing through a unit area of surface with the unit normal $\vec{n}$, while the integrand on the RHS,

$$\frac{2}{\pi}\vec{n} \cdot \left\{\Re\left[\widehat{\vec{S}}_c(\vec{r}, \omega)\right]\right\},$$

is the energy passing through a unit area of this surface per unit frequency, $(\mathrm{J/m^2})/\mathrm{Hz}$ and is known as energy spectral density (Lang and Kohn, 1971).

Chapter 2

Wave optics and polarization

2.1 Electromagnetic theory of propagation

Optical phenomena may be divided mainly into four areas, such as geometrical optics, wave optics, quantum optics, and statistical optics. Geometrical optics is the study of optics when the propagation of light is described using ray tracing. It deals with the image formation and related phenomena that can be discussed within the framework of the laws concerning reflection, refraction, and rectilinear propagation. This can be applied in those cases where interference and diffraction phenomena are ignored. It has found an application in the method of ray tracing widely used in optical design. Physical (or wave) optics that helps in describing diffraction and interference phenomena at length, is founded on Maxwell's equations, according to which, light is composed of electromagnetic waves of different frequencies.

Light is an electromagnetic radiation, propagating disturbance involving space and time variation. A wave may be described as a periodic disturbance that transports energy from one point to another. Its direction of propagation is the direction in which energy is carried. The wave properties are described by the quantities: period, wavenumber, speed, amplitude, and phase and then are interrelated. It is important to note that unlike radio waves, the phase of optical waves cannot be measured directly due to the quantum nature of light dealt by quantum optics.

With the discovery of the quantized nature of light, and particularly with the statistical interpretation of quantum mechanics introduced by Max Born, statistical optics became a new branch of optics. The problem of photon[1] statistics in fluctuating light fields, which may have different

[1]Photon is a quantum or packet of an electromagnetic radiation. It travels at the

frequencies or may be generated by scattering at a rough surface, arises. Wolf (1954) introduced a broad framework for considering the coherence properties of waves.

The harmonic wave equations represent any type of wave disturbance that varies in a sinusoidal manner. This includes, wave on a string, water waves[2], sound waves[3] etc. For electromagnetic waves, the propagation of light, $V(\vec{r}, t)$, in which $\vec{r}$ is the position vector of a point (x, y, z), stands for either of the varying field vectors that together constitute the wave. The quantity $V(\vec{r}, t)$ may refer to vertical displacements of a string or pressure variations due to a sound wave propagating in a gas.

2.1.1 *Intensity of a light wave*

Electromagnetic theory interprets the light intensity as the energy flux of the field. Poynting showed that the intensity is the direct consequences of the Maxwell's equations. For classical waves that result from the wave-equation based on the concepts of classical physics, such as sound waves, water waves etc., intensity is proportional to the square of the amplitude, and for a spherical wave it is inversely proportional to the square of the distance from the source. The time average is specified as,

$$\langle E(t) \rangle = \lim_{T \to \infty} \frac{1}{2T} \int_{-T}^{T} E_T(t) dt. \tag{2.1}$$

The quantity within the sharp brackets is due to the assumed ergodicity[4]

celerity (speed) of light. Photons have properties like energy $\hbar\omega$, zero rest mass, momentum, spin $\hbar/2$, where $\hbar$ is the Planck's constant, $h = 6.626196 \times 10^{34}$ Joules (J), divided by 2π and they are non interacting Bosons.

[2]Water waves involve a combination of both longitudinal (where the particle displacement is parallel to the direction of wave propagation) and transverse (where the particle displacement is perpendicular to the direction of wave propagation) motions. It is restored by gravity.

[3]A sound wave is the pattern of disturbance caused by the movement of energy traveling through a medium as it propagates away from the source. The characteristic speed of acoustic disturbance is the sound speed, v_s, given by,

$$v_s = \sqrt{\frac{\gamma P}{\rho}},$$

where $\gamma(= C_P/C_v)$ is the ratio of specific heat at constant pressure (C_P) to the specific heat at constant volume (C_v), P the pressure, and ρ the density of medium.

[4]Ergodicity implies that each ensemble average is equal to the corresponding time average involving a typical member of the ensemble, while the stationary field implies that all the ensemble averages are independent of the origin of time.

of the field. The quantity, $< \vec{E}^2 >$ is taken as measured intensity, since the intensities are compared in the same condition. In the monochromatic fields, the electric vector, $E(\vec{r}, t)$, is expressed as,

$$\begin{aligned} E(\vec{r}, t) &= \Re\left\{A(\vec{r})e^{-i\omega t}\right\} \\ &= \frac{1}{2}\left[A(\vec{r})e^{-i\omega t} + A^*(\vec{r})e^{i\omega t}\right], \end{aligned} \tag{2.2}$$

where the complex vector function $A(\vec{r})$ of position $\vec{r} = x, y, z$ is denoted by,

$$A(\vec{r}) = a_j(\vec{r})e^{i\psi_j(\vec{r})}, \tag{2.3}$$

in which the amplitudes, $a_j(\vec{r})$, and the phases, $\psi_j(\vec{r})$, are real functions, $j = 1, 2, 3$, and $A^*(\vec{r})$ represents the complex conjugate of $A(\vec{r})$.

By taking the square of equation (2.2), one obtains,

$$|E(\vec{r}, t)|^2 = \frac{1}{4}\left[A^2(\vec{r})e^{-2i\omega t} + A^{*2}(\vec{r})e^{2i\omega t} + 2A(\vec{r})A^*(\vec{r})\right]. \tag{2.4}$$

Since the intensities are compared in the same condition, the quantity $\left\langle|E(\vec{r}, t)|^2\right\rangle$ is used to measure intensity. The average of this equation (2.4) can be recast into,

$$\begin{aligned} \left\langle|E(\vec{r}, t)|^2\right\rangle &= \frac{1}{4}\left\{\frac{1}{2T}\int_{-T}^{T}\left[A^2(\vec{r})e^{-2i\omega t} + A^{*2}(\vec{r})e^{2i\omega t} + 2A(\vec{r})A^*(\vec{r})\right]dt\right\} \\ &= \frac{1}{4}\left\{\left[\frac{A^2(\vec{r})}{2i\omega T} + \frac{A^{*2}(\vec{r})}{2i\omega T}\right]\left[e^{i\omega T} - e^{-i\omega T}\right] + 2A(\vec{r})A^*(\vec{r})\right\}. \end{aligned} \tag{2.5}$$

The photocurrent at the detector[5] is proportional to the light intensity, and the detector circuitry averages over a time, which is longer than a cycle of oscillation. A detector receives an average of the effects produced by the different values of the amplitude, $a(t)$, which is sensitive to the square of $E(\vec{r}, t)$, therefore the time average of the intensity tends to a finite value as the averaging interval is increased indefinitely. The intensity of the wave averaged over the time interval $2T$ is needed to make an observation. Assuming a stationary wave field, taking the time average of the energy

[5]A detector transforms the light intensity into electrical signal, which provides the number of photo-events collected during the time of measurement and an additive random noise.

over an interval much greater than the period, $T = 2\pi/\omega$, the intensity is deduced as,

$$I(\vec{r},t) = \lim_{T\to\infty} \frac{1}{2T}\int_{-T}^{T} |E(\vec{r},t)|^2\, dt = \langle EE^*\rangle\,. \tag{2.6}$$

When T is large, the first two terms of the equation (2.5) are negligible in comparison with the last term, $2A(\vec{r})A^*(\vec{r})$. The intensity I at the same point is derived as,

$$\begin{aligned} I &\propto \left\langle |E(\vec{r},t)|^2\right\rangle = \frac{1}{2}A(\vec{r})A^*(\vec{r}) \\ &= \frac{1}{2}\left(|A_x|^2 + |A_y|^2 + |A_z|^2\right) = \frac{1}{2}\left(a_1^2 + a_2^2 + a_3^2\right). \end{aligned} \tag{2.7}$$

The intensity for a stationary wave as defined in equation (2.7) cannot be described adequately for the fluctuations or pulse type phenomena, thus,

$$I(\vec{r},t) = \frac{1}{2T}\int_{t-T}^{t+T} |E(\vec{r},t)|^2\, dt. \tag{2.8}$$

Equation (2.8) corresponds to the moving average with a window of width T centered at t.

2.1.2 *Harmonic plane waves*

A harmonic plane wave represents a wave field spread out periodically in space and time. The notable features are:

- the harmonic variations of the electric and magnetic fields are always perpendicular to each other and to the direction of propagation, $\vec{\kappa}$,
- the cross product $\vec{E} \times \vec{B}$ provides the direction of travel, and
- the field always vary sinusoidally, and vary with the same frequency and are in phase with each other.

In an homogeneous medium[6] in a region free of currents and charges, each rectangular component $V(\vec{r},t)$ of the field vectors, obeys the homogeneous wave equation,

$$\left(\nabla^2 - \frac{1}{v^2}\frac{\partial^2}{\partial t^2}\right) V(\vec{r},t) = 0, \tag{2.9}$$

[6]Homogeneous optical medium through which light energy passes partially or completely has a uniform composition throughout.

in which $V(\vec{r},t)$ is the wave function describing the propagation of the wave, t the time, $\vec{r} = x\vec{i} + y\vec{j} + z\vec{k}$ the position vector, and v the propagation velocity of the wave and is dependent on the physical features of the medium.

This is a linear second-order differential equation of a hyperbolic signature. In principle, the differential equation is nonlinear, but in this case the initial approximation is considered to be linearized. A plane wave may be a simple two (2-D) or three-dimensional (3-D) wave. The characteristic of such a wave is that all the points on a plane that is perpendicular to direction of propagation, have same phase value. Waves propagating in 2- or 3-dimensions are analyzed using the concept of wavefronts similar to ripples created by a stone when dropped in a pool of water. The wavefront is the locus of points of constant phase and the successive wavefronts are separated by one wavelength. For 2-D wave, the wavefronts are parallel lines, separated by a distance equal to the wavelength of the wave, while for 3-D wave, the wavefronts are parallel planes. They are perpendicular to the direction of the wave at every point.

The general solution of the wave equation (2.9) in free space, in the form of $V = V(\vec{r}\cdot\vec{s},t)$ represents a plane wave in the direction given by the unit vector, $\vec{s}$, since at each instant of time, V is constant over each of the planes, $\vec{r}\cdot\vec{s} =$ constant, which are perpendicular to $\vec{s}$.

$$V(\vec{r},t) = V(\vec{r}\cdot\vec{s} - vt), \tag{2.10}$$

where V is an arbitrary function of its arguments, which can be differentiated twice, $V(\vec{r}\cdot\vec{s} - vt)$ describes a disturbance propagating along the $\vec{s}$ direction at a constant velocity v, with shape unchanged.

The velocity is defined to be the distance traveled of one wavelength by a wave in unit time period, T, i.e., $v = \lambda/T$. By simplifying the consideration to one dimensional scalar waves in a vacuum, the equation (2.9) takes the form,

$$\left(\frac{\partial^2}{\partial z^2} - \frac{1}{v^2}\frac{\partial^2}{\partial t^2}\right) V(\vec{r},t) = 0. \tag{2.11}$$

On setting $z - vt = p$ and $z + vt = q$, the equation (2.11) translates into,

$$\frac{\partial^2 V}{\partial p \partial q} = 0. \tag{2.12}$$

In uni-dimensional space, the general solution of this equation (2.12) in free, boundary less space is,

$$\begin{aligned} V &= V_1(p) + V_2(q) \\ &= V_1(\vec{r}\cdot\vec{s} - vt) + V_2(\vec{r}\cdot\vec{s} + vt), \end{aligned} \tag{2.13}$$

where $V_1(\vec{r}\cdot\vec{s} - vt)$ is right moving wave that propagates along the positive direction at velocity v and $V_2(\vec{r}\cdot\vec{s} + vt)$ is left moving wave.

The most general solution of equation (2.9) in uni-dimensional free space is that of a superposition of arbitrary independent right-moving and left-moving waves. The wave disturbance $V(\vec{r}_0, t)$ is a function of time, t, at a point $\vec{r}_0$ in space,

$$V(\vec{r}_0, t) = a\cos(2\pi\nu t + \psi). \tag{2.14}$$

where $a > 0$ is the amplitude of the wave, ν the frequency of the wave which is the reciprocal of its period T, ψ the phase constant, and the argument of the cosine is known as the phase.

A wave equation representing a harmonic plane wave that propagates in the direction specified by a unit vector $\vec{s}$ satisfies the homogeneous wave equation (2.9). On replacing t by $(t - \vec{r}\cdot\vec{s}/v)$ in equation (2.14), one finds,

$$\begin{aligned} V(\vec{r}, t) &= a\cos\left[\omega\left(t - \frac{\vec{r}\cdot\vec{s}}{v}\right) + \psi\right] \\ &= a\cos(\omega t - \vec{\kappa}\cdot\vec{r} + \psi), \end{aligned} \tag{2.15}$$

in which the phase term is given by,

$$\omega t - \vec{\kappa}\cdot\vec{r} = \omega\left(t - \frac{\vec{r}\cdot\vec{s}}{v}\right) = \text{constant}, \tag{2.16}$$

where $\cos(\omega t - \vec{\kappa}\cdot\vec{r})$ is the oscillatory term, $\vec{\kappa}$ the wave vector, $\omega = 2\pi\nu$ the angular frequency, and $\vec{\kappa}\cdot\vec{r}$ represents planes in space of constant phase.

Equation (2.15) remains unchanged when $\vec{r}\cdot\vec{s}$ is replaced by $\vec{r}\cdot\vec{s} + \lambda$, where $\lambda = v(2\pi/\omega) = vT$, is the wavelength of the wave which is the minimum distance between the two points on the wave profile that are oscillating in phase at a given instant of time. Unlike a plane harmonic wave, the most general wave is not periodic in space. The velocity of the planar wavefronts can be found from the condition, $\vec{\kappa}\cdot\vec{r} - \omega t + \psi = \text{constant}$. This implies that the points on the wavefront move at a velocity $d\vec{r}/dt$, satisfying $d\psi/dt = 0$, i.e.,

$$\vec{\kappa} \cdot \frac{d\vec{r}}{dt} = \omega, \qquad \text{or,} \tag{2.17}$$

$$\frac{d\vec{r}}{dt} = v = \frac{\omega}{|\vec{\kappa}|} = \frac{\lambda}{T} = \lambda\nu. \tag{2.18}$$

The wavelength propagating in free space corresponding to harmonic wave of the same frequency is represented by λ_0 and is given by,

$$\lambda_0 = cT = n\lambda. \tag{2.19}$$

The wave number vector, $\vec{\kappa}$ is given by the relation,

$$\vec{\kappa} = \frac{\omega}{c}\vec{s} = \frac{2\pi}{\lambda}\vec{s}, \tag{2.20}$$

where $\vec{s}$ is the unit vector in the direction of propagation of the wave.

In most general case, ω is dependent on the direction of propagation, $\vec{\kappa}$, as well as on the magnitude $\kappa = |\vec{\kappa}|$, in which

$$\kappa = \frac{\omega}{c} = \frac{2\pi}{\lambda} = \frac{2\pi\nu}{c}. \tag{2.21}$$

is called the wavenumber and is defined as the number of wavelength in vacuum per unit length (m).

The wave number for propagation in free space is defined as, $\kappa_0 = 2\pi/\lambda_0$. This quantity is widely used in spectroscopy[7] and the spectroscopic wave number is expressed as, $\mathrm{k} = 1/\lambda_0$. The vector $\vec{\kappa} = \kappa\vec{s}$, is called wave vector in the medium in the direction of propagation, $\vec{s}$ and the wave vector in vacuum is given by $\vec{\kappa}_0 = \kappa_0\vec{s}$, where $\kappa = n\kappa_0 = 2\pi/\lambda = n\omega/c = \omega/v$ and n the index of refraction.

The path length l is the distance through which a wavefront recedes when the phase increases by δ and is expressed as,

$$l = \frac{v}{\omega}\delta = \frac{\lambda}{2\pi}\delta = \frac{\lambda_0}{2\pi n}\delta. \tag{2.22}$$

[7]Spectroscopy analyzes the lines of light emitted from excited atoms as the electrons drop back through their orbitals. It can be used to understand the characteristics such as velocities, redshifts, abundances, magnetic field, etc.

2.1.3 *Harmonic spherical waves*

In a homogeneous medium, harmonic wave disturbances emanating from a point source[8] spread out in all directions. These waves are spherical waves. The wavefronts corresponding to such waves are concentric surfaces and are perpendicular to the direction of propagation at any point. In contrast to the plane wave, where the amplitude remains constant as it propagates away from the source, the spherical wave decreases in amplitude.

The strength of a wave is described by the intensity, I, of the wave. If an isotropic source radiating its energy equally in all directions, the energy emitted per second by the source S passes through the surface of a sphere of area $4\pi r^2$ (see Figure 2.1), in which r is the radius of the sphere. If the energy is not absorbed, the energy flowing through the sphere per second is $I \times 4\pi r^2$. For a spherical wave, the intensity I is proportional to the distance from the source by an inverse square law i.e., $1/r^2$. An energy flux at a distance r from a point source is distributed over an area A is spread over an area $4A$ at a distance $2r$. The energy per second reaching a detector of fixed area decreases inversely proportional to the distance squared.

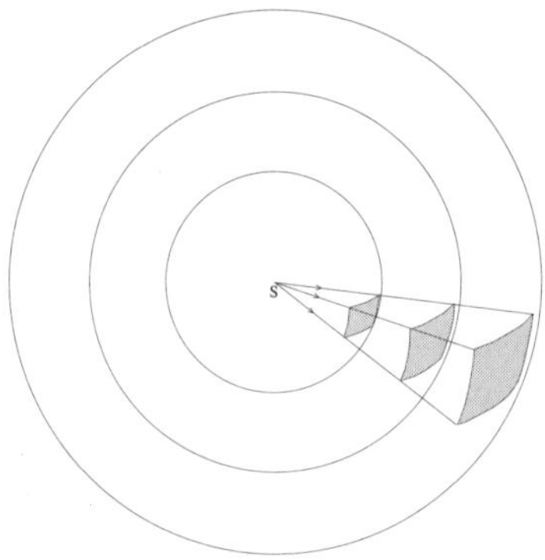

Fig. 2.1 Propagation of a spherical wave.

For the solutions representing spherical waves, making the assumption that the function $V(\vec{r}, t)$ has spherical symmetry about the origin, i.e., $V(\vec{r}, t) = V(r, t)$, where $r = |\vec{r}| = \sqrt{x^2 + y^2 + z^2}$, and

$$x = r \sin\theta \cos\phi; \quad y = r \sin\theta \sin\phi; \quad z = r \cos\theta. \tag{2.23}$$

[8]A point source is a source of light which is of the size of pin head of a common pin. Stars may be considered as a point source since they are sufficiently distant and isolated; an unresolved binary or a multiple system can also be considered as a point source.

The Lapacian operator in the case of spherical coordinates reads,

$$\nabla^2 = \frac{1}{r^2}\frac{\partial}{\partial r}\left(r^2\frac{\partial}{\partial r}\right) + \frac{1}{r^2\sin\theta}\frac{\partial}{\partial\theta}\left(\sin\theta\frac{\partial}{\partial\theta}\right) + \frac{1}{r^2\sin^2\theta}\frac{\partial^2}{\partial\phi^2}. \tag{2.24}$$

Since the spherical wave is spherically symmetric showing no dependence on θ and ϕ, the Lapacian operator reduces to the first term of the RHS of the equation, i.e.,

$$\nabla^2 = \frac{1}{r^2}\frac{\partial}{\partial r}\left(r^2\frac{\partial}{\partial r}\right) = \frac{\partial^2}{\partial r^2} + \frac{2}{r}\frac{\partial}{\partial r}. \tag{2.25}$$

By using this relation,

$$\nabla^2 V(r,t) = \frac{1}{r}\frac{\partial^2}{\partial r^2}\left[rV(r,t)\right], \tag{2.26}$$

thus, the wave equation (2.9) attains the form,

$$\left(\frac{\partial^2}{\partial r^2} - \frac{1}{v^2}\frac{\partial^2}{\partial t^2}\right) rV(r,t) = 0, \tag{2.27}$$

The general solution may be written as,

$$V(\vec{r},t) = \frac{1}{\vec{r}}V_1(r - vt) + \frac{1}{\vec{r}}V_2(r + vt), \tag{2.28}$$

The first term on the right hand side (RHS) of this equation (2.28) represents a spherical wave diverging from the origin while the second term converging towards the origin. The amplitude of the disturbance falls off as $1/\vec{r}$. An outgoing spherical wave is obtained as,

$$V(\vec{r},t) = \frac{a}{\vec{r}}\cos\left[\omega\left(t - \frac{r}{v}\right) + \psi\right]. \tag{2.29}$$

2.2 Complex representation of monochromatic light waves

The term, $V(\vec{r},t)$, in equation (2.15) is periodic both in $\vec{\kappa}\cdot\vec{r}$ and t, and it describes a wave propagating along the $\vec{\kappa}$ direction at a velocity $v = \omega/\kappa$. A general time harmonic wave of frequency ω may be defined from the real solution of the wave equation, $V(\vec{r},t)$ at a point $\vec{r}$ of the form,

$$V(\vec{r},t) = \Re\left\{a(\vec{r},\nu)e^{-i[2\pi\bar{\nu}t - \psi(\vec{r},\nu)]}\right\}, \tag{2.30}$$

where $\bar{\nu}$ the mean optical frequency, $\psi(\vec{r},\nu)$ the phase functions, and $a(\vec{r},\nu)$ the phasor amplitude of the field, which rotates with angular frequency, $\omega = 2\pi\bar{\nu}$ and its projection on the real axis provides the physical amplitude.

The phase component gains significance in the case of superimposing several waves. The oscillations of V in the equation (2.30) are bounded by $0 \leq |V| \leq a$ or, $-a \leq V \leq a$. The most physically relevant information is embodied in the relative phase differences among superimposed waves and these relative amplitude ratios. This information is encoded in the complex exponential representation. Let the complex amplitude of the vibration, $A(\vec{r},\nu)$ be,

$$A(\vec{r},\nu) = a(\vec{r},\nu)e^{i\psi(\vec{r},\nu)}, \tag{2.31}$$

where $a(\vec{r},\nu)$ and $\psi(\vec{r},\nu)$ are real functions, which provide the amplitude and phase respectively, of each monochromatic[9] component of frequency ν.

From a quantum-mechanical point, the term, $e^{i\psi}$ is a propagator of a probability amplitude from one place and time to another, in which ψ represents the change in phase (modulo 2π) of the probability amplitude along the minimum path. The complex representation of the analytic[10] signal of a plane wave, $U(\vec{r},t)$, is recast into,

$$U(\vec{r},t) = a(\vec{r},\nu)e^{-i[2\pi\bar{\nu}t - \psi(\vec{r},\nu)]} \tag{2.32}$$

$$= A(\vec{r},\nu)e^{-i2\pi\bar{\nu}t}. \tag{2.33}$$

This complex representation is preferred for linear time invariant systems, because the eigenfunctions of such systems are of the form $e^{-i\omega t}$. The complex representation of the analytic signal of a spherical wave is represented by,

$$U(\vec{r},t) = \left[\frac{a(\vec{r},\nu)}{\vec{r}}\right] e^{-i[2\pi\bar{\nu}t - \psi(\vec{r},\nu)]}. \tag{2.34}$$

Equation (2.34) is the solution to the wave equation and represents spherical wave propagating outward from the origin. The irradiance of the spherical wave is proportional to the square of the amplitude $a(\vec{r},\nu)/\vec{r}$ at a distance $\vec{r}$.

The complex amplitude is a constant phasor in the monochromatic case, therefore, the Fourier transform (FT) of the complex representation of the

---

[9] Monochromatic radiation is the radiation of single precise energy; the energy is related with its wavelength which is often used to specify the color of the visible radiation.

[10] A function is analytic if its components are harmonic conjugates.

signal, $U(\vec{r},t)$, is given by,

$$\widehat{U}(\vec{r},\nu) = a(\vec{r},\nu)e^{i\psi}\delta(\nu - \bar{\nu}), \tag{2.35}$$

in which ^ stands for a Fourier transform.

The spectrum is equal to twice the positive part of the instantaneous spectrum, $\widehat{V}(\vec{r},\nu)$.

2.2.1 *Superposition of waves*

The superposition principle (sum of the amplitudes of the contributions from each point on the surface) states the most important property of wave phenomena. The effects of interference and diffraction are the corollaries of this principle (Pedrotti and Pedrotti, 1987). If two or more waves meet at a point in space, the net disturbance at the point at each instant of time is defined by the sum of the disturbances created by each of the waves individually, i.e.,

$$V = V_1 + V_2 + \cdots, \tag{2.36}$$

where $V_{j=1,2,\cdots}$ are the constituent waves.

Since the free wave equation is a linear homogeneous differential equation (equation 2.9), any linear combination of its solution is a solution as well. The superposition of electromagnetic waves in terms of their electric field, $\vec{E}$ and magnetic field, $\vec{B}$, may also be expressed as,

$$\begin{aligned}\vec{E} &= \vec{E}_1 + \vec{E}_2 + \cdots,\\ \vec{B} &= \vec{B}_1 + \vec{B}_2 + \cdots.\end{aligned} \tag{2.37}$$

This type of linear superposition stands valid in the presence of matter. Deviations from linearity are observed at high intensities produced by lasers[11] when the electric fields approach the electric fields comparable to atomic fields (non-linear optics). Let two waves, E_1, and E_2 of the same frequency, but different in amplitude and phase, combine to form a resultant wave, E; the orientation of the electric or magnetic fields must be

[11] A laser (Light Amplification by Stimulated Emission of Radiation) is an optical source that makes use of mechanisms such as absorption, stimulation, and spontaneous emission. The laser output can be continuous or pulsed and is contributing significantly to the science, communications, and technology. Solid state lasers emits the ultrafast pulses, whose time-durations are of the order of picoseconds (ps; 10^{-12} s) or femtoseconds (fs; 10^{-15} s). These lasers are used to map the sequence of the events, while tunable lasers are useful in strong-field interactions.

taken into account. The time variations of these waves are expressed as,

$$E_1(x,t) = a_1 \cos(\kappa x - \omega t + \psi_1) = \Re\left[a_1 e^{i\psi_1} e^{\kappa x - \omega t}\right], \quad (2.38)$$

$$E_2(x,t) = a_2 \cos(\kappa x - \omega t + \psi_2) = \Re\left[a_2 e^{i\psi_2} e^{\kappa x - \omega t}\right]. \quad (2.39)$$

These two waves (equations 2.38, 2.39) travelling in the same direction and intersecting at a fixed point in space may differ in phase, $\vec{\kappa}\cdot(\vec{r}_2 - \vec{r}_1) + (\psi_2 - \psi_1)$. The difference in phase is due to a path difference provided by the first term and an initial phase difference in the range $-\pi < \delta \leq \pi$ is given by the second term, $\delta = \psi_2 - \psi_1$. The real part is avoided for convenience while deriving the resultant wave,

$$\begin{aligned} E_0(x,t) &= E_1(x,t) + E_2(x,t) \\ &= a_1 e^{i\psi_1} e^{\kappa x - \omega t} + a_2 e^{i\psi_2} e^{\kappa x - \omega t} \\ &= a_0 e^{i\psi_0} e^{i(\kappa x - \omega t)}. \end{aligned} \quad (2.40)$$

Thus, the superposition of two harmonic waves of given frequency produces a harmonic wave of same frequency with a given amplitude and phase. The amplitude and phase of the resultant wave are derived from,

$$a_0 e^{i\psi_0} = a_1 e^{i\psi_1} + a_2 e^{i\psi_2}, \quad (2.41)$$

or,

$$a_0 \cos\psi_0 + i a_0 \sin\psi_0 = a_1 \cos\psi_1 + a_2 \cos\psi_2 + i\left(a_1 \sin\psi_1 + a_2 \sin\psi_2\right). \quad (2.42)$$

Here a_0 is the resultant amplitude and ψ_0 the resultant phase. The intensity of the resultant wave is proportional to,

$$\begin{aligned} I_0 = \left|a_0 e^{i\psi_0}\right|^2 &= a_1^2 + a_2^2 + 2a_1 a_2 \cos(\psi_2 - \psi_1) \\ &= a_1^2 + a_2^2 + 2a_1 a_2 \cos\delta. \end{aligned} \quad (2.43)$$

Equation (2.43) is proportional to the intensity distribution in a two-beam[12] interference pattern. If $a_1 = a_2$, one gets,

$$I_0 = a_0^2 = 2a_1^2\left(1 + \cos\delta\right) = 4a_1^2 \cos^2\left(\frac{\delta}{2}\right). \quad (2.44)$$

Thus the linear combination provides a cosine intensity distribution or equivalently a cosine squared distribution. Since the real and imaginary

[12]Beam of light is a collection of large number of rays of light, the path along which the light energy travels in a given direction.

parts on the two sides of equation (2.42) must be equal, the phase angle is determined by,

$$\tan\psi_0 = \frac{a_0 \sin\psi_0}{a_0 \cos\psi_0} = \frac{a_1 \sin\psi_1 + a_2 \sin\psi_2}{a_1 \cos\psi_1 + a_2 \cos\psi_2}. \tag{2.45}$$

For the combination of several sine waves, the resultant wave is given by,

$$E(x,t) = \sum_{n=1}^{N} a_n e^{i\psi_n} e^{i(\kappa x - \omega t)} = a_0 e^{i\psi_0} e^{i(\kappa x - \omega t)}, \qquad \text{or,} \tag{2.46}$$

$$a_0 e^{i\psi_0} = \sum_{n=1}^{N} a_n e^{i\psi_n}. \tag{2.47}$$

The phase of this resultant wave is derived as,

$$\tan\psi_0 = \frac{\sum_{n=1}^{N} a_n \sin\psi_n}{\sum_{n=1}^{N} a_n \cos\psi_n}. \tag{2.48}$$

The resultant intensity is proportional to,

$$\begin{aligned} I_0 &= \left| a_0 e^{i\psi_0} \right|^2 = \left| \sum_{n=1}^{N} a_n e^{i\psi_n} \right|^2 = \sum_{n=1}^{N} \sum_{m=1}^{N} a_n e^{i\psi_n} a_m e^{-i\psi_m} \\ &= \sum_{n=1}^{N} a_n^2 + \sum_{n=1}^{N} \sum_{m=1, n\neq m}^{N} a_n a_m e^{i(\psi_n - \psi_m)} \\ &= \sum_{n=1}^{N} a_n^2 + \sum_{n=1}^{N} \sum_{m=1, n>m}^{N} a_n a_m \left(e^{i(\psi_n - \psi_m)} + e^{-i(\psi_n - \psi_m)} \right) \\ &= \sum_{n=1}^{N} a_n^2 + 2 \sum_{n=1}^{N} \sum_{m=1, n>m}^{N} a_n a_m \cos(\psi_n - \psi_m). \end{aligned} \tag{2.49}$$

Equation (2.49) expresses a harmonic waves of same frequency in which the last term is the interference term.

If the number of randomly phased sources of equal amplitude is large, the waves are incoherent, and the sum of cosine term approaches zero. The resultant irradiance of N identical but randomly phased sources is the sum of the individual irradiance, i.e.,

$$a_0^2 = \sum_{n=1}^{N} a_n^2. \tag{2.50}$$

If all the a_n are equal, one finds

$$a_0^2 = N a_0^2. \tag{2.51}$$

While for the in-phase coherent case, the equation (2.49) turns out to be,

$$a_0^2 = \sum_{n=1}^{N} \sum_{m=n}^{N} a_n a_m = \left[\sum_{n=1}^{N} a_n \right]^2. \tag{2.52}$$

Here all ψ are assumed to be equal. If all the amplitudes are equal, we find,

$$a_0^2 = N^2 a_0^2. \tag{2.53}$$

The resultant irradiance of N identical coherent sources, radiating in phase with each other, is equal to the N^2 times the irradiance of the individual sources. For both coherent and in-coherent cases, the total energy does not change, but the distribution of the energy change.

2.2.2 *Standing waves*

The standing wave is produced by the superposition of the two traveling waves of same amplitudes and frequency which propagate in a uni-dimensional space, in opposite directions. From the principle of superposition, the resultant wave, $U(x,t)$ is given by,

$$U(x,t) = ae^{i(\kappa x - \omega t + \psi_1)} + ae^{i(-\kappa x - \omega t + \psi_2)} \tag{2.54}$$

$$= 2ae^{-i\omega t} \cos(\kappa x + \psi). \tag{2.55}$$

For $\psi = \pi/2$,

$$V(x,t) = \Re[U(x,t)] = 2a \sin \kappa x \cos \omega t. \tag{2.56}$$

Equation (2.56) represents the standing wave and provides a solution of the uni-directional wave equation for waves which propagate along a bounded uni-directional region, and interfere destructively at its end points culminating to vanish. The wave function vanishes at the points,

$$x_n = \frac{nx}{\kappa} = \frac{n\lambda}{2}, \qquad n = 0, \pm 1, \pm 2, \cdots, \tag{2.57}$$

for any value of t. These points are called the nodes of the wave and are separated by half a wavelength. At a given time t, the function E obtains its extreme at the points,

$$x_m = \left(m + \frac{1}{2}\right)\frac{\pi}{\kappa} = \left(m + \frac{1}{2}\right)\frac{\lambda}{2}. \tag{2.58}$$

These points are known as the antinodes of the wave. Unlike the traveling waves, standing waves transmits no energy.

2.2.3 *Phase and group velocities*

Superposition of two waves with wave crests moving at different speeds, exhibits periodically large and small amplitudes. This raises the important notion of the 'group velocity' and its relationship to the 'phase velocity'. Let a wave formed by the superposition of two plane monochromatic waves of equal amplitude, differing in frequency and wavenumber, ω_1, κ_1 and ω_2, κ_2, be propagated in the direction of the z-axis. Differences in frequency imply differences in wavelength and velocity. The resultant wave is expressed as,

$$\begin{aligned} U(z,t) &= Ae^{i(\kappa_1 z - \omega_1 t)} + Ae^{i(\kappa_2 z - \omega_2 t)} \\ &= A\left(e^{i\{[(\kappa_1 - \kappa_2)/2]z - [(\omega_1 - \omega_2)/2]t\}}\right. \\ &\qquad \left. + e^{i\{[(\kappa_2 - \kappa_1)/2]z - [(\omega_2 - \omega_1)/2]t\}}\right) \\ &\qquad \times e^{i[(\kappa_2 - \kappa_1)/2]z - i[(\omega_2 - \omega_1)/2]t} \\ &= 2A\cos\left(\frac{\kappa_1 - \kappa_2}{2}z - \frac{\omega_1 - \omega_2}{2}t\right) \\ &\qquad \times e^{i\{[(\kappa_1 + \kappa_2)/2]z - [(\omega_1 + \omega_2)/2]t\}} \\ &= 2A\cos(\kappa_g z - \omega_g t)e^{i(\bar{\kappa} z - \bar{\omega} t)}, \end{aligned} \tag{2.59}$$

with

$$\begin{aligned} \bar{\kappa} &= \frac{\kappa_1 + \kappa_2}{2}, \qquad & \bar{\omega} &= \frac{\omega_1 + \omega_2}{2}, \\ \kappa_g &= \frac{\kappa_1 - \kappa_2}{2}, \qquad & \omega_g &= \frac{\omega_1 - \omega_2}{2}. \end{aligned} \tag{2.60}$$

By this transformation the wave is split into an amplitude factor slowly oscillating at $\omega_1 - \omega_2$ and a phase factor rapidly oscillating at $\omega_1 + \omega_2$. Figure (2.2) displays the cosine function plot. In this the high frequency carrier wave is 300 MHz, while the low frequency is 10 MHz. The low frequency wave serves as an envelop modulating the high frequency wave. The curve depicting the envelope of the resulting wave disturbance exhibits the phenomenon of beats[13]. Let the period of first and seconds signals respectively be, $\kappa_1 = 2\pi/\lambda_1$ and $\kappa_2 = 2\pi/\lambda_2$, therefore,

$$2\kappa_g = (\kappa_1 - \kappa_2) = 2\pi\left(\frac{1}{\lambda_1} - \frac{1}{\lambda_2}\right) = \frac{2\pi}{\lambda_g}, \tag{2.61}$$

with

$$\frac{1}{\lambda_g} = \frac{1}{\lambda_1} - \frac{1}{\lambda_2}.$$

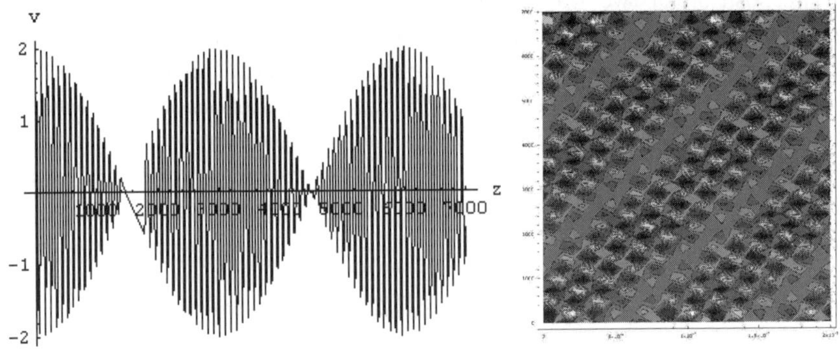

(a) (b)

Fig. 2.2 (a) 1-D plot of the cosine factor of $V(z,t) = 2a\cos(\bar{\kappa}z - \bar{\omega}t)\cos(\kappa_g z - \omega_g t)$ in space domain; (b) 2-D contour plot of the same in time and space domains.

It is to be noted that modulated waves, familiar to radio physicists and engineers, are encountered throughout physics. The amplitude of a cosinusoidal wave varies with time either periodically as in the curve depicting the envelope of the resulting wave disturbance exhibits the phenomenon of beats, or aperiodically as in the isolated group of waves. A cosine modulation has a spectrum with 'sidebands' on either side of the components at

[13]The superposition of two equal amplitude harmonic waves of different frequency produces a beat, which is perceived as periodic variations in volume whose rate is the difference between the two frequencies.

$\pm\kappa_1$. The beat frequency, ω_b is derived as,

$$\omega_b = 2\omega_g = 2\left[\frac{\omega_1 - \omega_2}{2}\right] = \omega_1 - \omega_2. \tag{2.62}$$

Substituting equation (2.59) into the wave equation (2.9), the following relationship emerges,

$$\begin{aligned}\left(\nabla^2 - \frac{1}{v^2}\frac{\partial^2}{\partial t^2}\right) Ae^{i(\vec{\kappa}\cdot\vec{r} - \omega t)} &= -\left(\vec{\kappa}^2 - \frac{\omega^2}{v^2}\right) Ae^{i(\vec{\kappa}\cdot\vec{r} - \omega t)} \\ &= -\left(\vec{\kappa}^2 - \frac{\omega^2}{v^2}\right) U = 0. \end{aligned}\tag{2.63}$$

Thus the dispersion relation is given by,

$$\vec{\kappa}^2 = \frac{\omega^2}{v^2}. \tag{2.64}$$

The wave velocity in equation (2.18), i.e., $v = \omega/|\vec{\kappa}|$, provides the dispersion relation. This is a functional relation between ω and $\vec{\kappa}$, and is given by,

$$\omega = \omega(\vec{\kappa}). \tag{2.65}$$

Since the wavefront is a plane normal to $\vec{\kappa}$, $d\vec{r}/dt$ is directed along $\vec{\kappa}$, hence,

$$v = \left|\frac{d\vec{r}}{dt}\right| = v_p, \tag{2.66}$$

with v_p as the phase velocity of the wave, which can be derived from the equation (2.60),

$$v_p = \frac{\bar{\omega}}{\bar{\kappa}} = \frac{\omega_1 + \omega_2}{\kappa_1 + \kappa_2} \cong \frac{\omega}{\kappa}, \tag{2.67}$$

where the final term, ω/κ, is an approximation in the case for neighboring frequency and wavelength components in a continuum.

The ensemble of waves travel as a group or packet whose mean progress is described by a group velocity, v_g of the wave train in the superposition,

$$v_g = \frac{\omega_g}{\kappa_g} = \frac{\omega_1 - \omega_2}{\kappa_1 - \kappa_2} \cong \frac{d\omega}{d\kappa}. \tag{2.68}$$

In this case, one assumes the differences between the frequencies and propagation constants are small. The group velocity is the velocity at which physical information is carried away by the wave train. The energy of the wave is determined by its amplitude, so in general, the group velocity provides the velocity of the energy transport. Let the equation (2.67) be plugged into equation (2.68), thus the relation between group and phase velocities is derived as,

$$\begin{aligned} v_g &= \frac{d\omega}{d\kappa} = \frac{d}{d\kappa}(dv_p) \\ &= v_p + \kappa\left(\frac{dv_p}{d\kappa}\right) = v_p - \lambda\left(\frac{dv_p}{d\lambda}\right). \end{aligned} \tag{2.69}$$

In a dispersion-free medium, $dv_p/d\kappa = 0$, the equation (2.69) turns out to be $v_g = v_p$. This is the case of light propagating in vacuum, where $v_g = v_p = c$. In amplitude modulation (AM) of radio waves, the carrier waves are modulated to contain information. Here the group velocity, known as signal velocity is normally less than the phase velocity of the carrier waves. If the light pulses are transmitted through a dispersive medium, the group velocity is the velocity of the pulses, and will be different from the velocity of the individual harmonic waves. In the case of a non-trivial dispersion, for example, the propagation of electromagnetic radiation through an ionized plasma[14], the relation is provided by,

$$\omega^2 = c^2\kappa^2 + \bar{\omega}^2. \tag{2.70}$$

It describes the relation between the energy (ω) and the momentum (κ) of massive relativistic particles in relativistic quantum mechanics. In this equation (2.70), ω never drops below ω_p, which serves as cutoff frequency.

2.3 Complex representation of non-monochromatic fields

In practice, the optical fields are never completely monochromatic. The electromagnetic waves emitted by the atoms are discharged as wave trains. Owing to the finite length of these wave trains, the radiation forms a frequency spectrum. The realistic light beam (polychromatic), $U^{(r)}$, that is regarded as a member of an ensemble consisting of all realizations of the field, fluctuates as a function of time. At optical frequencies, the fluctuating field components are not observable quantities, but quadratic averages of them are. Since as a rule, the stochastic field is treated as ergodic, the ensemble average can be replaced by a time average.

[14]A plasma is typically an ionized gas, the fourth state of matter, and is considered to be a distinct phase of matter in contrast to solids, liquids, and gases because of its unique characteristic that deals with electric charges. It consists of a collection of free moving electrons and ions. The free electric charges make the plasma electrically conductive. The term ionized means that at least one electron has been dissociated from an atom.

Let $V(\vec{r},t)$ be a non-monochromatic wave, with an associated analytic signal $U(\vec{r},t)$. Following conventional Fourier analysis, one may represent the wave form of such wave as the superposition of a large number of sinusoidal components (of monochromatic waves) of different frequencies, each component having a constant amplitude and phase over the period of observation. The amplitude and phase being random with respect to each other, one may express,

$$V(\vec{r},t) = \Re \int_0^\infty a(\vec{r},\nu)e^{-i[2\pi\nu t - \psi(\vec{r})]}d\nu. \tag{2.71}$$

The expression for the complex amplitude in the case of polychromatic wave can be derived by analogy to the monochromatic case (equation 2.25) yielding,

$$U(\vec{r},t) = 2\int_0^\infty \widehat{U}(\vec{r},\nu)e^{-i2\pi\nu t}d\nu, \tag{2.72}$$

where the Fourier transform (FT) of the complex representation of the signal, $U(\vec{r},t)$, is $\widehat{U}(\vec{r},\nu)$.

The disturbance produced by a real physical source is calculated by the integration of the monochromatic signals over an optical band pass. For a real non-monochromatic vibration, $U^{(r)}(\vec{r},t)(-\infty \leq t \leq \infty)$ is expressed as,

$$U^{(r)}(\vec{r},t) = \int_0^\infty a(\vec{r},\nu)\cos\left[\psi(\vec{r},\nu) - 2\pi\nu t\right]d\nu. \tag{2.73}$$

Equation (2.73) is the Fourier cosine integral representation of the real valued signal $U^{(r)}(\vec{r},t)$. The associated Fourier sine integral function is given by,

$$U^{(i)}(\vec{r},t) = \int_0^\infty a(\vec{r},\nu)\sin\left[\psi(\vec{r},\nu) - 2\pi\nu t\right]d\nu. \tag{2.74}$$

Invoking Euler's formula, one derives the complex analytic signal $U(\vec{r},t)$ associated with the real function, $U^{(r)}(\vec{r},t)$ as,

$$U(\vec{r},t) = U^{(r)}(\vec{r},t) + iU^{(i)}(\vec{r},t). \tag{2.75}$$

The imaginary part $U^{(i)}(\vec{r},t)$ contains no new information about the optical field. Therefore, the complex function, $U(\vec{r},t)$ is derived in the form of a Fourier integral as,

$$U(\vec{r},t) = \int_0^\infty a(\vec{r},\nu)e^{i[\psi(\vec{r},\nu) - 2\pi\nu t]}d\nu, \tag{2.76}$$

where the signal in equation (2.76) contains positive frequencies.

The functions $U^{(i)}(\vec{r},t)$ and $U(t)$ are uniquely specified by $U^{(r)}(\vec{r},t)$. $U^{(i)}$ is obtained from $U^{(r)}$ by replacing the phase $\psi(\nu)$ of each Fourier component by $[\psi(\nu) - \pi/2]$. The integrals of the equations (2.73 and 2.74) are allied integrals. Hence, the conjugate functions, $U^{(r)}$ and $U^{(i)}$ are described by Hilbert transform (see Appendix B),

$$U^{(r)}(t) = \frac{1}{\pi}\int_{-\infty}^\infty \frac{U^{(i)}(\tau)d\tau}{\tau - t}, \tag{2.77}$$

$$U^{(i)}(t) = -\frac{1}{\pi}\int_{-\infty}^\infty \frac{U^{(r)}(\tau)d\tau}{\tau - t}. \tag{2.78}$$

Let $\widehat{U}^{(r)}(\vec{r},\nu)$ be the Fourier transform of $U^{(r)}(\vec{r},t)$, therefore, the latter represents as,

$$U^{(r)}(\vec{r},t) = \int_{-\infty}^\infty \widehat{U}^{(r)}(\vec{r},\nu)e^{-i2\pi\nu t}d\nu. \tag{2.79}$$

By splitting the integral on the right hand side of this equation (2.79) into two integrals with limits of $-\infty \to 0$ and $0 \to \infty$, one deduces,

$$\begin{aligned} U^{(r)}(\vec{r},t) &= \int_0^\infty \widehat{U}^{(r)*}(\vec{r},\nu)e^{i2\pi\nu t}d\nu + \int_0^\infty \widehat{U}^{(r)}(\vec{r},\nu)e^{-i2\pi\nu t}d\nu, \\ &= 2\Re\left[\int_0^\infty \widehat{U}^{(r)}(\vec{r},\nu)e^{-i2\pi\nu t}d\nu\right]. \end{aligned} \tag{2.80}$$

Comparing equations (2.72) and (2.73), one obtains,

$$\widehat{U}^{(r)}(\vec{r},\nu) = \frac{1}{2}a(\vec{r},\nu)e^{i\psi(\vec{r},\nu)} \qquad \nu \geq 0. \tag{2.81}$$

Therefore, the equation (2.76) takes the form,

$$U(\vec{r},t) = \int_0^\infty a(\vec{r},\nu)e^{i[\psi(\vec{r},\nu) - 2\pi\nu t]}d\nu \tag{2.82}$$

$$= 2\int_0^\infty \widehat{U}^{(r)}(\vec{r},\nu)e^{-i2\pi\nu t}d\nu. \tag{2.83}$$

Hence $U(\vec{r},t)$ may be derived from $U^{(r)}(\vec{r},t)$ if the operations on the latter are linear. The real part provides the real valued wave field according to the equation (2.72) by representing $U^{(r)}$ as a Fourier integral; U is also known as the complex half-range function associated with $U^{(r)}$. The following relationships follow from the equations (2.72), and (2.83) by Parseval's theorem (see Appendix B) and by the use of the relation $\widehat{U}(-\nu) = \widehat{U}^*(\nu)$,

$$\int_{-\infty}^{\infty} \left|U^{(r)}(\vec{r},t)\right|^2 dt = \int_{-\infty}^{\infty} \left|U^{(i)}(\vec{r},t)\right|^2 dt = 2\int_{0}^{\infty} \left|\widehat{U}(\vec{r},\nu)\right|^2 d\nu. \quad (2.84)$$

2.3.1 *Convolution relationship*

Let the Fourier transform representation of the real function $U^{(r)}$ be represented by,

$$U^{(r)}(t) = \int_{-\infty}^{\infty} \widehat{U}^{(r)}(\nu) e^{-i2\pi\nu t} d\nu, \quad (2.85)$$

and by Fourier inverse transform,

$$\widehat{U}^{(r)}(\nu) = \int_{-\infty}^{\infty} U^{(r)}(t) e^{i2\pi\nu t} dt. \quad (2.86)$$

The real field variable at a point represented by a position vector $\vec{r}(= x, y, z)$, at time t. It is convenient to carry out analysis in terms of associated analytic signal $U(\vec{r},t)$ instead of the real field variable $V^{(r)}(\vec{r},t)$. With the help of correlation of two functions, i.e.,

$$h(t) = \int_{-\infty}^{\infty} f(t) g(t+\tau) dt,$$

in which $f(t)$ represents an input curve, $g(t)$ a blurring function, and $h(t)$ the output value, one may deduce the correlation of two analytical signals, is expressed as (Francon, 1966),

$$\begin{aligned}\int_{-\infty}^{\infty} U_1(t+\tau) U_2^*(t) dt &= \int_{-\infty}^{\infty} U_2^*(t) \left[2\int_{0}^{\infty} \widehat{U}_1(\nu) e^{-i2\pi\nu(t+\tau)} d\nu\right] dt \\ &= 2\int_{0}^{\infty} \widehat{U}_1(\nu) e^{-i2\pi\nu\tau} \left[\int_{-\infty}^{\infty} U_2^*(t) e^{i2\pi\nu t} dt\right] d\nu.\end{aligned} \quad (2.87)$$

where $U_1(t)$ and $U_2(t)$ are the analytic signals associated with the real vibrations $U_1^{(r)}(t)$ and $U_2^{(r)}(t)$.

According to the equation (2.83),

$$2\widehat{U}_1(\nu) = \int_{-\infty}^{\infty} U_2(t)e^{i2\pi\nu t}dt, \tag{2.88}$$

$$2\widehat{U}_2^*(\nu) = \int_{-\infty}^{\infty} U_2^*(t)e^{-i2\pi\nu t}dt. \tag{2.89}$$

Therefore, the equation (2.87) turns out to be,

$$\int_{-\infty}^{\infty} U_1(t+\tau)U_2^*(t)dt = 4\int_{0}^{\infty} \widehat{U}_1(\nu)\widehat{U}_2^*(\nu)e^{-i2\pi\nu\tau}d\nu. \tag{2.90}$$

The correlation of the two functions $U_1^{(r)}(t)$ and $U_2^{(r)}(t)$, as well as the correlation of the functions $U_1^{(i)}(t)$ and $U_2^{(i)}(t)$, or $U_1^{(i)}(t)$ and $U_2^{(r)}(t)$ can be deduced as,

$$\int_{-\infty}^{\infty} U_1^{(r)}(t+\tau)U_2^{(r)}(t)dt = \int_{-\infty}^{\infty} U_1^{(i)}(t+\tau)U_2^{(i)}(t)dt, \tag{2.91}$$

$$\int_{-\infty}^{\infty} U_1^{(r)}(t+\tau)U_2^{(i)}(t)dt = -\int_{-\infty}^{\infty} U_1^{(i)}(t+\tau)U_2^{(r)}(t)dt. \tag{2.92}$$

In order to determine the convolutions of $U_1^{(i)}(t)$ and $U_2^{(i)}(t)$, we may pass from $U^{(r)}(t)$ and $U^{(i)}(t)$ by introducing the factor $e^{i\pi/2}$ in the right hand side of the equation (2.79) since a phase difference of $\pi/2$ exists between each component of $U^{(r)}(t)$ and $U^{(i)}(t)$. Therefore, according to the equation (2.76),

$$\begin{aligned}\int_{-\infty}^{\infty} U_1(t+\tau)U_2^*(t)dt = \int_{-\infty}^{\infty} &\left[U_1^{(r)}(t+\tau) + iU_1^{(i)}(t+\tau)\right] \\ &\times \left[U_2^{(r)}(t) - iU_2^{(i)}(t)\right] dt.\end{aligned} \tag{2.93}$$

By using equations (2.91) and (2.92), one obtains,

$$\begin{aligned}\int_{-\infty}^{\infty} U_1(t+\tau)U_2^*(t)dt = 2&\int_{-\infty}^{\infty} U_1^{(r)}(t+\tau)U_2^{(r)}(t)dt \\ -2i&\int_{-\infty}^{\infty} U_1^{(r)}(t+\tau)U_2^{(i)}(t)dt,\end{aligned} \tag{2.94}$$

Equation (2.94) relates that the real part of the correlation of two analytic signals equals up to a factor of 2, the correlation of the real functions associated with the analytic signals. By setting $U_1(t) = U_2(t)$ in the equation (2.94),

$$\int_{-\infty}^{\infty} U_1(t+\tau)U_1^*(t)dt = 2\int_{-\infty}^{\infty} U_1^{(r)}(t+\tau)U_1^{(r)}(t)dt$$
$$-2i\int_{-\infty}^{\infty} U_1^{(r)}(t+\tau)U_1^{(i)}(t)dt. \qquad (2.95)$$

From the equation (2.92), one finds,

$$\int_{-\infty}^{\infty} U_1^{(r)}(t+\tau)U_1^{(i)}(t)dt = -\int_{-\infty}^{\infty} U_1^{(i)}(t+\tau)U_1^{(r)}(t)dt, \qquad (2.96)$$

and for $\tau = 0$,

$$\int_{-\infty}^{\infty} U_1^{(r)}(t)U_1^{(i)}(t)dt = 0. \qquad (2.97)$$

It has been observed from this equation (2.97) that the functions $U^{(r)}(t)$ and $U^{(i)}(t)$ are orthogonal. By setting $\tau = 0$ in equation (2.95), we find,

$$\int_{-\infty}^{\infty} U_1(t)U_1^*(t)dt = 2\int_{-\infty}^{\infty} \left|U_1^{(r)}(t)\right|^2 dt. \qquad (2.98)$$

Hence the time integral of the square of the modulus of the analytic signal is equal to twice the time integral of the real functions with which the analytic signal is associated. By adding $\tau = 0$ to the equation (2.90), and using Parseval's one obtains,

$$\int_{-\infty}^{\infty} U_1(t)U_1^*(t)dt = 2\int_{-\infty}^{\infty} \left|\widehat{U}(\nu)\right|^2 d\nu = 4\int_{0}^{\infty} \left|\widehat{U}(\nu)\right|^2 d\nu. \qquad (2.99)$$

2.3.2 *Case of quasi-monochromatic light*

For a monochromatic wave field, the amplitude of the wave at any point is constant and the phase varies linearly with time. Conversely, the amplitude and phase in the case of quasi-monochromatic is wave field, undergo irregular fluctuations (Born and Wolf, 1984). The fluctuations arise since the real valued wave field $U^{(r)}$ consists of a large number of contributions that are independent of each other, the superposition of which gives rise to a fluctuating field. The rapidity of fluctuations depends on the light crossing time of the emitting region.

For a quasi-monochromatic waves, the field possesses a narrowband power spectrum of width $\Delta\nu \ll \bar{\nu}$, in which $\bar{\nu}$ is its mean frequency, therefore, the analytic signal is expressed as (Born and Wolf, 1984),

$$U(\vec{r},t) = A(\vec{r},t)e^{i[\psi(\vec{r},t) - 2\pi\bar{\nu}t]}, \tag{2.100}$$

where the field is characterized by the complex amplitude, $A(\vec{r},t)$.

Both the amplitude $A(\geq 0)$ and the phase ψ are the real functions. By invoking equation (2.83), the equation (2.100) is expressed as,

$$U(\vec{r},t) = A(\vec{r},t)e^{i\psi(\vec{r},t)}e^{-2\pi\bar{\nu}t} = 2\int_0^\infty \widehat{U}(\vec{r},\nu)e^{-i2\pi\nu t}d\nu. \tag{2.101}$$

In most of the applications, the spectral amplitudes will have appreciable values in a frequency interval of width $\Delta\nu = \nu - \bar{\nu}$ which is small compared to the mean frequency $\bar{\nu}$. The equation (2.101) can be recast in,

$$A(\vec{r},t)e^{i\psi(\vec{r},t)} = 2\int_0^\infty \widehat{U}(\vec{r},\nu)e^{-i2\pi(\nu - \bar{\nu})t}d\nu. \tag{2.102}$$

The field is characterized by the complex amplitude $a(t) = A(t)e^{i\psi(t)}$, but this phasor is time dependent. Since $\Delta\nu/\bar{\nu} \ll 1$, $A(\vec{r},t)$ and $\psi(\vec{r},t)$ vary slowly with respect to the exponential frequency term $e^{-i2\pi\bar{\nu}t}$. In this case the light that is emitted is known to be quasi-monochromatic. It is possible to derive that this complex amplitude obeys the same propagation law as the instantaneous amplitude. The spectral amplitudes were assumed to differ appreciably from zero nearly in the neighborhood of $\nu = \bar{\nu}$, hence, the integral (equation 2.101) represents a superposition of harmonic components of low frequencies. Since $U^{(r)}$ and $U^{(i)}$ are the real and imaginary parts of U, therefore, in terms of A and ψ,

$$\begin{aligned} U^{(r)}(\vec{r},t) &= A(\vec{r},t)\cos\left[\psi(\vec{r},t) - 2\pi\bar{\nu}t\right], \\ U^{(i)}(\vec{r},t) &= A(\vec{r},t)\sin\left[\psi(\vec{r},t) - 2\pi\bar{\nu}t\right]. \end{aligned} \tag{2.103}$$

These formulae (equation 2.103) express $U^{(r)}$ and $U^{(i)}$ in the form of normalized[15] signals of carrier frequency $\bar{\nu}$. The complex analytic signal is intimately connected with the envelope of the real signal. By squaring equations (2.103) followed by the additions, one gets,

$$A^2(\vec{r},t) = \left|U^{(r)}(\vec{r},t)\right|^2 + \left|U^{(i)}(\vec{r},t)\right|^2. \tag{2.104}$$

[15]Normalization is a process of reducing measured results as nearly as possible to a common scale. It is essential to insure that appropriate comparisons are made.

Thus, in terms of the analytic signal U, the envelope $A(\vec{r},t)$ is deduced as,

$$\begin{aligned} A(\vec{r},t) &= \sqrt{\left|U^{(r)}(\vec{r},t)\right|^2 + \left|U^{(i)}(\vec{r},t)\right|^2} \\ &= \sqrt{U(\vec{r},t)U^*(\vec{r},t)} = |U(\vec{r},t)|. \end{aligned} \tag{2.105}$$

Dividing $U^{(i)}(\vec{r},t)$ by $U^{(r)}(\vec{r},t)$, one gets,

$$\frac{U^{(i)}(\vec{r},t)}{U^{(r)}(\vec{r},t)} = \tan\left[\psi(t) - 2\pi\bar{\nu}t\right], \tag{2.106}$$

thus, the associated phase factor $\psi(\vec{r},t)$ is obtained,

$$\begin{aligned} \psi(\vec{r},t) &= 2\pi\bar{\nu}t + \tan^{-1}\frac{U^{(i)}(\vec{r},t)}{U^{(r)}(\vec{r},t)} \\ &= 2\pi\bar{\nu}t + \tan^{-1}\left(i\frac{U^*(\vec{r},t) - U(\vec{r},t)}{U^*(\vec{r},t) + U(\vec{r},t)}\right). \end{aligned} \tag{2.107}$$

Here, $A(\vec{r},t)$ is independent of $\bar{\nu}$ and that of $\psi(\vec{r},t)$ dependent on ν only through the additive term $2\pi\bar{\nu}t$.

2.3.3 *Successive wave-trains emitted by an atom*

Light emitted by an atom in a small spectral band is not strictly monochromatic, but is made up of wave-trains of finite length. A large number of such wave-trains pass at random time intervals during the time necessary to make an observation. The coherence of two interfering light beams is linked with the duration and consequently with the length of the wave-trains. This length determines the bandwidth of the radiations emitted by the atoms (Born and Wolf, 1984). Let $F(t)$ be the vibration at a given point at a time, t, due to a single wave train and is represented by the integral,

$$F(t) = \int_{-\infty}^{\infty} \widehat{F}(\nu)e^{-2i\pi\nu t}d\nu. \tag{2.108}$$

Assuming that the atom emits complex vibrations $F_1(t), F_2(t)\cdots$ at times $t_1, t_2, \cdots$ distributed at random, the expressions $F_1(t), F_2(t)\cdots$ are the analytic signals associated with real valued waves. The vibration emitted at times t_1 is represented by $F_1(t-t_1)$ at time t. If N such wave trains pass a point during the time required to make an observation, the total light disturbance involved in the observation may be written as,

$$U(t) = \sum_{n=1}^{N} F(t - t_n) = \sum_{n=1}^{N} \int_{-\infty}^{\infty} \widehat{F}(\nu) e^{-i2\pi\nu(t - t_n)} d\nu, \tag{2.109}$$

where t_n's denote the time of arrival of wave trains, and from which one gets,

$$\begin{aligned} U(t) &= \int_{-\infty}^{\infty} \left[\widehat{F}_1(\nu) e^{i2\pi\nu t_1} + \widehat{F}_2(\nu) e^{i2\pi\nu t_2} + \cdots \right] e^{-i2\pi\nu t} d\nu \\ &= \int_{-\infty}^{\infty} \widehat{U}(\nu) e^{-i2\pi\nu t} d\nu, \end{aligned} \tag{2.110}$$

with

$$\widehat{U}(\nu) = \sum_{n=1}^{N} \widehat{F}_n(\nu) e^{2i\pi\nu t_n}. \tag{2.111}$$

The phase of each wave-trains is variable and there is no phase relation between the different wave-trains. The complex vibration, $U(t)$, provided by one or other of these equations represents a succession of wave-trains emitted by an atom. Such a vibration represents the disturbance emitted at the time, t, by an extended incoherent source[16]. If the light is quasi-monochromatic one uses the equation (2.100). It is reiterated that the response of a detector is governed by the intensity of the optical wave falling on the surface. The value of coherence time, τ_c is, in general, very small for thermal light sources with respect to the time required for an observation. For a monochromatic thermal source, the value of such a time is of the order of 10^{-8} seconds, while for a laser source, it may be of the order of 10^{-2} seconds. Therefore the light intensity can be derived following equation (2.68),

$$I = \frac{1}{2T} \int_{-T}^{T} \left| U^{(r)}(t) \right|^2 dt \sim \frac{1}{2T} \int_{-\infty}^{\infty} \left| U^{(r)}(t) \right|^2 dt. \tag{2.112}$$

The truncated functions,

$$U_T^{(r)}(t) = \begin{cases} U^{(r)}(t) & \text{if } |t| \le T, \\ 0 & \text{if } |t| > T, \end{cases} \tag{2.113}$$

[16]Source is a source of light which is bigger than a point source. It may be considered as a mosaic of point sources.

may be analyzed by using Fourier method. Each truncated functions is assumed to be square integrable, thus in the form of a Fourier integral, one may express,

$$U_T^{(r)}(t) = \int_{-\infty}^{\infty} \widehat{U}_T(\nu) e^{-i2\pi\nu t} d\nu. \tag{2.114}$$

Let $U_T^{(i)}$ be the associated function and U_T the corresponding analytic signal, i.e.,

$$U_T(t) = U_T^{(r)} + iU_T^{(i)}(t) = 2\int_0^{\infty} \widehat{U}_T(\nu) e^{-i2\pi\nu t} d\nu. \tag{2.115}$$

Dividing each expression by $2T$, a basic property of the spectral density follows from the Parseval's theorem,

$$\begin{aligned}\frac{1}{2T}\int_{-\infty}^{\infty} \left|U_T^{(r)}(t)\right|^2 dt &= \frac{1}{2T}\int_{-\infty}^{\infty} \left|U_T^{(i)}(t)\right|^2 dt \\ &= \frac{1}{2}\cdot\frac{1}{2T}\int_{-\infty}^{\infty} U_T(t)U_T^*(t)dt = 2\int_0^{\infty} \widehat{G}_T(\nu)d\nu,\end{aligned} \tag{2.116}$$

in which

$$\widehat{G}_T(\nu) = \frac{\left|\widehat{U}_T(\nu)\right|^2}{2T}, \tag{2.117}$$

is known as periodogram, does not tend to a limit, but fluctuates with increasing T. The average of $\widehat{G}_T(\nu)$ taken over the ensemble of the functions $U^{(r)}(t)$ tends to a definite limit and $T \to \infty$. Thus, the smoothed (the smoothness of a function is the number of its continuous successive derivatives) periodogram, $\overline{\widehat{G}_T(\nu)}$, posses a limit,

$$\widehat{G}(\nu) = \lim_{T\to\infty} \overline{\widehat{G}_T(\nu)} = \lim_{T\to\infty} \frac{\overline{\left|\widehat{U}_T(\nu)\right|^2}}{2T}, \tag{2.118}$$

in which the bar denotes the ensemble average and in the limit as $T \to \infty$, one finds,

$$\left\langle \left|U_T^{(r)}(t)\right|^2 \right\rangle = \left\langle \left|U_T^{(i)}(t)\right|^2 \right\rangle = \frac{1}{2}\langle U_T(t)U_T^*(t)\rangle = 2\int_0^{\infty} \widehat{G}(\nu)d\nu. \tag{2.119}$$

The function $\widehat{G}(\nu)$ is called the 'Power spectrum' of the random process characterized by the ensemble of the function $U^{(r)}(t)$ and is also referred as the spectral density of the light vibrations. The term $\widehat{G}(\nu)d\nu$ is proportional to the contribution to the intensity from the frequency range $(\nu, \nu+d\nu)$, in which the truncated function, $U^{(r)}(t)$, is considered to be the light vibration.

2.3.4 *Coherence length and coherence time*

It is stated in the preceding section (2.4.3) that emitted waves are discharged as wave-trains. The coherence of two interfering beams is linked with the duration and consequently with the length of the wave-trains. Such length determines the bandwidth of the radiations emitted by the atoms (Born and Wolf, 1984). If all the wave-trains have the same duration during which $U^{(r)}(t)$ is a simple harmonic of the frequency $\bar{\nu}$, i.e,

$$U^{(r)}(t) = \begin{cases} \cos 2\pi\nu t, & \text{if } |t| \le \tau_c/2, \\ 0 & \text{if } |t| > \tau_c/2. \end{cases} \tag{2.120}$$

From equations (2.108) and (2.109) one gets,

$$U^{(r)}(t) = \int_{-\infty}^{\infty} \widehat{U}^{(r)}(\nu)e^{-2i\pi\nu t}d\nu, \tag{2.121}$$

the Fourier transform of which,

$$\widehat{U}^{(r)}(\nu) = \int_{-\infty}^{\infty} U^{(r)}(t)e^{2i\pi\nu t}d\nu, \tag{2.122}$$

Parseval's theorem (see Appendix B) provides the following relation,

$$\begin{aligned}\int_{-\infty}^{\infty} \left|U^{(r)}(t)\right|^2 dt &= \int_{-\infty}^{\infty} \left|U^{(r)}(\nu)\right|^2 d\nu \\ &= \int_{-\infty}^{\infty} \left|\widehat{F}^{(r)}(\nu)\right|^2 \sum_{n=1}^{N}\sum_{m=1}^{N} e^{2i\pi\nu(t-t_n)}d\nu, \end{aligned} \tag{2.123}$$

we have,

$$\begin{aligned}\sum_{n=1}^{N}\sum_{m=1}^{N} e^{2\pi i\nu(t-t_n)} &= N + \sum_{n\neq m} e^{2i\pi\nu(t_n-t_m)} \\ &= N + 2\sum_{n<m} \cos 2\pi\nu(t_n - t_m), \end{aligned} \tag{2.124}$$

where t_n's are distributed at random.

From the equations (2.112) and (2.123), the mean intensity, I, is recast as,

$$I = \frac{N}{2T}\int_{-\infty}^{\infty}\left|\widehat{F}^{(r)}(\nu)\right|^2 d\nu. \tag{2.125}$$

Equation (2.125) states that the mean intensity is proportional to the integral of the intensities $\widehat{I}(\nu) = |\widehat{F}^{(r)}(\nu)|^2$ for incoherent superposition of the monochromatic components of which a single wave train is made up.

According to equation (1.122), the Fourier transform of $U^{(r)}(t)$ is written as,

$$\begin{aligned}\widehat{U}^{(r)}(\nu) &= \frac{1}{2}\int_{-\tau_c/2}^{\tau_c/2}\left[e^{-2i\pi\nu_0 t} + e^{2i\pi\nu_0 t}\right]e^{2i\pi\nu t}dt \\ &= \frac{\tau_c}{2}\left[\frac{\sin\pi(\nu-\nu_0)\tau_c}{\pi(\nu-\nu_0)\tau_c} + \frac{\sin\pi(\nu+\nu_0)\tau_c}{\pi(\nu+\nu_0)\tau_c}\right],\end{aligned} \tag{2.126}$$

where τ_c is the coherence time of the light.

The frequency interval, $\nu_0 - \Delta\nu/2 \le \nu \le \nu_0 + \Delta\nu/2$, over which the intensity may be appreciable but the first zero corresponds to $\nu - \nu_0 = \pm 1/\tau_c$, it is clear that,

$$\Delta\nu = \nu - \nu_0 \sim \frac{1}{\tau_c}, \tag{2.127}$$

and if the light is quasi-monochromatic, the frequency interval turns out to be, $\Delta\nu \ll \nu$.

In the case of an idealization of light from a real source, the effective frequency range of the Fourier spectrum turns out to be of the order of the reciprocal of the duration of a single wave-train. According to the atomic theory, the loss of energy by atoms during emission gives rise to the damping of the wave-train. A very weakly damped wave-train is almost a sinusoidal and therefore monochromatic, while a highly damped wave-train corresponds to a non-monochromatic. During the time of one observation, the complex amplitude, $a(t)$, assumes a large number of values that spread over at random. In limited sinusoidal vibration one has a large number of points distributed at random, while in damped harmonic vibrations, the segments of straight lines that are distributed at random pass through the origin.

Atoms are in random thermal motion relative to the observer, therefore the observed spectra are changed by an effect, called Doppler effect[17]. The

[17] Doppler effect is known to be the apparent shift in frequency and wavelength of

emitting atoms are disturbed by their neighbor as well yielding irregular modifications of the wave-trains. Thus it is not advisable to say the terms such as duration of the wave-train or frequency range of the Fourier spectrum. However, the averages of these terms satisfy the reciprocity inequality, which is analogous to the Heisenberg uncertainty relation in quantum mechanics.

$$\tau_c \Delta\nu \geq \frac{1}{4\pi}. \tag{2.128}$$

If the spectrum becomes too narrow it prevents overlapping of two terms on RHS of equation (2.126). Using this expression (equation 2.126), the equation (2.121) can be recast as,

$$U^{(r)}(t) = \frac{\tau_c}{2}\int_{-\infty}^{\infty} \widehat{U}^{(r)}(\nu - \nu_0)e^{-2i\pi\nu t}d\nu, \qquad \text{with} \tag{2.129}$$

$$\widehat{U}^{(r)}(\nu - \nu_0) = \frac{\sin\pi(\nu - \nu_0)\tau_c}{\pi(\nu - \nu_0)\tau_c}. \tag{2.130}$$

After certain manipulations, one derives,

$$U^{(r)}(t) \simeq \tau_c \int_0^{\infty} \widehat{U}^{(r)}(\nu - \nu_0)\cos[-2\pi\nu t]d\nu. \tag{2.131}$$

---

a light wave that is perceived by an observer moving with respect to the object. The Doppler shift of the wavelength of light due to the velocity of the source is given by,

$$\lambda_o = \lambda_r\sqrt{\frac{1 + v_r/c}{1 - v_r/c}},$$

in which λ_o is the observed wavelength, λ_r the wavelength one would observe if the object is at rest relative to the observer, v the velocity of the object, and c the speed of light. The resulting redshift is commonly expressed in terms of the z-parameter, which is the fractional shift in the spectral wavelength,

$$z = \frac{\Delta\lambda}{\lambda_r} = \frac{\lambda_o - \lambda_r}{\lambda_r},$$

so that

$$\frac{v_r}{c} = \frac{(z+1)^2 - 1}{(z+1)^2 + 1} \approx z \qquad \text{if } z \ll 1.$$

In such a situation, one observes the wavelengths stretched out from the receding object and $\Delta\lambda_r$ becomes positive number; the spectral lines are shifted to longer wavelength (a red-shift). On the other hand, wavelengths appear to be compressed from the approaching object and $\Delta\lambda$ turns out to be negative; the spectral lines are shifted to shorter wavelength (a blue shift).

The distribution of energy in the spectrum of the vibration, $U^{(r)}(t)$, can be deduced by comparison with the equation (2.74),

$$a^2(\nu) = (\tau_c)^2 \left[\frac{\sin \pi(\nu - \nu_0)\tau_c}{\pi(\nu - \nu_0)\tau_c}\right]^2, \tag{2.132}$$

in which $a^2(\nu)$ is the spectral distribution of the energy.

As stated earlier, the time necessary to make one observation, T, is very large in comparison with the coherence time, τ_c, therefore the complex amplitude $a(t)$ assumes a large number of values which are distributed at random during the time of one observation. With c the velocity of light, the coherence length, l_c, is defined as,

$$l_c = c\tau_c, \tag{2.133}$$

thus the mean wavelength, $\bar{\lambda}_0 = c/\nu$, the coherence length is given by,

$$l_c \sim \frac{c}{\Delta\nu} = \frac{\bar{\lambda}_0^2}{\Delta\lambda_0}. \tag{2.134}$$

2.4 Polarization of plane monochromatic waves

The three-dimensional plane of wave motion has a random distribution of electric fields vibrating in all directions. If the directions of electric fields in the plane perpendicular to the direction of propagation are not evenly distributed, the radiation is polarized. The measurement of polarization parameters is important in understanding of the emission mechanisms. Processes such as electric and magnetic fields, scattering, chemical interactions, molecular structure, and mechanical stress cause changes in the polarization state of an optical beam. Applications relying on the study of these changes cover a vast area, among them are astrophysics and molecular biology.

The polarization state of the waves is an important characteristic of the nature of light. The behavior of the electric field vector, $\vec{E}$, at a particular time in space and its changes from point to point of the field with time are captured by the state of polarization of a wave. This state of polarization enters many physical phenomena, such as reflection, transmission etc. The amount of reflected and transmitted light depends on the state of polarization. Augustin Jean Fresnel in 1823 had derived the reflection and transmission formulae for a plane wave that is incident on a static and plane interface between two dielectric isotropic media.

As stated earlier in section (1.5.1), the plane waves described by the equations (1.86 and 1.87) are transverse implying electric and magnetic field vectors oscillate in a plane perpendicular to the wave number vector, $\vec{\kappa}$. Let an electric field for the plane wave be propagate in the direction of the $\pm z$ axis in free space,

$$\begin{aligned} E(z,t) &= E_x(z,t)\vec{\epsilon}_1 + E_y(z,t)\vec{\epsilon}_2 \\ &= a_x \cos(\kappa_0 z \mp \omega t + \psi_1)\vec{\epsilon}_1 + a_y \cos(\kappa_0 z \mp \omega t + \psi_2)\vec{\epsilon}_2, \end{aligned} \tag{2.135}$$

in which a_x and a_y are the instantaneous amplitudes of the two orthogonal components, E_x, and E_y of the electric vector along x- and y- axes respectively, ψ_1 and ψ_2 the respective instantaneous phases at a fixed point in space as a function of time, and $\vec{\epsilon}_1$ and $\vec{\epsilon}_2$ the two orthogonal real unit vectors perpendicular to each other in directions of $\vec{B}_0$ and $\vec{E}_0$.

The plane defined by the electric field vector $\vec{E}$ and the propagation vector $\vec{\kappa}$ is known as plane of polarization. The unit vectors, $\vec{\epsilon}_{j=1,2}$, span the plane normal to polarization plane $\vec{\kappa}$. If z is held fixed, as t evolves, the tip of the vector $\vec{E}^{(r)}$ draws a trajectory in the $x - y$ plane, which is the result of composing two perpendicular, equal frequency oscillations. This curve is the locus points whose coordinates $(E_x^{(r)}, E_y^{(r)})$ are given by,

$$E_x^{(r)} = a_x \cos(\psi_1 - \omega t), \tag{2.136}$$

$$E_y^{(r)} = a_y \cos(\psi_2 - \omega t). \tag{2.137}$$

Since according to the wave equation, the field is transverse, the x and y components of this electric field are different from zero. The expression for the trajectory parametrized by these equations (2.136 and 2.137) is obtained by eliminating ωt,

$$\begin{aligned} \left(\frac{E_x^{(r)}}{a_x}\right)^2 + \left(\frac{E_y^{(r)}}{a_y}\right)^2 &= \cos^2(\psi_1 - \omega t) + \cos^2(\psi_2 - \omega t) \\ &= \sin^2(\psi_1 - \psi_2) + 2\frac{E_x^{(r)}}{a_x}\frac{E_y^{(r)}}{a_y}\cos(\psi_1 - \psi_2) \\ &= \sin^2\delta + 2\frac{E_x^{(r)}}{a_x}\frac{E_y^{(r)}}{a_y}\cos\delta, \end{aligned} \tag{2.138}$$

where $\delta = \psi_1 - \psi_2$, $-\pi < \delta \leq \pi$, is the phase difference between the x- and y- vibrations.

Light can be polarized in ordinary circumstances under natural conditions if the incident light strikes the surface, say water, at an angle equal to the polarizing angle of that substance. At other angles of incidence, the light is said to be partially polarized and consist of mixture of polarized and non-polarized components. On such cases, the plane of polarization of the reflected light is parallel with the surface of the reflecting material. By rearranging the equation (2.138), one obtains,

$$\left(\frac{E_x^{(r)}}{a_x}\right)^2 + \left(\frac{E_y^{(r)}}{a_y}\right)^2 - 2\frac{E_x^{(r)}}{a_x}\frac{E_y^{(r)}}{a_y}\cos\delta = \sin^2\delta. \tag{2.139}$$

Equation (2.139) is an expression for the polarization ellipse of the electric field for a monochromatic light in which the amplitudes and phases are constant. When $\delta = \pi/2$, equation (2.139) reduces to,

$$\left(\frac{E_x^{(r)}}{a_x}\right)^2 + \left(\frac{E_y^{(r)}}{a_y}\right)^2 = 1, \tag{2.140}$$

which means the reference axes coincide with the major and minor axes of the ellipse.

The general form of polarization is elliptical polarization in which the end points of the instantaneous electric vectors lie on an ellipse. The other states of polarization are its corollaries. As the monochromatic wave propagates through space in a fixed plane perpendicular to the direction of light, the end point of the electric vector at a fixed point traces out an ellipse. The shape of the ellipse changes continuously. When the ellipse maintains a constant orientation, ellipticity, and sense in the ellipse, the wave is known to be completely polarized at that point. In rest of the cases, the reference axes do not correspond to the major and minor axes of the ellipse. This state is called an elliptic polarization. The cross-term $E_x^{(r)}E_y^{(r)}$ of the equation (2.139) implies that the polarization ellipse of the electric field rotates through an angle θ.

Figure 2.3 displays the different Lissajous representations of polarization ellipses for different phase angles. The components are in phase and the polarization ellipse reduces to a segment of straight line, known as linear polarization or plane polarized. With $\delta = \pi$, one gets again linear polarization. For $0 < \delta < \pi$, the polarization ellipse is traced with a left hand sense, while for $-\pi < \delta < 0$, it is traced with right hand sense. When δ turns out to be $n\pi/2$, where $n = \pm 1, \pm 3, \cdots$, the polarization ellipse becomes circularly polarized, implying that the ellipse degenerates to a circle. Intermediate states, where there is some jiggling around an average direction are called partially polarized. The amount of order is specified by

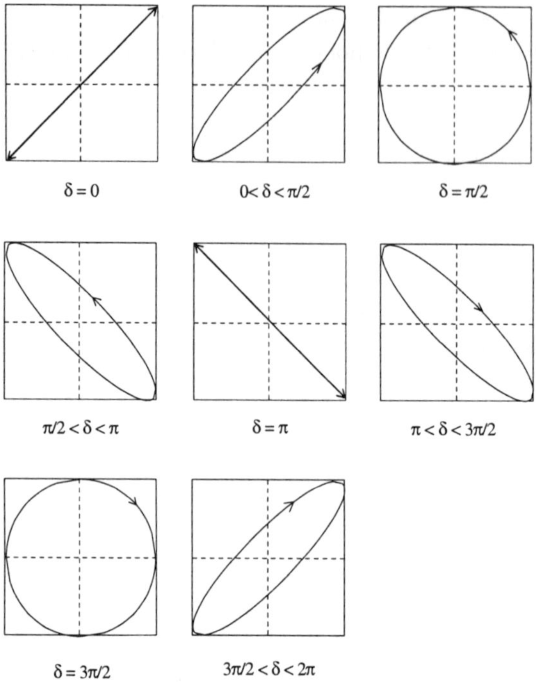

Fig. 2.3 The elliptical polarization with various values of the phase difference δ.

the degree of polarization. Plane polarization is usually caused by scattering[18], while circular polarization is due to the strong magnetic field. In the case of quasi-monochromatic approximation, the equation (2.139) may be expressed as,

$$\left(\frac{E_x^{(r)}(t)}{a_x}\right)^2 + \left(\frac{E_y^{(r)}(t)}{a_y}\right)^2 - 2\frac{E_x^{(r)}(t)E_y^{(r)}(t)}{a_x a_y}\cos\delta = \sin^2\delta. \tag{2.141}$$

For $\delta = 0$, one obtains linear polarization; the field components E_x and E_y are in phase and the polarization ellipse reduces to a straight line. For a linearly polarized wave, the amplitudes are constant with respect

[18]Scattering is a physical process in which a beam of particles or radiation disperse into a range of directions as a result of physical interactions. The trajectory of the scattered particle is a hyperbola. The term, scattering, is also used for the diffusion of electromagnetic waves by the atmosphere. There are two broad types of scattering such as (i) elastic scattering where the photon energies of the scattered photons is unaltered that involves hardly any loss or gain of energy by the radiation and (ii) inelastic scattering which involves some changes in the energy of the radiation.

to magnitude and direction and both E_x and E_y have the same phase. Assuming that $\psi_1 = \psi_2$, the equation (2.141) takes the form,

$$\left(\frac{E_x^{(r)}(t)}{a_x} - \frac{E_y^{(r)}(t)}{a_y}\right)^2 = 0, \qquad \text{or,} \tag{2.142}$$

$$a_y E_x^{(r)}(t) - a_x E_y^{(r)}(t) = 0. \tag{2.143}$$

Equation (2.143) is an expression for a straight line in the E_x, E_y plane. Since the amplitudes a_x and a_y are non-negative, this line lies in the first and the third quadrants, thus $E(z,t)$ oscillates along this line. Similarly, if $\psi_1 - \psi_2 = \pi$, the equation (2.141) translates into,

$$a_y E_x^{(r)}(t) + a_x E_y^{(r)}(t) = 0, \tag{2.144}$$

which lies in second and fourth quadrants.

If the magnitudes of a_1 and a_2 are equal, but exhibit a phase difference of $\delta = \pm\pi/2$, the situation corresponds to that of a state of circular polarization. The relative amplitude of the components E_x and E_y of the field is described by the angle γ, which is given by,

$$\tan\gamma = \frac{a_y}{a_x} \qquad 0 < \gamma < \frac{\pi}{2}. \tag{2.145}$$

The shape and the orientation of the polarization ellipse and the senses in which the said ellipse is traced out are determined by the angles γ and δ. The size of the ellipse depends on the amplitude of the electric field.

2.4.1 *Stokes vector representation*

A major advance to the study of polarization came from the introduction of four measurable parameters of polarized waves by George Gabriel Stokes in 1852, known as Stokes parameters that are parameterized by I, Q, U, and V. A general radiation field is generally described by these parameters specifying intensity of the field, the degree of polarization, the plane of polarization and the ellipticity of the radiation at each point and in any given direction.

A stationary quasi-monochromatic plane wavefield is considered, in which $\Delta\nu \ll \bar{\nu}$, propagating in the direction characterized by the unit vector $\vec{\epsilon}_3$. The plane wave components of the optical field at a point $\vec{r}$ at time t in terms of complex quantities can be written as,

$$E_x(\vec{r}, t) = a_x(\vec{r}, t) e^{i[\psi_1(\vec{r}, t) - 2\pi\bar{\nu}t]},$$

$$E_y(\vec{r},t) = a_y(\vec{r},t)e^{i[\psi_2(\vec{r},t) - 2\pi\bar{\nu}t]}. \tag{2.146}$$

The Stokes parameters measure the average over an ensemble or over time of fields, therefore, the equation (2.141) can be recast in,

$$\frac{\left\langle E_x^{(r)^2}(t)\right\rangle}{a_x^2} + \frac{\left\langle E_y^{(r)^2}(t)\right\rangle}{a_y^2} - 2\frac{\left\langle E_x^{(r)}(t)E_y^{(r)}(t)\right\rangle}{a_x\, a_y}\cos\delta = \sin^2\delta. \tag{2.147}$$

At a particular point in space, the four instantaneous Stokes parameters, are defined by the realizations at time t of the following combinations of the complex analytical signals, $E_x(t)$ and $E_y(t)$. These are associated with the real-valued components of the electric vector in mutually orthogonal directions perpendicular to $\vec{\epsilon}_3$. By using the Stokes parameters (Van der Hulst, 1957), the state of polarization of a beam of light, in the case of quasi-monochromatic case is described as,

$$\begin{aligned}
I &= |E_x(t)|^2 + |E_y(t)|^2 = \langle a_x^2\rangle + \langle a_y^2\rangle,\\
Q &= |E_x(t)|^2 - |E_y(t)|^2 = \langle a_x^2\rangle - \langle a_y^2\rangle,\\
U &= \frac{1}{2}\left(|E_x(t) + E_y(t)|^2 - |E_x(t) - E_y(t)|^2\right)\\
&= E_x(t)E_y^*(t) + E_x^*(t)E_y(t) = 2\,|E_x(t)|\,|E_y(t)|\cos\delta = 2\,\langle a_x\, a_y\rangle\cos\delta,\\
V &= \frac{1}{2}\left(|E_x(t) + iE_y(t)|^2 - |E_x(t) - iE_y(t)|^2\right)\\
&= i\left(-E_x^*(t)E_y(t) + E_x(t)E_y^*(t)\right) = 2\,|E_x(t)|\,|E_y(t)|\sin\delta\\
&= 2\,\langle a_x\, a_y\rangle\sin\delta.
\end{aligned} \tag{2.148}$$

In the case of monochromatic wave, a_1, a_2, and δ are independent of the time, thus, the aforementioned equations reduces to the monochromatic Stokes parameters, such as,

$$\begin{aligned}
I &= a_x^2 + a_y^2,\\
Q &= a_x^2 - a_y^2,\\
U &= 2a_x\, a_y\cos\delta,\\
V &= 2a_x\, a_y\sin\delta.
\end{aligned} \tag{2.149}$$

A vector $\vec{a}_0$ is defined here whose amplitude is $[a_x^2 + a_y^2]^{1/2}$ and which makes an angle θ with the positive x- axis such that $a_x = a_0\cos\theta$ and $a_x = a_0\sin\theta$. Here θ is the position angle of polarization. Both a_x and a_y may take a positive (+ve) or a negative (-ve) sign defining the quadrant in which θ lies. In terms of the positive angle of polarization and the phase difference δ, the Stokes parameters are deduced as,

$$\begin{aligned} I &= a_0^2, \\ Q &= I\cos 2\theta, \\ U &= I\sin 2\theta\cos\delta, \\ V &= I\sin 2\theta\sin\delta. \end{aligned} \tag{2.150}$$

In the case of $\delta = 0$, the vibration lie in a plane which makes an angle θ with the $x - y$ plane. When $\delta = \pm\pi$, the light turns out to be plane polarized. If the magnitudes of a_1 and a_2 are equal, but the phase difference of $\delta = \pm\pi/2$, the major and minor axes of the ellipse traced by the instantaneous electric vectors coincide with the x- and y- axis. Such a state is said to be circularly polarized. When δ has any value other than the afore-mentioned values, the resultant electric vector traces an ellipse in $x - y$ plane with the major axis arbitrarily inclined to the x- axis,

$$\tan 2\chi = \tan 2\theta\cos\delta \tag{2.151}$$

$$\cos 2\theta = \cos 2\chi\cos 2\beta, \tag{2.152}$$

where $\tan\beta = \mathcal{B}/\mathcal{A}$, in which $\mathcal{A}$ and $\mathcal{B}$ are the semi-major and semi-minor axes of the ellipse, and χ the angle between the major axis of the ellipse and the x- axis.

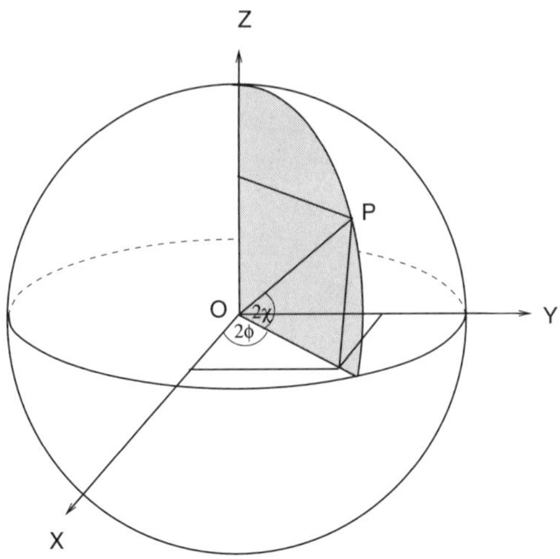

Fig. 2.4 The Poincaré sphere that represents the state of polarization of a monochromatic wave.

The state of polarization is indicated by the location of a point P on the surface of a sphere of radius I in the subspace of Q, U, and V, known as Poincaré sphere (see Figure 2.4) that provides a convenient graphical tool for utilizing changes in the state of polarization. Each point on the sphere represents a definite state of polarization. For partially polarized light, the radius is smaller than I and it becomes zero in the case of unpolarized light. There is one to one correspondence between the state of polarization and the points on this sphere. The parameter I is proportional to the intensity of the wave. The other parameters, Q, U, and V are related to the angles β $(0 \leq \beta < \pi)$ and χ $(-\pi/4 \leq \phi \leq \pi/4)$. These angles β and χ specifies the orientation of the ellipse and characterizes the ellipticity and sense in which the ellipse is being described respectively. (Q, U, V) is regarded as the Cartesian co-ordinates of a point P of radius I, such that 2β and 2χ are the spherical angular co-ordinates of the point, i.e., the longitude and latitude, respectively. The lines of longitude and latitude represent the equi-azimuth and equi-ellipticity contours respectively. The three co-ordinates (Q, U, V) of a point (I, χ, β) on this Poincaré sphere are given by,

$$\begin{aligned} Q &= I \cos 2\chi \cos 2\beta, \\ U &= I \sin 2\chi \cos 2\beta, \\ V &= I \sin 2\beta. \end{aligned} \tag{2.153}$$

These four Stokes parameters have the physical dimensions of time-average power per unit area. These parameters are not independent, since,

$$I^2 = Q^2 + U^2 + V^2. \tag{2.154}$$

The column vector formed by the Stokes parameters is referred as the Stokes vector, $\vec{S}$,

$$\vec{S} = \begin{bmatrix} I \\ Q \\ U \\ V \end{bmatrix}. \tag{2.155}$$

The special degenerate form of elliptically polarized waves characterized by normalized states of polarization (Jones vector), $\vec{E}$, as well as their corresponding Stokes vector, $\vec{S}$, and the density matrix, $\vec{D}$ are given in Table II (Appendix A).

2.4.2 *Optical elements required for polarimetry*

In order to determine the characteristic matrix for polarizers, retarders or wave plates, and rotators in the Jones matrix calculus, it is assumed that the components of a beam emerging from a polarizing element are linearly related to the components of the incident light that propagates as a polarized monochromatic plane wave. These elements change the state of polarization of the incident beam of light appropriately, but do not affect the direction of propagation. This relation is expressed as,

$$\begin{aligned} E'_x &= J_{xx}E_x + J_{xy}E_y, \\ E'_y &= J_{yx}E_x + J_{yy}E_y, \end{aligned} \tag{2.156}$$

where E'_x and E'_y are the components of the emerging beam and E_x and E_y are the components of the incident beam and the quantities, J_{ij}, and $i, j(= x, y)$ are the transforming factors (elements).

For a linear optical device, the relation between the input and output Jones vectors, $\vec{E}$ and $\vec{E}'$ is expressible in the form,

$$\begin{aligned} \vec{E}' = \vec{J} \cdot \vec{E} &= \begin{bmatrix} E'_x \\ E'_y \end{bmatrix} \\ &= \begin{bmatrix} J_{xx} & J_{xy} \\ J_{yx} & J_{yy} \end{bmatrix} \begin{bmatrix} E_x \\ E_y \end{bmatrix}, \end{aligned} \tag{2.157}$$

in which the Jones matrix or coherent matrix,

$$\vec{J} = \begin{bmatrix} \langle E_x E_x^* \rangle & \langle E_x E_y^* \rangle \\ \langle E_y E_x^* \rangle & \langle E_y E_y^* \rangle \end{bmatrix} = \begin{bmatrix} J_{xx} & J_{xy} \\ J_{yx} & J_{yy} \end{bmatrix}. \tag{2.158}$$

couples the incident wave of the system to the transmitted wave of that system.

A polarizer is an optical element which changes the orthogonal amplitudes unequally. Such an element can be linear, circular, or in general, elliptical, depending on the type of polarization that emerges. There is a large variety of polarizers with their respective advantages and disadvantages. Dichroism[19] and birefringence[20] are two properties of materials,

[19]Dichroism is defined as the difference in absorption co-efficients for the two polarization components. The size distribution of grains responsible for producing dichroic scattering may be obtained by studying the wavelength dependence of multiwavelength.

[20]Birefringence is the property by which certain materials have two different refractive indices for two orthogonal polarization components. This property is observed mostly in an anisotropic crystal structure having axis of symmetry, known as principal axis, like quartz and calcite.

which have been employed to produce optical elements with specific polarization behavior. Such a polarizer is characterized by the relations,

$$\begin{aligned} E'_x &= p_x E_x, \\ E'_y &= p_y E_y \qquad 0 \le p_{x,y} \le 1. \end{aligned} \tag{2.159}$$

For complete transmission $p_{x,y} = 1$, and for complete attenuation $p_{x,y} = 0$. Figure 2.5(a) depicts an input wave passing through an optical element, say a polaroid sheet, with an electric field lying in the plane of the sheet. Usually a polaroid sheet (Land, 1951) is used as linear polarizer and retarder.

The most common types of sheet polarizer are made by processes involving unidirectional stretching of large plastic sheets such as a polymer. These polarizers are thin and do not cause appreciable lateral shift of the image[21], therefore, it is possible to construct simple astronomical polarimeter using these as analyzer. These polarizers are efficient in restricted wavelength region, but have poor transmittance. The typical ratio of $|J_{yy}/J_{xx}|$ is in the range 5×10^{-3} to 2×10^{-1} at $0.4\mu m \le 0.7\mu m$. In an ideal polarizer, this ratio is zero.

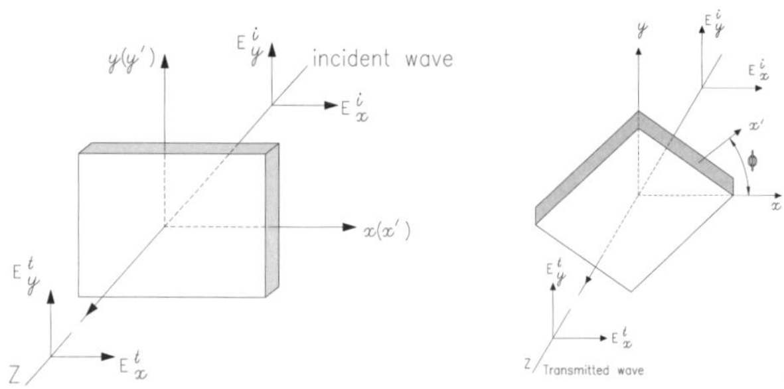

Fig. 2.5 Schematic representation for optical element, (a) element aligned with axes (x, y) and (b) element rotated (right-handed) through angle ϕ about z axis.

If the field is decomposed with components parallel to and perpendicular to the stretched direction y, the field component in the case of the former suffers severe attenuation, while in the case of the latter it suffers little attenuation. Consequently, the emerging wave is linearly polarized with the electric field perpendicular to the stretched direction. In the Figure 2.5(b), the optical element is rotated by an angle ϕ about the $z-$ axis, so

[21]An image is a recording of the intensity of an optical wavefront.

that $0 \leq \phi < \pi$. At the output of the rotator, the orthogonal components of the electric field, one may write,

$$E'_x = E_x \cos\phi + E_y \sin\phi, \tag{2.160}$$
$$E'_y = -E_x \sin\phi + E_y \cos\phi, \tag{2.161}$$

where ϕ denotes the angle through which the field is rotated.

Equations (2.160 and 2.161) are the amplitude equations for rotation. Rotators are primarily used to change the orientation angle of the polarization ellipse. The Jones matrix for a rotator is represented by a unitary rotation matrix defined for the linear polarization-state basis set,

$$\vec{J}(\phi) = \begin{bmatrix} \cos\phi & \sin\phi \\ -\sin\phi & \cos\phi \end{bmatrix}, \tag{2.162}$$

where $\vec{J}(\phi)$ is the rotation matrix in one direction.

The action of rotating a linear polarizer through this angle with respect to the one direction results,

$$\vec{J}' = \vec{J}(-\phi)\cdot\vec{J}\cdot\vec{J}(\phi). \tag{2.163}$$

The rotation matrix, $\vec{J}(-\phi)$ in other direction is,

$$\vec{J}(-\phi) = \begin{bmatrix} \cos\phi & -\sin\phi \\ \sin\phi & \cos\phi \end{bmatrix}. \tag{2.164}$$

Here $\vec{R}_l^{-1} = \vec{R}_r$, where $\vec{R}_l$ is the matrix introduced for rotating the linear polarizer through the angle ϕ with respect to the other direction.

For cascaded elements, the transmission matrix is the product of the transmission matrices for the individual elements,

$$\vec{J}' = \vec{J}_1\cdot\vec{J}_2\cdots\vec{J}_{m-1}\cdot\vec{J}_m. \tag{2.165}$$

With additional constraints that an input wave linearly polarized along the x- axis is transmitted unattenuated and along the y-axis normal to the transmission axis is completely absorbed. When the transmission axis ($x-$) of the linear polarizer is at the angle ϕ, as in Figure 2.5(b), the normalized Jones matrix, the equation (2.163) translates into,

$$\begin{aligned} \vec{J}' &= \begin{bmatrix} \cos\phi & -\sin\phi \\ \sin\phi & \cos\phi \end{bmatrix} \begin{bmatrix} 1 & 0 \\ 0 & 0 \end{bmatrix} \begin{bmatrix} \cos\phi & \sin\phi \\ -\sin\phi & \cos\phi \end{bmatrix} \\ &= \begin{bmatrix} \cos^2\phi & \cos\phi\sin\phi \\ \cos\phi\sin\phi & \sin^2\phi \end{bmatrix}. \end{aligned} \tag{2.166}$$

The Jones vector for the transmitted wave (equation 2.160) is deduced as,

$$\vec{E}' = \vec{J} \cdot \vec{E} = E'_x e^{-i\xi} \cos\phi \begin{bmatrix} \cos\phi \\ \sin\phi \end{bmatrix}. \tag{2.167}$$

Thus, the Malus (Éthene Louis Malus, 1775-1812) law for the linear polarizer is determined by the ratio of the time average powers per unit area for the transmitted and incident wave, i.e.,

$$\frac{I'}{I} = \frac{\left|\vec{E}'\right|^2}{\left|\vec{E}\right|^2} = \cos^2\phi. \tag{2.168}$$

Another polarizing element of importance is the retarder, also known as a compensator. It introduces a differential phase shift $\delta = \delta_1 - \delta_2$ between the two orthogonally linearly polarized components of the incident field. This is accomplished by causing a phase shift of $+\delta/2$ along the x-axis and a phase shift of $-\delta/2$ along the y-axis. These axes are referred as the fast and slow axes respectively. It divides the incident field into two components, viz., E_x and E_y and retards the phase of one of these components relative to the other. When the two components of the wave are reunited to form transmitted wave, the state of polarization is changed. In an ideal retarder, the phase of component E_y is retarded relative to the phase of the component E_x. In this case, the components of the emerging beam are related to the incident beam by,

$$\begin{aligned} E'_x(z,t) &= E_x(z,t)e^{i\delta/2}, \\ E'_y(z,t) &= E_y(z,t)e^{-i\delta/2}. \end{aligned} \tag{2.169}$$

The Jones matrix of a compensator, $\vec{C}$ is,

$$\vec{C} = \begin{bmatrix} e^{i\delta/2} & 0 \\ 0 & e^{-i\delta/2} \end{bmatrix}. \tag{2.170}$$

The retarder, in the simplest case, is a plane parallel plate of uniaxial crystal cut parallel to the optical axis. An ideal retarder neither changes the intensity of the light, nor the degree of polarization. Any retarder may be characterized by the two (not identical) Stokes vectors of incoming light that are not changed by the retarder. These two vectors are sometimes referred as eigenvectors of the retarder. Depending on whether these vectors describe linear, circular, or elliptical polarization, the retarder is known as a linear, circular, or elliptical retarder.

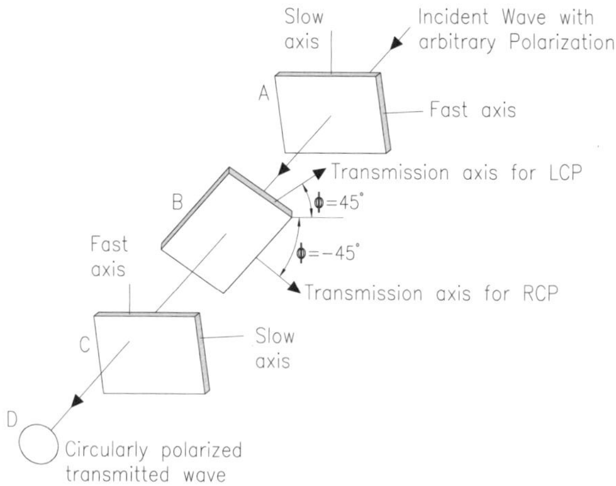

Fig. 2.6 Three element, circular polarizer formed from two quarter wave plates A, C and a linear polarizer, B. Angle of linear polarizer; $\phi = 45°$ for left-handed circular polarizer (LCP) and $\phi = -45°$ for right-handed circular polarizer (RCP).

The most common types of wave plates are the half-wave and quarter-wave plates. The half-wave plate was checked for depolarization (which means the conversions of plane polarized light to any other polarization state) produced by dispersion in the retardance and the orientation of the fast axis[22]. It can also be used for testing non-parallelism of the surfaces, which can lead to image motion, if rotated. The position angle of the fast axis of such a plate may vary with wavelength even within the operating wavelength band. The retardance can deviate as well from π within the band width. These cause depolarization of light as it passes through the half-wave plate. The ordinary and the extra-ordinary[23] rays travel with

[22]Fast axis is the direction in which light travels faster, while the orthogonal direction is known as slow axis. For a positive uniaxial crystal the ellipsoid touches the spheroid along the fast axis and for a negative uniaxial crystal along the slow axis.

[23]Huygens suggested that the propagation of light in a uniaxial crystal may be understood by assuming that the light is split into two components. The one with the plane of polarization perpendicular to the plane containing the optical axis and the incident beam obeys the Snell's law of refraction at the surface of the prism, that acts a dispersing element spreading out the different wavelengths of light into a spectrum, is called ordinary ray and travels along the direction of the initial ray path. The second wavefront is an ellipsoid of revolution and is called as extra-ordinary beam. Such a beam

different velocities within a birefringent material and hence acquire a phase difference between them as they propagate. A plate made of such a material with optically flat surfaces and having a thickness such that the path difference between the ordinary and extra-ordinary beams acts as a half-wave plate at the wavelength $\lambda/2$. The equation governing the half-wave plate is $(n_o - n_e)t = \lambda/2$, in which n_o and n_e are the refractive indices of ordinary and extra-ordinary rays respectively, and t the thickness of the plate.

When light passes through the half-wave plate the phase difference of two orthogonally polarized components is $\delta = \pi$ radian[24] and for a quarter-wave device $\delta = \pi/2$. If the incident light is plane polarized at angle, θ, the emerging light is polarized at an angle $-\theta$ with respect to the fast axis. Crystalline quartz and magnesium fluoride are commonly used for making half-wave plates. For a quarter-wave plate, the Jones matrix turns out to be,

$$\vec{C} = \begin{bmatrix} e^{i\pi/4} & 0 \\ 0 & e^{-i\pi/4} \end{bmatrix}. \tag{2.171}$$

When the fast axis, x'- axis of the linear retarder is at the angle ϕ as depicted in Figure 2.5(b), the normalized Jones matrix in the case of quarter-wave plate translates into,

$$\vec{C} = \begin{bmatrix} \cos(\pi/4) + i\cos(2\phi)\sin(\pi/4) & \sin(\pi/4) + i\sin(2\phi)\sin(\pi/4) \\ \mp i\sin(2\phi)\sin(\pi/4) & \cos(\pi/4) - i\cos(2\phi)\sin(\pi/4) \end{bmatrix}. \tag{2.172}$$

An ideal circular polarizer produces either a left-handed (LCP) or right-handed (RCP) circularly polarized transmitted field from an incident field of arbitrary polarization. A circular polarizer is generally made from a linear polarizer placed in between two quarter-wave plates as in Figure 2.6, where the angles of linear polarizer, $\phi = 45°$ and $\phi = -45°$ are maintained for the respective left- and right-handed circular polarizer. Figure (2.6) depicts the three element ideal polarizer formed from a linear polarizer and a couple of quarter wave plates. By multiplying the Jones matrices for

---

with vibrations lying in the above plane does not obey the Snell's law, albeit travels in the prism with a different speed and with different direction.

[24]The radian is the plane angle between two radii of a circle that cuts off on the circumference an arc equal in length to the radius.

these elements, one finds,

$$\vec{C}' = \begin{bmatrix} e^{i\pi/4} & 0 \\ 0 & e^{-i\pi/4} \end{bmatrix} \frac{1}{2} \begin{bmatrix} 1 & \pm 1 \\ \pm 1 & 1 \end{bmatrix} \begin{bmatrix} e^{-i\pi/4} & 0 \\ 0 & e^{i\pi/4} \end{bmatrix} = \frac{1}{2} \begin{bmatrix} 1 & \pm i \\ \mp i & 1 \end{bmatrix}, \tag{2.173}$$

where $\pm$ refers to right-handed and left-handed circular polarization.

This matrix is the normalized Jones matrix for right-handed (left-handed) circular polarizer.

2.4.3 *Degree of polarization*

In order to study the interaction of polarized light with elements which can change its state of polarization, the derivation of the matrix representation of the Stokes parameters is required. Let the incident beam be interacting with a polarizing element. Such a beam is characterized by its Stokes parameters, I, Q, U, and V. The input polarized beam interacts with the polarizing element, and hence, the output beam is characterized by a new set of Stokes parameters, I', Q', U', and V'.

$$\begin{aligned} I' &= m_{00}I + m_{01}Q + m_{02}U + m_{03}V, \\ Q' &= m_{10}Q + m_{11}Q + m_{12}U + m_{13}V, \\ U' &= m_{20}U + m_{21}Q + m_{22}U + m_{23}V, \\ V' &= m_{30}V + m_{31}Q + m_{32}U + m_{33}V. \end{aligned} \tag{2.174}$$

In terms of the Stokes column matrix, it is written as,

$$\vec{S}' = \begin{bmatrix} I' \\ Q' \\ U' \\ V' \end{bmatrix} = \begin{bmatrix} m_{00} & m_{01} & m_{02} & m_{03} \\ m_{10} & m_{11} & m_{12} & m_{13} \\ m_{20} & m_{21} & m_{22} & m_{23} \\ m_{30} & m_{31} & m_{32} & m_{33} \end{bmatrix} \begin{bmatrix} I \\ Q \\ U \\ V \end{bmatrix} = \vec{M} \cdot \vec{S}, \tag{2.175}$$

where

$$\vec{M} = \begin{bmatrix} m_{00} & m_{01} & m_{02} & m_{03} \\ m_{10} & m_{11} & m_{12} & m_{13} \\ m_{20} & m_{21} & m_{22} & m_{23} \\ m_{30} & m_{31} & m_{32} & m_{33} \end{bmatrix}, \tag{2.176}$$

is known as the Mueller matrix[25] (Brosseau, 1998) of the optical element,

[25] A Mueller matrix is a 4×4 matrix whose components define the relations between the Stokes parameter vector, $\vec{S}$ and $\vec{S'}$, where it represent light before and after it passes through the corresponding optical component. It is a generalization of the Jones matrix.

which depends on the frequency. The Mueller matrix for an ideal homogeneous linear compensator can be derived from the following expressions that arise from the definition of the Stokes parameters (equation 2.149),

$$
\begin{aligned}
I' &= |E'_x|^2 + |E'_y|^2 = |E_x|^2 + |E_y|^2 = I, \\
Q' &= |E'_x|^2 - |E'_y|^2 = |E_x|^2 - |E_y|^2 = Q, \\
U' &= E'_x E'^*_2 + E'_y E'^*_1 = U \cos\delta - V \sin\delta, \\
V' &= i(E'_x E'^*_y - E'_y E'^*_x) = U \sin\delta + V \cos\delta,
\end{aligned}
\tag{2.177}
$$

Equation (2.177) is expressed in matrix form as,

$$
\begin{bmatrix} I' \\ Q' \\ U' \\ V' \end{bmatrix} = \begin{bmatrix} 1 & 0 & 0 & 0 \\ 0 & 1 & 0 & 0 \\ 0 & 0 & \cos\delta & -\sin\delta \\ 0 & 0 & \sin\delta & \cos\delta \end{bmatrix} \begin{bmatrix} I \\ Q \\ U \\ V \end{bmatrix}. \tag{2.178}
$$

For an ideal phase shifter (retarder), there is no loss in intensity, $I' = I$, thus, the Mueller matrix for a retarder with a phase shift δ is,

$$
\vec{C}(\delta) = \begin{bmatrix} 1 & 0 & 0 & 0 \\ 0 & 1 & 0 & 0 \\ 0 & 0 & \cos\delta & -\sin\delta \\ 0 & 0 & \sin\delta & \cos\delta \end{bmatrix}. \tag{2.179}
$$

In the case of a rotator, the equations (2.160 and 2.161) is plugged into equation (2.149), therefore,

$$
\begin{aligned}
I' &= |E'_x|^2 + |E'_y|^2 = |E_x|^2 + |E_y|^2 = I, \\
Q' &= |E'_x|^2 - |E'_y|^2 = |E_x|^2 - |E_y|^2 = Q\cos 2\phi + U \sin 2\phi, \\
U' &= E'_x E'^*_y + E'_y E'^*_x = -U \sin 2\phi + V \cos 2\phi, \\
V' &= i(E'_x E'^*_y - E'_y E'^*_x) = V.
\end{aligned}
\tag{2.180}
$$

A Mueller matrix exhibits an angular dependence in the case of light scattering. The effect of an optical device that transforms the aberrations into the light intensity variations, on the polarization of light is characterized by a Mueller matrix. The description of optical systems in terms of such matrices is applicable to more general situation than the description in terms of Jones matrices. Light which is unpolarized or partially polarized must be treated using Mueller calculus, while fully polarized light may be treated with Jones calculus since the latter works with amplitude rather than intensity of light.

Thus the Mueller matrix of a rotator is expressed as,

$$\vec{R}(\phi) = \begin{bmatrix} 1 & 0 & 0 & 0 \\ 0 & \cos 2\phi & \sin 2\phi & 0 \\ 0 & -\sin 2\phi & \cos 2\phi & 0 \\ 0 & 0 & 0 & 1 \end{bmatrix}. \tag{2.181}$$

From the equations (2.149) and (2.166), the form of Mueller matrix of an ideal linear polarizer oriented at angle ϕ is expressed as (Collett, 1993),

$$\vec{P}(\phi) = \frac{1}{2} \begin{bmatrix} 1 & \cos 2\phi & \sin 2\phi & 0 \\ \cos 2\phi & \cos^2 2\phi & \frac{1}{2}\sin 4\phi & 0 \\ \sin 2\phi & \frac{1}{2}\sin 4\phi & \sin^2 2\phi & 0 \\ 0 & 0 & 0 & 0 \end{bmatrix} \tag{2.182}$$

The state of partially elliptically polarized light is described in terms of combination of two independent components, such as:

(1) natural unpolarized light of intensity $I(1 - p_e)$, and
(2) fully elliptically polarized light of intensity Ip_e, in which $p_e = Ip/I$ is the degree of polarization, Ip being the polarized part of the intensity.

Therefore, the Stokes parameters are modified to,

$$\begin{aligned} Q &= Ip_e \cos 2\chi \cos 2\beta, \\ U &= Ip_e \sin 2\chi \cos 2\beta, \\ V &= Ip_e \sin 2\beta, \qquad \text{with} \end{aligned} \tag{2.183}$$

$$p_e = \frac{\left[Q^2 + U^2 + V^2\right]^{1/2}}{I} \qquad 0 \le p_e \le 1, \tag{2.184}$$

and the factor $p_e \cos 2\beta = p$ is known as the degree of linear polarization.

Thus, the equations (2.183) translate into,

$$\begin{aligned} Q &= Ip \cos 2\chi, \\ U &= Ip \sin 2\chi, \\ V &= Ip_v, \end{aligned} \tag{2.185}$$

with $p_v = p_e \sin 2\beta$ as the degree of ellipticity, which is positive for right-handed elliptical polarization and negative for left-handed polarization.

For a beam of light, the quantities I, $(Q^2 + U^2)$, and V are invariant under a rotation of the reference coordinate system. For a celestial object, the equatorial coordinate system provides a system of reference with the positive x- axis in the direction of north defined by the meridian through the object and the positive y- axis in the direction of east, the direction of positive $z-$ axis being along the line of sight towards the observer. The degree of linear polarization and the angle which the axis of the resultant ellipse makes with the x- axis are expressed as,

$$p = \frac{\left[Q^2 + U^2\right]^{1/2}}{I}, \tag{2.186}$$

$$\chi = \frac{1}{2}\tan^{-1}\frac{U}{Q}. \tag{2.187}$$

Since $\delta = 0$ for plane polarized light, the equation (2.151) provides $\chi = \theta$, the position angle of polarization.

2.4.4 *Transformation of Stokes parameters*

The combination of half-wave plates and the analyzer modifies the state of polarization of the incident beam of light appropriately. The role of an analyzer is performed by the calcite block introduced in the light path immediately after the stationary half-wave plate. The optical components, dichroic filters[26] and diffraction grating change the state of polarization of the incident beam, hence the analyzer is placed in front of all such components. The Mueller matrix for a half-wave plate with its optical axis at a position angle, α, with respect to fiducial direction is given by,

$$\vec{H}(\alpha) = \begin{bmatrix} 1 & 0 & 0 & 0 \\ 0 & \cos 4\alpha & \sin 4\alpha & 0 \\ 0 & \sin 4\alpha & -\cos 4\alpha & 0 \\ 0 & 0 & 0 & -1 \end{bmatrix}, \tag{2.188}$$

From the equation (2.182), one finds that the Stokes parameters I', Q', U', and V' of light transmitted through a perfect analyzer with the principal plane containing the optical axis and the incident beam, at position angle, ϕ, are related to the Stokes parameters I, Q, U, and V of the incident light

[26] Filters allow electromagnetic radiation to pass in a limited range of wavelengths. In astronomy, two main types of filters, namely colored filters and interference filters, are used. Colored type filters employ chemical dyes to restrict the wavelengths that pass. In order to make a bandpass filter, two glasses are combined. Interference filters are composed of several layers of partially transmitting material separated by certain fractions of a wavelength of the light that the filter is designed to pass.

by the Mueller matrix of transformation equation (Serbowski, 1947),

$$\begin{bmatrix} I' \\ Q' \\ U' \\ V' \end{bmatrix} = \frac{1}{2} \begin{bmatrix} 1 & \cos 2\phi & \sin 2\phi & 0 \\ \cos 2\phi & \cos^2 2\phi & \frac{1}{2}\sin 4\phi & 0 \\ \sin 2\phi & \frac{1}{2}\sin 4\phi & \sin^2 2\phi & 0 \\ 0 & 0 & 0 & 0 \end{bmatrix} \begin{bmatrix} I \\ Q \\ U \\ V \end{bmatrix}, \tag{2.189}$$

Equation (2.189) reveals that the Stokes parameters I, Q, U, and V of the incident light can be determined by measuring the intensities of light transmitted by an analyzer with its optical axis at several position angles ϕ. If the analyzer is rotated at a frequency ν, the transmitted light modulates at a frequency 2ν.

In order to determine the degree of polarization, the intensity of light at different position angles may be measured either by rotating the analyzer, or by rotating the entire polarimeter about the telescope. Both the methods are not feasible at the same time as they involve a large amount of overhead time. The spatial shift[27] of the image causes a rotation of the image when the analyzer is rotated. This image rotation, in turn, introduces an instrumental polarization that varies with wavelength (Breger, 1979). A slight misalignment between the rotational axis and the normal to the incident direction contributes to the rotation of the image in the focal plane. The intensity of light falling on the detector is modulated during the observations dealing with the measurement of the Stokes parameters. If the thickness of the optical element is small, the spatial shift in the image becomes negligible.

The afore-mentioned problems can be avoided by introducing a rotating half-wave plate in the light path for modulating the starlight. If ν is the frequency of rotation of the half-wave plate, most of the disturbing instrumental effects, in particular, those caused by image motion on the photocathode modulate the signal ν and 2ν frequencies, while the linear polarization modulates the signal with a frequency 4ν. This reduces the risk of spurious polarization caused by the image motion effectively. It is ideal to have the modulating half-wave plate in front of the optical elements that isolate spectral regions for observations.

In order to create circularly polarized light and to rotate or reverse the polarization ellipse, the quarter-wave plates and half-wave plates respectively are most commonly used. In the case of the combination of half-wave plates and the analyzer, the Stokes parameters can be determined by mea-

[27]Spatial shift results from the inherent deviation in the direction of the emergent beam from the incident direction.

suring the intensity of light transmitted by the above combination. The Mueller matrix for the wave plate rotated through an angle α is,

$$\vec{P}(\tau,\alpha) = \begin{bmatrix} 1 & 0 & 0 & 0 \\ 0 & G + H\cos 4\alpha & H\sin 4\alpha & -\sin\tau\sin 2\alpha \\ 0 & H\sin 4\alpha & G - H\cos 4\alpha & \sin\tau\cos 2\alpha \\ 0 & \sin\tau\sin 2\alpha & -\sin\tau\cos 2\alpha & \cos\tau \end{bmatrix}, \qquad (2.190)$$

with

$$G = \frac{1}{2}(1+\cos\tau); \qquad H = \frac{1}{2}(1-\cos\tau), \qquad (2.191)$$

and τ is the retardance, the phase difference introduced by the retarder between the vibrations in the principal plane and those in the plane perpendicular to it.

The Stokes parameters I', Q', U', and V' of the light transmitted through a perfect retarder is transformed according to the following matrix,

$$\begin{bmatrix} I' \\ Q' \\ U' \\ V' \end{bmatrix} = \begin{bmatrix} 1 & 0 & 0 & 0 \\ 0 & G + H\cos 4\alpha & H\sin 4\alpha & -\sin\tau\sin 2\alpha \\ 0 & H\sin 4\alpha & G - H\cos 4\alpha & \sin\tau\cos 2\alpha \\ 0 & \sin\tau\sin 2\alpha & -\sin\tau\cos 2\alpha & \cos\tau \end{bmatrix} \begin{bmatrix} I \\ Q \\ U \\ V \end{bmatrix}, \qquad (2.192)$$

If two retarders are kept in series, the square matrix in the equation (2.192) is replaced by the product of two such matrices.

The determination of the Stokes parameters in the above cases reduces to the measurement of the intensity of light transmitted by the retarder and the analyzer in the path, for the various position angles of the optical axis of the modulating retarder. The intensity of light transmitted by two retarders with the optical axes at position angles, α_1 and α_2 followed by an analyzer,

$$\begin{aligned} I' = \frac{1}{2}\{&I \pm Q\left[G_1G_2 + H_1H_2\cos 4(\alpha_1-\alpha_2) + H_1G_2\cos 4\alpha_1\right. \\ &\left.+G_1H_2\cos 4\alpha_2 - \sin\tau_1\sin\tau_2\sin 2\alpha_1\sin 2\alpha_2\right] \\ &\pm U\left[H_1H_2\sin 4(\alpha_1-\alpha_2) + H_1G_2\sin 4\alpha_1 + G_1H_2\sin 4\alpha_2\right. \\ &\left.+\sin\tau_1\sin\tau_2\cos 2\alpha_1\sin 2\alpha_2\right] \mp V\left[H_2\sin\tau_1\sin(2\alpha_1-4\alpha_2)\right. \\ &\left.+G_2\sin\tau_1\sin 2\alpha_1 + \cos\tau_1\sin\tau_2\sin 2\alpha_2\right]\}, \end{aligned} \qquad (2.193)$$

in which τ_1 and τ_2 are the retardance of the two retarders. The upper and lower signs in equation (2.193) correspond to the principal plane of the analyzer at position angles $\phi = 0°$ and $\phi = 90°$ respectively.

The computed path difference of a super-achromatic half-wave plate is $\lambda/2 \pm 1.3\%$ in the wavelength range $0.3 - 1.1\mu$m, and, if the accuracy of the theoretical retardation is achieved to $\pm 3\%$, providing a value of $\tau = 180 \pm 6.5°$ for the half-value plates used. The stationary half-wave plate is mounted with its effective optical axis approximately parallel to the principal plane of the analyzer, thus all the terms proportional to $\sin 2\alpha_2$ and $\sin 4\alpha_2$ turn out to be negligibly small in the equation (2.193), and it reduces to,

$$I' = \frac{1}{2}\left[I \pm Q\cos 4(\alpha_1 - \alpha_2) \pm U \sin 4(\alpha_1 - \alpha_2) \pm CV \sin 2\alpha_1\right], \quad (2.194)$$

where $C = H_2 \sin \tau_1 \cos 2\alpha_2$.

For plane polarized light, $V = 0$, and therefore, the equation (2.194) turns out to be,

$$I' = \frac{1}{2}\left[I \pm Q\cos 4(\alpha_1 - \alpha_2) \pm U \sin 4(\alpha_1 - \alpha_2)\right], \quad (2.195)$$

With $\tau_1 = \tau_2 = 180 \pm 6.5°$, $|C| \approx 0.1$, and the increase in the probable errors in Q and U due to the 2α modulation is negligible unless V is appreciably large. If the retarders and the calcite block analyzer do not cause any attenuation in the two orthogonally polarized emerging beams, the intensity of the two beams emerging from the calcite block are effectively given by,

$$I' = \frac{1}{2}\left(I \pm Q\cos 4\alpha \pm U \sin 4\alpha\right), \quad (2.196)$$

where α is the angle between the effective optical axes of the two half-wave plates.

The upper and lower signs correspond to the beam with vibrations in the principal plane of the top plate of the calcite block and with vibrations perpendicular to the principal plane respectively. The Stokes parameters Q and U, hence the degree of linear polarization p (equation 2.186) and the position angle χ (equation 2.187) of polarization can be derived by measuring the intensity of the transmitted beam as a function of different orientations α of the rotating half-wave plate and calculating the amplitude and phase of the 4α modulation in the data.

2.4.4.1 *Polarimeter*

A Polarimeter is an instrument for measuring the state of polarization of a beam of light or other form of electromagnetic radiation. An astronomical polarimeter based on the principle that is illustrated in Figure (2.7) was first built by Pirola (1973). A beam displacement prism divides the incident light into two beams with mutually perpendicular planes of polarization. The incident light beam emerges from the prism as two spatially separated beams, ordinary beam and extra-ordinary beam. Unlike the star which is a point source, the background sky that acts as an extended object illuminates the entire top surface of the said prism, hence, produces two broad emergent beams whose centers are spatially separated by the equal amount as above. A considerable amount of overlap between these two beams occur about the geometrical axis of the prism. Therefore, the overlapping region in the focal plane of the telescope remains unpolarized by the prism and provides the background sky (unmodulated) brightness directly.

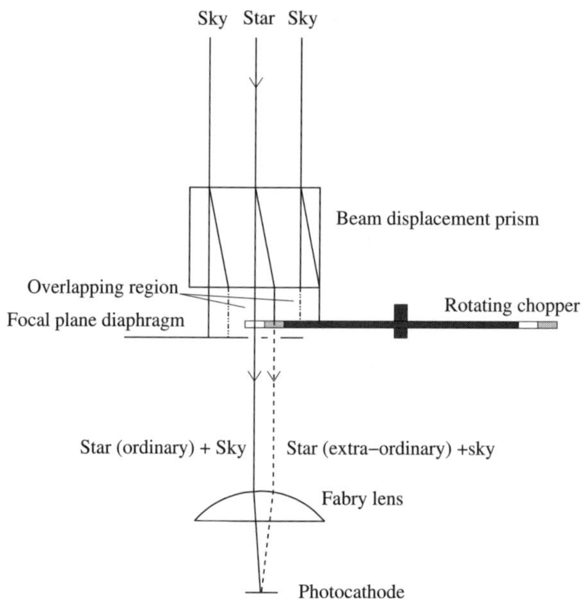

Fig. 2.7 Working principle of the polarimeter.

In the case of the star, the emergent beams from the prism do not overlap. The two images of the star with mutually perpendicular planes of polarization at the focal plane superposed on the unpolarized background sky are observed. Two identical apertures are used to isolate these images, and a rotating chopper is used to alternately block one of the images in order to allow the other to be detected by the same photomultiplier tube. The advantages of such a system are:

- the contribution of background sky polarization is eliminated from the data,
- the effect of any time-dependent variations in the sensitivity of the detector is negligible, since the same photomultiplier tube is used as detector, and
- the quasi-simultaneous detection of both the star beams using a fast rotating chopper eliminates the effects of variations in sky-transparency and reduces the errors due to atmospheric scintillation for bright stars significantly; scintillation noise is independent of the brightness of the star and dominates the photon noise for bright objects (Young, 1967).

The efficiency of a polarimeter at the telescope is determined by (i) the faintest magnitude that could be observed with a specified accuracy in a given time, and (ii) obtaining the maximum amount of information on wavelength dependence during the same time.

2.4.4.2 *Imaging polarimeter*

A conventional imaging polarimeter should be able to measure two orthogonal polarization components simultaneously to eliminate atmospheric scintillation effects. In order to minimize the image motion on the detector, one may employ one moving optical component. An accurate guiding mechanism is required to allow long exposure image to be recorded, which is necessary to obtain sufficient S/N ratio. It is also desirable to develop the instrument capable of making multi-wavelength observations in various wavelength bands in the optical and near-IR regions. An instrument of this kind has a high quality achromat lens[28] acting as a focal plane reducer, which reimages the telescope focal plane on the surface of a CCD detector. A Wollaston prism[29] is placed in the optical path just before the camera

[28] A lens collects light from a point of source and refracts that light to form an image. A simple lens is a single element, while a compound lens consists of a single group of two or more simple lenses. The complex lens is made up of multiple groups of lens elements. An Achromat is a single lens, in which two or more elements, usually of crown and flint glass, with differing dispersion are bonded together. They have been corrected for chromatic aberration with respect to two selected wavelengths.

[29] A Wollaston prism separates randomly polarized or unpolarized light into two orthogonal linearly polarized beams that exit the prism at an angle determined by the wavelength of the light and the length of the prism. It is essentially made of two orthogonal calcite prisms that are cemented together on their base in such a way that the combination forms a plane parallel plate. The optic axes in the two prisms are parallel to the external faces and are mutually perpendicular. Light striking the surface of incidence at right angles is refracted in the first prism into an ordinary (o) and an extra-ordinary (e) ray. The angular splitting α is given by the relation, $\alpha = 2(n_e - n_o)\tan\theta$, where θ

lens. Light passing through such a prism gets split into two orthogonal components parallel (ordinary ray) and perpendicular (extra-ordinary ray) to the axis of the prism. These two beams travel in slightly different directions separated by about 0.5°. This results in two images being produced on the CCD for every point at the telescope focal plane.

The normalized Stokes parameter is defined in a coordinate system aligned with the axis of Wollaston prism be the ratio (R) of the difference between the intensities in the ordinary (O) and extra-ordinary (E) images to their sum.

The Mueller matrices corresponding to the ordinary and extra-ordinary rays emerging from the Wollaston prism are obtained by putting $\phi = 0°$ and $\theta = 90°$ in equation (2.182). By combining the equations (2.188) and (2.189), the Stokes vector of the light emerging from the half-wave plate, the orthogonal polarization components produced by the Wollaston prism analyzer are written as,

$$\frac{1}{2}\begin{bmatrix} 1 & \pm 1 & 0 & 0 \\ \pm 1 & 1 & 0 & 0 \\ 0 & 0 & 0 & 0 \\ 0 & 0 & 0 & 0 \end{bmatrix} \begin{bmatrix} I \\ Q\cos 4\alpha + U\sin 4\alpha \\ Q\sin 4\alpha - U\cos 4\alpha \\ -V \end{bmatrix} = [I'] \begin{bmatrix} 1 \\ \pm 1 \\ 0 \\ 0 \end{bmatrix}, \tag{2.197}$$

where $I' = 1/2(I \pm Q\cos 4\alpha \pm U\sin 4\alpha)$ (see equation 2.196) and α is the angle between the effective optical axis of the two half-wave plates.

At $\phi = 0°$, the (+ve) signs in equation (2.197) are for the polarization components that are ordinary, and at $\phi = 90°$, the (+ve) signs are for the extra-ordinary components.

is the prism angle, n_o and n_e are the ordinary and extra-ordinary indices of refraction respectively.

Chapter 3

Interference and diffraction

3.1 Fundamentals of interference

Principles of reflection and refraction are well explained using geometrical optics. Physical optics deals with light as a wave and the principle of linear superposition is particularly important. The most interesting cases of interference usually involve identical waves, with the same amplitude and wavelength, coming together. Consider the case of just two waves, although one may generalize to more than two. When these waves are in phase and travel together are superposed, the intensity at the point of superposition varies from point to point between maxima which exceed the sum of the intensities in the beams, and minima, which may be zero. This is known as interference. When the crest of one wave passes through the crest of another wave, it is referred as constructive interference. It also occurs when the trough of one wave is superpositioned upon the trough of another wave. The other extreme case occurs when the trough of one corresponds with the crest of the other and tend to cancel each other out, resulting in a flat or no wave while interfering. This type of interference is referred to as destructive interference.

Basic principles of optical interference has wide range of applications ranging from on-line real-time wavefront control in astronomy to experiments in relativity. In what follows, the principle of interference and diffraction and the necessary conditions in physical applications are elucidated.

3.2 Interference of two monochromatic waves

Let the two monochromatic waves $\vec{E}_1$ and $\vec{E}_2$ be superposed at the recombination point P. The correlator sums the instantaneous amplitudes of the fields. The total electric field at P at the output is,

$$\vec{E} = \vec{E}_1 + \vec{E}_2, \tag{3.1}$$

with

$$\begin{aligned}\vec{E}(\vec{r},t) &= \Re\left\{A(\vec{r})e^{-i\omega t}\right\}\\ &= \frac{1}{2}\left[A(\vec{r})e^{-i\omega t} + A^*(\vec{r})e^{i\omega t}\right],\end{aligned} \tag{3.2}$$

where $\vec{r} = x, y, z$ is the position vector of a point, $A(\vec{r}) = a_j(\vec{r})e^{i\psi_j(\vec{r})}$ is the complex vector function of position $\vec{r}$, $j = 1, 2, 3$, $\psi_j(\vec{r},\omega)(= \vec{\kappa}\cdot\vec{r} - \delta_j)$ the phase functions, δ_j's the phase constant, which specify the state of polarization, and $\vec{\kappa}$ the propagation vector.

The vector components of the amplitude, $A(\vec{r})$, are

$$A_x = a_1 e^{i\vec{\kappa}\cdot\vec{r} - \delta_1}, \quad A_y = a_2 e^{i\vec{\kappa}\cdot\vec{r} - \delta_2}, \quad A_z = a_3 e^{i\vec{\kappa}\cdot\vec{r} - \delta_3}. \tag{3.3}$$

The intensity of light, I, is,

$$I \propto \left\langle \vec{E}^2 \right\rangle = \frac{1}{2}A(\vec{r})A^*(\vec{r}), \tag{3.4}$$

in which $\langle\,\rangle$ stands for the time average of the energy within the bracket and A^* represents for the complex conjugate of A.

By squaring the equation (3.1), one gets,

$$\vec{E}^2 = \vec{E}_1^2 + \vec{E}_2^2 + 2\vec{E}_1\cdot\vec{E}_2. \tag{3.5}$$

The intensity at the same point, P is given by,

$$I = I_1 + I_2 + J_{12}, \tag{3.6}$$

in which

$$I_1 = \left\langle \vec{E}_1^2 \right\rangle, \quad I_2 = \left\langle \vec{E}_2^2 \right\rangle, \quad J_{12} = \left\langle \vec{E}_1\cdot\vec{E}_2 \right\rangle. \tag{3.7}$$

Here $I_{j=1,2}$ are the intensities of the two waves, and J_{12} the interference term.

This interference term J_{12} is dependent on the relative phase, thus the intensities of superimposed waves cannot be added. The complicated intensity distribution of a diffracted wave is a direct result of the mutual interference of an infinite number of spherical wavelets[1] emanating from

[1] Wavelets (little waves) can be envisaged as a series of tiny spherical waves which repeatedly generate themselves at all points across the wavefront, and propagate outwards.

the aperture.

Let $A_1(\vec{r})$ and $A_2(\vec{r})$ be the complex amplitudes of the two waves, where

$$A_1(\vec{r}) = a_1(\vec{r})e^{i\psi_1(\vec{r})}; \qquad A_2(\vec{r}) = a_2(\vec{r})e^{i\psi_2(\vec{r})}. \tag{3.8}$$

The (real) phases, $\psi_1(\vec{r})$, $\psi_2(\vec{r})$ of the two waves are in general be different since the waves will have to travel to P by different paths, but if the experimental conditions are such that the same phase difference, δ, is introduced between the corresponding components, we have,

$$\psi_1(\vec{r}) - \psi_2(\vec{r}) = \delta = \frac{2\pi}{\lambda_0}\Delta\varphi, \tag{3.9}$$

where $\delta = 2\pi\Delta\varphi/\lambda_0$, in which $\Delta\varphi$ is the optical path difference (OPD) between two waves from the common source to the intersecting point, and λ_0 the wavelength in vacuum.

In terms of $A_1(\vec{r})$ and $A_2(\vec{r})$,

$$\begin{aligned}\vec{E}_1 \cdot \vec{E}_2 &= \frac{1}{4}\left[A_1(\vec{r})e^{-i\omega t} + A_1^*(\vec{r})e^{i\omega t}\right] \cdot \left[A_2(\vec{r})e^{-i\omega t} + A_2^*(\vec{r})e^{i\omega t}\right] \\ &= \frac{1}{4}\left[A_1(\vec{r}) \cdot A_2(\vec{r})e^{-2i\omega t} + A_1^*(\vec{r}) \cdot A_2^*(\vec{r})e^{2i\omega t}\right. \\ &\quad \left. + A_1(\vec{r}) \cdot A_2^*(\vec{r}) + A_1^*(\vec{r}) \cdot A_2(\vec{r})\right].\end{aligned} \tag{3.10}$$

So that

$$\begin{aligned}J_{12} = 2\left\langle \vec{E}_1 \cdot \vec{E}_2 \right\rangle &= \frac{1}{2}\left[A_1(\vec{r}) \cdot A_2^*(\vec{r}) + A_1^*(\vec{r}) \cdot A_2(\vec{r})\right] \\ &= a_1(\vec{r})a_2(\vec{r})\cos(\psi_1(\vec{r}) - \psi_2(\vec{r})) \\ &= a_1(\vec{r})a_2(\vec{r})\cos\delta.\end{aligned} \tag{3.11}$$

Equation (3.11) relates the dependence of the interference term on the amplitude components, as well as on the phase difference of the two waves. The correlation term, $A_1^*(\vec{r}) \cdot A_2(\vec{r})$ can take significant values, albeit for short period of time, much shorter than the response time of any optical detector, and $< A_1^*(\vec{r})A_2(\vec{r}) >= 0$.

Fresnel-Arago made an extensive study of the conditions under which the interference of polarized light occurs. Their conclusions, known as 'Fresnel-Arago law', are:

- two waves that are linearly polarized in the same plane can interfere and
- two waves, linearly polarized with perpendicular polarizations, cannot interfere and no fringes yield.

In the case of the latter, the situation remains same even if they are derived from perpendicular components of unpolarized light and subsequently brought into the same plane, but interfere when they are derived from the same linearly polarized wave and subsequently brought into the same plane (Collett, 1993).

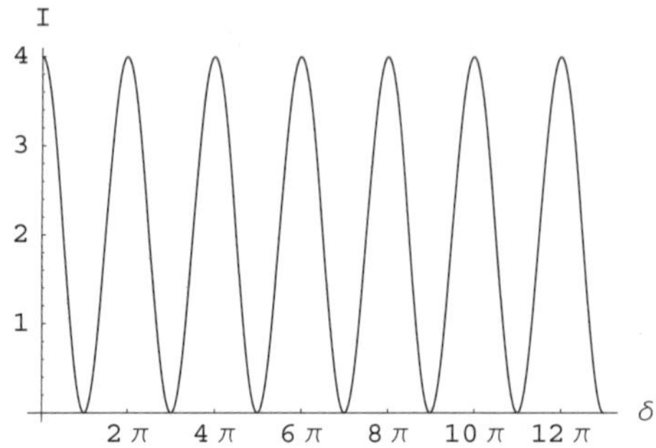

(a)

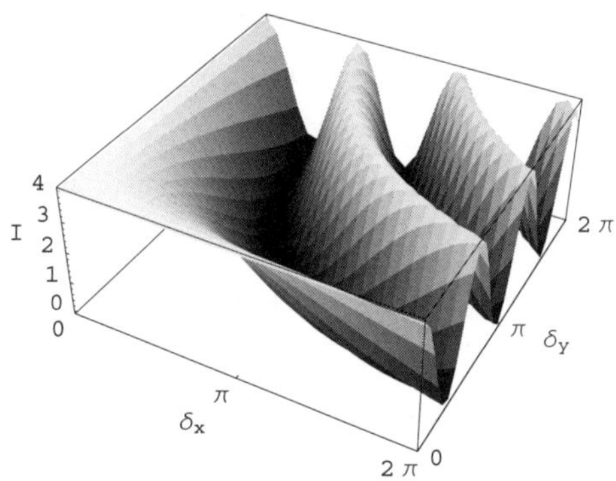

(b)

Fig. 3.1 Interference of the two plane waves with $I_1 = I_2$, in which $I_{\max} = 4I_1$; variation of intensity with phase difference. (a) $I = 4I_1 \cos^2(\delta/2)$ and (b) $I = 4I_1 \cos^2(\delta x \delta y/4)$.

The distribution of intensity resulting from the superposition of the two waves propagates in the z-direction, and linearly polarized with their $\vec{E}$ vectors in the x-direction. On using equations (3.3, 3.6, and 3.11), one obtains,

$$I_1 = \frac{1}{2}a_1^2, \qquad I_2 = \frac{1}{2}a_2^2,$$
$$J_{12} = a_1 a_2 \cos\delta = 2\sqrt{I_1 I_2}\cos\delta. \tag{3.12}$$

The intensity of illumination at P is derived as,

$$I = I_1 + I_2 + 2\sqrt{I_1 I_2}\cos\delta. \tag{3.13}$$

The interference term enables the positions of the fringe intensity maxima and minima to be calculated. The intensity of illumination at P attains its maximal value when the ψ_1 and ψ_2 are in phase. They enhance by each other by interfering, and their interference is called constructive interference. When both ψ_1 and ψ_2 are out of phase by half a cycle and are of equal amplitude, destructive interference takes place. Mathematically both these interferences are respectively expressed as,

$$\begin{aligned} I_{max} &= I_1 + I_2 + 2\sqrt{I_1 I_2} \qquad \text{when } |\delta| = 0, 2\pi, 4\pi, \cdots, \\ I_{min} &= I_1 + I_2 - 2\sqrt{I_1 I_2} \qquad \text{when } |\delta| = \pi, 3\pi, 5\pi, \cdots. \end{aligned} \tag{3.14}$$

When $I_1 = I_2$ the equation (3.13) can be recast as,

$$I = 2I_1(1 + \cos\delta) = 4I_1\cos^2\frac{\delta}{2}. \tag{3.15}$$

Equation (3.15) reveals that the intensity varies between a maximum value $I_{max} = 4I_1$ and a minimum value $I_{min} = 0$. Figure (3.1) depicts the interference of the two beams of equal intensity.

The contrast or visibility, $\mathcal{V}$, is defined by,

$$\mathcal{V} = \frac{I_{max} - I_{min}}{I_{max} + I_{min}} = \frac{2\sqrt{I_1 I_2}}{I_1 + I_2}. \tag{3.16}$$

The visibility of the fringe is a dimensionless number between zero and one that indicates the extent to which a source is resolved on the baseline being used. It contains information about both the spatial and spectral nature of the source. The visibility equals 1 when $I_1 = I_2$.

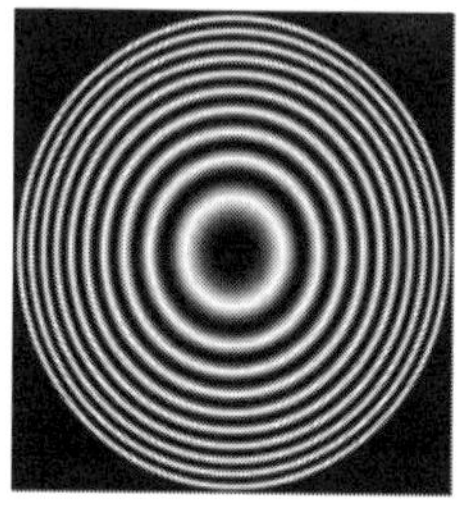

Fig. 3.2 Newton's rings.

From the historical interest in connection with Newton's views on the nature of light, an example of fringes, known as 'Newton's rings' is displayed in Figure (3.2). These fringes are computer simulated, but can be observed in the air film between the convex spherical surface of a lens and a plane glass surface in contact, and illuminated at normal incidence with quasimonochromatic light.

3.2.1 *Young's double-slit experiment*

The key discovery to understand the wave theory of light was the double-slit optical interference experiment of Young performed in 1801 (Young, 1802). According to which he established that beams of light can interfere constructively, as well as destructively. This experiment is based on wave-front division which is sensitive to the size and bandwidth of the source. This interferometer is generally used to measure spatial coherence of a source which is never truly a point source. High visibility of the fringes are discernible on the observation screen if such an interferometer is fed by a monochromatic light source. If a second source is placed in the same plane, but shifted slightly, the condition of conservation of the OPD allows to derive the spatial shift of the new set of fringes on the screen. This leads to the loss of the fringe contrast resulting from blurring due to the superposition of the shifted interferograms. This stresses the importance of the size of the source.

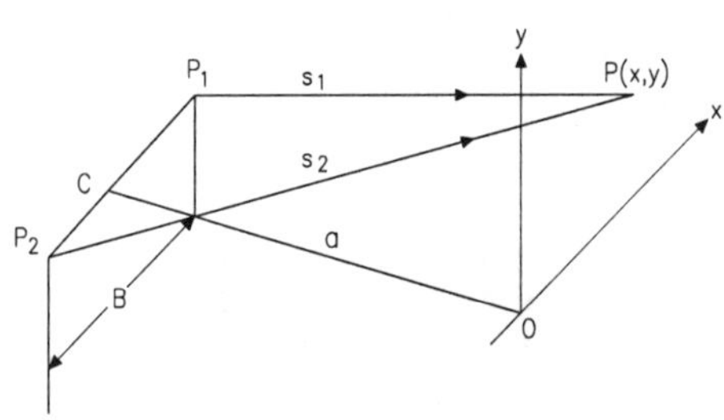

Fig. 3.3 Illustration of interference with two point sources.

For a point $\mathsf{P}(x, y)$ in the plane of observation, let a plane monochromatic wave emanating from a point source be falling on two pinholes $\mathtt{P}_1$ and $\mathtt{P}_2$ in an opaque screen[2] and equidistant from the source. Here B is the separation (baseline) of the pinholes and is assumed to be an order of magnitude λ. These pinholes act as monochromatic point sources which are in phase, and the interference pattern is obtained on a remote screen over a plane $x\mathtt{O}y$ normal to a perpendicular bisector $\mathtt{CO}$ of $\mathtt{P}_1\mathtt{P}_2$ and with the $x-$axis parallel to $\mathtt{P}_1\mathtt{P}_2$ (see Figure 3.3). Assume that a is the distance between the aperture mask and interference point at $\mathtt{P}$, where $a \gg B$.

$$\begin{aligned} s_1 = \mathtt{P}_1\mathtt{P} &= \left[a^2 + y^2 + \left(x - \frac{B}{2}\right)^2\right]^{1/2}, \\ s_2 = \mathtt{P}_2\mathtt{P} &= \left[a^2 + y^2 + \left(x + \frac{B}{2}\right)^2\right]^{1/2}, \end{aligned} \tag{3.17}$$

and by squaring these sub-equations (3.17) followed by subtracting one obtains,

$$s_2^2 - s_1^2 = 2xB. \tag{3.18}$$

The geometrical path difference between the spherical waves reaching the observation point, $\mathtt{P}$, is caused by the difference of propagation distances of the waves from the pinholes, $\mathtt{P}_2$ and $\mathtt{P}_1$ to $\mathtt{P}$ and is expressed as,

$$\Delta s = \frac{xB}{a} = B\sin\theta, \tag{3.19}$$

in which θ is the angle $\mathtt{OCP}$.

The observed intensity along the observation screen is given by,

$$I = I_{max}\cos^2\left(\frac{\kappa B\sin\theta}{2}\right), \tag{3.20}$$

The phase difference, δ, resulting from the difference in propagation distance is of the form,

$$\delta = \kappa\Delta\varphi = \kappa B\sin\theta = 2\pi\left(\frac{B\sin\theta}{\lambda}\right). \tag{3.21}$$

If n is the refractive index of the homogeneous medium, the different optical path from $\mathtt{P}_1$ and $\mathtt{P}_2$ to the point $\mathtt{P}$, the optical path difference

[2] Opaque screens do not allow the light energy to pass through.

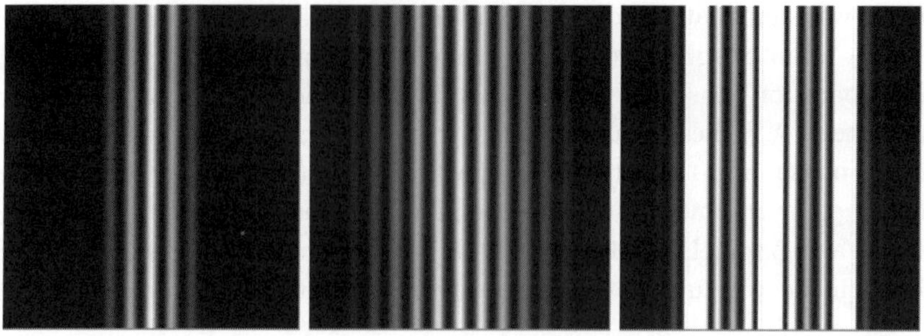

(a) (b) (c)

Fig. 3.4 Computer simulated double slit Young's fringe patterns (a) and (b) and multislit fringe patterns (c). There are two input parameters considered in the program, i.e., size of the slit and the size of the gap between two slits. By varying these two parameters different patterns were obtained in which diffraction phenomenon is taken into consideration.

(OPD), $\Delta\varphi$, is given by,

$$\Delta\varphi = n\Delta s = \frac{nxB}{a}, \tag{3.22}$$

and are corresponding phase difference is,

$$\delta = \frac{2\pi}{\lambda_0}\frac{nxB}{a}. \tag{3.23}$$

Adding other phase differences arising as a result of propagation through different media or initial phase differences. All these phase differences are required to be summed up into a total phase difference $\delta(\theta)$. Thus the intensity observed at P is derived as,

$$I = I_{max}\cos^2\left(\frac{\delta}{2}\right). \tag{3.24}$$

Since the angle P_1PP_2 is very small, one may consider the waves from P_1 and P_2 to be propagated in the same direction at P, so that the intensity can be calculated from the equation (3.13). If the two waves arriving from the same source, or sources that are emitting waves in phase, they interfere constructively at a certain point if the distance traveled by one wave is the same as, or differs by an integral number of wavelengths from, the path length traveled by the second wave. For the waves to interfere destructively, the path lengths must differ by an integral number of wavelengths plus half a wavelength. According to the equations (3.14) and (3.23), there are maxima and minima of intensities respectively when,

$$x = \begin{cases} ma\lambda_0/nB & |m| = 0, 1, 2, \cdots, \\ ma\lambda_0/nB & |m| = 1/2, 3/2, 5/2, \cdots. \end{cases} \tag{3.25}$$

The interference pattern in the immediate vicinity of O thus consists of bright and dark bands, called interference fringes equidistant (see Figure 3.4), and are at right angles to the line P_1P_2 joining the two sources. The separation of adjacent bright fringes, Δ, is proportional to the wavelength, λ_0 and inversely proportional to the baseline between the apertures B, i.e.,

$$\Delta = \frac{a\lambda_0}{nB}. \tag{3.26}$$

The order of the interference at any point is given by,

$$m = \frac{\delta}{2\pi} = \frac{\Delta\varphi}{\lambda_0}. \tag{3.27}$$

If the contact of the two waves is like the left side of the Figure (3.5), the corresponding fringes can be seen from the right side of the same figure. The formula for $\cos(x)^2$ fringes used in this case is expressed as,

$$\left| Ae^{2\pi ux^2} + Be^{2\pi vx} \right|^2 = A^2 + B^2 + 2AB\cos\left[2\pi(ux^2 + vx)\right], \tag{3.28}$$

in which the first term of the LHS represents a cylindrical wave, while the second term of the same side represent an inclined plane wave. Here u and v are the spatial frequencies; unit should be lines per mm if the unit of x is in mm.

It is worthwhile to note that the interference of the two tilted plane waves provides straight line fringes - more the tilt thinner (slimmer) the fringes. If the plane wave is assumed to be not inclined one, v turns out to be zero, $v = 0$ and the nature of the fringe becomes $\cos(2\pi ux^2)$. The term $\cos^2(2\pi ux^2)$ is plotted as one sees always the intensity pattern. Two types of $\cos x^2$ fringes have been drawn. In one type where the central order fringe is the thickest, the nature of the wave is like symmetrical half cylinder. In the other type at the corner the phase front is like x-square curve starting from the left corner.

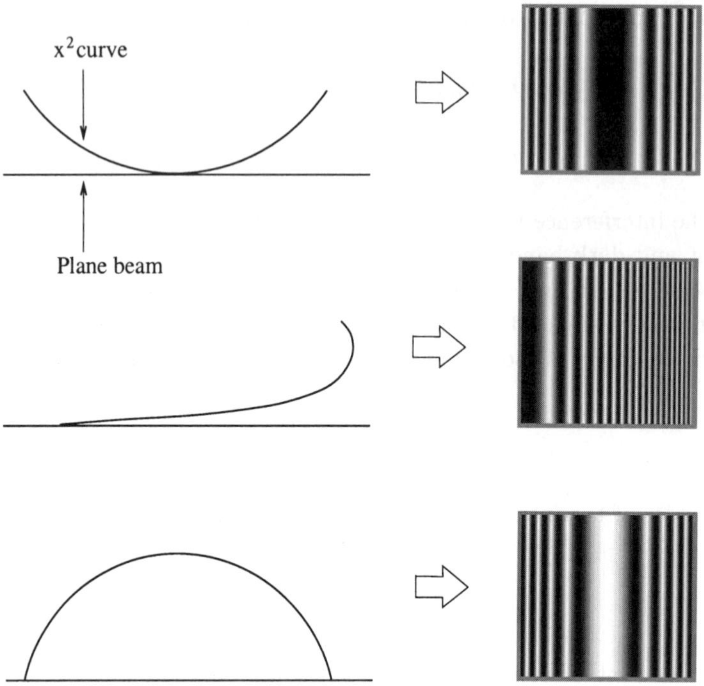

Fig. 3.5 2-D patterns of $\cos x^2$ fringes when the contacts of the two waves are like the left side of the figure.

3.2.2 *Michelson's interferometer*

Michelson's interferometer is based on amplitude division and is generally used to measure the temporal coherence of a source which is never strictly monochromatic. Results from the Michelson interferometer were used to verify special relativity. They are also being used in possible gravity-wave detection.

Let a monochromatic light source be placed at the focus of a collimating lens. The incident beam is divided at the semi-reflecting surface of a plane parallel glass plate, D, into two beams at right angles. One of these beams is reflected back by a fixed mirror[3] kept at one arm and the other beam

[3] A mirror is an object whose surface is smooth enough to form an image. A plane mirror has a flat surface, in which a parallel beam of light changes its direction as a whole; the images formed by such a mirror are virtual images of the same size as the original object. Curved mirrors are used to produce magnified or demagnified images. In a concave mirror, a parallel beam of light becomes a convergent beam, whose rays intersect in the focus of the mirror, while in a convex mirror, a parallel beam becomes

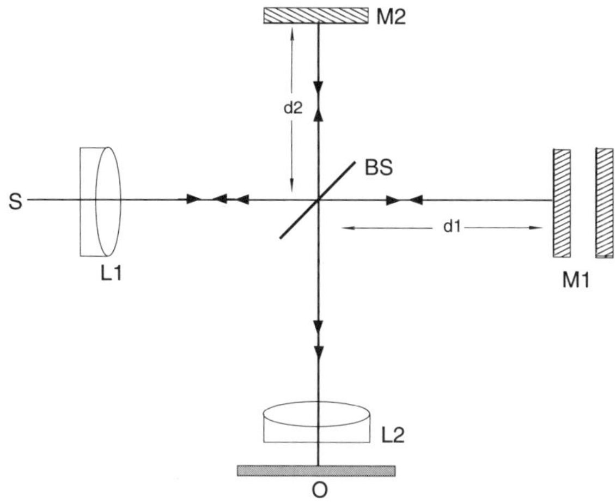

Fig. 3.6 Schematic diagram of a classical Michelson interferometer.

that is transmitted through the beam splitter, is reflected back by another mirror kept at the movable arm (see Figure 3.6). The extra path traversed by one of the beams is compensated by translating the latter. Both these reflected beams are divided again by the beam splitter, wherein one beam from each mirror propagates to a screen. Successive maxima and minima of the fringes are observed at the output with a periodicity governed by ratio of the OPD to the wavelength. All the wavelengths add in phase at zero OPD. The loss of visibility away from zero OPD refers to the blurring due to stretched interference fringes. This stresses the importance of the spectrum of the source.

Let $U(\vec{r},t) = A(\vec{r},t)e^{-i2\pi\nu_0 t}$ be the analytic signal of light emitted by the source. The observed complex disturbance at the focus of the lens is determined by,

$$U(\vec{r},\tau) = K_1 U(\vec{r},t) + K_2 U(\vec{r},t+\tau), \tag{3.29}$$

where $K_{j(=1,2)}$ are real numbers determined by the losses for each light paths, $\tau(= 2h/c)$ the relative time delay suffered by light in the arm with the movable mirror, c the velocity of light, ν_0 the frequency of light in vacuum, and h the mirror displacement from the position of equal pathlength.

If both the fields are sent to a quadratic detector, it yields the desired cross-term (time average due to time response). The measured intensity at

divergent, with the rays appearing to diverge from a common intersection behind the mirror.

the detector is deduced as,

$$\begin{aligned}I(\vec{r},\tau) &= \left\langle |K_1 U(\vec{r},t) + K_2 U(\vec{r},t+\tau)|^2 \right\rangle \\ &= K_1^2 \left\langle |U(\vec{r},t)|^2 \right\rangle + K_2^2 \left\langle |U(\vec{r},t+\tau)|^2 \right\rangle \\ &\quad + K_1 K_2 \left\langle |U(\vec{r},t) U^*(\vec{r},t+\tau)| \right\rangle + K_1 K_2 \left\langle |U^*(\vec{r},t) U(\vec{r},t+\tau)| \right\rangle .\end{aligned} \tag{3.30}$$

If the field $U(\vec{r},t)$ is stationary, that is,

$$\left\langle |U(\vec{r},t)|^2 \right\rangle = \left\langle |U(\vec{r},t+\tau)|^2 \right\rangle = I_0(\vec{r}),$$

the equation (3.30) is recast as,

$$I(\vec{r},\tau) = I_0(\vec{r}) \left(K_1^2 + K_2^2 \right) + 2K_1 K_2 \Re \left[\Gamma(\vec{r},\tau) \right], \tag{3.31}$$

in which $\Gamma(\vec{r},\tau)$ is the autocorrelation (see Appendix B) of the signal $U(\vec{r},t)$.

The autocorrelation can be expressed as an ensemble average over all possible realizations, known as coherence function. Here the complex self-coherence function is given by,

$$\begin{aligned}\Gamma(\vec{r},\tau) &= \langle U^*(\vec{r},t) U(\vec{r},t+\tau) \rangle \\ &= \lim_{T_m \to \infty} \frac{1}{2T_m} \int_{-T_m}^{T_m} U^*(\vec{r},t) U(\vec{r},t+\tau) dt.\end{aligned} \tag{3.32}$$

For a harmonic wave, $U(\vec{r},t) = A(\vec{r},t)e^{-i\omega_0 t}$, the self-coherence function, $\Gamma(\vec{r},\tau)$, takes the form,

$$\begin{aligned}\Gamma(\vec{r},\tau) &= \lim_{T_m \to \infty} \frac{1}{2T_m} \int_{-T_m}^{T_m} U^*(\vec{r},t) U(\vec{r},t+\tau) dt \\ &= \lim_{T_m \to \infty} \frac{1}{2T_m} \int_{-T_m}^{T_m} |A(\vec{r},t)|^2 e^{i\omega_0 t} e^{-i\omega_0 (t+\tau)} dt \\ &= |A(\vec{r},t)|^2 e^{-i\omega_0 \tau}.\end{aligned} \tag{3.33}$$

Equation (3.33) implies that the self-coherence function harmonically depends on the time delay, τ. The normalized form of $\Gamma(\vec{r},\tau)$, known as the complex degree of (mutual) coherence of light, $\gamma(\vec{r},\tau)$, can be derived as,

$$\gamma(\vec{r},\tau) = \frac{\Gamma(\vec{r},\tau)}{\Gamma(\vec{r},0)}. \tag{3.34}$$

This normalized degree of coherence is given by,

$$\gamma(\vec{r},\tau) = |\gamma(\vec{r},\tau)|\, e^{-i2\pi\nu_0\tau \;-\; \psi(\vec{r},\tau)}. \tag{3.35}$$

Since $\Gamma(\vec{r},0) = I_1$ is real and is the largest value that occurs when the modulus of the autocorrelation function, $\Gamma(\vec{r},\tau)$, is taken as, $|\gamma(\vec{r},\tau)| \leq 1$. The observed interferogram has the form,

$$I(\vec{r},\tau) = I_0(\vec{r})\left(K_1^2 + K_2^2\right)\left[1 + \frac{2K_1K_2}{K_1^2 + K_2^2}\,|\gamma(\vec{r},\tau)|\cos[2\pi\nu_0\tau - \psi(\vec{r},\tau)]\right]. \tag{3.36}$$

By assuming equal losses in the two arms of the interferometer i.e., $K_1 = K_2 = K$, the equation (3.36) reduces to,

$$I(\vec{r},\tau) = 2I_0(\vec{r})K^2\left[1 + |\gamma(\tau)|\cos[2\pi\nu_0\tau - \psi(\vec{r},\tau)]\right]. \tag{3.37}$$

The interferogram consists of a sinusoidal fringe term $\cos 2\pi\nu_0\tau$, modulated by the coherence term $|\gamma(\vec{r},\tau)|e^{i\psi(\vec{r},\tau)}$, varying from $4K^2I_0$ to zero about a mean level $2K^2I_0$. As the OPD increases, the amplitude modulation $\gamma(\vec{r},\tau)$ falls from unity towards zero, and the fringes suffer a phase modulation $\psi(\vec{r},\tau)$. From the measurement of the fringe visibility, the temporal coherence of the source can be determined,

$$\mathcal{V} = \frac{I_{max} - I_{min}}{I_{max} + I_{min}} = \frac{|\Gamma(\vec{r},\tau)|}{\Gamma(\vec{r},0)} = |\gamma(\vec{r},\tau)|. \tag{3.38}$$

Equation (3.38) implies that the visibility function, $\mathcal{V}$, equals the modulus of the complex degree of coherence, $|\gamma(\vec{r},\tau)|$. It is found that the fringe visibility $\mathcal{V}$ is a function of time delay, τ, between light waves. The temporal coherence can be expressed in terms of the spectrum of the source radiation (Goodman, 1985),

$$\gamma(\vec{r},t) = \int_{-\infty}^{\infty} B(\vec{r},\nu)e^{-i2\pi\nu\tau}\,d\tau, \tag{3.39}$$

in which $B(\vec{r},\nu)$ is the normalized power spectral density of the radiation.

With the interferometer each monochromatic component produces an interference pattern as the path difference is increased from zero; two component patterns show increasing mutual displacement, because of the difference of wavelength. The visibility of the fringes therefore decreases and they disappear when the OPD is sufficiently large. The maximum transit time difference for good visibility of the fringes is known as coherence time of the field. In order to keep the time correlation close to unity, the delay

τ must be limited to a small fraction of the temporal width or coherence time, $\tau_c \Delta\nu \simeq 1$, where $\Delta\nu$ is the spectral width. The coherence time is expressed as,

$$\tau_c = \int_{-\infty}^{\infty} |\gamma(\vec{r}, t)|^2 \, d\tau, \tag{3.40}$$

The coherence time is the characteristic evolution time of the complex amplitude. A factor less than unity affects the degree of coherence. The corresponding limit for the OPD between two fields is the coherence length and is determined by $l_c = c\tau_c = c/\Delta\nu$. Such a length measures the maximum OPD for which the fringes are still observable.

The angular dimension of the source producing fringes can be determined simply by observing the smallest value of d for which the visibility of the fringes is minimum. This condition occurs when,

$$d = \frac{A\lambda_0}{\theta}, \tag{3.41}$$

where $A = 0.5$ for two point sources of angular separation θ, and $A = 1.22$ for a uniform circular disc source of angular diameter, θ.

A variation of the Michelson's interferometer was developed by Twyman and Green in which a point source of quasi-monochromatic light at the focus of a a well corrected collimating lens is employed, so that all rays enter the interferometer parallel to the optical axis. The parallel rays emerging from the interferometer are brought to a focus by a second well corrected lens. The fringes of equal thickness appear on the observing screen which reveals imperfections in the optical system that cause variations in OPD. The difference of optical path between the emergent rays at the virtual intersecting point is $\Delta\varphi = nh$ and the corresponding phase difference would be $\delta = 2\pi nh/\lambda_0$. This interferometer made with high quality optical components is used at the laboratory to test the quality of optical component under that.

3.2.3 *Mach-Zehnder interferometer*

A more radical variation of the Michelson's interferometer is the Mach-Zehnder interferometer, which is used for analyzing the temporal coherence of a collimated beam of light. It is also employed to measure variations of refractive index, and density in compressible gas. For aerodynamic research, in which the geometry of air flow around an object in a wind tunnel is required to be determined through local variations of pressure and refractive index, such an interferometer is used. The usefulness of such an equipment for high angular resolution[4] stellar spectroscopy can be envisaged by obtaining the objective prism spectra of a few close binary stars (Kuwamura et al., 1992).

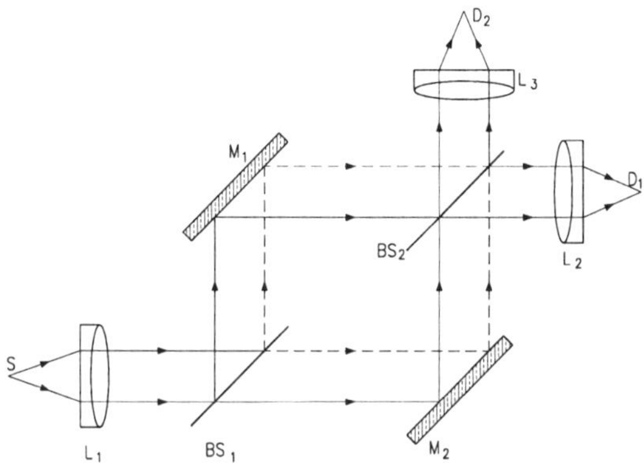

Fig. 3.7 The Mach-Zehnder interferometer.

Figure (3.7) depicts the schematic diagram of a Mach-Zehnder interferometer. This instrument enables the spatial separation of the interfering beams and therefore the use of test objects in one of them. Light from a quasi-monochromatic point source in the focal plane of a well corrected lens, L_1, making the beam collimated is divided into two beams by a beam splitter. Each beam is reflected by the two mirrors, M_1, M_2, of which one is fixed and the other is movable that is used to adjust the optical path-length difference between them, kept at diagonally opposite and the beams are made to coincide again by another semi-transparent[5] mirror. The four reflecting surfaces are arranged to be approximately parallel, with their centers at the corners of a parallelogram. The geometry of this equipment depicts that both shear and tilts can be introduced independently, without introducing shifts. Path-lengths of beam 1 and 2 around the rectangular system and through the beam splitter are identical. In such a situation, the beams interfere constructively in channel 1 and deliver their energy to

[4]Resolution is defined as the ability to discern finer detail; greater the resolution greater the ability to distinguish objects or features.

[5]Transparent medium allows most of the light energy to pass through, while translucent energy partially allows such an energy to pass through.

the detector 1. Any deviations from the path-length equality are sent to channel 2 either the entire beam or a fraction of it.

The relative phase between the two beams can be varied continuously by adjusting the length of one of the interferometer's arms. At equal path-length the output of the detector 1, D_1 reaches maximum. With the increase of the length of the movable arm by a quarter of a wavelength, the phase shift becomes 180°, thus D_1 reaches minimum. Such a situation is periodically repeated as long as the two beams remain coherent or partially coherent and therefore, the output of each detector oscillates between a maximum and a minimum. The oscillations ceases when the path-lengths differ by more than the coherence length of the beam, and both channels receive equal amount of light irrespective of the path-length difference. The time-averaged detector output, $\mathcal{P}_{1,2}$ is written as (Mansuripur, 2002),

$$\begin{aligned}\mathcal{P}_{1,2}(\tau) &= \frac{1}{T}\int_0^T \left\{\frac{1}{2}\left[A(\vec{r},t) \pm A(\vec{r},t-\tau)\right]\right\}^2 dt \\ &= \frac{1}{2T}\int_0^T |A(\vec{r},t)|^2\, dt \pm \frac{1}{2T}\int_0^T A(\vec{r},t)A(\vec{r},t-\tau)dt \\ &= \frac{1}{2}\left[I \pm \frac{1}{2}\sum_n |a_n(\vec{r},\nu)|^2\, \Delta\nu \cos(2\pi\nu_n\tau)\right], \qquad (3.42)\end{aligned}$$

in which the intensity I is independent of time delay, τ, and is given by,

$$I(\vec{r}) = \frac{1}{T}\int_0^2 |A(\vec{r},t)|^2\, dt = \frac{1}{2}\sum_n |a_n(\vec{r},\nu)|^2\, \Delta\nu, \qquad (3.43)$$

and the amplitude of the waveform is,

$$A_n(\vec{r},t) = \sum_n a_n(\vec{r},\nu)\,(\Delta\nu)^{1/2}\cos\left[2\pi\nu_n(\tau - t) + \psi_n\right], \qquad (3.44)$$

a_n and ψ_n are the amplitude and phase of the spectral component whose frequency is ν_n, T the time period, and τ the time delay.

The second term of the equation (3.42) coincides with the first order coherence function of the field in the case of a stationary process. This term is also known as the autocorrelation function of the waveform $A(\vec{r},t)$.

3.3 Interference with quasi-monochromatic waves

The Figure (3.8) is a sketch of a Young's set up where the wave field is produced by an extended polychromatic source. Assuming that the respective distances of a typical point $\mathtt{P}(\vec{r})$ on the screen B from the pinhole positions, $\mathtt{P}_1(\vec{r}_1)$ and $\mathtt{P}_2(\vec{r}_2)$ are s_1 and s_2. Here, the quantity of interest is the correlation factor $< U(\vec{r}_1, t-\tau)U^*(\vec{r}_2, t) >$. The analytic signal obtained at the screen, $\mathtt{P}(\vec{r})$, is expressed as,

$$U(\vec{r}, t) = K_1 U(\vec{r}_1, t - \tau_1) + K_2 U(\vec{r}_2, t - \tau_2), \tag{3.45}$$

where $\vec{r}(= x, y, z)$ the position vector at time t. $\tau_1 = s_1/c$ and $\tau_2 = s_2/c$ are the transit time of light from $\mathtt{P}_1(\vec{r}_1)$ to $\mathtt{P}(\vec{r})$, and from $\mathtt{P}_2(\vec{r}_2)$ to $\mathtt{P}(\vec{r})$ respectively, and $K_{j=1,2}$ the constant factors that depend on the size of the openings and on the geometry of the arrangement, i.e., the angle of incident and diffraction at $\mathtt{P}_1(\vec{r}_1)$ and $\mathtt{P}_2(\vec{r}_2)$.

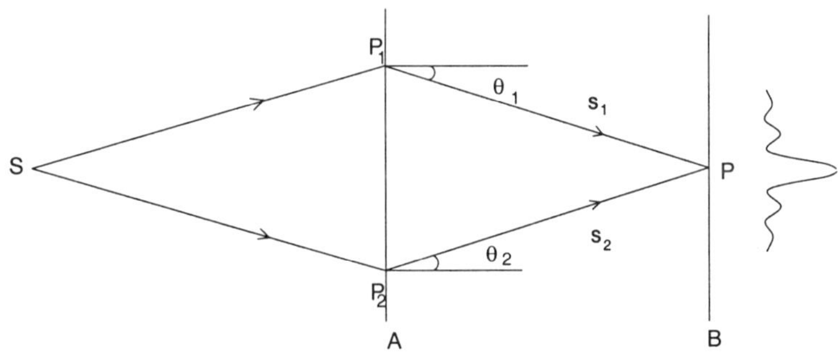

Fig. 3.8 Coherence of the two holes $\mathtt{P}_1(\vec{r}_1)$ and $\mathtt{P}_2(\vec{r}_2)$ illuminated by source σ.

If the pinholes at $\mathtt{P}_1(\vec{r}_1)$ and $\mathtt{P}_1(\vec{r}_1)$ are small and the diffracted fields are considered to be uniform, the values $|K_j|$ satisfy $K_1^* K_2 = K_1 K_2^* = K_1 K_2$. The diffracted fields are approximately uniform, that is, K_1 and K_2 do not depend on θ_1 and θ_2. In order to derive the intensity of light at $\mathtt{P}(\vec{r})$, by neglecting the polarization effects, one assumes that the averaging time is effectively infinite which is a valid assumption for true thermal light. The desired intensity, $I(\vec{r}, t)$ at $\mathtt{P}(\vec{r})$ is defined by the formula,

$$I(\vec{r}, t) = \langle U(\vec{r}, t) U^*(\vec{r}, t) \rangle . \tag{3.46}$$

It follows from the two equations (3.45 and 3.46),

$$\begin{aligned} I(\vec{r}, t) = {} & |K_1|^2 \left\langle |U(\vec{r}_1, t - \tau_1)|^2 \right\rangle + |K_2|^2 \left\langle |U(\vec{r}_2, t - \tau_2)|^2 \right\rangle \\ & + K_1 K_2^* \langle U(\vec{r}_1, t - \tau_1) U^*(\vec{r}_2, t - \tau_2) \rangle \\ & + K_1^* K_2 \langle U(\vec{r}_2, t - \tau_2) U^*(\vec{r}_1, t - \tau_1) \rangle . \end{aligned} \tag{3.47}$$

The field was assumed to be stationary. One may shift the origin of time in all these expressions. Therefore,

$$\langle U(\vec{r}_1, t-\tau_1)U^*(\vec{r}_1, t-\tau_1)\rangle = \langle U(\vec{r}_1, t)U^*(\vec{r}_1, t)\rangle = I(\vec{r}_1, t). \tag{3.48}$$

Similarly,

$$\langle U(\vec{r}_2, t-\tau_2)U^*(\vec{r}_2, t-\tau_2)\rangle = \langle U(\vec{r}_2, t)U^*(\vec{r}_2, t)\rangle = I(\vec{r}_2, t), \tag{3.49}$$

and if one sets $\tau = \tau_2 - \tau_1$, hence

$$\begin{aligned} \langle U(\vec{r}_1, t-\tau_1)U^*(\vec{r}_2, t-\tau_2)\rangle &= \langle U(\vec{r}_1, t+\tau)U^*(\vec{r}_2, t)\rangle\,, \\ \langle U^*(\vec{r}_1, t-\tau_1)U(\vec{r}_2, t-\tau_2)\rangle &= \langle U^*(\vec{r}_1, t+\tau)U(\vec{r}_2, t)\rangle\,. \end{aligned} \tag{3.50}$$

These two sub-equations (3.50) show that the terms, $U(\vec{r}_1, t+\tau)U^*(\vec{r}_2, t)$ and $U^*(\vec{r}_1, t+\tau)U(\vec{r}_2, t)$ are conjugate, therefore,

$$U(\vec{r}_1, t+\tau)U^*(\vec{r}_2, t) + U^*(\vec{r}_1, t+\tau)U(\vec{r}_2, t) = 2\Re\left[U(\vec{r}_1, t+\tau)U^*(\vec{r}_2, t)\right]. \tag{3.51}$$

The values $|K_j|$ satisfy $K_1^*K_2 = K_1K_2^* = K_1K_2$ for smaller pinholes. By denoting $I_j(\vec{r}, t) = |K_j|^2 < |U(\vec{r}_j, t-\tau_j)|^2 >$, in which $j = 1, 2$, one derives the intensity at $\mathtt{P}(\vec{r})$,

$$I(\vec{r}, t) = I_1(\vec{r}, t) + I_2(\vec{r}, t) + 2\,|K_1K_2|\,\Re\left[\Gamma\left(\vec{r}_1, \vec{r}_2, \frac{s_2 - s_1}{c}\right)\right], \tag{3.52}$$

where the time delay $(s_2 - s_1)/c$ may be denoted by τ.

By introducing a normalization of the coherence function, a further simplification yields. According to the inequality after Schwarz,

$$|\Gamma(\vec{r}_1, \vec{r}_2, \tau)| \leq \sqrt{\Gamma(\vec{r}_1, \vec{r}_1, 0)\Gamma(\vec{r}_2, \vec{r}_2, 0)},$$

in which $\Gamma(\vec{r}_1, \vec{r}_1, \tau)$ and $\Gamma(\vec{r}_2, \vec{r}_2, \tau)$ are the self coherence functions of light at the pinholes, $\mathtt{P}_1(\vec{r}_1)$ and $\mathtt{P}_2(\vec{r}_2)$ respectively, and $\Gamma(\vec{r}_1, \vec{r}_1, 0)$ and $\Gamma(\vec{r}_2, \vec{r}_2, 0)$ represent the intensities of light incident on the two aforementioned pinholes, $\mathtt{P}_1(\vec{r}_1)$ and $\mathtt{P}_2(\vec{r}_2)$, respectively.

If the last term of equation (3.52) does not vanish, the intensity at $\mathtt{P}(\vec{r})$ is not equal to the sum of the intensities of the two beams that reach the point from the two pinholes. It differs from their sum by the term $2|K_1K_2|\Re[\Gamma(\vec{r}_1, \vec{r}_2, \tau)]$. Since $K_j \neq 0$, it follows that if $\Gamma \neq 0$, the superposition of the two beams gives rise to interference. The mutual coherence function, $\Gamma(\vec{r}_1, \vec{r}_2, \tau)$, of the fields $U(\vec{r}_1, t)$ and $U(\vec{r}_2, t)$ at the two points

$\mathsf{P}_1(\vec{r}_1)$ and $\mathsf{P}_2(\vec{r}_2)$ is given by,

$$\Gamma(\vec{r}_1, \vec{r}_2, \tau) = \langle U(\vec{r}_1, t+\tau) U^*(\vec{r}_2, t) \rangle . \tag{3.53}$$

When the two points coincide ($P_1 = P_2$) one obtains,

$$\Gamma(\vec{r}_1, \vec{r}_1, \tau) = \langle U(\vec{r}_1, t+\tau) U^*(\vec{r}_2, t) \rangle , \tag{3.54}$$

At a point where both the points coincide, the self coherence reduces to ordinary intensity. When $\tau = 0$,

$$\begin{aligned} \Gamma(\vec{r}_1, \vec{r}_1, 0) &= I(\vec{r}_1, t), \\ \Gamma(\vec{r}_2, \vec{r}_2, 0) &= I(\vec{r}_2, t). \end{aligned} \tag{3.55}$$

The term $|K_1|^2 I(\vec{r}_1, t)$ is the intensity observed at $\mathsf{P}(\vec{r})$ when the pinhole at $\mathsf{P}_1(\vec{r}_1)$ alone is opened ($K_2 = 0$) and the term $|K_2|^2 I(\vec{r}_2, t)$ has similar interpretation. The intensities at $I_1(\vec{r}, t)$ and $I_2(\vec{r}, t)$ respectively are:

$$\begin{aligned} I_1(\vec{r}, t) &= |K_1|^2 I(\vec{r}_1, t) = |K_1|^2 \Gamma(\vec{r}_1, \vec{r}_1, 0), \\ I_2(\vec{r}, t) &= |K_2|^2 I(\vec{r}_2, t) = |K_2|^2 \Gamma(\vec{r}_2, \vec{r}_2, 0). \end{aligned} \tag{3.56}$$

The normalized complex degree of (mutual) coherence, $\gamma(\vec{r}_1, \vec{r}_2, \tau)$, is defined as,

$$\begin{aligned} \gamma(\vec{r}_1, \vec{r}_2, \tau) &= \frac{\Gamma(\vec{r}_1, \vec{r}_2, \tau)}{\sqrt{\Gamma(\vec{r}_1, \vec{r}_1, 0)}\sqrt{\Gamma(\vec{r}_2, \vec{r}_2, 0)}} = \frac{\Gamma(\vec{r}_1, \vec{r}_2, \tau)}{\sqrt{I_1(\vec{r}, t) I_2(\vec{r}, t)}} \\ &= \frac{\langle U(\vec{r}_1, t+\tau) U^*(\vec{r}_2, t) \rangle}{\sqrt{I_1(\vec{r}, t) I_2(\vec{r}, t)}}, \end{aligned} \tag{3.57}$$

where $\gamma(\vec{r}_1, \vec{r}_2, \tau)$ is the complex degree of coherence of light vibrations at the points $\mathsf{P}_1(\vec{r}_1)$ and $\mathsf{P}_2(\vec{r}_2)$.

The degree of mutual coherence, $\gamma(\vec{r}_1, \vec{r}_2, \tau)$, obeys two wave equations. In Cartesian coordinates, they read (Goodman, 1985),

$$\left(\nabla_j^2 - \frac{1}{c^2} \frac{\partial^2}{\partial \tau^2} \right) \gamma(\vec{r}_1, \vec{r}_2, \tau) = 0, \tag{3.58}$$

where

$$\nabla_j^2 = \frac{\partial^2}{\partial x_j^2} + \frac{\partial^2}{\partial y_j^2} + \frac{\partial^2}{\partial z_j^2},$$

and $\vec{r}_{j=1,2} = x_j, y_j, z_j$.

The term $\gamma(\vec{r}_1, \vec{r}_2, \tau)$ measures both the spatial and the temporal coherence. The degree of coherence of the vibration is given by the Schwarz's inequality, $0 \leq |\gamma(\vec{r}_1, \vec{r}_2, \tau)| \leq 1$. In general, two light beams are not correlated but the correlation term, $U(\vec{r}_1, t)U^*(\vec{r}_2, t)$, takes on significant values for a short period of time and $< U(\vec{r}_1, t)U^*(\vec{r}_2, t) > = 0$. Time variations of $U(\vec{r}, t)$ for a thermal source are statistical in nature (Mandel and Wolf, 1995). Hence, one seeks a statistical description of the field (correlations) as the field is due to a partially coherent source. Depending upon the correlations between the phasor amplitudes at different object points, one would expect a definite correlation between the two points of the field. The effect of $|\gamma(\vec{r}_1, \vec{r}_2, \tau)|$ is to reduce the visibility of the fringes. When $|\gamma(\vec{r}_1, \vec{r}_2)| = 1$, the averaged intensity around the point P in the fringe pattern undergoes periodic variation, between the values $4I_1(\vec{r})$ and zero. This case represents complete coherence, while in the case of incoherence, i.e., when when $|\gamma(\vec{r}_1, \vec{r}_2, \tau)|$ turns out to be zero; no interference fringes are formed. The intermediate values $(0 < |\gamma(\vec{r}_1, \vec{r}_2, \tau)| < 1)$ characterize partial coherence.

Finally, the equation (3.52) can be recast as,

$$I(\vec{r}, t) = I_1(\vec{r}, t) + I_2(\vec{r}, t) + 2\sqrt{I_1(\vec{r}, t)}\sqrt{I_2(\vec{r}, t)}\Re\left[\gamma(\vec{r}_1, \vec{r}_2, \tau)\right]. \quad (3.59)$$

Such an equation (3.59) is known as the general interference law for stationary optical fields. In order to determine light intensity at $P(\vec{r})$ when two light waves are superposed, the intensity of each beam and the value of the real term, $\gamma(\vec{r}_1, \vec{r}_2, \tau)$, of the complex degree of coherence must be available.

The visibility of the fringes $\mathcal{V}(\vec{r})$, at a point $P(\vec{r})$ in terms of the intensity of the two beams and of their degree of coherence may be expressed as,

$$\begin{aligned}\mathcal{V}(\vec{r}) &= \frac{I_{max} - I_{min}}{I_{max} + I_{min}} = |\gamma(\vec{r}_1, \vec{r}_2, \tau)| \frac{2\sqrt{\Gamma(\vec{r}_1, \vec{r}_1, 0)}\sqrt{\Gamma(\vec{r}_2, \vec{r}_2, 0)}}{\Gamma(\vec{r}_1, \vec{r}_1, 0) + \Gamma(\vec{r}_2, \vec{r}_2, 0)} \\ &= |\gamma(\vec{r}_1, \vec{r}_2, \tau)| \frac{2\sqrt{I_1(\vec{r}, t)}\sqrt{I_2(\vec{r}, t)}}{I_1(\vec{r}, t) + I_2(\vec{r}, t)} \\ &= |\gamma(\vec{r}_1, \vec{r}_2, \tau)| \qquad \text{if } I_1(\vec{r}, t) = I_2(\vec{r}, t). \end{aligned} \quad (3.60)$$

If the differential time delay, $\tau_2 - \tau_1$, is very small compared to the coherence time τ_c, $\gamma(\vec{r}_1, \vec{r}_2, \tau)$ is no longer sensitive to the temporal coherence. This case occurs under quasi-monochromatic field conditions. Assuming $\Delta\nu \ll \bar{\nu}$, i.e., $\tau_2 - \tau_1 \ll \tau_c$, one expresses the complex degree of coherence as,

$$\begin{aligned}\gamma(\vec{r}_1, \vec{r}_2, \tau) &= |\gamma(\vec{r}_1, \vec{r}_2, \tau)|\, e^{\Phi(\vec{r}_1, \vec{r}_2, \tau) - 2\pi\bar{\nu}\tau} \\ &= \gamma(\vec{r}_1, \vec{r}_2, 0)e^{-i2\pi\bar{\nu}\tau}. \end{aligned} \quad (3.61)$$

The exponential term is nearly constant and $\gamma(\vec{r}_1, \vec{r}_2, 0)$ measures the spatial coherence only. If $|\gamma(\vec{r}_1, \vec{r}_2, \tau)| = 0$, the equation (3.59) takes the form,

$$I(\vec{r}, t) = I_1(\vec{r}, t) + I_2(\vec{r}, t). \tag{3.62}$$

The intensity at a point $\mathtt{P}(\vec{r})$, in the interference pattern in the case of $|\gamma(\vec{r}_1, \vec{r}_2, \tau)| = 1$,

$$\begin{aligned} I(\vec{r}, t) &= I_1(\vec{r}, t) + I_2(\vec{r}, t) + 2\sqrt{I_1(\vec{r}, t)}\sqrt{I_2(\vec{r}, t)} \\ &\quad \times |\gamma(\vec{r}_1, \vec{r}_2, \tau)| \cos\left[\Phi(\vec{r}_1, \vec{r}_2, \tau) - \delta\right], \end{aligned} \tag{3.63}$$

in which

$$\begin{aligned} \delta = 2\pi\bar{\nu}\tau &= \frac{2\pi(s_2 - s_1)}{\bar{\lambda}} \\ &= \bar{\kappa}(s_2 - s_1), \end{aligned}$$

where $\bar{\kappa} = 2\pi\bar{\nu}/c = 2\pi/\bar{\lambda}$, with $\bar{\kappa}$ as the mean wave number, $\bar{\nu}$ the mean frequency, and $\bar{\lambda}$ the mean wavelength.

The relative coherence of the two beams diminishes as the difference in path length increases, culminating in lower visibility of the fringes. Let $\psi(\vec{r}_1, \vec{r}_2)$, be the argument of $\gamma(\vec{r}_1, \vec{r}_2, \tau)$, thus,

$$\begin{aligned} I(\vec{r}_1, \vec{r}_2, \tau) &= I_1(\vec{r}, t) + I_2(\vec{r}, t) + 2\sqrt{I_1(\vec{r}, t)}\sqrt{I_2(\vec{r}, t)} \\ &\quad \times \Re\left[|\gamma(\vec{r}_1, \vec{r}_2, 0)|\, e^{i[\Phi(\vec{r}_1, \vec{r}_2) - 2\pi\bar{\nu}\tau]}\right]. \end{aligned} \tag{3.64}$$

Equation (3.64) directly illustrate the Young's pin-hole experiment. The measured intensity at a distance x from the origin (point at zero OPD) on a screen at a distance, s, from the apertures is,

$$\begin{aligned} \mathcal{I}(x) &= I_1(\vec{r}, t) + I_2(\vec{r}, t) + 2\sqrt{I_1(\vec{r}, t)}\sqrt{I_2(\vec{r}, t)} \\ &\quad \times |\gamma(\vec{r}_1, \vec{r}_2, 0)| \cos\left[\frac{2\pi d(x)}{\lambda} - \Phi(\vec{r}_1, \vec{r}_2)\right], \end{aligned} \tag{3.65}$$

where $d(x) = Bx/a\lambda$, is the OPD corresponding to x, B the distance between the two apertures, and a the distance between the aperture plane and focal plane.

On introducing two quantities, $J(\vec{r}_1, \vec{r}_2)$ and $\mu(\vec{r}_1, \vec{r}_2)$, which are known respectively as the equal-time correlation function and the complete coherence factor, one obtains,

$$J(\vec{r}_1, \vec{r}_2) = \Gamma(\vec{r}_1, \vec{r}_2, 0) = \langle U(\vec{r}_1, t)U^*(\vec{r}_2, t)\rangle. \tag{3.66}$$

The equal-time correlation function, $J(\vec{r}_1, \vec{r}_2)$, is also called the mutual intensity of light at the apertures $\mathtt{P}_1(\vec{r}_1)$ and $\mathtt{P}_2(\vec{r}_2)$. The complex coherence factor of light $\mu(\vec{r}_1, \vec{r}_2)$ may be written in terms of mutual intensity,

$$\begin{aligned}\mu(\vec{r}_1, \vec{r}_2) = \gamma(\vec{r}_1, \vec{r}_2, 0) &= \frac{\Gamma(\vec{r}_1, \vec{r}_2, 0)}{\sqrt{\Gamma(\vec{r}_1, \vec{r}_1, 0)}\sqrt{\Gamma(\vec{r}_2, \vec{r}_2, 0)}} \\ &= \frac{J(\vec{r}_1, \vec{r}_2)}{\sqrt{J(\vec{r}_1, \vec{r}_1)}\sqrt{J(\vec{r}_2, \vec{r}_2)}} = \frac{J(\vec{r}_1, \vec{r}_2)}{\sqrt{I_1(\vec{r}, t)}\sqrt{I_2(\vec{r}, t)}}. \end{aligned} \tag{3.67}$$

When $I_1(\vec{r}, t)$ and $I_2(\vec{r}, t)$ are constant, in the case of quasi-monochromatic light, the observed interference pattern has constant visibility and constant phase across the observation region. The visibility in terms of the complex coherence factor is,

$$\mathcal{V} = \begin{cases} \dfrac{2\sqrt{I_1(\vec{r}, t)}\sqrt{I_2(\vec{r}, t)}}{I_1(\vec{r}, t) + I_2(\vec{r}, t)}\mu(\vec{r}_1, \vec{r}_2) & \text{if } I_1(\vec{r}, t) \neq I_2(\vec{r}, t), \\ \\ \mu(\vec{r}_1, \vec{r}_2) & \text{if } I_1(\vec{r}, t) = I_2(\vec{r}, t). \end{cases} \tag{3.68}$$

When the complex coherent factor, $\mu(\vec{r}_1, \vec{r}_2)$, turns out to be zero, the fringes vanish and the two lights are known to be mutually incoherent, and in the case of $\mu(\vec{r}_1, \vec{r}_2)$ being 1, the two waves are called mutually coherent. For an intermediate value of $\mu(\vec{r}_1, \vec{r}_2)$, the two waves are partially coherent.

3.4 Propagation of mutual coherence

Analogous to the detailed structure of an optical wave, which undergoes changes as the wave propagates through space, the detailed structure of the mutual coherence function undergoes changes. This coherence function is said to propagate. In what follows, the basic laws obeyed by mutual coherence followed by the mutual coherence function obeys a pair of scalar wave equations are elucidated.

3.4.1 *Propagation laws for the mutual coherence*

Let a wavefront with arbitrary coherence properties be propagating to a surface B from an optical system lying on a surface, A (see Figure 3.9). The mutual coherence function at the surface A is expressed as,

$$\Gamma(\vec{r}_1, \vec{r}_2, \tau) = \langle U(\vec{r}_1, t + \tau)U^*(\vec{r}_2, t)\rangle\,. \tag{3.69}$$

Knowing the values for the mutual coherence function, all points $\mathtt{Q}(\vec{r}_1)$ and $\mathtt{Q}(\vec{r}_2)$ on the surface B, one estimates,

$$\Gamma(\vec{r}_1', \vec{r}_2', \tau) = \langle U(\vec{r}_1', t+\tau) U^*(\vec{r}_2', t) \rangle . \tag{3.70}$$

The analytic signals at $\mathtt{Q}(\vec{r}_1)$ and $\mathtt{Q}(\vec{r}_2)$ on the surface B are obtained by applying the Huygens-Fresnel principle (see section 3.6.1) with narrowband conditions (Goodman, 1985), that is,

$$\begin{aligned} U(\vec{r}_1', t+\tau) &= \int_{\mathtt{A}} \frac{\chi_1(\kappa)}{s_1} U\left(\vec{r}_1, t+\tau-\frac{s_1}{c}\right) d\mathtt{S}_1, \\ U^*(\vec{r}_2', t) &= \int_{\mathtt{A}} \frac{\chi_2^*(\kappa)}{s_2} U^*\left(\vec{r}_2, t-\frac{s_2}{c}\right) d\mathtt{S}_2, \end{aligned} \tag{3.71}$$

in which χ_1 and χ_2 are the inclination factors, s_1 and s_2 the distances $\mathtt{P}_1\,\mathtt{Q}_1$ and $\mathtt{P}_2\,\mathtt{Q}_2$ respectively, $\kappa = 2\pi\nu/c$, and ν the frequency of light.

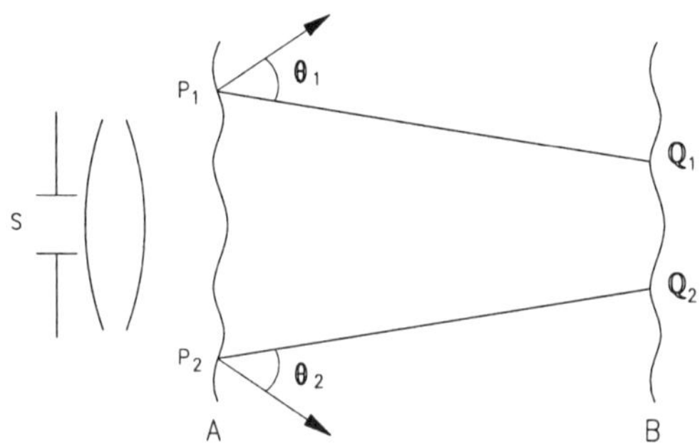

Fig. 3.9 Geometry for propagation laws for the cross-spectral density and for the mutual coherence.

By plugging the sub-equations (3.71) into the equation (3.69) one finds,

$$\Gamma(\vec{r}_1', \vec{r}_2', \tau) = \int_{\mathtt{A}} \int_{\mathtt{A}} \frac{\chi_1}{s_1} \frac{\chi_2^*}{s_2} \Gamma\left(\vec{r}_1, \vec{r}_2, \tau - \frac{s_2 - s_1}{c}\right) d\mathtt{S}_1 d\mathtt{S}_2. \tag{3.72}$$

Such an equation (3.72) is regarded as the propagation law for the mutual coherence function at points $\mathtt{Q}_1(\vec{r}_1')$ and $\mathtt{Q}_2(\vec{r}_2')$ of the surface B. Under quasi-monochromatic conditions, one may write,

$$\Gamma\left(\vec{r}_1, \vec{r}_2, \tau - \frac{s_2 - s_1}{c}\right) = J(\vec{r}_1, \vec{r}_2) e^{i\bar{\kappa}(s_2 - s_1)}. \tag{3.73}$$

in which $\bar{\kappa} = 2\pi\bar{\nu}/c = 2\pi/\bar{\lambda}$ is the mean wavenumber, $\bar{\nu}$ the mean frequency of light, and $J(\vec{r}_1, \vec{r}_2)$ the mutual intensity of light.

Hence, the equation (3.72) can be recast as,

$$J(\vec{r}_1', \vec{r}_2') = \int_{\mathtt{A}}\int_{\mathtt{A}} \frac{\chi_1}{s_1}\frac{\chi_2^*}{s_2} J(\vec{r}_1, \vec{r}_2) e^{i\bar{\kappa}(s_2 - s_1)} d\mathtt{S}_1 d\mathtt{S}_2. \tag{3.74}$$

Equation (3.74) relates the general law for propagation of mutual intensity. When the two points $\mathtt{Q}_1(\vec{r}_1')$ and $\mathtt{Q}_2(\vec{r}_2')$ coincide at $\mathtt{Q}(\vec{r}')$ and $\tau = 0$, the intensity distribution on the surface $\mathtt{B}$ is deduced as,

$$I(\vec{r}') = \int_{\mathtt{A}}\int_{\mathtt{A}} \frac{\sqrt{I(\vec{r}_1)I(\vec{r}_2)}}{s_1 s_2} \gamma\left(\vec{r}_1, \vec{r}_2, \frac{s_2 - s_1}{c}\right) \chi_1\chi_2^* d\mathtt{S}_1 d\mathtt{S}_2, \tag{3.75}$$

in which γ is the correlation function and

$$\Gamma(\vec{r}_1, \vec{r}_2, \tau) = \sqrt{I(\vec{r}_1)I(\vec{r}_2)}\gamma(\vec{r}_1, \vec{r}_2, \tau). \tag{3.76}$$

Equation (3.75) expresses the intensity at an arbitrary point $\mathtt{Q}(\vec{r}')$ as the sum of surface contributions from all pairs of elements of the arbitrary surface $\mathtt{A}$.

3.4.2 *Wave equations for the mutual coherence*

In a scalar wave equation governing the propagation of fields and the mutual coherence function obeys a pair of wave equations (Wolf 1955). Let $U^{(r)}(\vec{r}, t)$ represents the real wave disturbance in free space at the point $\vec{r}$, at time t. It obeys the partial differential equation,

$$\left(\nabla^2 - \frac{1}{c^2}\frac{\partial^2}{\partial t^2}\right) U^{(r)}(\vec{r}, t) = 0, \tag{3.77}$$

in which c is the velocity of light in vacuum, and ∇^2 the Lapacian operator with respect to the Cartesian rectangular coordinates.

It is possible to show that the complex analytic signal $U(\vec{r}, t)$ associated with $U^{(r)}(\vec{r}, t)$ also obeys the equation (3.77), i.e.,

$$\left(\nabla^2 - \frac{1}{c^2}\frac{\partial^2}{\partial t^2}\right) U(\vec{r}, t) = 0, \tag{3.78}$$

In vacuum, let $U(\vec{r}_1, t)$ and $U(\vec{r}_2, t)$ represent the disturbances at points $\vec{r}_1$ and $\vec{r}_2$ respectively. The mutual coherence function is given by,

$$\Gamma(\vec{r}_1, \vec{r}_2, \tau) = \langle U(\vec{r}_1, t + \tau) U^*(\vec{r}_2, t)\rangle\,.$$

For a stationary field Γ depends on t_1 and t_2 only through the difference $t_1 - t_2 = \tau$. Therefore, one may write $\Gamma(\vec{r}_1, \vec{r}_2; t_1, t_2) = \Gamma(\vec{r}_1, \vec{r}_2, \tau)$. On applying the Lapacian operator, ∇_1^2 with respect to the Cartesian rectangular coordinates of $\vec{r}_1(= x_1, y_1, z_1)$ to the definition of $\Gamma(\vec{r}_1, \vec{r}_2, \tau)$, one derives,

$$\begin{aligned}\nabla_1^2\Gamma(\vec{r}_1, \vec{r}_2, \tau) &= \left\langle \nabla_1^2 U(\vec{r}_1, t+\tau) U^*(\vec{r}_2, t)\right\rangle \\ &= \left\langle \frac{1}{c^2}\frac{\partial^2 U(\vec{r}_1, t+\tau)}{\partial (t+\tau)^2} U_2^*(\vec{r}_2, t)\right\rangle \\ &= \frac{1}{c^2}\frac{\partial^2}{\partial \tau^2}\langle U(\vec{r}_1, t+\tau) U^*(\vec{r}_2, t)\rangle \\ &= \frac{1}{c^2}\frac{\partial^2}{\partial \tau^2}\Gamma(\vec{r}_1, \vec{r}_2, \tau). \end{aligned} \tag{3.79}$$

Similarly, the Lapacian operator, ∇_2^2 with respect to the Cartesian rectangular coordinates of $\vec{r}_2(= x_2, y_2, z_2)$ to the definition of $\Gamma(\vec{r}_1, \vec{r}_2, \tau)$, is applied, therefore, yields a second equation,

$$\begin{aligned}\nabla_2^2\Gamma(\vec{r}_1, \vec{r}_2, \tau) &= \left\langle \nabla_2^2 U(\vec{r}_2, t+\tau) U^*(\vec{r}_1, t)\right\rangle \\ &= \frac{1}{c^2}\frac{\partial^2}{\partial \tau^2}\Gamma(\vec{r}_1, \vec{r}_2, \tau), \end{aligned} \tag{3.80}$$

which $\Gamma(\vec{r}_1, \vec{r}_2, \tau)$ satisfies.

Thus, one finds that in free space the second order correlation function, $\Gamma(\vec{r}_1, \vec{r}_2, \tau)$, of an optical field propagates in accordance with a pair of wave equations (3.79 and 3.80). τ represents a time difference between the instants at which the correlation at two points is considered. When τ is small compared to the coherence time, then $\Gamma(\vec{r}_1, \vec{r}_2, \tau) \sim J(\vec{r}_1, \vec{r}_2) e^{-i2\pi\bar{\nu}\tau}$. The mutual intensity $J(\vec{r}_1, \vec{r}_2)$ in vacuum (within the range of validity of the quasi-monochromatic theory) obeys the Helmholtz equations,

$$\begin{aligned}\nabla_1^2 J(\vec{r}_1, \vec{r}_2) + \bar{\kappa}^2 J(\vec{r}_1, \vec{r}_2) &= 0, \\ \nabla_2^2 J(\vec{r}_1, \vec{r}_2) + \bar{\kappa}^2 J(\vec{r}_1, \vec{r}_2) &= 0. \end{aligned} \tag{3.81}$$

In order to derive the propagation of cross-spectral density, the equation for light disturbance, $U(\vec{r}, t)$ is written in terms of generalized Fourier integral,

$$U^{(r)}(\vec{r}, t) = \int_{-\infty}^{\infty} \widehat{U}(\vec{r}, \nu) e^{-i2\pi\nu t} d\nu. \tag{3.82}$$

On taking the Fourier transform of the equation (3.77), one gets,

$$\nabla^2 \widehat{U}(\vec{r}, \nu) + \kappa^2 \widehat{U}(\vec{r}, \nu) = 0, \tag{3.83}$$

in which $\kappa = 2\pi\nu/c$ and $\widehat{}$ stands for the Fourier transform.

Equation (3.83) obeys the Helmholtz equation. By applying the operator, $\nabla^2 - (1/c^2)\partial^2/\partial t^2$ to both of the equations (3.78) and (3.83) one finds,

$$\nabla^2 U(\vec{r}, t) - \frac{1}{c^2}\frac{\partial^2 U(\vec{r}, t)}{\partial t^2} = \int_0^\infty \left[\nabla^2 \widehat{U}(\vec{r}, \nu) + \kappa^2 \widehat{U}(\vec{r}, \nu)\right] e^{-i2\pi\nu t} d\nu. \tag{3.84}$$

The expression for the mutual coherence function can also be written as an inverse Fourier transform of the cross-spectral density function,

$$\Gamma(\vec{r}_1, \vec{r}_2, \tau) = \int_0^\infty \widehat{\Gamma}(\vec{r}_1, \vec{r}_2, \nu) e^{-i2\pi\nu\tau} d\nu, \tag{3.85}$$

where $\widehat{\Gamma}(\vec{r}_1, \vec{r}_2)$ is noted to be zero for negative frequencies.

Since the propagation equations (3.79 and 3.80) obeyed by the mutual intensity, and on applying them to the equation (3.85), the laws for cross-spectral density are deduced,

$$\begin{aligned} \int_0^\infty \left[\nabla_1^2 - \frac{1}{c^2}\frac{\partial^2}{\partial \tau^2}\right] \widehat{\Gamma}(\vec{r}_1, \vec{r}_2, \nu) e^{-i2\pi\nu\tau} d\nu = 0, \\ \int_0^\infty \left[\nabla_2^2 - \frac{1}{c^2}\frac{\partial^2}{\partial \tau^2}\right] \widehat{\Gamma}(\vec{r}_1, \vec{r}_2, \nu) e^{-i2\pi\nu\tau} d\nu = 0. \end{aligned} \tag{3.86}$$

On applying the τ derivatives to the exponentials, a pair of Helmholtz equations that must be satisfied by the cross-spectral density are obtained,

$$\begin{aligned} \nabla_1^2 \widehat{\Gamma}(\vec{r}_1, \vec{r}_2, \nu) + \kappa^2 \widehat{\Gamma}(\vec{r}_1, \vec{r}_2, \nu) = 0, \\ \nabla_2^2 \widehat{\Gamma}(\vec{r}_1, \vec{r}_2, \nu) + \kappa^2 \widehat{\Gamma}(\vec{r}_1, \vec{r}_2, \nu) = 0. \end{aligned} \tag{3.87}$$

The sub-equations (3.87) state that the cross-spectral density obey the same propagation laws as do mutual intensities.

3.5 Degree of coherence from an extended incoherent source: partial coherence

The theory of partial coherence was formulated by van Cittert (1934) and in a more general form by Zernike (1938), which subsequently known as the van Cittert-Zernike theorem. This theorem is the basis for all high angular resolution interferometric experiments, which deals with the relation between the mutual coherence and the spatial properties of an extended incoherent source. It states that, the modulus of the complex degree of coherence, in a plane illuminated by an incoherent quasimonochromatic source is equal to the modulus of the normalized spatial Fourier transform of its brightness distribution (Born and Wolf, 1984, Mandel and Wolf, 1995). The observed image is the Fourier transform (FT) of the mutual coherence function or the correlation function.

3.5.1 *The van Cittert-Zernike theorem*

The geometrical factors required for derivation of the van Cittert-Zernike theorem are depicted in Figure (3.10). The field points, $\mathtt{P}_1(\vec{r}_1')$ and $\mathtt{P}_2(\vec{r}_2')$ on a screen $\mathtt{A}$ are illuminated by an extended quasi-monochromatic source, σ, whose dimensions are small compared to the distance to the screen. Here $\vec{r}_{j=1,2} = x_j, y_j$ is the 2-D position vector. If the source is divided into small elements $d\sigma_1, d\sigma_2, \cdots,$, centered on points $S_1, S_2, \cdots$, which are mutually incoherent, and of linear dimensions small compared to the mean wavelength $\bar{\lambda}$, the complex disturbance due to element $d\sigma_m$ at $\mathtt{P}_{j=1,2}$ in the screen is,

$$U_{mj}(t) = A_m\left(t - \frac{s_{mj}}{c}\right)\frac{e^{-i2\pi\bar{\nu}(t - s_{mj}/c)}}{s_{mj}}, \tag{3.88}$$

where the strength and phase of the radiation coming from element $d\sigma_m$ are characterized by the modulus of A_m and its argument, respectively, and s_{mj} is the distance from the element $d\sigma_m$ to the point $\mathtt{P}_j$.

The extended astronomical source is spatially incoherent because of an internal physical process. Any two elements of the source are assumed to be uncorrelated. The distance $s_{m2} - s_{m1}$ is small compared to the coherence length of the light. Hence, the mutual coherence function (see equation 3.53) of $\mathtt{P}_1$ and $\mathtt{P}_2$ becomes,

$$\Gamma(\vec{r}_1, \vec{r}_2, 0) = \sum_m \langle A_m(t)A_m^*(t)\rangle \frac{e^{i2\pi\bar{\nu}(s_{m1} - s_{m2})/c}}{s_{m1}s_{m2}}. \tag{3.89}$$

Considering a source with a total number of elements so large that it can be regarded as continuous, the sum in equation (3.89) is replaced by the integral,

$$\Gamma(\vec{r}_1, \vec{r}_2, 0) = \int_\sigma I(\vec{r})\frac{e^{i\bar{\kappa}(s_1 - s_2)}}{s_1 s_2} d\mathtt{S}. \tag{3.90}$$

in which s_1 and s_2 are the distances of $\mathtt{P}_1(\vec{r}_1')$ and $\mathtt{P}_2(\vec{r}_2')$ from a typical

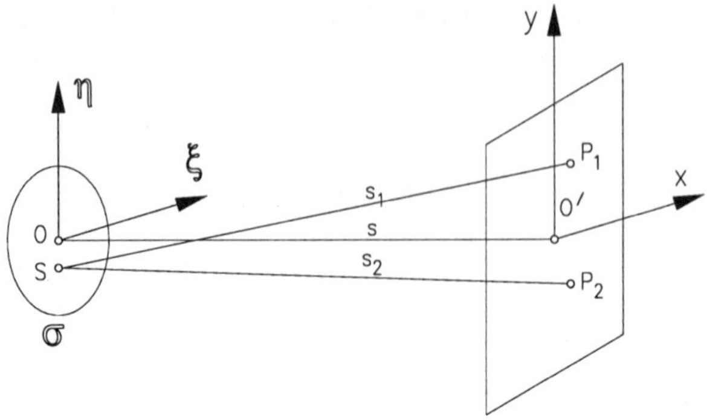

Fig. 3.10 Calculation of degree of coherence of an extended object.

point $S(\vec{r})$ on the source respectively, $\bar{\kappa} = 2\pi/\bar{\lambda}$ the wave number, and $I(\vec{r})$ the intensity per unit area of the source.

The complex degree of coherence $\mu(\vec{r}_1, \vec{r}_2)$ is, according to equations (3.66) and (3.90), we find,

$$\mu(\vec{r}_1', \vec{r}_2') = \frac{1}{\sqrt{I(\vec{r}_1)I(\vec{r}_2)}} \int_\sigma I(\vec{r}) \frac{e^{i\bar{\kappa}(s_1 - s_2)}}{s_1 s_2} dS, \tag{3.91}$$

where

$$I(\vec{r}_j) = \Gamma(\vec{r}_j', \vec{r}_j', 0) = \int_\sigma \frac{I(\vec{r})}{s_j^2} dS, \tag{3.92}$$

with $I(\vec{r}_j)$ as the corresponding intensities at $P_j(\vec{r}_j')$ and $j = 1, 2$.

This equation (3.91) is known as van Cittert-Zernike theorem. It expresses the complex degree of coherence at two fixed points, $P_1(\vec{r}_1')$ and $P_2(\vec{r}_2')$ in the field illuminated by an extended quasi-monochromatic source in terms of the intensity distribution $I(\vec{r})$ across the source and the intensity $I(\vec{r}_1)$ and $I(\vec{r}_2)$ at the corresponding points, $P_1(\vec{r}_1')$ and $P_2(\vec{r}_2')$.

Let the planar geometry (Figure 3.10) be adopted where the source and observation regions are assumed to be in a parallel plane, separated by distance s. If the linear dimensions of the source and the distance $P_1(\vec{r}_1')$ and $P_2(\vec{r}_2')$ are small compared to the distance of these points from the source, the degree of coherence, $|\mu(\vec{r}_1', \vec{r}_2')|$, is equal to the absolute value of the normalized Fourier transform of the intensity function of the source. Let (ξ, η) be the coordinates of the source plane, $\mathtt{S}(\vec{r})$, referred to axes at $\mathtt{O}$,

$$s_1 = \sqrt{s^2 + (x_1 - \xi)^2 + (y_1 - \eta)^2} \simeq s + \frac{(x_1 - \xi)^2 + (y_1 - \eta)^2}{2s}, \quad (3.93)$$

$$s_2 = \sqrt{s^2 + (x_2 - \xi)^2 + (y_2 - \eta)^2} \simeq s + \frac{(x_2 - \xi)^2 + (y_2 - \eta)^2}{2s}, \quad (3.94)$$

where (x_j, y_j) are the coordinates in the observation plane, and the term in $x_j/s, y_j/s, \xi/s$, and η/s are retained.

On setting,

$$p = \frac{(x_1 - x_2)}{s}, \qquad q = \frac{(y_1 - y_2)}{s}, \quad (3.95)$$

$$\psi(\vec{r}_1, \vec{r}_2) = \bar{\kappa}\left[\frac{(x_1^2 + y_1^2) - (x_2^2 + y_2^2)}{2s}\right]. \quad (3.96)$$

The quantity $\psi(\vec{r}_1, \vec{r}_2)$ represents the phase difference $2\pi(\mathrm{OP}_1 - \mathrm{OP}_2)/\bar{\lambda}$ and may be neglected if $(\mathrm{OP}_1 - \mathrm{OP}_2) \ll \bar{\lambda}$. By normalizing the equation (3.91), the van Cittert-Zernike theorem yields,

$$\mu(\vec{r}_1', \vec{r}_2') = e^{i\psi(\vec{r}_1, \vec{r}_2)} \frac{\displaystyle\iint_{-\infty}^{\infty} I(\xi, \eta) e^{-i\bar{\kappa}(p\xi + q\eta)} d\xi d\eta}{\displaystyle\iint_{-\infty}^{\infty} I(\xi, \eta) d\xi d\eta}. \quad (3.97)$$

Equation (3.97) states that for an incoherent, quasi-monochromatic, circular source, the complex coherence factor far from the source is equal to the normalized Fourier transform of its brightness distribution. This form of the van Cittert-Zernike theorem is widely used in stellar interferometry, since the stellar sources are supposed to be at a distance very large compared to the separation of the telescopes and the size of the source itself, and are also supposed to be two-dimensional objects. However, this result calls for important remarks:

- $\mu(\vec{r}_1, \vec{r}_2)$ and I are second order quantities which are proportional to the irradiances and
- the van Cittert-Zernike theorem holds good wherever the quadratic wave approximation is valid (Mariotti, 1988).

In the case of a heterogeneous medium[6], when the angle where SP makes with the normal to $d\sigma$ is sufficiently small, equation (3.90) is written as,

$$J(\vec{r}_1', \vec{r}_2') = \bar{\lambda}^2 \int_\sigma I(\vec{r}) K(\vec{r}, \vec{r}_1', \bar{\nu}) K^*(\vec{r}, \vec{r}_1', \bar{\nu}) d\text{S}. \tag{3.98}$$

in which $K(\vec{r}, \vec{r}', \nu)$ is the transmission function of the medium, representing the complex disturbance at $\text{P}(\vec{r}')$ due to the monochromatic point source of frequency ν, of unit strength and of zero phase, situated at the element $d\sigma$ at $\text{S}(\vec{r})$.

Equation (3.97) is known as Hopkins' theorem. Since $\mu(\vec{r}_1', \vec{r}_2') = \text{J}(\vec{r}_1', \vec{r}_2')/\sqrt{I_1 I_2}$, one may write,

$$\mu(\vec{r}_1', \vec{r}_2') = \frac{\bar{\lambda}^2}{\sqrt{I(\vec{r}_1) I(\vec{r}_2)}} \int_\sigma I(\vec{r}) K(\vec{r}, \vec{r}_1', \bar{\nu}) K^*(\vec{r}, \vec{r}_1', \bar{\nu}) d\text{S}. \tag{3.99}$$

where $I(\vec{r}_1) = J(\vec{r}_1', \vec{r}_1')$ and $I(\vec{r}_2) = J(\vec{r}_2', \vec{r}_2')$ are the intensities at $\text{P}_1(\vec{r}_1')$ and $\text{P}_2(\vec{r}_2')$ respectively.

Defining,

$$U(\vec{r}, \vec{r}_1') = i\bar{\lambda} K(\vec{r}, \vec{r}_1', \bar{\nu}) \sqrt{I(\vec{r})}, \qquad U(\vec{r}, \vec{r}_2') = i\bar{\lambda} K(\vec{r}, \vec{r}_2', \bar{\nu}) \sqrt{I(\vec{r})}, \tag{3.100}$$

equations (3.98) and (3.99) can be recast as,

$$J(\vec{r}_1', \vec{r}_2') = \int_\sigma U(\vec{r}, \vec{r}_1') U^*(\vec{r}, \vec{r}_2') d\text{S}, \tag{3.101}$$

$$\mu(\vec{r}_1', \vec{r}_2') = \frac{1}{\sqrt{I(\vec{r}_1) I(\vec{r}_2)}} \int_\sigma U(\vec{r}, \vec{r}_1') U^*(\vec{r}, \vec{r}_2') d\text{S}. \tag{3.102}$$

The term $U(\vec{r}, \vec{r}')$ is proportional to the disturbance which would arise at P from a monochromatic source of frequency $\bar{\nu}$ strength $\sqrt{I(\vec{r})}$ and zero phase, situated at S. Equations (3.101) and (3.102) express the mutual intensity, $J(\vec{r}_1', \vec{r}_2')$, and complex degree of coherence, $\mu(\vec{r}_1', \vec{r}_2')$, due to an extended quasi-monochromatic source in terms of the disturbances produced at $\text{P}_1(\vec{r}_1')$ and $\text{P}_2(\vec{r}_2')$ by each source point of an associated monochromatic source.

3.5.2 *Coherence area*

In most of the practical cases, the phase factor $\psi(\vec{r}_1, \vec{r}_2)$ that appears in the equation (3.97) may be neglected, in particular, when $\text{P}_1(\vec{r}_1')$ and $\text{P}_2(\vec{r}_2')$

[6]Heterogeneous medium has different composition at different points.

lie at the same distance from the origin, i.e., $\rho_1 = \rho_2$, with $\rho = x^2 + y^2$, or if $\bar{\lambda} \gg 2(\rho_2^2 - \rho_1^2)/s$, holds. On introducing the coordinates $(\alpha, \beta) = (\xi/s, \eta/s)$ and the spatial frequencies, $(u, v) = (p/\bar{\lambda}, q/\bar{\lambda})$, the equation (3.97) becomes,

$$\mu(u, v) = \left(\frac{e^{i\psi(\vec{r}_1, \vec{r}_2)}}{\displaystyle\int_\sigma I d\text{S}} \right) \cdot \widehat{I}(\alpha, \beta). \tag{3.103}$$

If the normal $\text{O}\,\text{O}'$ to the plane of the source passes through $\text{P}_2(\vec{r}_2')$, a classical illustration of the van Cittert-Zernike theorem for an incoherent circular source of diameter, θ, is shown as,

$$I(\alpha, \beta) = I_0 \Pi \left(\frac{\sqrt{\alpha^2 + \beta^2}}{\theta/2} \right). \tag{3.104}$$

The modulus of the corresponding coherence factor is given by,

$$|\mu(\vec{u})| = \left| \frac{2J_1(Z)}{Z} \right|, \tag{3.105}$$

where $\vec{u}(= u, v)$ is the 2-D spatial vector, and $J_1(Z)$ the first order Bessel function[7] of the variable Z,

$$Z = \pi \theta \vec{u}. \tag{3.106}$$

Thus with the increase of separation between the points $\text{P}_1(\vec{r}_1')$ and $\text{P}_2(\vec{r}_2')$, the degree of coherence decreases, and complete incoherence, i.e., $\mu(\vec{u}) = 0$, results when the spacing between these points $\text{P}_1(\vec{r}_1')$ and $\text{P}_2(\vec{r}_2')$ equals to $1.22\bar{\lambda}/\theta$.

[7] Bessel function is used as solution to differential equation dealing with problems in which the boundary conditions bear circular symmetry. The higher order Bessel functions $J_n(x)$ are:

$$x\frac{dJ_n}{dx} + nJ_n = xJ_{n-1},$$

and its recurrence relation:

$$x^{n+1}J_n(x) = \frac{d}{dx}\quad x^{n+1}J_{n=1}(x) \quad .$$

The property of the first order Bessel function, i.e.,

$$\lim_{x=0} \frac{J_1(x)}{x} = \frac{1}{2}.$$

The coherence area, $\mathtt{A}_c$, of the radiation is defined as,

$$\mathtt{A}_c = \iint\limits_{-\infty}^{\infty} |\mu(p,q)|^2 \, dpdq. \tag{3.107}$$

From the van Cittert-Zernike theorem, the similarity and the power theorem, one derives,

$$\mathtt{A}_c = (\bar{\lambda}s)^2 \frac{\iint\limits_{-\infty}^{\infty} I^2(\xi,\eta)d\xi d\eta}{\left[\iint\limits_{-\infty}^{\infty} I(\xi,\eta)d\xi d\eta\right]^2}. \tag{3.108}$$

If the brightness distribution inside a contour is uniform,

$$I(\xi,\eta) = \begin{cases} I_0 & \text{inside the contour,} \\ 0 & \text{otherwise,} \end{cases} \tag{3.109}$$

one finds

$$\mathtt{A}_c = \frac{(\bar{\lambda}s)^2}{\mathtt{A}_s} = \frac{\bar{\lambda}^2}{\Omega}, \tag{3.110}$$

in which $\mathtt{A}_s$ is the area of the source and Ω the solid angle subtended by the contour from the observing plane.

By introducing a diaphragm of area $\mathtt{A}_c$ in the incident wavefront, the throughput becomes,

$$\mathtt{A}_c\Omega = \lambda^2. \tag{3.111}$$

Equation (3.111) states that if the throughput is known, the degree of spatial coherence of a beam can be checked. The spatio-temporal coherence of a beam is represented by the coherence volume $\mathtt{A}_c l_c$. The number of photons with the same polarization states inside the coherence volume of an optical field is known as degeneracy.

3.6 Diffraction

Diffraction is the apparent bending of light waves around small obstacles in its path. A close inspection of a shadow under a bright source reveals that it is made up of finely spaced bright and dark regions. The obstacle alters the amplitude or phase of the light waves such that the regions of the wavefront that propagate beyond the obstacle interfere with each other. The effects of diffraction can be envisaged on a CD or a DVD, which act as a diffraction grating[8] to form rainbow pattern.

If the medium in which the waves propagate is of finite extension and is bounded, the boundaries of that medium affect the propagating waves. The waves bounce against these boundaries and partly reflect back into the medium and partly transmit outside the medium. The reflected waves superimpose on incident waves, which may, in turn, cancel each other. If the volume of the medium is finite, the propagation vector, $\vec{\kappa}$, becomes one of a discrete set of values.

Another boundary effect is the diffraction of waves from apertures and opaque screens. Let a plane wave of frequency ω and wave vector $\vec{\kappa}$ be incident on an infinite opaque screen, containing a finite aperture; the wave decays at infinity on the other side of the opaque plane. These data form a well defined boundary value problem for the wave equation (2.9). The boundary values define the wave at any point and at any time on the rear side of the opaque screen. The diffracted wave on the other side of the screen is constructed by considering each point of the aperture of the opaque screen as a point source of waves, as well as of same ω, and then, superimposing these waves on the other side of the screen. Point sources for waves obeying the wave equation (2.9) generate spherical wavefronts; the form of these spherical waves in three-space dimensions is represented by the equation (2.24). The total sum over all the emerging spherical waves at a point on the observation plane provides the diffracted wave at that point.

---

[8]A diffraction grating is a reflecting or transparent element, which is commonly used to isolate spectral regions in multichannel instruments. A typical grating has fine parallel and equally spaced grooves or rulings, typically of the order of several hundreds per millimeter and has higher dispersion or ability to spread the spectrum than a prism. If a light beam strikes such a grating, diffraction and mutual interference effects occur, and light is reflected or transmitted in discrete directions, known as diffraction orders. Because of the dispersive properties of gratings, they are employed in monochromators and spectrometers. A grating creates monochromatic light from a white light source, which can be achieved by utilizing the grating's ability of spreading light of different wavelengths into different angles. The relation between the incidence and diffraction angles is given by,

$$\sin\theta_m(\lambda) + \sin\theta_i = -m\lambda/d,$$

in which θ_i is the incident angle, $\theta_m(\lambda)$ the diffracted angle, m the order number of the diffracted beam, which may be positive or negative, resulting in diffracted orders on both sides of the zero order beam, d the groove spacing of the grating, $\lambda(=\lambda_0/n)$ the wavelength, λ_0 the wavelength in vacuum, and n the refractive index.

3.6.1 *Derivation of the diffracted field*

Huygens-Fresnel theory of diffraction has played a major role in elaborating the wave theory of light since the description of its effects in 17th century. The operative theory of diffraction by Fresnel, based on Huygens' principle and Young's principle of interference, followed by the experimental confirmation of some unexpected predictions, established the importance of this new theory. A firm mathematical foundation of the theory was obtained later by Kirchhoff (Born and Wolf, 1984). The rigorous treatment can be used on Maxwell's equations and the properties of the electromagnetic waves with suitable conditions. If one neglects the coupling between electric and magnetic fields and treats light as a scalar field, decisive simplifications arise. The validity conditions for this approximation are fulfilled in the practical cases.

Huygens'-Fresnel principle states that the complex amplitude at P can be calculated by considering each point within the aperture to be a source of spherical waves. As a wave propagates, its disturbance is given by the superposition and interference of secondary spherical wavelets weighted by the amplitudes at the points where they originate on the wave. Kirchhoff had shown that the amplitude and phase ascribed to the secondary sources by Fresnel were indeed logical consequences of the wave nature of light.

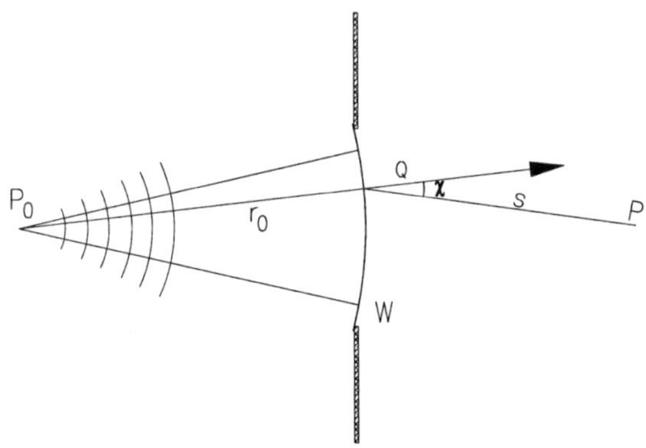

Fig. 3.11 Fresnel zone construction.

Let a monochromatic wave emitted from a point source P_0 fall on an aperture W, and S be the instantaneous position of a spherical monochromatic wavefront of radius r_0 (see Figure 3.2). The disturbance at a point $\mathtt{Q}(\vec{r}')$ on the wavefront is represented by,

$$U(\vec{r}', t) = \frac{Ae^{i\kappa r_0}}{r_0}, \tag{3.112}$$

in which $\kappa = 2\pi/\lambda$, A is the amplitude at unit distance from the source.

The time periodic factor $e^{-i\omega t}$ is omitted in equation (3.112). The expression for the elementary contribution, $dU(\vec{r})$, due to the element $d\mathtt{S}$ at $\mathtt{Q}(\vec{r}')$ is given by,

$$dU(\vec{r}, t) = K(\chi)\frac{Ae^{i\kappa r_0}}{r_0}\frac{e^{i\kappa s}}{s}d\mathtt{S}, \tag{3.113}$$

where s is the distance between the points $\mathtt{Q}(\vec{r}')$ and $\mathtt{P}(\vec{r})$, $K(\chi)$ the obliquity factor which accounts for the properties of the secondary wavelet, χ the angle of diffraction between the normal at $\mathtt{Q}(\vec{r}')$, and the direction $\mathtt{P}(\vec{r})$.

Fresnel assumed the value of the obliquity factor as unity in the original direction of propagation, i.e., for $\chi = 0$ and that it decreases with increasing χ, i.e., when $\chi = \pi/2$. An unobstructed part of the primary wave contributes to the effect at $\mathtt{P}(\vec{r})$, hence, the total disturbance at $\mathtt{P}(\vec{r})$ is deduced as,

$$U(\vec{r}, t) = \frac{Ae^{i\kappa r_0}}{r_0}\int_{\mathtt{W}}\frac{e^{i\kappa s}}{s}K(\chi)d\mathtt{S}, \tag{3.114}$$

In order to derive an expression for $K(\chi)$, Fresnel has evaluated the integral by considering in the diffraction aperture successive zones of constant phase, i.e., for which the distance s is constant within $\lambda/2$. The field at $\mathtt{P}(\vec{r})$ yields from the interference of the contributions of these zones (Born and Wolf, 1984). The obliquity factor is given by,

$$K(\chi) = -\frac{i}{2\lambda}(1 + \cos\chi). \tag{3.115}$$

with $i\lambda K = 1$, one may write,

$$K = \frac{1}{i\lambda} = \frac{e^{-i\pi/2}}{\lambda}. \tag{3.116}$$

The factor $e^{-i\pi/2}$ is accounted by:

(1) the secondary wave oscillate a quarter of a wave out of phase with the primary and

(2) the amplitudes of the secondary wave and that of the primary wave are in the ratio $1 : \lambda$.

Kirchhoff and Summerfield also derived similar results using Helmholtz's equation and Green's theorem[9] (Born and Wolf, 1984). Considering a monochromatic scalar wave,

$$V(\vec{r}, t) = U(\vec{r}, \nu)e^{-i\omega t}, \tag{3.117}$$

where $\vec{r}$ is the position vector of a point (x, y, z), the complex function $U(\vec{r})$ must satisfy the time-independent wave equation,

$$\left(\nabla^2 + \kappa^2\right) U = 0, \tag{3.118}$$

where $\kappa = \omega/c = 2\pi\nu/c = 2\pi/\lambda$.

Equation (3.118) is known as Helmholtz equation. With rigorous mathematics, Kirchoff showed that the Huygens-Fresnel principle can be expressed as (Born and Wolf, 1984),

$$U(\vec{r}) = \frac{1}{4\pi}\int_{\mathtt{S}} \left\{ U\frac{\partial}{\partial n}\left(\frac{e^{i\kappa s}}{s}\right) - \frac{e^{i\kappa s}}{s}\frac{\partial U}{\partial n} \right\} d\mathtt{S}, \tag{3.119}$$

which is known as the integral theorem of Helmholtz and Kirchhoff.

Considering a monochromatic wave from a point source $\mathtt{P}_0$, propagated through an opening in a plane opaque screen, the light disturbance at a point, $\mathtt{P}$, one obtains,

$$U(\vec{r}) = -\frac{iA}{2\lambda}\int_{\mathtt{A}} \frac{e^{i\kappa(r+s)}}{rs} \left[\cos(n, r) - \cos(n, s)\right] d\mathtt{S}. \tag{3.120}$$

Equation (3.120) is the Fresnel-Kirchhoff's diffraction formula. If the radius of curvature of the wave is sufficiently large, $\cos(n, r_0) = 0$ on $\mathtt{W}$. On setting $\chi = \pi - (r_0, s)$, one obtains

$$U(\vec{r}) = \frac{Ae^{i\kappa r_0}}{2i\lambda r_0}\int_{\mathtt{W}} \frac{e^{i\kappa s}}{s}(1 + \cos\chi) d\mathtt{S}, \tag{3.121}$$

where r_0 is the radius of the wavefront $\mathtt{W}$.

[9]Green's theorem is a vector identity which is equivalent to the curl theorem in the plane. It provides the relationship between a line integral around a closed curve and a double integral over the plane region. It can be described by

$$\int_{\mathtt{V}} \phi\nabla^2\psi - \psi\nabla^2\phi \; d\mathtt{V} = \int_{\mathtt{S}} \phi\frac{\partial\psi}{\partial n} - \psi\frac{\partial\phi}{\partial n} \; d\mathtt{S}.$$

3.6.2 *Fresnel approximation*

The Fresnel or near-field approximation yields good results for near field diffraction region which begins at some distance from the aperture, hence, the curvature of the wavefront must be taken into account. Both the shape and the size of the diffraction pattern depend on the distance between aperture and screen.

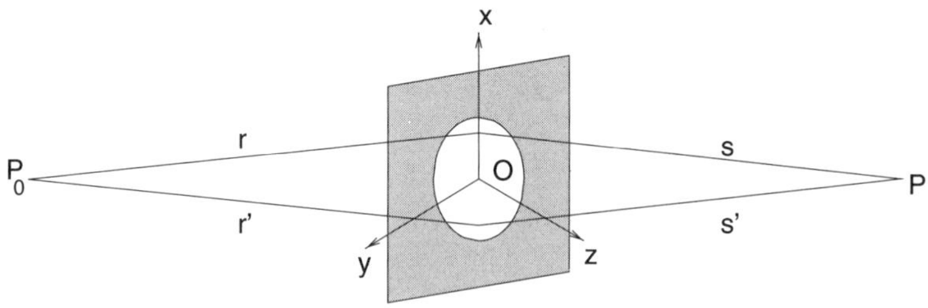

Fig. 3.12 Calculating the optical field at P from the aperture plane.

From the equation (3.120), one finds that as the element $d\mathsf{S}$ explores the domain of integration, $r + s$ changes by many wavelengths. Let O be any point in the aperture and assume that the angles which the lines $\mathsf{P_0O}$ and OP make with $\mathsf{P_0P}$ are not too large. Therefore, the factor, $\cos(n, r) - \cos(n, s)$, is replaced by $2\cos\delta$, in which δ is the angle between the line $\mathsf{P_0P}$ and the normal to the screen (see Figure 3.7). The factor $1/rs$ is also replaced by $1/r's'$, where r' and s' are the distances of $\mathsf{P_0}$ and P from the origin. Thus, the equation (3.120) takes the form,

$$U(\vec{r}) \sim \frac{A}{i\lambda}\frac{\cos\delta}{r's'}\int_{\mathsf{A}} e^{i\kappa(r+s)}d\mathsf{S}, \tag{3.122}$$

where A is the amplitude of the plane wave.

Let $\mathsf{P_0}(x_0, y_0, z_0)$ and $\mathsf{P}(x, y, z)$ respectively be the source and observation points, and $\mathsf{O}(\xi, \eta)$ is a point in the aperture plane. The diffraction angles are restricted to tiny, so that $K(\chi) \approx 1$. The size of the diffraction aperture and the region of observation of the diffracted rays should be small compared to their distance s', so that $1/s \approx 1/s'$. However, the wavelength, λ is small compared to both s and s', it is improbable to approximate $e^{i\kappa s}$ by $e^{i\kappa s'}$, thus,

$$U(x, y) = \frac{1}{i\lambda s'}\iint_{-\infty}^{\infty} P(\xi, \eta)e^{i\kappa s}d\xi d\eta. \tag{3.123}$$

where $P(\xi, \eta)$ is the pupil function, which is defined in absence of aberration and pupil absorption, as

$$P(\xi,\eta) = \begin{cases} 1 & \text{inside the aperture} \\ 0 & \text{otherwise,} \end{cases} \tag{3.124}$$

while, in the case of a transparent pupil with aberrations, $P(\xi, \eta)$ represents the aberration function. Furthermore, one has

$$\begin{aligned} r &= \left[(\xi - x_0)^2 + (\eta - y_0)^2 + z_0^2\right]^{1/2}, \quad r' = \left[x_0^2 + y_0^2 + z_0^2\right]^{1/2}, \\ s &= \left[(x - \xi)^2 + (y - \eta)^2 + z^2\right]^{1/2}, \quad s' = \left[x^2 + y^2 + z^2\right]^{1/2}, \end{aligned} \tag{3.125}$$

hence

$$\begin{aligned} r &= r'\left[1 + \left(\frac{\xi - x_0}{r'}\right)^2 + \left(\frac{\eta - y_0}{r'}\right)^2\right]^{1/2}, \\ s &= s'\left[1 + \left(\frac{x - \xi}{s'}\right)^2 + \left(\frac{y - \eta}{s'}\right)^2\right]^{1/2}. \end{aligned} \tag{3.126}$$

By using the Taylor series expansion,

$$(1+x)^{1/2} = 1 + \frac{1}{2}x - \frac{1}{8}x^2 + \cdots \approx 1 + \frac{1}{2}x,$$

the near-field (Fresnel) is approximated to the form,

$$r \approx r' + \frac{1}{2r'}\left[(\xi - x_0)^2 + (\eta - y_0)^2\right], \tag{3.127}$$

$$s \approx s' + \frac{1}{2s'}\left[(x - \xi)^2 + (y - \eta)^2\right]. \tag{3.128}$$

The first order development of s is valid if $x^2 + y^2 \ll s'^2$ and $\xi^2 + \eta^2 \ll s'^2$, which are not very stringent conditions. This approximation is equivalent to changing the emitted spherical wave into a quadratic wave. The diffracted field is derived and expressed as convolution equation,

$$\begin{aligned} U(x,y) &= \frac{Ae^{i\kappa s'}}{i\lambda s'} \iint_{-\infty}^{\infty} P(\xi,\eta) e^{i(\kappa/2s')\left[(x-\xi)^2 + (y-\eta)^2\right]} d\xi d\eta \\ &= \iint_{-\infty}^{\infty} P(\xi,\eta)\, U_{s'}(x - \xi, y - \eta)\, d\xi d\eta, \end{aligned} \tag{3.129}$$

in which

$$U_{s'}(x,y) = \frac{Ae^{i\kappa s'}}{i\lambda s'} e^{i\,(\kappa/2s')\,\left[x^2+y^2\right]}, \tag{3.130}$$

In $U_{s'}$, two phase factors appear (Mariotti, 1988):

(1) the first one corresponds to the general phase retardation as the wave travels from one plane to the other, and
(2) the second one is a quadratic phase term that depends on the positions, O and P.

Equation (3.129) is clearly a convolution equation, and the final result for the Fresnel diffraction is recast as,

$$U(x,y) = P\,(\xi,\eta) \star U_{s'}, \tag{3.131}$$

in which $\star$ denotes convolution operator.

3.6.3 *Fraunhofer approximation*

Fraunhofer or far-field approximation, takes place if the distances of the source of light, P_0 and observation screen, P are effectively large compared to the dimension of the diffraction aperture so that wavefronts arriving at the aperture and the observation screen may be considered as plane. In this approximation, the factor $[\cos(n,r) - \cos(n,s)]$ in equation (3.120) does not vary appreciably over the aperture. The diffraction pattern changes uniformly in size as the viewing screen is moved relative to the aperture. The far-field approximation is developed by simplifying equation (3.128) as,

$$s \approx s' + \frac{1}{2s'}\left[(x^2+y^2\right] - \frac{1}{s'}\left[x\xi + y\eta\right]. \tag{3.132}$$

The conditions of validity are $x^2+y^2 \ll s'^2$ and $\xi^2+\eta^2 \ll 2s'/\kappa = \lambda s'/\pi$, which are far more restrictive. The distance separating the Fresnel and the Fraunhofer regions is called as the Rayleigh distance $s'_R = D^2/\lambda$, in which D is the size of the diffracting aperture. In the Fraunhofer case,

$$U\,(x,y,s') = \frac{Ae^{i\kappa s'}}{i\lambda s'} e^{i\,(\kappa/2s')\,\left[x^2+y^2\right]} \times \iint_{-\infty}^{\infty} P\,(\xi,\eta)\, e^{-i\,(\kappa/s')\,\left[x\xi+y\eta\right]} d\xi d\eta. \tag{3.133}$$

The phase terms outside the integral of this equation (3.133) have no influence, and is therefore discarded. This integral in equation (3.133) is obtained from the Fourier transform (FT) of the distribution of field, $P(\xi,\eta)$ across the aperture. The result is that the amplitude diffracted at infinity is the Fourier transform of the amplitude distribution over the aperture. Retaining $(\xi^2+\eta^2)$ in expression (equation 3.128), i.e.,

$$(x-\xi)^2+(y-\eta)^2=\left(x^2+y^2\right)-2\left(x\xi+y\eta\right)+\xi^2+\eta^2,$$

the equation (3.129) may be written as,

$$U\left(x,y,s'\right)\propto\iint_{-\infty}^{\infty}P\left(\xi,\eta\right)e^{i\left(\kappa/2s'\right)\left[\xi^2+\eta^2\right]}e^{-i\left(\kappa/s'\right)\left[x\xi+y\eta\right]}d\xi d\eta. \tag{3.134}$$

Equation (3.134) is another approach to Fresnel diffraction. For far-field approximation, i.e., $s'\gg(\xi^2+\eta^2)/\lambda$, the phase factor is much smaller, and is therefore ignored. One may define a new coordinate system, (u,v), called spatial frequencies, by

$$u=\frac{x}{\lambda s'};\qquad v=\frac{y}{\lambda s'}, \tag{3.135}$$

and have the units of inverse distance. In the time domain, this is analogous to the spectra in which frequency is inverse of time. The Fraunhofer diffraction pattern for such a field is proportional to the FT of the pupil transmission function (PTF).

The conjugate coordinates are $(x/s',y/s')$, which are more convenient as it corresponds to the direction cosines of the diffracted rays. Assuming that the incident wavefront is plane and a lens is placed somewhere behind the diffracting screen. At the back focal plane of this lens, each set of rays diffracted in the direction $(x/s',y/s')$ may focus at a particular position. In other words, one obtains an image of the diffracted rays at infinity. Thus, the field distribution at a focus is provided by the Fraunhofer diffraction expression which is referred to as the Fourier transform property of lenses. This situation is common in imaging systems that will be discussed in the following chapter 4, which explains the importance of Fraunhofer diffraction[10].

[10]Quantum mechanics has thrown a light on the problem of diffraction by offering a unified approach to the wave-corpuscle dualism. A diffraction set-up may be interpreted as an experiment for localizing particles moving toward the screen. The deviation of the motion appears as a result of Heisenberg's uncertainty principle, which itself can be derived from the properties of conjugate Fourier transform pairs (see Appendix B).

The phase factors in the diffraction integral are,

$$e^{i\omega t}e^{-i\kappa(r+s)} = e^{i\psi_0}e^{-i2\pi(u\xi + v\eta)}, \tag{3.136}$$

in which $\psi_0 = \omega t - \kappa(r' + s')$ is independent of ξ and η (Klein and Furtak, 1986).

By introducing new coordinates, (u, v) the Fraunhofer diffraction integral (equation 3.133) may be expressed in the form,

$$\widehat{U}(u,v) = \mathcal{C}\iint_{-\infty}^{\infty} P(\xi,\eta)\, e^{-i2\pi\,(u\xi + v\eta)}d\xi d\eta, \tag{3.137}$$

with $\mathcal{C}$ as the constant and is defined in terms of quantities depending on the position of the source and of the point of observation.

Equation (3.137) states that the Fraunhofer diffraction pattern for the wave disturbance is proportional to the Fourier transform of the pupil function.

3.6.3.1 *Diffraction by a rectangular aperture*

Optical devices often use slits as aperture stops. Let O be the origin of a rectangular coordinate system at the center of a rectangular aperture of sides $2a$ and $2b$, and $\mathsf{O}\xi$ and $\mathsf{O}\eta$ are the axes parallel to the sides. Let the pupil function be,

$$P(\xi,\eta) = \begin{cases} 1 & -a\ |\xi| < a; \quad -b\ |\eta| < b, \\ 0 & \text{otherwise}. \end{cases} \tag{3.138}$$

The Fourier transform of the complex disturbance, $U(\xi, \eta)$ is evaluated according to equation (3.137),

$$\begin{aligned} \widehat{U}(u,v) &= \mathcal{C}\int_{-a}^{a}\int_{-b}^{b} e^{-i2\pi\,[u\xi + v\eta]}d\xi d\eta \\ &= \mathcal{C}\int_{-a}^{a} e^{-i2\pi u\xi}d\xi \int_{-b}^{b} e^{-i2\pi v\eta}d\eta \\ &= \mathcal{C}\widehat{U}(u)\widehat{U}(v). \end{aligned} \tag{3.139}$$

Fruitful experiments involving diffraction of electrons and neutrons can be considered as definitively settling the wave-corpuscle controversy.

where

$$\widehat{U}(u) = \int_{-a}^{a} e^{-i2\pi u\xi} d\xi; \qquad \widehat{U}(v) = \int_{-b}^{b} e^{-i2\pi v\eta} d\eta.$$

On integration, one finds,

$$\begin{aligned}\widehat{U}(u) &= -\frac{1}{i2\pi u}\left[e^{-i2\pi ua} - e^{i2\pi ua}\right] \\ &= \frac{\sin 2\pi ua}{2\pi u} = 2a\frac{\sin 2\pi ua}{2\pi ua}.\end{aligned} \tag{3.140}$$

Similarly,

$$\widehat{U}(v) = 2b\frac{\sin 2\pi vb}{2\pi vb}. \tag{3.141}$$

The diffraction pattern related to the field distribution of a rectangular aperture is given by the Fourier transform of a rectangular distribution. This varies as the so called sinc function, $sinc(u) = \sin \pi u/\pi u$.

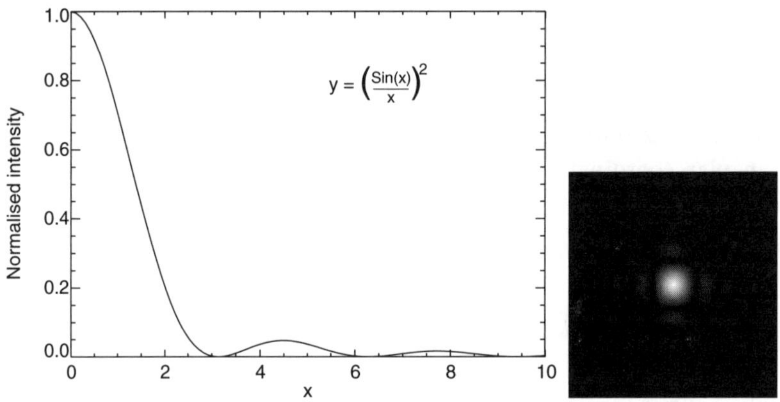

(a) (b)

Fig. 3.13 (a) 1-D Fraunhofer diffraction pattern of a rectangular aperture and (b) 2-D pattern of the same of a square aperture.

Therefore the intensity at the point, $\mathtt{P}(\vec{r})$, is given by,

$$\widehat{I}(u, v) = |\widehat{U}(u, v)|^2 = \widehat{I}(0, 0)\, sinc^2(2\pi ua)\, sinc^2(2\pi ub), \tag{3.142}$$

where $\widehat{I}(0, 0)$ is the intensity at the center.

The intensity distribution in the diffraction pattern is given by equation (3.142). In this case the curve is sharper, and of course, always positive.

The function $y = (\sin x/x)^2$, is displayed in Figure (3.17) with a strong central peak and small secondary peaks. The function is plotted in normalized units. It has a principal maximum $y = 1$ at $x = 0$ and zero minimum at $x = \pm\pi, \pm 2\pi, \pm 3\pi, \cdots$. The diffraction pattern in the case of a rectangular aperture consists of a central bright spot[11] and a number of equispaced bright spots along the $x-$ and $y-$ axes, but with decreasing intensity.

If b is very large so that rectangular aperture degenerates into a slit, thus there is little diffraction in $y-$ direction. The irradiance distribution in the diffraction pattern of a slit aperture is given by,

$$\widehat{I}(u) = \widehat{I}(0) sinc^2(2\pi ua). \tag{3.143}$$

The diffraction pattern of a slit aperture consists of a central bright line parallel to the slit and parallel bright lines of decreasing intensity on both sides.

3.6.3.2 *Diffraction by a circular pupil*

The Fraunhofer diffraction produced by the optical instruments that use circular pupils or apertures, say telescope lenses or mirrors, play an important role in the performances of these instruments. It is important to note that a telescope is an ideal condition for the image. When a continuum of wave components pass through such an aperture, the superposition of these components result in a pattern of constructive and destructive interference. For astronomical instruments, the incoming light is approximately a plane wave. In this far-field limit, Fraunhofer diffraction occurs and the pattern formed at the focal plane of a telescope may have little resemblance to the aperture.

Let ρ, θ be the polar coordinates of a point in a circular aperture of radius a. The pupil function represented by $P(\rho, \theta)$ is,

$$P(\rho, \theta) = \begin{cases} 1, & 0 < \rho < a \\ 0, & \text{otherwise.} \end{cases} \tag{3.144}$$

The polar coordinates are expressed in terms of u and v as,

$$\begin{aligned} \xi &= \rho\cos\theta; \qquad \eta = \rho\sin\theta; \\ u &= w\cos\phi; \qquad v = w\sin\phi, \end{aligned} \tag{3.145}$$

[11]When a light beam strikes an opaque surface, the spot of light takes any shape depending upon the apertures and the wavefront of the beam.

with

$$w = \sqrt{u^2 + v^2} = \frac{1}{\lambda}\left[\left(\frac{x_0}{r'} + \frac{x}{s'}\right)^2 + \left(\frac{y_0}{r'} + \frac{y}{s'}\right)^2\right]^{1/2},$$

as the radial distance of the point P from the origin in units of length^{-1}.

In most cases, one assumes that the source is on the optical axis, so that $x_0 = y_0 = 0$. Assuming circular symmetry for the optical field behind the aperture, $U(\rho, \theta) = CP(\rho)$, and the Fraunhofer diffraction integral, takes the form,

$$\widehat{U}(u, v) = \mathcal{C}\int_{\mathtt{W}} e^{-i2\pi(u\xi + v\eta)} d\xi d\eta. \tag{3.146}$$

Using the relations (equation 3.145) in the equation (3.146), one gets

$$\begin{aligned}\widehat{U}(w, \phi) &= \mathcal{C}\int_0^a \rho d\rho \int_0^{2\pi} e^{-i2\pi\rho w\cos(\theta - \phi)} d\theta \\ &= 2\pi\mathcal{C}\int_0^a \rho J_0(2\pi\rho w)d\rho,\end{aligned} \tag{3.147}$$

where

$$J_0(x) = \frac{1}{2\pi}\int_0^{2\pi} e^{i(x\cos\alpha)} d\alpha,$$

is the Bessel function of the first kind and order zero.

On using the relation

$$\int_0^x x' J_0(x')dx' = xJ_1(x),$$

the equation (3.147) may be written as,

$$\widehat{U}(w) = \mathcal{C}\pi a^2\left[\frac{2J_1(2\pi aw)}{2\pi aw}\right], \tag{3.148}$$

where $J_1(x)$ is Bessel function of the first kind and order one.

In the case of a circular aperture, the diffraction pattern is a central spot surrounded by concentric rings. The intensity distribution in this pattern is given by,

$$\widehat{I}(w) = |\widehat{U}(w)|^2 = \widehat{I}(0)\left[\frac{2J_1(2\pi aw)}{2\pi aw}\right]^2, \tag{3.149}$$

where $\widehat{I}(0)[= \mathcal{C}^2(\pi a^2)^2]$ is the intensity on-axis ($w = 0$).

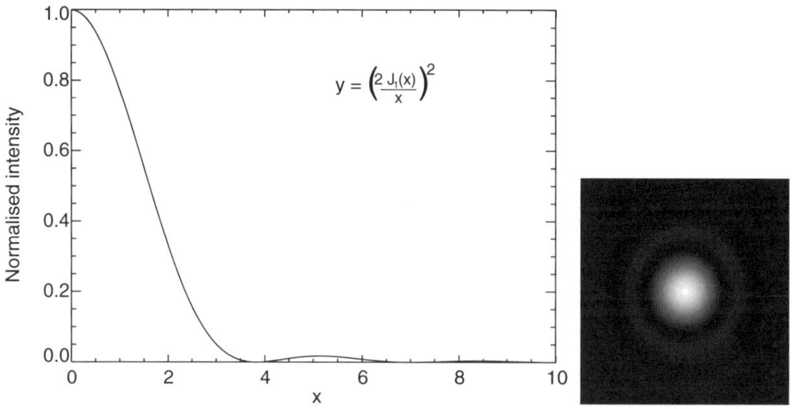

(a) (b)

Fig. 3.14 (a) 1-D intensity distribution of a circular aperture and (b) 2-D pattern of the same of a circular aperture.

Equation (3.149) was first derived by Airy. The intensity distribution consisting of constructive and destructive interference rings is called the Airy diffraction pattern. The central bright spot is known as Airy disc. Unlike Gaussian intensity profiles[12] (see Figure 3.15), which do not have sharp cutoffs at the edge, the Fraunhofer diffraction pattern of a circular aperture has easily defined edge. Such an edge of the circular spot where the Bessel function reaches its first zero is known as the first dark ring. This dark ring is called a region of destructive interference.

Figure 3.14 displays the intensity distribution in the Airy pattern where $x = 2\pi a w$. This distribution has its principal maximum $y = 1$ at $x = 0$, i.e., at $w = 0$. With increasing x (or w) it oscillates with gradually diminishing magnitude. The intensity is zero for values of x given by $J(x) = 0$; the intensity of the bright rings decreases rapidly with their radii. 84% of the total energy is located within the central circle or the Airy disc. The

[12] A light beam where the electric field profile in a plane perpendicular to the beam axis may be described with a Gaussian function. The transverse profile of the intensity of the beam is defined by the equation:

$$I(r) = I(0)e^{-2r^2/w^2},$$

where I is the intensity at a point r from the maximum intensity, $I(0)$, and w the beam radius, which is the distance from the beam axis to a point where the intensity drops to $I(0)/e^2 (\approx 13.5\%)$ of the maximum value.

As the beam propagates, the beam begins to diverge, causing the width of the beam to increase, which gives rise to a wider intensity profile and a smaller value for $I(0)$.

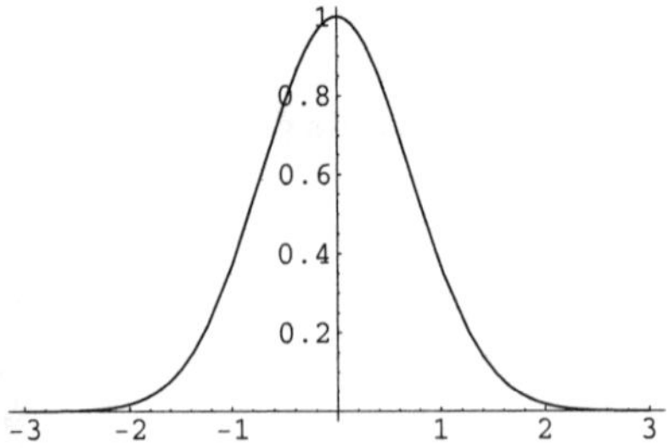

Fig. 3.15 Gaussian intensity profile.

dark destructive interference rings occur at the minima of $J_1(x)$, where $x = 3.83, 7.02, \cdots$. The first minimum null point at $x = 1.22\pi = 3.83$ provides the diameter of the central spot, $\theta \sim 1.22\lambda/2a$. The second null point at $x = 7.02$ is discernible when θ becomes $\sim 2.23\lambda/2a$.

The intensity distribution in diffraction pattern for pupils with or without aberrations is called the point spread function (PSF). For an imaging system, it is proportional to the modulus square of the Fourier transform of its pupil transmission function, thus the resolution at the image plane is determined by the width of this PSF.

Chapter 4

Image formation

4.1 Image of a source

The process of image formation is an important topic which helps in understanding the retrieval of high angular resolution information. An image of an extended object is essentially a convolution of the brightness distribution in the object with the diffraction pattern of a point source (impulse response) produced by the imaging system. Information on the object is retrieved from an analysis of this diffraction pattern, or of the interference properties of the incoming wavefront.

An optical system redirects the rays from an object point to a real image space, the wavefronts are contracting and the rays are converging to a common point, known as image point. A perfect image of a small object lying at a large distance from an imaging system is an exact replica of the object except for its magnification. Such an image is known to be the Gaussian image[1]. Let $\vec{X}_0(= X_0, Y_0)$ be the position vector of a point in the object plane, the position of its Gaussian image is given by,

$$\vec{x}_0 = \mathrm{M}\vec{X}_0, \tag{4.1}$$

where M is the lateral magnification and $\vec{x}_0 = (x_0, y_0)$ the position vector of the image point.

Considering that the object and the image lie in mutually parallel planes that are perpendicular to the optical axis of the system; the entrance and the exit pupils respectively lie in planes that are parallel to the object and the image planes. The propagation of light inside the system is limited

[1]Gaussian image of a sinusoidal object is sinusoidal with the same modulation and phase as the object, while the diffracted image is sinusoidal, but with a reduced contrast and changed phase depending on the spatial frequency.

by one aperture, which is known as pupil. The transfer of light from the object to the image plane, barring the effects of diffraction at the pupil, is treated by the laws of geometrical optics. The distribution of the complex disturbances in the image plane, in the case of monochromatic wave can be expressed as a superposition integral. Let the point $\mathrm{M}_1(\vec{x}_1)$ be the geometrical image of $\mathrm{M}_0(\vec{x}_0)$, in which $\vec{x}_1 = (x_1, y_1)$ is the position vector. In an imaging system consisting of a simple lens based telescope, in which the point spread function (PSF) is invariant under spatial shifts[2]. The distribution of the complex disturbances, $U(\vec{x})$, in the image plane for a monochromatic point source of wavelength λ, is formulated as,

$$dU_1(\vec{x}_1) = U_0(\vec{x}_0)K(\vec{x}_0;\vec{x}_1)d\vec{x}_0, \tag{4.2}$$

where $U_0(\vec{x}_0)$ represents the complex disturbance in the plane of the object, the element at $\vec{x}_0$ makes a contribution of $dU_1(\vec{x}_1)$ to the disturbance at the point $\vec{x}_1$ in the image plane, and the imaging properties of the system is characterized by means of transmission function $K(\vec{x}_0;\vec{x}_1)$, defined as the complex amplitude per unit area of $\vec{x}_0$ plane at $\vec{x}_1$ in the image plane, due to a disturbance of unit amplitude and zero phase at the object point $\vec{x}_0$.

The total disturbance in the image plane due to an extended object in a monochromatic light of wavelength λ at $\vec{x}_1$ is,

$$U_1(\vec{x}_1) = \int_{-\infty}^{\infty} U_0(\vec{x}_0)K(\vec{x}_0;\vec{x}_1)d\vec{x}_0. \tag{4.3}$$

The impulse response of the system is the Fraunhofer diffraction pattern due to the pupil, i.e.,

$$K(\vec{x}_0;\vec{x}_1) \propto \int_{-\infty}^{\infty} P(\vec{x})e^{-i2\pi(\vec{x}_1 - \vec{x}_0)\cdot\vec{x}/\lambda s}d\vec{x}, \tag{4.4}$$

in which $P(\vec{x})$ is the pupil function,

$$P(\vec{x}) = \begin{cases} U(\vec{x}) & \text{inside the pupil,} \\ 0, & \text{otherwise.} \end{cases} \tag{4.5}$$

Equation (4.4) provides the relationship between the transmission function (impulse response) K and the pupil function $P(\vec{x})$. The function, $P(\vec{x})$, is complex function. The phase disturbances introduced by aberrations are

[2]As in an electrical network where the response to a unit-impulse must be independent of the time at which that input is applied, known as time-invariant system, in an optical imaging system the PSF must be the same over the entire field, called space-invariant system.

incorporated in it. $K(\vec{x}_1)$ has a sharp maximum, at or near the Gaussian image point $\vec{x}_1 = \vec{x}_0$ and falls off sharply with increasing distance from this point. In a well corrected system, the size of K is of the order of the first dark ring of the corresponding Airy pattern. The transmission function, $K(\vec{x}_0)$ varies slowly as that point explores the object surface. In a circular pupil, the spot has a sharp edge, hence it is possible to specify spot diameter. The spot size of an image patch can be assessed by taking the area of the smallest spot that has 84% of the energy and by computing the diameter of an equivalent circular bright patch. Figure 4.1 depicts the image spot size at the focus and away from the focus.

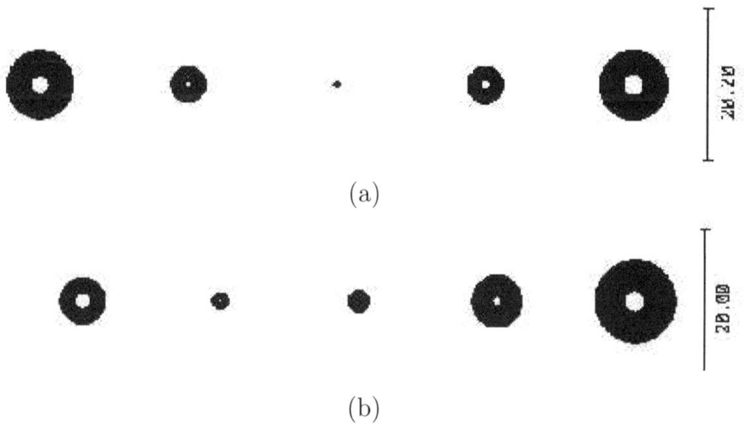

Fig. 4.1 Image spot size at the focus and away from the focus; (a) for an ideal telescope and (b) for an aberrated one; the spot size enlarges on either side of the focus compared to the ideal case. The units are given in micrometers (μm).

The size of the Airy disc of diffraction pattern is inversely proportional to the diameter of the pupil. The deviation of light does not occur if the diameter of the aperture increases. But with the increase of the incident angle of the beam, the entrance pupil may become occulted partially. This is known as vignetting effect which modifies the distribution of light in the diffraction pattern.

It is implicitly assumed in this case that the impulse response or the PSF is independent of the image-plane coordinates. The PSF is shift-invariant and the Fourier transform of which is the coherent transfer function. In order to make the system space invariance, the following conditions are required to be satisfied:

- the object plane coordinate should be normalized in a way that the object magnification between the object and image plane coordinate systems is unity,
- the object coordinate axes must be directed so as to enable one to remove the effects of image inversion from the mathematical calculation, and
- the amplitude spread function must be free of space-variant phase factors.

In optics, the condition of spatial invariance, known as iso-planatism condition, is an important requirement since the field dependent geometrical aberrations change the impulse response with respect to the distance to the optical axis. Aberrations are contributions arise due to imperfections in the spherical wavefronts or rays. In an ideal optical system, all rays of light from a point in the object plane would converge to the same point in the image plane, forming an image. The influences which cause different rays to converge to different points are called aberrations. They can be depicted in the wavefront, which is directly related to determine image quality. It can also be depicted as ray form offering convenient graphical interface for evaluating basic quality level of optical systems. Rather than a true point image as determined by diffraction, the optical system produces a blur[3]. Typically, such a system suffers with following primary aberrations:

(1) Spherical aberration: The on axis ray bundle emanating from a point source shows spherical aberration if the light rays from the center and edges converge at different points. These rays are imaged by a spherical optical element as a bright dot surrounded by a halo of light. The spherical aberration is an image imperfection that is due to the spherical shape of the optical element, since the outer parts of such an element have a different focal length than does the central area. Such aberration is uniform over the field, in the sense that the difference in longitudinal focus between the lens margins and center does not depend on the obliquity of the incident light. A concave mirror, being a portion of a sphere, is a section of a paraboloid where parallel rays incident on all ares of the surface are reflected to a point free of such aberration. Combination of convex and concave optics may eliminate spherical aberration.
(2) Coma: It is the manifestation of differences in lateral magnification for

[3]When the beam is slightly out of focus, the intensity pattern becomes blurred.

rays coming from an object point not on the optic axis. If the centroid is shifted slightly from the chief ray, coma appears. Light from a point source is spread out a family of circles that fit into a cone, and in a plane perpendicular to the optical axis. This may be minimized (and in some cases eliminated) by a suitable choice of curvature of the optical surfaces to match the application. Corresponding or conjugate object and image points, with freedom from both spherical and coma are referred to aplanatic points, and a optics possessing such a points is called aplanatic optics.

(3) Astigmatism: It is an off-axis point wavefront aberration, caused by the inclination of incident wavefronts relative to the optical surface. It occurs when an optical element has different focii for rays that propagate in two perpendicular planes; more de-centered subset of rays may produce astigmatism. Usually centroid shifts with change in focus.

(4) Chromatic aberration: This is caused by the dispersion of the optical material, the variation of its refractive index with the wavelength of light. The focal length of such a system also varies causing longitudinal and axial chromatic variations. Such aberration, envisaged as fringes of color around the image, can be minimized by using an achromat. This may reduce the amount of chromatic aberration over a certain range of wavelengths, albeit unable to provide perfect correction.

In order to ensure space-invariance, the iso-planatism condition imposes that the supports of the functions is restricted to the iso-planatic (space invariant) patch; in the space invariant case, each amplitude spread function depends on only two independent variables. In a well corrected system the transmission function represents Airy pattern centered on the Gaussian image point, apart from the constant factor. Let the object be small, which forms an iso-planatic region of the system, for all points on it $K(\vec{x}_0;\vec{x}_1)$ may be replaced to a good approximation by a function depending on the coordinate differences $(\vec{x}_1-\vec{x}_0)$,

$$K(\vec{x}_0;\vec{x}_1)=K(\vec{x}_1-\vec{x}_0). \tag{4.6}$$

Equation (4.4) for the transfer function in the space-invariant case, is written as,

$$K(\vec{x}_1-\vec{x}_0)=\int_{-\infty}^{\infty}P(\lambda\vec{u})e^{-2\pi(\vec{x}_1-\vec{x}_0)\cdot\vec{u}}d\vec{u}, \tag{4.7}$$

where the dimensionless variable $\vec{u}$ is equal to $\vec{x}/\lambda$ and $\vec{u}(=u,v)$ the 2-D spatial frequency vector with magnitude $|\vec{u}|$.

The complex disturbances exhibit a time-dependency in the case of quasi-monochromatic case; therefore, the field in the image plane is expressed as,

$$U_1(\vec{x}_1, t) = \int_{-\infty}^{\infty} U_0(\vec{x}_0, t) K(\vec{x}_0; \vec{x}_1) d\vec{x}_0. \tag{4.8}$$

The detectors in the visual band are not sensitive to the field, but to the intensity as given by,

$$I(\vec{x}_1) = \langle U_1(\vec{x}_1, t) U_1^*(\vec{x}_1, t) \rangle, \tag{4.9}$$

where <> denotes the time-average.

The general expression for the image intensity distribution is,

$$I(\vec{x}_1) = \iint_{-\infty}^{\infty} K(\vec{x}_1 - \vec{x}_0') K^*(\vec{x}_1 - \vec{x}_0'') \langle U_0(\vec{x}_0', t) U_0^*(\vec{x}_0'', t) \rangle \, d\vec{x}_0' d\vec{x}_0'', \tag{4.10}$$

in which $\vec{x}_0' = x_0', y_0'$, and $\vec{x}_0'' = x_0'', y_0''$ are the position vectors.

The quantity, $< U_{0'}, U_{0''}^* >$, measures the correlation of the complex disturbances at two points of the object.

4.1.1 *Coherent imaging*

In the case of a coherent illuminated object, the complex amplitude distribution of its image is obtained by adding the complex amplitude distributions of the images of its infinitesimal elements. Application of coherent illumination can be found in optical processing, such as image smoothening, enhancing the contrast, correction of blur, extraction of the features etc. So one can act directly on the the transfer function by placing amplitude and phase screen in the pupil of the system. Phase contrast microscopy is one of the applications of coherent optical processing.

Considering the limiting form of equation (4.3), when the source reduces to a vanishingly small object (point source) of unit strength and zero phase at the vector, $\vec{x}_0 = \vec{x}_0'$, i.e.,

$$U_0(\vec{x}_0) = \delta(\vec{x}_0 - \vec{x}_0'), \tag{4.11}$$

where δ represents delta function[4].

[4]The Dirac delta function, referred to as the unit impulse function, is defined by the

Equation (4.3) provides,

$$U_1(\vec{x}_1) = K(\vec{x}_0; \vec{x}_1). \tag{4.12}$$

If the fields at $\vec{x}_0'$ and $\vec{x}_0''$ are fully correlated, that is, have the same instantaneous amplitudes and a constant phase delay, the situation is known to be a coherent case. The irradiance distribution in the corresponding image is obtained by taking modulus square of the disturbance and $\langle U_{0'}, U_{0''}^* \rangle$ equals to $U_0(\vec{x}_0') \cdot U_0^*(\vec{x}_0'')$. Thus the equation (4.10) translates into,

$$I(\vec{x}_1) = \int_{-\infty}^{\infty} K(\vec{x}_1 - \vec{x}_0')U_0(\vec{x}_0')d\vec{x}_0' \int_{-\infty}^{\infty} K^*(\vec{x}_1 - \vec{x}_0'')U_0(\vec{x}_0'')d\vec{x}_0''. \tag{4.13}$$

The complex disturbance $U_1(\vec{x}_1)$ is expressed as,

$$U_1(\vec{x}_1) = \int_{-\infty}^{\infty} K(\vec{x}_1; \vec{x}_0)U_0(\vec{x}_0)d\vec{x}_0 \tag{4.14}$$

According to the equation (4.14), U_1 is a convolution of U_0 and K. For an iso-planatic coherent object, the spatial frequency spectrum of its disturbance is given by the product of the spectrum of its Gaussian disturbance obtain by Fourier inversion method,

$$\widehat{U}_1(\vec{u}) = \widehat{K}(\vec{u})\widehat{U}_0(\vec{u}), \tag{4.15}$$

in which $\widehat{K}(\vec{u}) = \widehat{P}(-\lambda\vec{u})$ (from equation 4.7) and $\widehat{\ }$ stands for a Fourier transformation carried out on the variable $\vec{x}$.

Equation (4.15) implies that the disturbances in the object plane and in the image plane are considered as a superposition of space-harmonic components of the spatial frequencies $\vec{u}$. Each component of the image depends on the corresponding component of the object and the ratio of the components is $\widehat{K}$. Thus the transmission from the object to the image is equivalent to the action of a linear filter.

The frequency response function, $\widehat{K}(\vec{u})$, for coherent illumination is equal to the value of the pupil function P. In an aberration-free system with a circular pupil of geometrical diameter $2a$, all the spatial frequencies

---

property,

$$\delta(x) = \begin{array}{ll} 0 & x \neq 0, \\ \infty & x = 0, \end{array}$$

with the integral of the Dirac delta from any negative limit to any positive limit as 1, i.e.,

$$\int_{-\infty}^{\infty} \delta(x)dx = 1.$$

are transmitted by the system up to a limit, known as the cutoff frequency of the system,

$$f_c^{coh} = \frac{2a}{2\lambda R} = \frac{a}{\lambda R}, \tag{4.16}$$

where R is the radius of the reference sphere.

4.1.2 *Incoherent imaging*

For a monochromatic point source, the distinction between a coherently and an incoherently illuminated object disappears, but it stands out for an extended source. The irradiance distribution of the image of an incoherent object is obtained by adding the irradiance distributions of the images of its elements. When incoherent light propagates, the wave becomes partially coherent. According to van Cittert-Zernike theorem (see section 3.5.1), the degree of coherence as a function of spatial separation is the same as the diffraction pattern due to an aperture of the same size and shape as source. The implications of such a theorem are that light from small source, such as a star, is spatially coherent at the telescope aperture, while light from an extended source, such as Sun, is coherent only over a region of the aperture.

The optical systems indeed use incoherent detection based on received power level rather than on actual electric field amplitude and phase. If the fields at two different points are incoherent, the quantity $< U_0', U_0''^* >$ averages out, and

$$\left\langle U_0', U_0''^* \right\rangle = O(\vec{x}_0')\delta(\vec{x}_0' - \vec{x}_0''), \tag{4.17}$$

in which $O(\vec{x}_0')$ represents the spatial intensity distribution in the object plane, where the light is assumed to be incoherent (as in an actual source).

The intensity of light reaches the point $\vec{x}_1$ in the plane of the image from the element $d\vec{x}_0'$. The intensities from the different elements are additive since the object is incoherent. For sufficiently small objects, the total intensity, $I(\vec{x}_1)$ is of the form,

$$I(\vec{x}_1) = \int_{-\infty}^{\infty} O(\vec{x}_0')|K(\vec{x}_1 - \vec{x}_0')|^2 d\vec{x}_0'. \tag{4.18}$$

Equation (4.18) is a convolution of the intensity distribution in the object with the squared modulus of the transmission function,

$$I(\vec{x}) = O(\vec{x}) \star |K(\vec{x})|^2. \tag{4.19}$$

Taking the Fourier transform of both sides of equation (4.19),

$$\widehat{I}(\vec{u}) = \widehat{O}(\vec{u})\widehat{T}(\vec{u}), \tag{4.20}$$

in which $\widehat{T}(\vec{u})$ is the optical transfer function (OTF) representing the complex factor applied by imaging system to the frequency components of object intensity distribution with frequency, $\vec{u}$, relative to the factor applied to the zero frequency component.

4.1.3 *Optical transfer function*

Optical transfer function (OTF) is a measure of the imaging quality and represents how each spatial frequency component in an object intensity is transferred to the image. It describes the change of the modulus and phase of the object Fourier transform in the imaging process, which is the Fourier transform of the point spread function (PSF). Every sinusoidal component in the object distribution is transferred to the image distribution without changing its frequency, but the intensity and phase are subject to alteration by the imaging system. The fact that the OTF of an optical system is the Fourier transform of its PSF, may be used to express OTF in terms of the auto-correlation of its pupil function. For an object point at the origin, the complex amplitude distribution function in its image may be written from equation (4.4) as,

$$K(\vec{x}_1) = \int_{-\infty}^{\infty} P(\vec{x})e^{-i2\pi\vec{x}_1 \cdot \vec{x}/\lambda s}d\vec{x}. \tag{4.21}$$

Using the definition of spatial frequency from (3.135), one may write,

$$u = \frac{x}{\lambda s'}; \qquad v = \frac{y}{\lambda s'}; \qquad \vec{q} = \frac{\vec{x}}{\lambda\, s'}, \tag{4.22}$$

where $q(u, v)$ is the spatial frequency vector in the pupil plane changing $\vec{x}$ to $\vec{q}$, one writes the equation (4.21) as,

$$K(\vec{x}_1) = \int_{-\infty}^{\infty} P(\vec{q})e^{-i2\pi\vec{q} \cdot \vec{x}_1}d\vec{q}. \tag{4.23}$$

Therefore, the PSF is,

$$|K(\vec{x}_1)|^2 = \iint_{-\infty}^{\infty} P(\vec{q})P^*(\vec{q}\,')e^{-i2\pi(\vec{q} - \vec{q}\,') \cdot \vec{x}_1}d\vec{q}d\vec{q}\,'. \tag{4.24}$$

Putting $\vec{q}-\vec{q}' = \vec{f}$ (a frequency variable), one gets

$$|K(\vec{x}_1)|^2 = \iint_{-\infty}^{\infty} P(\vec{q}' + \vec{f})P^*(\vec{q}')e^{-i2\pi\vec{f}\cdot\vec{x}_1}d\vec{q}'d\vec{f}$$
$$= \int_{-\infty}^{\infty}\left[\int_{-\infty}^{\infty} P(\vec{q}' + \vec{f})P^*(\vec{q}')d\vec{q}'\right]e^{-i2\pi\vec{f}\cdot\vec{x}_1}d\vec{f}. \quad (4.25)$$

Using Fourier inversion transform theorem, one finds

$$\int_{-\infty}^{\infty} P(\vec{q}' + \vec{f})P^*(\vec{q}')d\vec{q}' = \int_{-\infty}^{\infty} |K(\vec{x}_1)|^2 e^{i2\pi\vec{f}\cdot\vec{x}_1}d\vec{x}_1, \quad (4.26)$$

with f, as the spatial frequency.

Equation (4.26) shows that the autocorrelation of the pupil function in terms of frequency variables is the OTF of the optical system. Changing q' to q , one may write OTF,

$$\widehat{\mathcal{T}}(\vec{f}) = \int_{-\infty}^{\infty} P(\vec{q}+\vec{f})P^*(\vec{q})d\vec{q}. \quad (4.27)$$

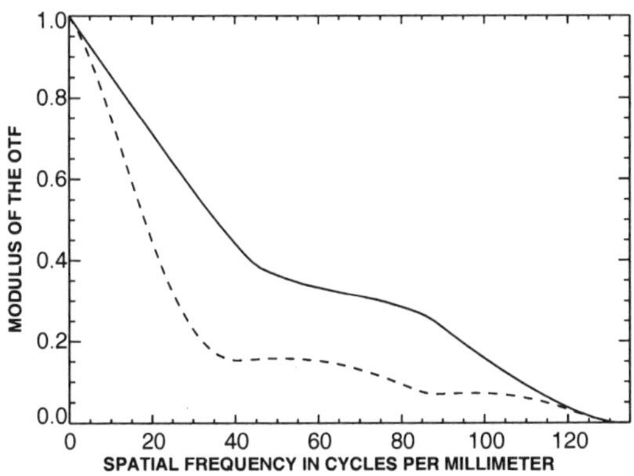

Fig. 4.2 Polychromatic diffraction MTF of a Cassegrain telescope. The solid line represents for an ideal diffraction-limited telescope, while the dashed line for a non-ideal case.

The spatial frequency spectrum of the diffracted image of an iso-planatic incoherent object is equal to the product of the spectrum of its Gaussian image and the OTF of the system. The magnitude of the OTF, $|\widehat{\mathcal{T}}(\vec{f})|$, is the ratio of the intensity modulation in the image to that in the object.

Figure (4.2) displays the normalized values of the MTF for an optical telescope. The modulation transfer function (MTF) is the modulus of generally the complex OTF. It is essentially an index of the efficiency of an optical system. The MTF contains no phase information, but is a real quantity. For incoherent imaging, the MTF, $|\widehat{\mathcal{T}}(\vec{f})| \leq 1$. Typically MTF decreases with increasing frequencies, hence the high frequency details in the image are weakened and eventually lost. The MTF is equivalent to the modulus of the Fourier transform of the PSF:

$$\left|\widehat{\mathcal{T}}(\vec{f})\right| = \left|\int_{-\infty}^{\infty} |K(\vec{x}_1)|^2 e^{i2\pi\vec{f}\cdot\vec{x}_1} d\vec{x}_1\right|. \tag{4.28}$$

or taking inverse transform

$$|K(\vec{x}_1)|^2 = \int_{-\infty}^{\infty} \widehat{\mathcal{T}}(\vec{f}) e^{-i2\pi\vec{f}\cdot\vec{x}_1} d\vec{f}. \tag{4.29}$$

From the equation (4.29), one obtains

$$|K(0)|^2 = \int_{-\infty}^{\infty} \widehat{\mathcal{T}}(\vec{f}) d\vec{f}. \tag{4.30}$$

Any frequency dependent phase changes introduced by the system would result in the image as lateral shifts of the sinusoidal frequency components comprising the image (Steward, 1983). This phenomenon is called phase transfer function (PTF) and its relation with the OTF is shown mathematically as,

$$OTF = MTF e^{i(PTF)}. \tag{4.31}$$

The contribution of this phase transfer function is often negligible. The normalized form of $\widehat{\mathcal{T}}(\vec{f})$ is,

$$\widehat{\mathcal{T}}_n(\vec{f}) = \frac{\widehat{\mathcal{T}}(\vec{f})}{\widehat{\mathcal{T}}(0)}. \tag{4.32}$$

For a perfect optical imaging system with a uniformly illuminated circular pupil of diameter D, the pupil function may be expressed as,

$$P(\vec{x}) = \begin{cases} 1 & \text{inside the pupil} \\ 0 & \text{otherwise,} \end{cases} \tag{4.33}$$

and the normalized OTF is given by,

$$\widehat{\mathcal{T}}_n(\vec{f}) = \frac{1}{\mathrm{A}_p}\int_{-\infty}^{\infty} P(\vec{\xi})P(\vec{\xi}+\vec{f})d\vec{\xi} = \frac{1}{\mathrm{A}_p}\int_{\mathrm{A}_o} d\vec{\xi}, \tag{4.34}$$

where $\vec{\xi}(=\xi,\eta)$ is the position vector of a point in the aperture, A_p the total area of the pupil, and A_o the overlap area of two replicas of the pupil displaced by an amount $\vec{f}$. It is to be noted that D is the reduced diameter of the pupil given by $D = 2a/\lambda s$, where a is the geometrical radius of the pupil. D has the same dimension as $\vec{f}$.

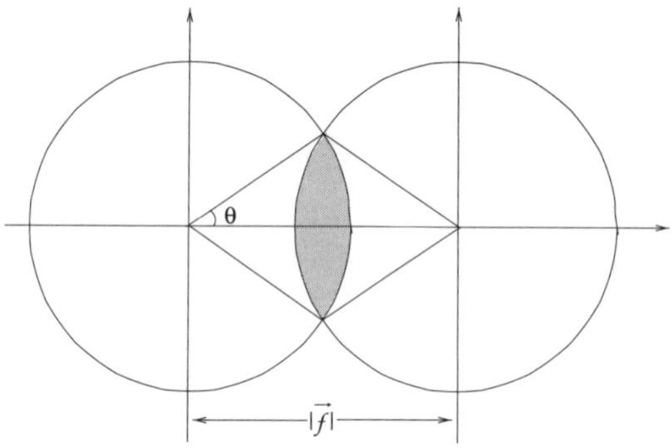

Fig. 4.3 Aberration free OTF as the fractional area of overlap of two circles for incoherent illumination.

Figure (4.3) depicts an area of overlap of two circles of unit diameter, whose centers are separated by an amount $|\vec{f}|$. Thus,

$$\widehat{\mathcal{T}}_n(|\vec{f}|) = \frac{2}{\pi}\left[\theta - \sin\theta\cos\theta\right], \tag{4.35}$$

with $\cos\theta = |\vec{f}|/D$.

The normalized optical transfer function (OTF) steadily decreases from 1 when $|\vec{f}| = 0$, i.e., $\theta = \pi/2$ to 0 when $|\vec{f}| = D$, i.e., $\theta = 0$. So that the cutoff frequency becomes,

$$|\vec{f}|_c = D = \frac{2a}{\lambda\mathcal{T}} = \frac{2a}{\lambda R}, \tag{4.36}$$

where $\mathcal{T} = R$, is the radius of the reference sphere, D the diameter of the aperture, and λ the wavelength of interest.

Comparing equation (4.16) with equation (4.36), one finds that incoherent cutoff frequency is twice the coherent cutoff frequency. For any optical system, $|\widehat{\mathcal{T}}(\vec{f})| = 0$ for $|\vec{f}| \geq f_c$. This states that the information at the spatial frequencies above the cutoff frequencies, f_c is irrevocably lost.

4.1.4 *Image in the presence of aberrations*

Aberrations affect the performance of optical imaging system. In order to obtain a quantitative estimate of the quality of images produced by such an imaging system, Karl Strehl introduced a criterion in 1902, known as Strehl's criterion, $\mathcal{S}_r$. In optics, the image quality is determined by such a criterion, which is defined as the ratio of the maximum intensity at a point of the image due to an aberrated optical system to the on axis intensity in the image due to the same but unaberrated system. In other words, it is the ratio of the current peak intensity due to an aberrated wavefront to the peak intensity in the diffraction-limited image where there are no phase fluctuations.

Denoting R as the radius of the Gaussian reference sphere with focus at $\mathtt{P}(\vec{r})$ in the region of image and s be the distance between the point, $\mathtt{Q}(\vec{r}')$, in which a ray in the image space intersects the wavefront through the center of the pupil and $\mathtt{P}(\vec{r})$. The disturbance at $\mathtt{Q}(\vec{r}')$ is represented by, $Ae^{i\kappa(\psi - R)}/R$, where A/R is the amplitude at $\mathtt{Q}(\vec{r}')$. From the Huygens-Fresnel principle, the disturbance, U, at $\mathtt{P}(\vec{r})$ at a distance z is given by (Born and Wolf, 1984),

$$U(\vec{r}) = -\frac{Aa^2}{i\lambda R^2} e^{i(R/a)^2 u} \int_0^1 \int_0^{2\pi} e^{i\left[\kappa\psi - v\rho\cos(\theta - \phi) - \frac{1}{2}u\rho^2\right]} \rho d\rho d\theta, \tag{4.37}$$

where a is the radius of the circular pupil on which ρ, θ are polar coordinates, r, ϕ the polar coordinates at the image plane, $\kappa = 2\pi/\lambda$ the wave number, and $\kappa\psi$ the deviation due to aberration in phase from a Gaussian sphere about the origin of the focal plane, and u, v the optical coordinates, i.e.,

$$u = \frac{2\pi}{\lambda}\left(\frac{a}{R}\right)^2 z, \qquad v = \frac{2\pi}{\lambda}\left(\frac{a}{R}\right)^2 \sqrt{x^2 + y^2} \tag{4.38}$$

The intensity at $\mathtt{P}(\vec{r})$ is expressed as,

$$I(\vec{r}) = \left(\frac{Aa^2}{\lambda R^2}\right)^2 \left| \int_0^1 \int_0^{2\pi} e^{i\left[\kappa\psi - v\rho\cos(\theta - \phi) - \frac{1}{2}u\rho^2\right]} \rho d\rho d\theta \right|^2. \tag{4.39}$$

In the absence of aberrations, the intensity, I_G, is a maximum on-axis ($\vec{r} = 0$) known as Gaussian image point, then according to the equation (4.39), one finds that,

$$I_G = \pi^2 \left(\frac{Aa^2}{\lambda R^2}\right)^2. \tag{4.40}$$

The Strehl intensity ratio, $\mathcal{S}_r$, is expressed as,

$$\mathcal{S}_r = \frac{I(\vec{r})}{I_G} = \frac{1}{\pi^2}\left|\int_0^1\int_0^{2\pi} e^{i\left[\kappa\psi - v\rho\cos(\theta-\phi) - \frac{1}{2}u\rho^2\right]}\rho d\rho d\theta\right|^2. \tag{4.41}$$

In the case of small aberrations, when tilt (a component of ψ linear in the coordinates $\rho\cos\theta$ or $\rho\sin\theta$ in the pupil plane) is removed and the focal plane is displaced to its Gaussian focus, the linear and quadratic terms in the exponential of the equation (4.41) vanish. The Strehl ratio simplifies to,

$$\mathcal{S}_r = \frac{I(0)}{I_G} = \frac{1}{\pi^2}\left|\int_0^1\int_0^{2\pi} e^{i\kappa\psi_{ab}(\rho,\theta)}\rho d\rho d\theta\right|^2, \tag{4.42}$$

in which ψ_{ab} represents the wave aberration referred to a reference sphere centered on the point P(0) in the image plane.

It is clear from the equation (4.42) that the Strehl ratio, $\mathcal{S}_r$, is bounded by $0 \leq \mathcal{S}_r \leq 1$. For strongly varying ψ_{ab}, $\mathcal{S}_r \ll 1$. The Strehl ratio tends to become larger for smaller wavenumber, κ, in the case of any given varying ψ_{ab}. For an unaberrated beam at the pupil, $\psi_{ab}(\rho,\theta) = 0$, such a ratio turns out to be unity, thus the intensity at the focus becomes diffraction-limited. By making the hypothesis that the aberrations are small, the phase error term is expanded as,

$$e^{i\kappa\psi_{ab}(\rho,\theta)} \simeq 1 + i\kappa\psi_{ab}(\rho,\theta) - \frac{\kappa^2}{2}\psi_{ab}^2(\rho,\theta) + \cdots, \tag{4.43}$$

in which $\psi_{ab}(\rho,\theta)$ is the optical path error introduced by the aberrations.

Since the aberrations are small, the third and higher powers of $\kappa\psi_{ab}(\rho,\theta)$ are neglected in this equation (4.43). Substituting equation (4.43) in equation (4.42), Strehl's intensity ratio is obtained, under the said condition, as

$$\mathcal{S}_r = \frac{I_{ab}(0)}{I(0)} = 1 + \kappa^2\left[\int_0^1\int_0^{2\pi}\psi_{ab}(\rho,\theta)\rho d\rho d\theta\right]^2 - \kappa^2\int_0^1\int_0^{2\pi}\psi_{ab}^2(\rho,\theta)\rho d\rho d\theta. \tag{4.44}$$

When $\psi_{ab} = 0$, $\mathcal{S}_r \sim 1$. The quality of image forming beam is directly related to the root-mean-square (RMS) phase error. The Strehl ratio for such an error less than $\lambda/2\pi$ is expressed as,

$$\mathcal{S}_r = 1 - \left(\frac{2\pi}{\lambda}\right)^2 \langle\sigma\rangle^2 \simeq e^{-(2\pi/\lambda)^2 \langle\sigma\rangle^2}, \tag{4.45}$$

where σ, is the RMS phase error or RMS wavefront error, and $\langle\sigma\rangle^2$ the variance (see Appendix B) of the aberrated wavefront with respect to a reference perfect wavefront,

$$\langle\sigma\rangle^2 = \frac{\int_0^1\int_0^{2\pi} \left(\psi_{ab}(\rho,\theta) - \bar{\psi}_{ab}(\rho,\theta)\right)^2 \rho d\rho d\theta}{\int_0^1\int_0^{2\pi} \rho d\rho d\theta}. \tag{4.46}$$

In equation (4.46), $\bar{\psi}_{ab}$ represent the average value of ψ_{ab}:

$$\bar{\psi}_{ab} = \frac{1}{\pi}\int_0^1\int_0^{2\pi} \psi_{ab}\rho d\rho d\theta. \tag{4.47}$$

For a good image quality, the Strehl's criterion should be $\mathcal{S}_r \geq 0.8$.

4.2 Imaging with partially coherent beams

In order to understand the quantitative relationship between an object and the image, it is necessary to know the coherence properties of the light being radiated by the object. These coherence properties have a profound influence on the character of the observed image (Goodman, 1985). This section is of paramount importance to develop an understanding of certain interferometric types of imaging systems, which measure the coherence of the light. It would explore the concept of speckle in coherent imaging systems as well.

4.2.1 *Effects of a transmitting object*

In an optical system where the object is illuminated from behind (trans-illuminated), the image is formed from the transmitted light. The no true object can be perfectly thin[5], hence an incident ray exits at slightly different transverse coordinates. For a non-uniform thickness of the object, the refractive index varies from point to point, refraction within the object modifies the position at which a given ray exits. Assuming that the field $U_1(\vec{x},t)$ enters at one end of a thin lens and exits at opposite side, where the field is $U_1'(\vec{x},t)$, the relationship between these two fields are given by,

$$U_1'(\vec{x},t) = B(\vec{x})U_1(\vec{x},t-\delta(\vec{x})), \tag{4.48}$$

where $B(\vec{x})$ reduces the amplitude of the transmitted field, $\vec{x} = x, y$ the 2-D space vector, $\delta(\vec{x})$ the delay suffered by the wave at the coordinates $\vec{x}$.

The relationship between the incident and transmitted fields in the form of mutual coherence functions is given by,

$$\begin{aligned}\Gamma_1'(\vec{x}_1,\vec{x}_1',\tau) &= \langle U_1'(\vec{x}_1,t+\tau)U_1'^*(\vec{x}_1',t)\rangle \\ &= B(\vec{x}_1)B(\vec{x}_1')\left\langle U_1(\vec{x}_1,t+\tau-\delta(\vec{x}_1))U_1^*(\vec{x}_1',t-\delta(\vec{x}_1'))\right\rangle \\ &= B(\vec{x}_1)B(\vec{x}_1')\Gamma_i(\vec{x}_1,\vec{x}_1',\tau-\delta(\vec{x}_1)+\delta(\vec{x}_1')).\end{aligned} \tag{4.49}$$

For a quasi-monochromatic light, the analytic signal representation for the fields in terms of a time varying phasor is,

$$U_1(\vec{x},t) = A_1(\vec{x},t)e^{-i2\pi\bar{\nu}t}, \tag{4.50}$$

in which $\bar{\nu}$ denotes the center frequency of the disturbance.

The mutual coherence function of the incident field is written as,

$$\Gamma_1(\vec{x}_1,\vec{x}_1',\tau) = \langle A_1(\vec{x}_1,t+\tau)A_1^*(\vec{x}_1',t)\rangle\, e^{-i2\pi\bar{\nu}t}. \tag{4.51}$$

Thus the equation (4.49) is recast into,

$$\begin{aligned}\Gamma_1'(\vec{x}_1,\vec{x}_1',\tau) &= B(\vec{x}_1)e^{i2\pi\bar{\nu}\delta(\vec{x}_1)}B(\vec{x}_1')e^{-i2\pi\bar{\nu}\delta(\vec{x}_1')} \\ &\quad\times\langle A_1(\vec{x}_1,t+\tau-\delta(\vec{x}_1)+\delta(\vec{x}_1'))A_1^*(\vec{x}_1',t)\rangle\, e^{-i2\pi\bar{\nu}t}.\end{aligned} \tag{4.52}$$

The time average is independent of $\delta(\vec{x}_1)$ and $\delta(\vec{x}_1')$ when,

$$|\delta(\vec{x}_1)-\delta(\vec{x}_1')| \ll \frac{1}{(\Delta\nu)} = \tau_c, \tag{4.53}$$

in which τ_c is the coherence time.

[5]A transmitting object such as a lens may be considered 'thin' whose thickness (distance along the optical axis between the two surfaces of the object) is negligible compared to its aperture.

Thus the relationship between the incident and transmitted mutual coherence function takes the form,

$$\Gamma_1'(\vec{x}_1, \vec{x}_1', \tau) = t_r(\vec{x}_1) t_r^*(\vec{x}_1') \Gamma_1(\vec{x}_1, \vec{x}_1', \tau), \tag{4.54}$$

in which $t_r(\vec{x}) = B(|\vec{x}|)e^{i2\pi\bar{\nu}\delta(\vec{x})}$ is the amplitude transmittance of the object at $\mathtt{P}(\vec{x})$.

Since in a physical experiment $\tau < \tau_c$, the equation (4.54) simplifies to,

$$J_1'(\vec{x}_1, \vec{x}_1') = t_r(\vec{x}_1) t_r^*(\vec{x}_1') J_1(\vec{x}_1, \vec{x}_1'). \tag{4.55}$$

4.2.2 *Transmission of mutual intensity*

The transmission of mutual intensity through an optical system can be envisaged from the geometry of object-image coherence relation of a thin lens. The object and image planes of the lens are at a distances f behind and in front of it. These planes are perpendicular to a line passing through the optical axis of the lens (see Figure 4.4).

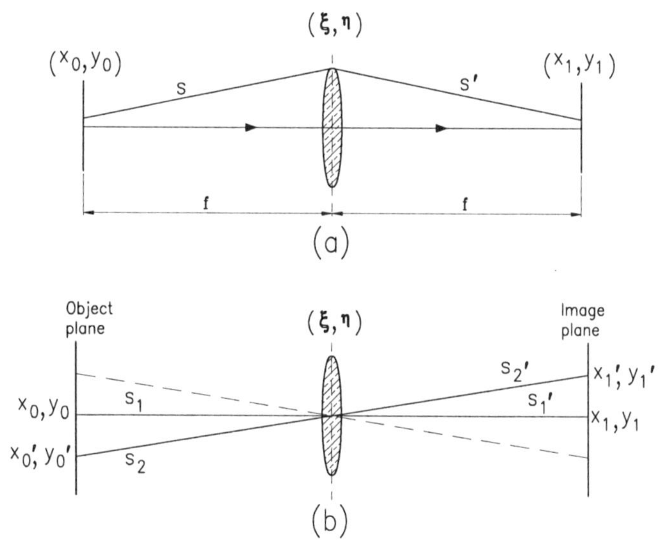

Fig. 4.4 Geometry for calculation of object-image coherence relationship for a single thin lens.

Let $J_0(\vec{x}_0; \vec{x}_0')$ be the mutual intensity for the points, $\vec{x}_0 = (x_0, y_0)$, and

$\vec{x}_0' = (x_0', y_0')$ in the object plane. By using the law for the propagation of mutual intensity (equation 3.74), the mutual intensity leaving the lens $J_1(\vec{x}_1; \vec{x}_1')$, in which $\vec{x}_1 = (x_1, y_1)$, and $\vec{x}_1' = (x_1', y_1')$, can be derived (Goodman, 1985). By employing four-dimensional (4-D) approach, the mutual intensity in the image plane is given by,

$$J_1\left(\vec{x}_1; \vec{x}_1'\right) = \iint_{-\infty}^{\infty} J_0\left(\vec{x}_0; \vec{x}_0'\right) K\left(\vec{x}_0; \vec{x}_1\right) K^*\left(\vec{x}_0'; \vec{x}_1'\right) d\vec{x}_0 d\vec{x}_0'. \tag{4.56}$$

where $K(\vec{x}_0; \vec{x}_1)$ is the transmission function (equation 4.4) of the system.

The quantity $K(\vec{x}_0; \vec{x}_1)K^*(\vec{x}_0; \vec{x}_1)$ may be regarded as the impulse response of the system. Equation (4.56) is known as a four-dimensional superposition integral and is characteristic of a linear system. Since J_0 is zero for all points in the object plane from which no light proceeds to the image plane, the integration extends only over an infinite domain. By setting $\vec{x}_1 = \vec{x}_1' = \vec{x}$ in equation (4.56), the intensity distribution in the image plane is derived:

$$I_1\left(\vec{x}\right) = \iint_{-\infty}^{\infty} J_0\left(\vec{x}_0; \vec{x}_0'\right) K\left(\vec{x}_0; \vec{x}\right) K^*\left(\vec{x}_0'; \vec{x}\right) d\vec{x}_0 d\vec{x}_0'. \tag{4.57}$$

Let the object be small forming an iso-planatic region of the system, hence the transmission function for all points on it is replaced as, $K(\vec{x}_0; \vec{x}_1) = K(\vec{x}_1 - \vec{x}_0)$. Thus the equation (4.56) becomes,

$$J_1\left(\vec{x}_1; \vec{x}_1'\right) = \iint_{-\infty}^{\infty} J_0\left(\vec{x}_0; \vec{x}_0'\right) K\left(\vec{x}_1 - \vec{x}_0\right) K^*\left(\vec{x}_1' - \vec{x}_0'\right) d\vec{x}_0 d\vec{x}_0'. \tag{4.58}$$

Equation (4.58) is a 4-D convolution equation and can be employed to represent the mapping of $J_0(\vec{x}_0; \vec{x}_0')$ into $J_1(\vec{x}_1; \vec{x}_1')$. The relationship in the Fourier domain, in which convolutions are represented by multiplications of transform, the 4-D Fourier spectra of the object and image mutual intensities are defined as,

$$\widehat{J}_0\left(\vec{u}; \vec{u}'\right) = \mathcal{F}\left[J_0\left(\vec{x}_0; \vec{x}_0'\right)\right], \qquad \widehat{J}_1\left(\vec{u}_1; \vec{u}_1'\right) = \mathcal{F}\left[J_1\left(\vec{x}_1; \vec{x}_1'\right)\right], \tag{4.59}$$

where the notation, $\mathcal{F}$, stands for Fourier transform.

The 4-D Fourier transform of $J(\vec{x};\vec{x}')$ is defined by,

$$\mathcal{F}\left[J\left(\vec{x};\vec{x}'\right)\right] = \widehat{J}\left(\vec{u};\vec{u}'\right) = \iint\limits_{-\infty}^{\infty} J\left(\vec{x};\vec{x}'\right) e^{i2\pi\left[\vec{u}\cdot\vec{x}+\vec{u}'\cdot\vec{x}'\right]} d\vec{x}d\vec{x}', \quad (4.60)$$

in which $\vec{x}$, $\vec{x}'$ are the space vectors and $\vec{u}$, $\vec{u}'$ the respective spatial frequency vectors.

Similarly, the 4-D transfer function of the space invariant, linear system is defined as,

$$\begin{aligned} \widehat{K}\left(\vec{u}_1;\vec{u}_1'\right) &= \mathcal{F}\left[K(\vec{x}_1)K^*\left(\vec{x}_1'\right)\right] \\ &= \widehat{K}\left(\vec{u}_1\right)\widehat{K}^*\left(-\vec{u}_1'\right). \end{aligned} \quad (4.61)$$

where $\widehat{K}$ represents the 2-D Fourier transform of the amplitude spread function.

In applying convolution theorem to equation (4.58), the effect of the imaging system is deduced in the 4-D Fourier domain,

$$\widehat{J}_1\left(\vec{u}_1;\vec{u}_1'\right) = \widehat{J}_0\left(\vec{u}_0;\vec{u}_0'\right)\widehat{K}\left(\vec{u}_1\right)\widehat{K}^*\left(-\vec{u}_1'\right), \quad (4.62)$$

Equation (4.62) shows that if the mutual intensity in the object and image planes are represented as superposition of 4-D space harmonic components of all possible spatial frequencies, $(\vec{u}_1;\vec{u}_1')$, then each component in the image depends on the corresponding component in the object. The ratio of the components is equal to the frequency response function, $\widehat{\mathcal{K}}(\vec{u}_1;\vec{u}_1')$, for partially coherent quasi-monochromatic illumination. Such a response function is related to the pupil function of the system.

$$\widehat{\mathcal{K}}\left(\vec{u}_1;\vec{u}_1'\right) = \widehat{K}\left(\vec{u}_1\right)\widehat{K}^*\left(-\vec{u}_1'\right). \quad (4.63)$$

It is observed that $\widehat{K}(\vec{u}_1)$ is equal to the value of the pupil function $P(\vec{\xi})$, in which $\vec{\xi} = (\xi,\eta)$, of the system at the point, $\vec{\xi} = \bar{\lambda}R\vec{u}_1$, on the Gaussian system of radius R. Therefore, the frequency response function for partially coherent quasi-monochromatic illumination is connected with the pupil function by the formula,

$$\widehat{\mathcal{K}}\left(\frac{\vec{\xi}}{\bar{\lambda}R},\frac{\vec{\xi}'}{\bar{\lambda}R}\right) = P\left(\vec{\xi}\right)P^*\left(-\vec{\xi}'\right), \quad (4.64)$$

in which $\bar{\lambda}$ denotes the mean wavelength in the image space.

The relation between the spectra of the mutual intensities (equation 4.62) is recast as,

$$\widehat{J}_1\left(\vec{u}_1;\vec{u}_1'\right) = \widehat{J}_0\left(\vec{u}_0;\vec{u}_0'\right)\widehat{P}\left(\bar{\lambda}R\vec{u}_1\right)\widehat{P}^*\left(-\bar{\lambda}R\vec{u}_1'\right), \tag{4.65}$$

where $\widehat{P}$ is the Fourier transform of the pupil function.

The pupil function is zero for points outside the area of the exit pupil, hence the spectral components belonging to frequencies above certain values do not transmit. In a circular exit pupil of radius a, $P(\vec{\xi})P^*(-\vec{\xi}')$ vanishes when $\vec{\xi}^2 > a^2$ or $\vec{\xi}'^2 > a^2$. Thus, the spectral components of the mutual intensity belonging to frequencies $(\vec{u}_1;\vec{u}_1')$ do not transmit for the following parameters,

$$\lambda^2\vec{u}_1^2 > \left(\frac{a}{R}\right)^2, \qquad \lambda^2\vec{u}_1'^2 > \left(\frac{a}{R}\right)^2. \tag{4.66}$$

4.2.3 *Images of trans-illuminated objects*

Let a portion of the object plane be occupied by a transparent or semi-transparent object which is illuminated from behind with a partially coherent quasi-monochromatic light originating from an incoherent source and reaches the object plane after passing through a condenser[6]. The transmission function of the object, $F(\vec{\xi})$, is expressed as,

$$F(\vec{\xi}) = \frac{V(\vec{\xi})}{V_0(\vec{\xi})}, \tag{4.67}$$

where

$$V_0(\vec{\xi}) = Ae^{i\kappa(\vec{l}_0\cdot\vec{\xi})},$$

is the disturbance in the $\vec{\xi}$-plane in the absence of any object, $\vec{l}_0(= l_0, m_0)$, $V(\vec{\xi})$ the disturbance in the presence of the object.

In general, the transmission function, F, depends on both $\vec{\xi}$-plane, as well as on the direction $\vec{l}_0$ of illumination. Since both amplitude and phase of the light may be altered on passing through the object, this transmission function is generally a complex function. Let $U_0'(\mathtt{S};\vec{x}_0)$ represent the disturbance at the point, $(\vec{x}_0)$, of the object plane due to a source point $\mathtt{S}$ of the associated monochromatic source. The disturbances from this source

[6] A condenser lens system collects energy from a light source. It consists of two plano-convex elements with short-focal lengths mounted convex sides together.

point after the passage through the object is of the form,

$$U_0\left(\mathtt{S};\vec{x}_0\right)=U_0'\left(\mathtt{S};\vec{x}_0\right)F\left(\vec{x}_0\right). \tag{4.68}$$

The mutual intensities of the light incident on the object and from the object are respectively given by the following equations,

$$\begin{aligned} J_0'\left(\vec{x}_0;\vec{x}_0'\right) &= \int_\sigma U_0'\left(\mathtt{S};\vec{x}_0\right)U_0'^*\left(\mathtt{S};\vec{x}_0'\right)d\mathtt{S}, \\ J_0\left(\vec{x}_0;\vec{x}_0'\right) &= \int_\sigma U_0\left(\mathtt{S};\vec{x}_0\right)U_0^*\left(\mathtt{S};\vec{x}_0'\right)d\mathtt{S}. \end{aligned} \tag{4.69}$$

The transmitted mutual intensity is given by,

$$\begin{aligned} J_0\left(\vec{x}_0;\vec{x}_0'\right) &= \int_\sigma U_0'\left(\mathtt{S};\vec{x}_0\right)F\left(\vec{x}_0\right)U_0'^*\left(\mathtt{S};\vec{x}_0'\right)F^*\left(\vec{x}_0'\right)d\mathtt{S} \\ &= J_0'\left(\vec{x}_0;\vec{x}_0'\right)F\left(\vec{x}_0\right)F^*\left(\vec{x}_0'\right). \end{aligned} \tag{4.70}$$

The mutual intensity of the incident light, $J_0'(\vec{x}_0;\vec{x}_0')$, depends on the coordinate differences, $\Delta\vec{x}=\vec{x}_0-\vec{x}_0'$, that is, $J_0'(\vec{x}_0;\vec{x}_0')=J_0'(\vec{x}_0-\vec{x}_0')$. The 4-D Fourier transform (see equation 4.60) of $J_0(\vec{x}_0;\vec{x}_0')$ takes the form,

$$\begin{aligned} \widehat{J}_0'(\vec{u}_0;\vec{u}_0') &= \iint_{-\infty}^{\infty} J_0'\left(\Delta\vec{x}\right)F\left(\vec{x}_0\right)F^*\left(\vec{x}_0'\right)e^{i2\pi\left(\vec{u}_0\cdot\vec{x}_0+\vec{u}_0'\cdot\vec{x}_0'\right)}d\vec{x}_0 d\vec{x}_0' \\ &= \int_{-\infty}^{\infty} F\left(\vec{x}_0\right)e^{i2\pi\left(\vec{u}_0+\vec{u}_0'\right)\cdot\vec{x}_0}d\vec{x}_0 \\ &\quad\times\int_{-\infty}^{\infty} J_0'\left(\Delta\vec{x}\right)F^*\left(\vec{x}_0+\Delta\vec{x}\right)e^{i2\pi\vec{u}_0'\cdot\Delta\vec{x}}d\Delta\vec{x}. \end{aligned} \tag{4.71}$$

The second integral of the equation (4.71) is the Fourier transform of the product of two functions. This may be evaluated as the convolution of their individual transforms. After proper manipulation, this integral is deduced as,

$$\int_{-\infty}^{\infty}\widehat{J}_0'\left(\vec{u}''\right)\widehat{F}^*\left(\vec{u}''-\vec{u}_0'\right)e^{-i2\pi\left[\left(\vec{u}''-\vec{u}_0'\right)\cdot\vec{x}_0\right]}d\vec{u}''.$$

By substituting this into equation (4.71), one gets,

$$\widehat{J}_0'\left(\vec{u}_0;\vec{u}_0'\right)=\int_{-\infty}^{\infty}\widehat{J}_0'\left(\vec{u}''\right)\widehat{F}\left(\vec{u}''+\vec{u}_0\right)\widehat{F}^*\left(\vec{u}''-\vec{u}_0'\right)d\vec{u}''. \tag{4.72}$$

By invoking the equation (4.62), the 4-D spectrum $\widehat{J}_1(\vec{u}_1;\vec{u}_1')$ of the image mutual intensity in terms of 2-D spectra of other quantities is recast into,

$$\widehat{J}_1(\vec{u}_1;\vec{u}_1') = \widehat{K}(\vec{u}_1)\widehat{K}^*(-\vec{u}_1') \int_{-\infty}^{\infty} \widehat{J}_0'(\vec{u}'')\,\widehat{F}(\vec{u}''+\vec{u}_0)\,\widehat{F}^*(\vec{u}''-\vec{u}_0')\,d\vec{u}'', \tag{4.73}$$

in which the function $\widehat{J}_1$ is known as transmission cross co-efficient of the system, while $\widehat{K}$ can be expressed in terms of the pupil function of the imaging lens, and $\widehat{J}_0'$ is expressed in terms of the pupil function of the condenser lens.

The intensity, $I_1(\vec{x}_1)$ at the image plane is given by,

$$\begin{aligned} I_1(\vec{x}_1) &= J_1(\vec{x}_1;\vec{x}_1') \\ &= \iint_{-\infty}^{\infty} J_0'(\vec{x}_0-\vec{x}_0')\,F(\vec{x}_0)\,F^*(\vec{x}_0')\,K(\vec{x}_1-\vec{x}_0)\,K^*(\vec{x}_1-\vec{x}_0')\,d\vec{x}_0 d\vec{x}_0'. \end{aligned} \tag{4.74}$$

The intensity at the image plane in terms of $\widehat{J}_1(\vec{u}_1;\vec{u}_1')$ is given by,

$$I_1(\vec{x}_1) = \iint_{-\infty}^{\infty} \widehat{J}_1(\vec{u}_1;\vec{u}_1')e^{-i2\pi\,[(\vec{u}_1+\vec{u}_1')\cdot\vec{x}_1]}d\vec{u}_1 d\vec{u}_1'. \tag{4.75}$$

It is observed in equation (4.73) that the influence of the object, $\widehat{F}$, and the combined effect of the illumination, $\widehat{J}_0'$, and of the system, $\widehat{K}$, are separated. The intensity of the light emerging from the object with a uniform illumination, is proportional to $|F|^2$. Equations (4.74) and (4.75) represent the true intensity. The ideal intensity represents as the sum of contributions from all pairs of frequencies $\vec{u}_1;\vec{u}_1'$ of spatial spectrum of the object. Thus the spatial spectrum $\widehat{I}(\vec{u}_1)$ of the intensity,

$$\widehat{I}_1(\vec{u}) = \int_{-\infty}^{\infty} \widehat{J}_1(\vec{u}_1;\vec{u}-\vec{u}_1)\,d\vec{u}_1. \tag{4.76}$$

With the changed variables $\vec{z} = \vec{u}''+\vec{u}_0$, as well as by putting the equation (4.72) in equation (4.76), the spectrum of the image intensity is recast in,

$$\begin{aligned} \widehat{I}_1(\vec{u}) &= \int_{-\infty}^{\infty} \widehat{F}(\vec{z})\,\widehat{F}^*(\vec{z}-\vec{u})\,d\vec{z} \\ &\quad \times \left[\int_{-\infty}^{\infty} \widehat{J}_0(\vec{u}'')\,\widehat{K}(\vec{z}-\vec{u}'')\,\widehat{K}^*(\vec{z}-\vec{u}''-\vec{u})\,d\vec{u}''\right]. \end{aligned} \tag{4.77}$$

The quantity in square brackets of the equation (4.77) describes the effects of the optical system from source to image plane.

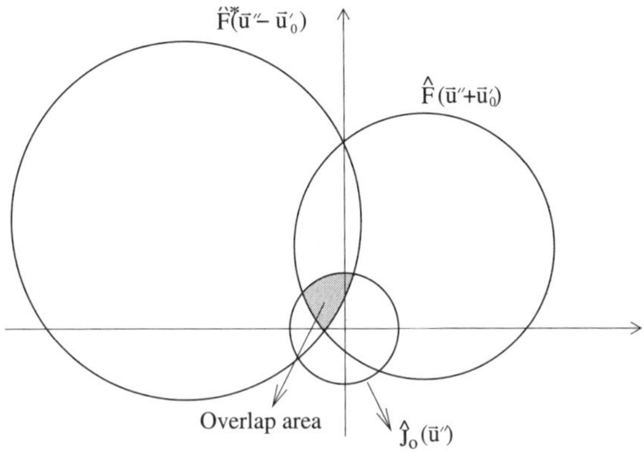

Fig. 4.5 Region of overlap

The response function, $\widehat{F}(\vec{u})$, of the object is non-zero within the exit pupil of radius a. When the object is trans-illuminated by a quasi-monochromatic light of mean wavelength, $\bar{\lambda}_0$, through a condenser of numerical aperture, $mn_0 \sin\theta_0$, in which $n_0 \sin\theta_0$ is the numerical aperture of the image forming system and m the number of times, the arguments imply that the function, $\widehat{J}_0(\vec{u})$, is a scaled version of the squared modulus of the pupil function of the condenser. The spectrum, $\widehat{J}'_0$, of the transmitted mutual intensity may be envisaged by integrating the product of the three partially overlapping functions, $\widehat{J}_0(\vec{u}'')$, $\widehat{F}(\vec{u}'' + \vec{u}'_0)$, and $\widehat{F}^*(\vec{u}'' - \vec{u}'_0)$ (see Figure 4.5). As the frequencies, $(\vec{u}_0; \vec{u}'_0)$ grow larger, the degree of overlap decreases, and consequently the value of the 4-D spectrum, $\widehat{J}'_0$, may drop.

4.3 The optical telescope

A telescope encodes the information about the source, which in most cases, contains in a specific energy distribution. Its purpose is to collect as many photons as possible from a given region of sky and directing them to a point. Larger the diameter of the objective, D, called telescope aperture, more the photons. The light gathering power of a telescope is proportional to the area of its aperture. For example, a 20 cm diameter telescope collects four times more photons than a 10 cm telescope.

An optical telescope collects light of any celestial object in the visible spectrum spanning from 300 nm (nanometers) to 800 nm. An aberration free telescope is used to image the celestial object on a detector and the digital data from the detector is processed in a computer to compute its required parameters. The first optical telescope was developed by Hans Lippershey in 1609, and Galileo made the first discovery. Such an instrument brought changes in understanding the universe. A telescope can be equipped to record light over a long period of time, by using photographic film or electronic detectors such as a photometer or a charge coupled device (CCD; will be discussed in chapter 8), while human eye has no capability to store light. A long-exposure image taken through a telescope can reveal objects too faint to be seen by the eye.

Optical telescope has passed through four major phases of development, each of which has caused a quantum jump in astronomical knowledge. Modern telescopes are high precision instruments that use sophisticated technology for its manufacture, testing, and deployment at sites best suited for astronomical observations. Since the telescope is required to point and track very faint light sources, accurate closed loop motor control systems employing computers, feedback devices with fast response is a necessity. In order to obtain good images of celestial objects, it is desired to have best design practices involving mechanical, electrical, optical, and thermal engineering methods.

Over the years, every effort has been made to incorporate latest technology and material available. Modern telescopes depend upon an active optics system, which maintains alignment with each other, as well as minimize gravitationally induced deformations. The optics need to be supported in some suitable structure. Each support is driven by some form of force drivers like stepper motors or electro-hydraulic support system to float the mirror. A computer monitors these forces by measuring the pressure between the load cells and the mirror in the actuators. Unlike passive optics[7], where lack of in-built corrective devices prevents to improve the quality of the star images during observations, active optics is capable of optimizing the image quality automatically by means of constant adjustments by in-built corrective optical elements. Such a technique corrects wavefront distortions[8] caused by the relatively slow mechanical, thermal, and optical effects in the telescope itself. This active optics technique has been devel-

[7] In passive mirror system, a set of springs and counter weights offer equal and opposite thrust to the mirror bottom, so that there is no net force acting on the mirror.

[8] Distortion is an aberration that affects the shape of an image rather than the sharpness.

oped for medium and large telescopes; the first of this kind is the European Southern Observatory (ESO) 3.5 m New Technology Telescope (NTT), entered into operation at La Silla, Chile, in 1989. At the onset of the fifth phase the very large diffraction-limited telescopes, which can be built from smaller or segments with active control of their alignment, may provide more insight of the universe.

The quantity, known as 'aperture ratio', $F\#$, is a number defined as the focal length of a mirror (primary) of a telescope divided by the effective diameter, D, of the aperture, $F\# = f/D$. Such a quantity is used to characterize the light gathering power of a telescope. If the aperture ratio is large, say near unity, one has a powerful fast telescope, while a small aperture ratio (the focal length is much greater than the aperture) provides a slow telescope. In the case of the former, the image is sharp and bright, therefore one takes photograph using shorter exposure, than that of the latter.

The exit pupil is the image of the objective lens, formed by the eyepiece, through which the light from the objective goes behind the eyepiece. The magnification, M, of a telescope depends on the focal length of a telescope, f, and can be determined by, $\mathrm{M} = f/f'$, in which f' is the focal length of the eyepiece that is employed to magnify the image produced by the telescope aperture.

Optical telescopes may be made out of a single or multi-element system depending on the uses that are required to perform. They may be employed in prime focus, Cassegrain, Nasmyth or Coudé configuration. A single lens based telescope, known as refractor, consists of two lenses made out of two different glass materials such as borosilicate (first lens and flint (second lens), but a reflector uses a mirror instead of lens (Sir Issac Newton had used a mirror for telescope). The former suffers from (i) the residual errors, (ii) loss of light due to transmission, and (iii) difficulties in fabrication and mounting, while the latter has advantages of (i) zero chromatic errors, (ii) maximum reflectivity, and (iii) easy fabrication and mounting.

A single element reflecting telescope has a prime focus; starlight from the paraboloid reflector is focused at a point. This mirror is mounted on a suitable structure that needs to be capable of following astronomical objects as they move across the sky. Most of the telescopes are made up of two components, namely, primary mirror and secondary mirror, though Nasmyth configuration requires to have an additional mirror called tertiary mirror.

The best combination of the mirrors is primary as parabolic that reflects

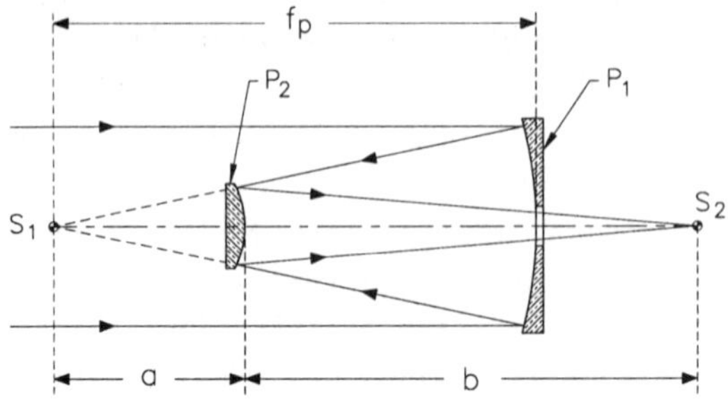

Fig. 4.6 Schematic diagram of a Cassegrain telescope.

the light rays towards the primary focus (prime focus) and a secondary as convex hyperbolic reflecting the rays back through a small hole through the center of the main mirror. Such an arrangement of a telescope is known as Cassegrain telescope (see Figure 4.6). The Richey-Chrétian type telescope is identical to the Cassegrain mode except that primary mirror is deepened to an hyperboloid and a stronger hyperboloid is used for the secondary. The nearer focus of the conic section which forms the surface of the secondary is coincident with the focus of the primary, and the Cassegrain focus is at the distant focus of the secondary mirror's surface. The light comes to focus behind the primary mirror. The secondary mirror is held in place by a spider (as in a Newtonian). Focussing is usually achieved by an external rack-and-pinion system similar to what a refractor would have.

Advantages of the Cassegrain system lies in its telephoto characteristics; the secondary mirror serves to expand the beam from the primary mirror so that the effective focal length of the whole system is several times that of the primary mirror. In general, the Cassegrain systems are designed to have long aperture ratios of $f/8, \cdots, f/15$, even though their primary mirrors may be $f/3$ or $f/4$. Thus the images remain tolerable over a field of view which may be several tenths of a degree across. The other noted advantages are:

- it provides much easier access to the focus position since the focus is at the back of primary mirror closer to the ground,
- it reduces optical distortion, since the focus lies on the optical axis of the primary, and

- the system supporting the rear end of the primary can be used for mounting instruments.

The focal length of secondary mirror is different from that of the primary. The effective focal length, F, of a Cassegrain telescope is determined as, $F = (bf_p)/a$, in which f_p is the focal length of the primary mirror, a the distance from the surface of secondary mirror to the focal point of the primary, and b the distance from the surface of secondary mirror to the Cassegrain focus. If $a \ll b$, the effective focal length turns out to be much larger than the prime focus, i.e., $F \gg f_p$. If a is small, that corresponds to a large magnification, a small diameter for the secondary and a view of a small area of the sky, while the large value of a corresponds to a small magnification, a large diameter for the secondary, and a large area of sky field. For an infrared telescope, it is essential to minimize the size of secondary in order to reduce the thermal radiation it emits.

The Coudé telescope is very closely related to the Cassegrain system. It is in effect a very long focal length Cassegrain or Ritchey-Chretian whose light beam is folded and guided by additional flat mirrors to give a focus whose position is fixed in space irrespective of the telescope position. The aperture ratio of such a telescope can be as large as $f/30, \cdots, f/100$. Such a focus is used mainly for high resolution spectroscopy.

The traditional solution of attaining Coudé focus is to use a chain of mirrors to direct the light to the Coudé room. After reflection from the secondary, the light is reflected down the hollow declination axis[9] by a diagonal flat mirror, and then down the hollow polar axis by a second diagonal. The light beam emerges from the end of the polar axis, whichever portion of the sky the telescope may be inspecting.

The Coudé room is generally situated in a separate location near the telescope, where bulky instruments, such as high dispersion spectrograph, are to be used. These instruments can be kept stationary and the temperature can be held accurately constant. The effective focal length is very large, so the image size is large, the region of sky imaged is the size of the star. The Coudé design has several disadvantages such as (i) loss of light in the process of reflections from several mirrors and (ii) rotation of field

[9]Equatorial mount for a telescope moves it along two perpendicular axes of motion, called right ascension (RA; α) and declination (Dec; δ). These two equatorial coordinate systems specify the position of a celestial object uniquely. They are comparable to longitude and latitude respectively projected onto the celestial sphere. The former is measured in hours (h), minutes (m), and seconds (s), while the latter is measured in degrees ($^\circ$), arc-minutes ($'$), and arc-seconds ($''$) north and south of the celestial equator.

of view as the telescope tracks an object across the sky. Another way of getting the light to the high resolution spectrograph is to use an optical fiber[10], which enables the instrument to be kept away from the former in a temperature-controlled room. This reduces the problems of flexure that occur within telescope-mounted spectroscopes[11] as gravitational loads change with the different telescope orientations. They can also be used to reformat stellar images. The light at the prime focus may be focused onto the end of the fiber, which is threaded over to instrument location. Sharp bends in the fiber may lose light, albeit a reasonable efficiency can be achieved with care.

4.3.1 *Resolving power of a telescope*

A large telescope helps in gathering more optical energy, as well as in obtaining better angular resolution; the resolution increases with the diameter of the aperture. But it has an inherent limitation to its angular resolution due to the diffraction of light at the telescope's aperture. According to diffraction theory, each point in the aperture may be considered as the center of an emerging spherical wave. In a far-field approximation, the spherical waves are equivalent to plane waves[12]. The incident star beam is a stream of photons arriving at random times from a range of random angles within its angular diameter. The photon senses the presence of all the details of the collecting aperture. However, it is prudent to think of wave, instead of photon, as a series of wavelets propagating outwards. The incident idealized photon is monochromatic in nature. The corresponding classical wave has the same extent as well.

The resolving power of a telescope refers to its ability to discern the two components of a binary star system, which is often used to gauze the spatial resolution. In the absence of any other effects such as the effect of

[10]Optical fibers are the convenient component to deliver light beam from one place to other, like coaxial cables that are used in radio frequencies, and are widely used to connect telescopes to spectroscopes. An optical fiber eliminates the need for mirrors or lenses, and alignment required for these elements. It exploits total internal reflection by having an inner region of low refractive index and a cladding of higher index; light is confined by repeated reflections.

[11]A spectroscope is a device that is used to separate light into its constituent wavelengths.

[12]The wavefronts from a distant point source, say a star are circular, because the emitted light travels at the same speed in all directions. After travelling a long time, the radius of the wavefront becomes so large that the wavefronts are planar over the aperture of a telescope.

the atmospheric turbulence (will be discussed in following chapter 5), the resolution of a telescope is limited by the diffraction of light waves. This is known as the 'diffraction-limit' of a telescope; the angular resolution θ according to Rayleigh criterion is,

$$\theta = 1.22\frac{\lambda}{D} \qquad \text{rad.} \tag{4.78}$$

The larger the aperture, the smaller is the Airy disc and the greater the maximum luminance at the center of the Airy disc. Lord Rayleigh introduced the afore-mentioned criterion, known as 'Rayleigh criterion', of resolving power for optical telescopes, which corresponds to the angular separation on the sky when one stellar component is centered on the first null in the diffraction pattern of the other; the binary star is said to be resolved. At this separation, the separate contributions of the two sources to the intensity at the central minimum is given by, $4/\pi^2 = 0.4053$ for a rectangular aperture; the resultant intensity at the center with respect to the two peaks is approximately 81% (see Figure 4.7). The Rayleigh criterion yields the result that the sources are resolved if the angle they subtend is, $\theta \sim 1.22\lambda/D$; the two maxima are completely separated at $2.33\lambda/D$ for the circular aperture and $2\lambda/D$ for the rectangular aperture.

In an ideal condition, the resolution that can be achieved in an imaging experiment, $\mathcal{R}$, is limited only by the imperfections in the optical system and according to Strehl's criterion, the resolving power, $\mathcal{R}$, of any telescope of diameter D is given, according to equation (4.30) by the integral of its transfer function,

$$\begin{aligned} \mathcal{R} &= \int_{-\infty}^{\infty} \widehat{\mathcal{S}}(\vec{u})d\vec{u} \int_{-\infty}^{\infty} \widehat{\mathcal{T}}(\vec{u})d\vec{u} \\ &= \frac{1}{\mathrm{A}_p} \iint\limits_{-\infty}^{\infty} \widehat{P}(\vec{u})\widehat{P}^*(\vec{u}+\vec{u}')d\vec{u}d\vec{u}' \\ &= \frac{1}{\mathrm{A}_p} \left| \int_{-\infty}^{\infty} \widehat{P}(\vec{u})d\vec{u} \right|^2, \end{aligned} \tag{4.79}$$

where A_p is the pupil area in wavelength squared units and $\widehat{\mathcal{T}}(\vec{u})$ the telescope transfer function.

The resolution of perfect telescope with diameter D is given by,

$$\mathcal{R} = \widehat{\mathcal{S}}(0) = \frac{\pi}{4}\left(\frac{D}{\lambda}\right)^2. \tag{4.80}$$

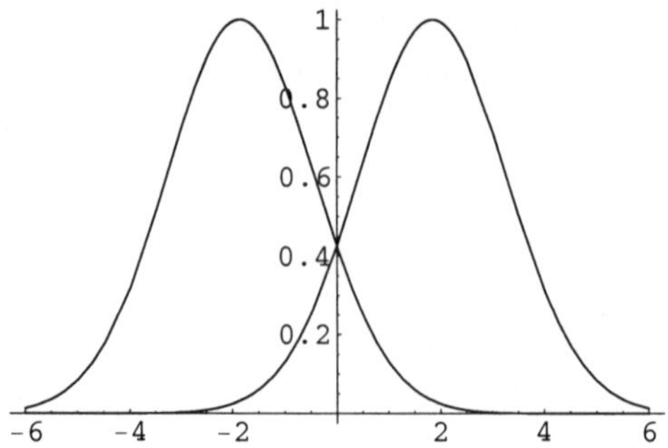

Fig. 4.7 The Rayleigh criterion of resolution.

From the Fourier transform properties, for a perfect non-turbulent atmosphere, normalizing $\widehat{S}(\vec{u})$ to unity at the origin, i.e.,

$$\widehat{S}(0) = 1. \tag{4.81}$$

This states that the Strehl ratio is proportional to the integral of OTF over frequencies. Typical ground-based observations with large telescopes in the visible wavelength range are made with a Strehl ratio ≤ 0.01 (Babcock, 1990), while a diffraction-limited telescope would, by definition, have a Strehl ratio of 1.

4.3.2 *Telescope aberrations*

Starlight collected by the telescope aperture do not image as a point, but distributes spatially in the image plane with the intensity falling off asymptotically as the inverse cube of the distance from its center, r^{-3}, in which r is the radial distance from the center. Diffraction phenomena occurring at its aperture causes Airy distribution. In practice, the aberrations of the diffracted waves are small (Mahajan, 2000); the depth of focus is determined by the amount of defocus aberration that can be tolerated. If the amplitudes of the small scale corrugations of the wavefront caused by the mirror aberrations are much smaller than the wavelength of the light, the instantaneous image of a star is sharp resembling the classical diffraction pattern taken through an ideal telescope, in which the PSF is invariant to spatial shifts. The PSF of a system with a radially symmetric pupil function behaves asymptotically as r^{-3} independent of the aberration. The centroid of the diffraction PSF is given by the slope of the imaginary part of its diffraction OTF at the origin.

The surface accuracy of a primary mirror must be specified in terms of the shortest wavelength at which the telescope is to be efficient. The reflector surface irregularities distort a plane wave into a wavefront with phase errors. If the errors are random, Gaussian, and uniformly distributed over the aperture, the surface efficiency, η_{surf}, is expressed according to Ruze (1966) as,

$$\eta_{surf} = e^{-(4\pi\sigma/\lambda)^2}, \tag{4.82}$$

where σ is the effective RMS surface errors.

It is worthwhile to mention that due to the inaccurate tracking of the telescope the star may move off the telescope axis by a certain angle, the input wavefronts are tilted by the same angle. The intensity patterns shift to a new center in the focal plane.

A Cassegrain telescope has a central obscuration, it is about 40 per cent of the full aperture. A partial obscuration of the entrance pupil may occur due to the existence of the secondary mirror and other structures in a telescope, and thus producing a deformation and an enlargement of the diffraction pattern. Let the annular aperture be bounded by two concentric circles of radii a and ϵa, in which ϵ is some positive number less than unity. The light distribution in the Fraunhofer pattern is represented by an integral of the form of equation (3.146), but with the ρ integration extending over the domain $\epsilon \leq \rho \leq a$ (Born and Wolf, 1984). Thus the equation (3.148) is recast as,

$$\widehat{U}(w) = C\pi a^2\left[\frac{2J_1(2\pi aw)}{2\pi aw}\right] - C\pi\epsilon^2 a^2\left[\frac{2J_1(2\pi a\epsilon w)}{2\pi a\epsilon w}\right], \tag{4.83}$$

thus,

$$I(w) = \frac{I(0)}{(1-\epsilon^2)^2}\left[\left(\frac{2J_1(2\pi aw)}{2\pi aw}\right) - \epsilon^2\left(\frac{2J_1(2\pi a\epsilon w)}{2\pi a\epsilon w}\right)\right]^2, \tag{4.84}$$

where $I(0) = |C|^2\pi^2 a^4(1-\epsilon^2)^2$ is the intensity at the center $w = 0$ of the pattern.

For a small ϵ, the intensity distribution is analogous, but it gets modified considerably with non-symmetrical structures; the spiders holding the secondary mirror produce long spikes on overexposed images.

Figure 4.8 displays the intensity distribution due to the diffraction effects at the focal plane of a 1 meter telescope with and without the central obscuration. The energy excluded as a function of the diaphragm radius

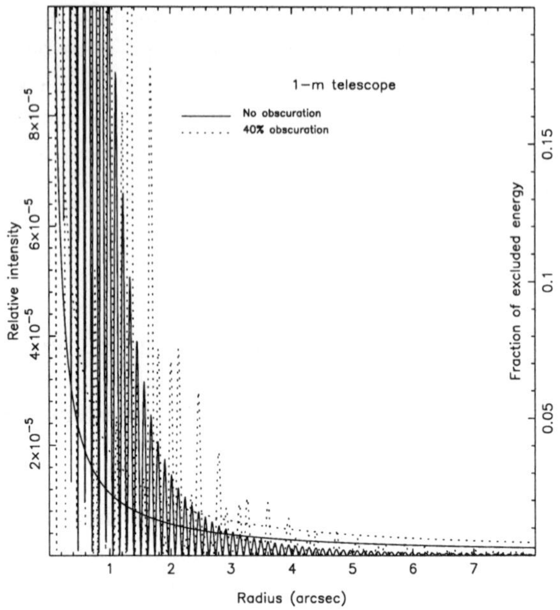

Fig. 4.8 Intensity profile and excluded energy due to diffraction at the focal plane of the 1 meter telescope, Vainu Bappu Observatory (VBO), Kavalur (Courtesy: A. V. Raveendran).

for the cases is shown as well. The main effect of the central obscuration is the redistribution of intensity in the outer rings, and thereby spreading out further in the image plane the light collected by the telescope. It is evident from this Figure (4.8) that even with an aperture of 16 arcsecond[13] diameter at the focal plane of a 1-m Cassegrain telescope, the excluded energy (the energy contained outside the aperture in the image plane) is more than 0.5 per cent. If the aperture is not exactly at the focal plane, the excluded energy becomes significantly larger than the above (Young, 1970).

[13]An arcsecond is the 1/3600 th of a degree of angle.

Chapter 5

Theory of atmospheric turbulence

5.1 Earth's atmosphere

The Earth's atmosphere is a mixture of gases and is primarily composed of nitrogen ($N_2; \sim 78\%$), oxygen ($O_2; \sim 21\%$), and argon ($Ar, \sim 1\%$). The notable other components are water ($H_2O; \sim 0-7\%$), ozone ($O_3; \sim 0-0.01\%$), and carbon dioxide ($CO_2; \sim 0.01 - 0.1\%$). The atmosphere is thickest near the surface of the Earth and thins out with height until it merges with interplanetary space; it reaches over 550 kilometers (km) from the ground. It is divided into five layers, depending on the thermal characteristics, chemical compositions, movement, and density that decays exponentially with altitude. These layers are:

(1) Troposphere: It begins at the Earth's surface and extends $\sim$ 8 to $\sim$ 14 km. Temperature falls down at the rate of $\approx 3°$ Centigrade (C) per km as one climbs up in this layer. It is generally known as the lower atmosphere. The tropopause separates it from the next layer called stratosphere.

(2) Stratosphere: It starts above the troposphere and extends to $\sim$50 km altitude. This layer is stratified in temperature, with warmer layers higher up and cooler layers farther down. The temperature increases gradually as altitude increases to a temperature of about 3° C, due to the absorption of solar ultraviolet (UV) radiation. The so called 'ozone layer' lies in this layer absorbing the longer wavelengths of the UV radiation. The stratopause separates stratosphere from the next layer called mesosphere.

(3) Mesosphere: It commences above the stratosphere and extends to $\sim$85 km, where the chemical substances are in excited state, as they absorbs energy from Sun. In this layer the temperature falls down again

as low as $\sim -90°$ C at the mesopause that separates it from the layer, called thermosphere.

(4) Thermosphere: This layer, also known as upper atmosphere, starts above the mesosphere and extends to as high as $\sim$ 600 km. The chemical reactions occur much faster than on the surface of the Earth. As the altitude increases, the temperature goes up to $\sim 1600°$ C and

(5) Exosphere: It commences at the top of the thermosphere and continues until it merges with space.

The atmosphere extends to great heights, with density declining by a factor of $e(2.72)$ over an altitude interval given by a quantity, called the scale height, H. If at a height of h the atmosphere has temperature, T, pressure, P, and density, ρ, considering cylinder with a length dh, the change in pressure dP, from the height, h to $h + dh$ is proportional to the mass of the gas in the cylinder. The equation of hydrostatic equilibrium is derived as,

$$dP = -g\rho dh. \tag{5.1}$$

with g as the acceleration due to gravity on the Earth's surface.

As a first approximation, one can assume that g does not depend on height. The error in case of the Earth is about 3%, if it is considered constant upto a height of 100 km from the surface. The equation of state for the ideal gas, $PV = Nk_BT$, in which N is the number of atoms or molecules, provides the expression for the pressure, P,

$$P = \frac{\rho k_B T}{\mu}, \tag{5.2}$$

where $\rho = \mu P/k_BT$, $k_B(= 1.38 \times 10^{-23}\text{JK}^{-1})$ the Boltzmann constant, $\mu(= 1.3\ \text{kgm}^{-3})$ is related to specific mass of air, and T the mean surface temperature; P and T are given in units of atmosphere (millibars) and degrees kelvins (K) respectively.

By using these two equations (5.1 and 5.2), one obtains,

$$\frac{dP}{P} = -\frac{\mu g}{k_B T} dh. \tag{5.3}$$

Integration of this equation (5.3) yields P as a function of height,

$$P = P_0 e^{-\int_0^h \frac{\mu g}{k_B T} dh} = P_0 e^{-\int_0^h dh/H}, \tag{5.4}$$

with $H = k_B T/(\mu g)$ as the scale height, which is a variable and has the dimension of length. It is used as a parameter in many formulae describing the structure of the atmosphere. If the change of the pressure or the density is known by a function of height, the mean molecular weight of the atmosphere can be computed. Although H is a function of height, one may consider it as constant here. With this approximation, therefore it is found,

$$-\frac{h}{H} = \log \frac{P}{P_0},$$

or using the equation of state (equation 5.2),

$$\frac{\rho T(h)}{\rho_0 T_0} = e^{-h/H}. \tag{5.5}$$

5.2 Basic formulations of atmospheric turbulence

Turbulence is caused by micro-thermal fluctuations in the atmosphere. The optical beam traversing through such turbulence is aberrated yielding in a blurred image. The resolution of conventional astro-photography is limited by the size of quasi-coherent areas (r_0) of the atmosphere. The density inhomogeneities appear to be created and maintained by the parameters such as thermal gradients, humidity fluctuations, and wind shears producing Kelvin-Helmoltz instabilities[1], which produce atmospheric turbulence and therefore refractive index inhomogeneities. The random fluctuations in the atmospheric motions occur predominantly due to

- the friction encountered by the air flow at the Earth's surface and consequent formation of a wind-velocity profile with large vertical gradients,
- differential heating of different portions of the Earth's surface by the Sun and the concomitant development of thermal convection,
- processes associated with formation of clouds involving release of heat of condensation and crystallization, and subsequent changes in the nature of temperature and wind velocity fields,
- convergence and interaction of airmasses with various atmospheric fronts, and
- obstruction of air-flows by mountain barriers that generate wave-like disturbances and rotor motions on their lee-side.

[1] Kelvin-Helmoltz instabilities are produced by shear at the interface between two fluids with different physical properties, typically different density

The gradients caused by the afore-mentioned environmental parameters warp the wavefront incident on the telescope pupil. If such distortions are the significant fraction of a wavelength across the aperture of a telescope, its resolution becomes limited.

Atmosphere is a non-stationary random process, which may be comparable to chaos[2], where the absence of predictable patterns and connections persist. The seeing conditions evolve with time, therefore, one needs to know the statistics of their evolution, mean value and standard deviation (see Appendix B) for a given telescope. In what follows, the properties of turbulence in the Earth's atmosphere, metrology of seeing, and its influence on the propagation of waves in the optical wavefield are enumerated.

5.2.1 *Turbulent flows*

Unlike the steady flows, also called laminar[3] flows, turbulent flows have a random velocity field. Reynolds formulated an approach to describe the turbulent flows using ensemble averages rather than in terms of individual components. He defined a dimensionless quantity, known as Reynolds number, which characterizes a turbulent flow. Such a quantity is obtained by equating the inertial and viscous forces, i.e.,

$$\mathrm{Re} = \frac{Lv}{\nu_v}, \tag{5.6}$$

where Re is the Reynolds number and is a function of the flow geometry, v the characteristic velocity of flow, L the characteristic size of the flow, and ν_v the kinematic viscosity of the fluid, the unit of which is m^2s^{-1}.

When the average velocity, v, of a viscous fluid of characteristic size, L, is gradually increased, two distinct states of fluid motion are observed (Tatarski, 1961, Ishimaru, 1978), viz., (i) laminar, at very low v, and (ii) unstable and random fluid motion at v greater than some critical value. Between these two extreme conditions, the flow passes through a series of unstable states. In the area of chaos theory, it is found that the final full blown turbulence may occur after a few such transitions. With high Reynolds number, the turbulence becomes chaotic in both space and time and exhibits considerable spatial structure, due to which it makes difficult to study atmosphere. Swirling water, puffs of smoke, and changing dust motion of sunlight exemplify such a chaotic condition as well. They are

[2]Chaos theory, a young branch of physics goes far beyond the neat stable mathematical model into the domain of constant change, where instability is the rule.

[3]Laminar flow is regular and smooth in space and time.

all unpredictable, random phenomena of patterns that emerge inside them. Such patterns dissolve as fast as they are created.

With the increase of Re, the velocity fluctuations increases and the inner Reynolds number corresponding to the size of fluctuations, Re_l, may exceed the certain critical Reynolds number (fixed), Re_{cr}. Therefore the 'first order' velocity fluctuations loose stability themselves and can transfer energy to new 'second order' fluctuations. As the number Re is increased further, the kinetic energy of the air motions at a given length scale is larger than the energy dissipated as heat by viscosity of the air at the same scale - the fluctuations become unstable and so on. The kinematic viscosity of air is of the order of $\nu_a = 1.5 \times 10^{-5}$ m^2 s^{-1}. With $v = 1$ m s^{-1} and $L = 15$ m, the Reynolds number turns out to be 10^6, which is considered to be turbulent.

In the standard model for atmospheric turbulence (Taylor, 1921, Kolmogorov, 1941b, 1941c), which states that energy enters the flow at scale length, L_0 and spatial frequency, $\kappa_{L_0} = 2\pi/L_0$, as a direct result of non-linearity of the Navier-Stokes equations governing fluid motion (Navier, 1823, Stokes, 1845),

$$\frac{\partial \vec{v}}{\partial t} + \vec{v}(\nabla \cdot \vec{v}) = -\frac{\nabla P}{\rho} + \nu_v \nabla^2 \vec{v}, \tag{5.7}$$

$$\nabla \cdot \vec{v}(\vec{r}, t) = 0, \tag{5.8}$$

in which

$$\nabla = \vec{i}\frac{\partial}{\partial x} + \vec{j}\frac{\partial}{\partial y} + \vec{k}\frac{\partial}{\partial z},$$

represents a vector differential operator, $P(\vec{r}, t)$ the dynamic pressure, and ρ the constant density, and $\vec{v}(\vec{r}, t)$ the velocity field, in which $\vec{r}$ is the position vector and t the time.

This forms the large-scale fluctuations, referred to as large eddies, which have size of the geometrically imposed outer scale length, L_0, generally the size of largest structure that moves with homogeneous speed. The large eddies also vary according to the local conditions, ranging from the distance to the nearest physical boundary. Measurements of outer scale length varying from 2 meters to 2 km have been reported (Colavita et al. 1987). Conan et al. (2000) derived a mean value $L_0 = 24$ m for a von Kármán spectrum from the data obtained at Cerro Paranal, Chile.

5.2.2 *Inertial subrange*

The velocity fluctuations occur on a wide range of space and time scales. The large eddies are not universal with respect to flow geometry, they are unstable and are broken up by turbulent flow and convection, spreading the scale of the inhomogeneities to smaller sizes, corresponding to a different scale length and higher spatial frequency. Energy is transported to smaller and smaller loss-less eddies until at a small enough Reynolds number. The motions at small scales would be statistically isotropic at the small scales, viscous dissipation would dominate the breakup process. In a stationary state, the energy flow from larger structures, L_0, to smaller structures must be constant. The amount of energy being injected into the largest structures must be equal to the energy that is dissipated as heat by friction in the smallest structures. Second order eddies are too unstable and may break up into smaller eddies and so on. Since the scale length associated with these eddies decreases, the Reynolds number associated with the flow defined in equation (5.6) decreases as well.

When the Reynolds number is low enough, the turbulent break up of the eddies stops and the kinetic energy of the flow is lost as heat via viscous dissipation resulting in a rapid drop in power spectral density, $\Phi(\vec{\kappa})$, for $\kappa > \kappa_0$, in which κ_0 is critical wave number, and $\vec{\kappa}$ the three dimensional (3-D) space wave numbers, κ_x, κ_y, κ_z. This imposes a highest possible spatial frequency on the flow beyond which hardly any energy is available to support turbulence (Tatarski, 1961). These changes are characterized by the inner scale length, l_0, at which viscous begins and spatial frequency, $\kappa_{l_0} = 2\pi/l_0$. This inner scale length varies from a few millimeters near the ground up to a centimeter high in the atmosphere.

In the smallest perturbations with sizes, l_0, the rate of dissipation of energy into heat is determined by the local velocity gradients in these smallest perturbations. By keeping the viscosity term which is dominant at l_0, the energy dissipated as heat, ε, is given by,

$$\varepsilon \sim \frac{\nu_v v^2}{l^2} \sim \frac{\nu_v v_0^2}{l_0^2}, \tag{5.9}$$

in which v_0 the velocity and l_0 the local spatial scale.

The unit of ε is expressed as per unit mass of the fluid per unit time, $\mathrm{m^2\,s^{-3}}$. Equation (5.9) gives rise to the scaling law, popularly known as the two-third law. Thus the energy, v^2, is given by,

$$v_0^2 \sim \varepsilon^{2/3} l_0^{2/3}. \tag{5.10}$$

Equation (5.10) states that in a turbulent flow at very high Reynolds number, the mean square difference of the velocities at two points in each fully developed turbulent flow behaves approximately as the two-third power of their distance. The law of finite energy dissipation can be envisaged in an experiment on turbulent flow where all the control parameters are kept same baring the viscosity, which is lowered as much as possible. The energy dissipation per unit mass, $d\mathcal{E}/dt$, in which $\mathcal{E}$ is the kinetic energy and t the time, behaves in a way consistent with a finite positive limit (Frisch, 1995). From the relationships, $v_0 \sim (\varepsilon l_0)^{1/3}$ and the equation (5.9), one obtains,

$$l_0 \sim \left(\frac{\nu_v^3}{\varepsilon}\right)^{1/4} \sim L_0(\mathrm{Re})^{-3/4}. \tag{5.11}$$

The distribution of turbule sizes ranges from millimeters to meters, with lifetimes varying from milliseconds to seconds. The quantity l_0 is expressed in terms of the dimensions of the largest eddies, L_0, in which $\varepsilon \sim v_{L_0}^3/L_0$. In other words, larger the Reynolds number, the smaller the size of the velocity inhomogeneities. All of the analysis to follow assumes are between these two lengths. If $\vec{r}$ is the vector between two points of the two scale lengths, the magnitude of which should be such that $l_0 < |\vec{r}| < L_0$. This is known as inertial subrange and is of fundamental importance to derive the useful predictions for turbulence within it.

It is worthwhile to note that the inertial range is the range of length scales over which energy is transferred and dissipation due to molecular viscosity is negligible. In 2-D hydrodynamic turbulence, the enstrophy (square of the vorticity) invariant, because of its stronger κ dependence compared to the energy invariant, dictates the large κ spectral behavior. The inertial range spectrum has two segments, such as the energy dominated low κ, and the enstrophy dominated high κ. The power spectrum has power law behavior over the inertial range. The inertial range kinetic energy spectrum is given by, $\mathcal{E}(\kappa) = \mathcal{C}_K\varepsilon^{2/3}\kappa^{-5/3}$, where κ is the wavenumber, ε the turbulent dissipation rate of total kinetic energy, and $\mathcal{C}_K$ the empirical Kolmogorov constant. The inertial subrange is an intermediate range of turbulence scales or wavelengths that is smaller than the energy containing eddies, but larger than the viscous eddies. In this, the net energy coming from the energy containing eddies is in equilibrium with net energy cascading to smaller eddies where it is dissipated. The small-scale fluctuations with sizes $l_0 < |\vec{r}| < L_0$, have universal statistics (scale-invariant behavior) independent of the flow geometry. This turbulence model was developed by Kolmogorov (1941a), widely known as Kolmogorov turbulence. The value of inertial subrange would be different at various locations on the site.

5.2.3 *Structure functions of the velocity field*

In turbulence theory, one uses the term, 'structure function', in place of correlation function. In the following sections the method is used to study ensemble averages rather than detailed properties, therefore the approach is to find a correlation or covariance function of refractive index. The effect of establishing a lower bound to spatial frequencies is that unknown contribution of the very low frequencies allows the variance to rise towards infinity. It appears to be a mathematical problem rather than a physical one since there are no observable consequences of such an infinite variance. Thus the structure functions are used, which do not suffer from this problem.

Structure function, $\mathcal{D}_f(\tau)$, is known to be the basic characteristic of a random process with stationary increments. Kolmogorov (1941a) states that the structure function in the inertial range for both homogeneous and isotropic random fields (see Appendix B) depends on the magnitude of $\vec{r} = |\vec{r}|$, in which $\vec{r} = \vec{\rho}' - \vec{\rho}$, as well as on the values of the rate of production or dissipation of turbulent energy ε and the rate of production or dissipation of temperature inhomogeneities η. The units of ε are $\mathrm{m^2\,s^{-3}}$ and those of η are expressed in $\mathrm{degree^2 s^{-1}}$.

The velocity structure function, $\mathcal{D}_v(\vec{r})$, due to the eddies of sizes r, i.e., $\mathcal{D}(\vec{r}) \sim v^2$ is defined as,

$$\begin{aligned}\mathcal{D}_v(\vec{r}) &= \left\langle |v(\vec{r}) - v(\vec{\rho}+\vec{r})|^2 \right\rangle, \\ &= 2\left[\left\langle v(\vec{r})^2\right\rangle - \langle v(\vec{r})v(\vec{\rho}+\vec{r})\rangle\right] \qquad l_0 \ll r \ll L_0.\end{aligned} \tag{5.12}$$

Here $\langle\,\rangle$ denotes an ensemble average over the repeated parameter ρ. Equation (5.12) expresses the variance at two points of distance $\vec{r}$ apart. The structure function for the range of values $l_0 \ll r \ll L_0$ is related to the covariance function, $\mathcal{B}_v(\vec{r})$, through

$$\mathcal{D}_v(\vec{r}) = 2[\mathcal{B}_v(\vec{0}) - \mathcal{B}_v(\vec{r})], \tag{5.13}$$

Here $\mathcal{B}_v(\vec{r}) =< v(\vec{\rho})\, v(\vec{\rho}+\vec{r}) >$ and the covariance is the 3-D FT of the spectrum, $\Phi_v(\vec{\kappa})$.

If turbulence is homogeneous, isotropic, and stationary, according to Kolmogorov, the velocity structure function can be expressed as a function of the normalized separation of two points, i.e.,

$$\mathcal{D}_v(r_1, r_2) = \alpha f\left(\frac{|r_1 - r_2|}{\beta}\right), \tag{5.14}$$

where f is the dimensionless function of a dimensionless argument, α and β are constants.

Dimensions of α are the units of velocity squared (v^2), while dimensions of β are distance (meter). Turbulent flow is governed only by the energy density, ε, and kinematic viscosity, ν_v. Combinations of ε and ν_v with right dimensions are,

$$\alpha = \nu_v^{1/2}\varepsilon^{1/2} \qquad \text{and} \qquad \beta = \nu_v^{3/4}\varepsilon^{-1/4}, \tag{5.15}$$

so,

$$\mathcal{D}_v(\mathbf{r}) = \nu_v^{1/2}\varepsilon^{1/2} f\left(\frac{|r_1 - r_2|}{\nu_v^{3/4}\varepsilon^{-1/4}}\right). \tag{5.16}$$

Above the inner scale of turbulence, l_0, according to Kolmogorov, the kinematic viscosity, ν_v plays no role in the value of the structure function. For f to be dimensionless, one must have $f(x) = x^{2/3}$, thus,

$$\mathcal{D}_v(\mathbf{r}) = \varepsilon^{2/3}\left|r_1 - r_2\right|^{2/3} \equiv \mathcal{C}_v^2\left|r_1 - r_2\right|^{2/3}. \tag{5.17}$$

in which $\mathcal{C}_v^2$ is the velocity structure constant.

By equation (5.9) $v \sim (\varepsilon r)^{1/3}$, one arrives at equation (5.12), which was derived by Kolmogorov (1941a) and Obukhov (1941) bears the name of the 'Two-Thirds law'. It is to be noted that the variations in velocity over very large distances have no effect on optical propagation. Asserting $\mathcal{D}_v(\mathbf{r})$ is a function of $\mathbf{r}$ and ε, it is written,

$$\mathcal{D}_v(\vec{r}) \propto \mathcal{C}_v^2 r^{2/3} \qquad l_0 \ll r \ll L_0. \tag{5.18}$$

5.2.4 *Kolmogorov spectrum of the velocity field*

In order to derive an expression for the spatial spectrum of turbulence, the procedure given by Tatarski (1961), runs as follows. Let the correlation function, $\mathcal{B}_f(\vec{r})$, of a locally isotropic scalar field be represented in the form,

$$\mathcal{B}_f(\vec{r}) = \int_{-\infty}^{\infty} \Phi(\vec{\kappa})\cos(\vec{\kappa}\cdot\vec{r})d\vec{\kappa}, \tag{5.19}$$

The function $\Phi(\vec{\kappa})$ is expressed in the form of $\mathcal{B}(\vec{r})$.

$$\Phi(\vec{\kappa}) = \frac{1}{(2\pi)^3}\int_{-\infty}^{\infty} \mathcal{B}_f(\vec{r})\cos(\vec{\kappa}\cdot\vec{r})d\vec{r}. \tag{5.20}$$

The function $\Phi(\vec{\kappa})$ and $\mathcal{B}_f(\vec{r})$ are Fourier transform of each other.

If the random field $f(\vec{r})$ is isotropic, the function $\mathcal{B}_f(\vec{r})$ depends on $|\vec{r}|$. Introducing spherical coordinates in equation (5.20), one obtains

$$\Phi(\vec{\kappa}) = \frac{1}{2\pi^2\kappa}\int_0^{\infty} r\mathcal{B}_f(r)\sin(\kappa r)dr, \tag{5.21}$$

where $\kappa = |\vec{\kappa}| = 2\pi/\lambda$ is the wave number.

Thus, the spectral density $\Phi(\vec{\kappa})$ is a function of the magnitude of $\vec{\kappa}$ alone, i.e., $\Phi(\vec{\kappa}) = \Phi(\kappa)$. Introducing spherical co-ordinates in the space of the vector, $\vec{\kappa}$, in equation (5.19), the following equation emerges,

$$\mathcal{B}_f(r) = \frac{4\pi}{r}\int_0^{\infty} \kappa\Phi(\kappa)\sin(\kappa r)d\kappa. \tag{5.22}$$

From the relation, $\mathcal{B}_f(r) = \int_{-\infty}^{\infty} V(\kappa)\cos(\kappa r)d\kappa$, the 1-D spectral density, $V(\kappa)$, is expressed as,

$$\begin{aligned} V(\kappa) &= \frac{1}{\pi}\int_0^{\infty} \mathcal{B}_f(r)\cos(\kappa r)dr, \\ \frac{dV(\kappa)}{d\kappa} &= -\frac{1}{\pi}\int_0^{\infty} \mathcal{B}_f(r)\sin(\kappa r)\cdot r dr. \end{aligned} \tag{5.23}$$

Comparing with equation (5.21), it is obtained,

$$\Phi(\kappa) = -\frac{1}{2\pi\kappa}\frac{dV(\kappa)}{d\kappa}. \tag{5.24}$$

It shows the relationship of 3-D spectral density $\Phi(\kappa)$ of an isotropic random field with the one-dimensional (1-D) spectral density $V(\kappa)$.

In order to study the spectrum of the velocity field in turbulent flow, the structure function, $\mathcal{D}_f(\vec{r})$, of a locally isotropic scalar field can be represented in the form,

$$\mathcal{D}_f(\vec{r}) = 2\int_{-\infty}^{\infty} \Phi_f(\vec{\kappa})\left[1 - \cos(\vec{\kappa}\cdot\vec{r})\right]d\vec{\kappa}. \tag{5.25}$$

The structure tensor, $\mathcal{D}_{i,k}(\vec{r}) = \langle(v_i - v_i')(v_k - v_k')\rangle$, in which v_i are the components with respect to x, y, z axes of the velocity vector at the point

$\vec{r}_1$ and v_i' are the components at the point $\vec{r}_1' = \vec{r}_1 + \vec{r}$, is written as,

$$\mathcal{D}_{i,k}(\vec{r}) = 2\int_{-\infty}^{\infty} \Phi_{i,k}(\vec{\kappa})\left[1 - \cos(\vec{\kappa}\cdot\vec{r})\right] d\vec{\kappa}, \tag{5.26}$$

where $\Phi_{i,k}(\vec{\kappa})$ is the spectral tensor of the velocity field and $i, k = 1, 2, 3$.

This tensor is expressed in terms of the vector κ and the unit tensor, $\delta_{i,k}$.

$$\Phi_{i,k}(\vec{\kappa})d\vec{\kappa} = \mathcal{G}(\kappa)\kappa_i\kappa_k + \mathcal{E}(\kappa)\delta_{i,k}, \tag{5.27}$$

where $\mathcal{G}(\kappa)$ and $\mathcal{E}(\kappa)$ are the scalar functions of a single argument.

It is essential to derive an expression for this energy, $\mathcal{E}(\vec{\kappa})d\vec{\kappa}$, between the spatial frequencies, $\vec{\kappa}$ and $\vec{\kappa} + d\vec{\kappa}$. This energy is proportional to the velocity squared and the spatial frequency is inversely proportional to $\vec{r}$. Therefore,

$$\Phi_{i,k}(\vec{\kappa}) = \mathcal{E}(\kappa)\left(\delta_{i,k} - \frac{\kappa_i\kappa_k}{\kappa^2}\right). \tag{5.28}$$

In order to convert the 3-D spectra, $\mathcal{E}(\vec{\kappa})$ to its one dimensional equivalent, $\mathcal{E}(\kappa)$, it is required to integrate over all directions. In the case of local isotropy, $\mathcal{E}(\kappa) = 4\pi\kappa^2\mathcal{E}(\vec{\kappa})$. Thus equation (5.20) can be recast into,

$$\mathcal{D}_{i,k}(\vec{r}) = 2\int_{-\infty}^{\infty} \mathcal{E}(\kappa)\left[1 - \cos(\vec{\kappa}\cdot\vec{r})\right]\left(\delta_{i,k} - \frac{\kappa_i\kappa_k}{\kappa^2}\right) d\vec{\kappa}. \tag{5.29}$$

In the case of the isotropic velocity field in which the correlation tensor $\mathcal{B}_{i,k}(\vec{r})$ exists as well as the structure tensor $\mathcal{D}_{i,k}(\vec{r})$. Therefore,

$$\mathcal{B}_{i,k}(\vec{r}) = \int_{-\infty}^{\infty} \mathcal{E}(\kappa)\cos(\vec{\kappa}\cdot\vec{r})\left(\delta_{i,k} - \frac{\kappa_i\kappa_k}{\kappa^2}\right) d\vec{\kappa}. \tag{5.30}$$

With $\delta_{ii} = 3$, and $\kappa_i\kappa_i = \kappa^2$,

$$\mathcal{B}_{ii}(\vec{r}) = \int_{-\infty}^{\infty} \cos(\vec{\kappa}\cdot\vec{r})2\mathcal{E}(\kappa)d\vec{\kappa}. \tag{5.31}$$

On setting $r = 0$ in equation (5.31), one gets

$$\frac{1}{2}\overline{v}'^2 = \int_{-\infty}^{\infty} \mathcal{E}(\vec{\kappa})d\vec{\kappa}. \tag{5.32}$$

By contracting equation (5.29) with respect to the indices i and k.

$$\mathcal{D}_{ii}(r) = 4\int_{-\infty}^{\infty} \mathcal{E}(\kappa)\left[1 - \cos(\vec{\kappa}\cdot\vec{r}\right] d\vec{\kappa}. \tag{5.33}$$

Using dimensional arguments and properties of the structure function, $\mathcal{D}_{ij}(r)$, one can show that,

$$\mathcal{D}_{ii}(r) = \frac{11}{3} C\varepsilon^{2/3} r^{2/3}, \tag{5.34}$$

with $\mathcal{D}_{ii}(r) = \mathcal{D}_{rr}(r) + 2\mathcal{D}_{tt}(r)$,

$$\mathcal{D}_{rr}(r) = C\varepsilon^{2/3} r^{2/3}, \qquad \mathcal{D}_{tt}(r) = \frac{4}{3} C\varepsilon^{2/3} r^{2/3}, \tag{5.35}$$

where $\mathcal{D}_{rr} = \langle (v_r - v_r')^2 \rangle$, in which v_r is the projection of the velocity at the point $\vec{r}_1$ along the direction of $\vec{r}$ and v_r' the same quantity at the point $\vec{r}_1'$, and $\mathcal{D}_{tt} = \langle (v_t - v_t')^2 \rangle$, v_t the projection of the velocity at the point $\vec{r}_1$ along some direction perpendicular to the vector $\vec{r}$ and v_t' the same quantity at the point $\vec{r}_1'$.

Therefore, the spectral density, $\mathcal{E}(\kappa)$ is derived as,

$$\mathcal{E}(\kappa) = A\varepsilon^{2/3} \kappa^{-11/3}, \tag{5.36}$$

with

$$A = \frac{11\Gamma(8/3) \sin \pi/3}{24\pi^2} C = 0.061C. \tag{5.37}$$

Equation (5.36) holds for any conserved passive additive including refractive index.

5.2.5 *Statistics of temperature fluctuations*

Turbulent flows produce temperature inhomogeneities by mixing adiabatically atmospheric layers at different temperatures. In such a case buoyancy becomes a source of atmospheric instability. The atmospheric stability may be measured by another dimensionless quantity, called Richardson number that is expressed as,

$$Ri = \frac{g}{T} \frac{\partial \bar{\theta}/\partial z}{(\partial \bar{u}/\partial z)^2}, \tag{5.38}$$

where g is the acceleration due to gravity, T the mean absolute temperature, $\partial \bar{\theta}/\partial z$ the gradient of the mean potential temperature, and $\partial \bar{u}/\partial z$ the gradient of the flow velocity.

When the term, $\partial \bar{\theta}/\partial z$, becomes negative, a parcel of air brought upward becomes warmer than surrounding air so that its upward motion would be maintained by buoyancy producing an instability, while in the reverse condition, i.e., if $\partial\bar{\theta}/\partial z$ is positive buoyancy produces stability. It is important to note that flows with $Ri > 0.25$ are stable, while flows with $Ri < 0.25$ are unstable.

The mixing of warm and cool air comes as a result of the turbulent measure of the atmosphere. But the exact pattern of the temperature distribution varies with location season and time of day. A small temperature fluctuation of one tenth of a degree would generate strong wavefront perturbations over a propagation distance of a few hundred meters. Naturally occurring variations in temperature ($< 1°$C) cause random changes in the wind velocity (eddies). Further, the changes in temperature give rise to small changes in atmospheric density and, hence, to the refractive index. These index changes of the order of 10^{-6} can accumulate. The cumulative effect can cause significant inhomogeneities in the index profile of the atmosphere.

The temperature structure function is,

$$\mathcal{D}_T(\vec{r}) = \left\langle |T(\vec{r}) - T(\vec{\rho}+\vec{r})|^2 \right\rangle. \tag{5.39}$$

The relationship between the structure function and the covariance,

$$\mathcal{D}_T(\vec{r}) = 2[\mathcal{B}_T(\vec{0}) - \mathcal{B}_T(\vec{r})]. \tag{5.40}$$

The expression $\mathcal{D}_T(\vec{r})$ is finite as long as $|\vec{r}|$ finite. It is to reiterate that according to Kolmogorov theory, the structure function depends on $r = |\vec{r}|$ for both homogeneous and isotropic field and on the values of ε and η. It follows from the simple dimension considerations that,

$$\mathcal{D}_T(\vec{r}) \propto \eta\varepsilon^{-1/3}r^{2/3}, \qquad \text{or} \tag{5.41}$$

$$\mathcal{D}_T(\vec{r}) \propto \mathcal{C}_T^2 r^{2/3}, \tag{5.42}$$

where $\mathcal{C}_T^2$ is known as the temperature structure constant. It is a measure of the local intensity of the temperature fluctuations. This is related to the average structure of the flow.

$$\mathcal{C}_T^2 = \alpha L_0^{4/3}\left(\frac{\partial\bar{\theta}}{\partial z}\right)^2 f(Ri), \tag{5.43}$$

in which α is the constant, L_0 the turbulence outerscale, $f(Ri)$ the function of the Richardson number.

5.2.6 *Refractive index fluctuations*

Difference in the refractive index of atmosphere through which the light propagates aberrates the wavefront; the velocity of air flow does not affect directly the light path. However, the velocity distribution is needed to couple with refractive index distribution via the variations of temperature, density, as well as water vapor (Tatarski, 1961). Water vapour above an observatory site is the major absorbent of infrared radiation. Aerosols absorb in the entire optical region, and also add to the background emission in the infrared. They are tiny particles, such as dust particles, rain drops, ice crystals etc., suspended in the atmosphere; the large particles can extinguish light through scattering and absorption. Information on ambient temperature and relative humidity can be used to obtain the surface water vapour pressure. This value, in conjunction with measured water vapour column, provides an estimate of water vapour scale height. These statistics are of interest towards an astronomical site characterization.

In order to determine the refractive index structure function, the idea of a 'conserved passive additive' that is introduced by Tatarski (1961) is required. A passive additive is a quantity, which does not affect the dynamics of the flow. It does not disappear through some chemical reaction in the flow. It was asserted that if the fluids in the atmosphere contain an irregular distribution of heat, the nature of turbulent flow results in a temperature structure function with a two-thirds power dependence on separation. Since there is no pressure induced variation in density within a small region, it follows that the density depends on the inverse of the absolute temperature.

The refractive index, $n(\vec{r})$, values fluctuate with time, t, due to the fluctuations in temperature, T, and pressure, P; the mean values of such meteorological variables change over minutes to hours. Dealing with small fluctuations in the absolute temperature, and since density and therefore refractive index are inversely proportional to temperature,

$$\frac{\partial n}{\partial T} = 80 \times 10^{-6} \frac{P}{T^2}. \tag{5.44}$$

The corresponding structure function of the refractive index, $\mathcal{D}_n$, is computed as,

$$\mathcal{D}_n(\vec{r}) = \left(\frac{\partial n}{\partial T}\right)^2 \mathcal{D}_T(\vec{r}). \tag{5.45}$$

From equation (5.44), one finds that the refractive index structure constant, $\mathcal{C}_n^2$ is related to the temperature structure constant as well,

$$\mathcal{C}_n^2 = \left[\frac{80 \times 10^{-6} P}{T^2}\right]^2 \mathcal{C}_T^2. \tag{5.46}$$

The quantity, $\mathcal{C}_n^2$, is called the structure constant of the refractive index fluctuations. Such a constant is a measure of the strength of the turbulence and has units of $\mathrm{m}^{-2/3}$, the value of which is always positive, and it varies with height. It characterizes the strength of the refractive index fluctuations. The structure function is given by,

$$\begin{aligned} \mathcal{D}_n(\vec{r}) &= \left\langle |n(\vec{r}) - n(\vec{\rho} + \vec{r})|^2 \right\rangle \\ &= 2[\mathcal{B}_n(\vec{0}) - \mathcal{B}_n(\vec{r})], \qquad l_0 \ll r \ll L_0. \end{aligned} \tag{5.47}$$

Therefore in terms of refractive index structure constant, $\mathcal{C}_n^2$, the refractive index structure function, $\mathcal{D}_n(\vec{r})$, is defined as,

$$\mathcal{D}_n(\vec{r}) = \mathcal{C}_n^2 r^{2/3}, \qquad l_0 \ll r \ll L_0, \tag{5.48}$$

The function $\Phi_n(\vec{\kappa})$ is the spectral density of the structure function $\mathcal{D}_n(\vec{r})$,

$$\begin{aligned} \mathcal{D}_n(\vec{r}) &= 2\int_{-\infty}^{\infty} (1 - \cos\vec{\kappa} \cdot \vec{r}) \Phi_n(\vec{\kappa}) d\vec{\kappa} \\ &= 8\pi \int_0^{\infty} \Phi_n(\kappa)\kappa^2 \left[1 - \frac{\sin \kappa r}{\kappa r}\right] d\kappa, \end{aligned} \tag{5.49}$$

where $\Phi_n(\kappa)$ is the spectral density in 3-D space wave numbers of the distribution of the amount of inhomogeneity in a unit volume.

The form of this function corresponding to two-third law for the concentration $\vec{n}$. By noting,

$$\int_0^{\infty} x^a \left[1 - \frac{\sin bx}{bx}\right] dx = -\frac{\Gamma(a)\sin(\pi a/2)}{b^{a+1}}, \qquad -3 < a < -1.$$

The power spectral density for the wave number, $\kappa > \kappa_0$, in the case of inertial subrange, can be equated as,

$$\Phi_n(\kappa) = \frac{\Gamma(8/3)\sin\pi/3}{4\pi^2} \mathcal{C}_n^2 \kappa^{-11/3}, \tag{5.50}$$

and

$$\Phi_n(\kappa) = \begin{cases} 0.033C_n^2\kappa^{-11/3} & \kappa < \kappa_0 \\ 0 & \kappa > \kappa_0, \end{cases} \tag{5.51}$$

where $\kappa = (\kappa_x, \kappa_y, \kappa_z)$ is the scalar wavenumber vector, $\kappa_0 \sim 1/L_0$, and L_0 is the outer scale of turbulence.

This spectrum for refractive index changes for a given structure constant and is valid within the inertial subrange, $(\kappa_{L_0} < \kappa < \kappa_{l_0})$. This model describing the power-law spectrum for the inertial intervals of wave numbers, known as Kolmogorov-Obukhov model of turbulence, is widely used for astronomical purposes (Tatarski, 1993). Owing to the non-integrable pole at $\kappa = 0$, mathematical problems arise to use this equation for modeling the spectrum of the refractive index fluctuations when, $\kappa \to 0$. The integral over $\Phi_n(\kappa) \propto \kappa^{-11/3}$ is infinite, i.e., the variance of the turbulent phase infinite. This is a well known property of Kolmogorov turbulence of the atmosphere. Since the Kolmogorov spectrum is not defined outside the inertial range, for a finite outer scale, the von Kármán spectrum can be used to perform the finite variance (Ishimaru, 1978). In order to accommodate the finite inner and outer scales, the power spectrum can be written as,

$$\Phi_n(\kappa) = 0.033C_n^2\left(\kappa^2 + \kappa_0^2\right)^{-11/6} e^{-\kappa^2/\kappa_i^2}, \tag{5.52}$$

with $\kappa_0 = 2\pi/L_0$ and $\kappa_i = 5.9/l_0$.

The root-mean-square (RMS) fluctuation of the difference between the refractive index at any two points in earth's atmosphere is often approximated as a power law of the separation between the points. The structure functions of the refractive index and phase fluctuations are the main characteristics of light propagation through the turbulent atmosphere, influencing the performance of the imaging system. The quantity C_n is a function of altitude and consequently depends on length, z, along the path of propagation, which may vary. The refractive index n is a function of $n(T, \mathrm{H})$ of the temperature, T and humidity, H. and therefore, the expectation value of the variance of the fluctuations about the average of the refractive index is given by,

$$\langle dn\rangle^2 = \left(\frac{\partial n}{\partial T}\right)^2 \langle dT\rangle^2 + 2\left(\frac{\partial n}{\partial T}\right)\left(\frac{\partial n}{\partial \mathrm{H}}\right)\langle dT\rangle\langle d\mathrm{H}\rangle + \left(\frac{\partial n}{\partial \mathrm{H}}\right)^2 \langle d\mathrm{H}\rangle^2. \tag{5.53}$$

It has been argued that in optical propagation, the last term is negligible, and that the second term is negligible for most astronomical observations. It could be significant, however, in high humidity situation, e.g., a marine boundary layer (Roddier, 1981). Most treatments ignore the contribution from humidity and express the refractive index structure function (Tatarski, 1961) as in equation (5.48).

As the temperature T, and humidity H, are both functions of height in the atmosphere, turbulent mixing creates inhomogeneities of temperature and humidity at scales comparable to eddy size. It has been argued that, to a good approximation, the power spectra of the temperature and humidity fluctuations are Kolmogorovian as well (Roddier, 1981). Thus the optically important property of turbulence is that the fluctuations in temperature and humidity, and therefore refractive index, are largest for the largest turbulent elements, upto the outer scale of turbulence. The spatial power spectrum $\Phi_T(\kappa)$ of temperature fluctuations and the power spectrum $\Phi_H(\kappa)$ of humidity fluctuations as functions of the wave number $\vec{\kappa}$ are described by,

$$\Phi_T(\kappa) \propto \kappa^{-5/3}, \qquad \Phi_H(\kappa) \propto \kappa^{-5/3}. \tag{5.54}$$

Equation (5.54) states that for the turbulent elements of sizes below the outer scale, the one-dimensional (1-D) power spectrum of the refractive index fluctuations falls off with (-5/3) power of frequency and is independent of the direction along the fluctuations are measured, i.e., the small-scale fluctuations are isotropic (Young, 1974). In the isotropic case, the 3-D power spectrum, Φ_n, for the wave number, $\kappa > \kappa_0$, in the case of inertial subrange, can be obtained by integrating over all direction, i.e.,

$$\Phi_T(\vec{\kappa}) \propto \kappa^{-11/3}, \qquad \Phi_H(\vec{\kappa}) \propto \kappa^{-11/3}. \tag{5.55}$$

In the inertial range, it is also given by,

$$\Phi_T(\vec{\kappa}) = 0.033 C_T^2 \kappa^{-11/3}. \tag{5.56}$$

Similarly, for humidity fluctuations one gets the power spectrum, $\Phi_H(\kappa)$,

$$\Phi_H(\vec{\kappa}) = 0.033 C_H^2 \kappa^{-11/3}. \tag{5.57}$$

It is to be noted here that the temperature structure constant, C_T^2, is proportional to the local vertical temperature gradient but is not related to the velocity. In the presence of large wind velocity, C_T^2 can be negligible, that is, the two structure constants are not strongly related.

5.2.7 *Experimental validation of structure constants*

Several experiments confirm this two-thirds power law in the atmosphere (Wyngaard et al., 1971, Coulman, 1974, Lopez, 1991). Robbe et al. (1997) reported from observations using a long baseline optical interferometer (LBOI) called Interféromètre à deux Télescopes (I2T; Labeyrie, 1975) that most of the measured temporal spectra of the angle of arrival exhibit a behavior compatible with the said power law.

The structure constants are thought to be a function of height above the ground, h (the altitude and all units of measure are in MKS units), and are constant within a given layer of the atmosphere. Various techniques, namely micro-thermal studies, radar and acoustic soundings, balloon and aircraft experiments have been used to measure the values of $\mathcal{C}_T^2$ (Tsvang, 1969, Lawrence et al. 1970, Coulman, 1974).

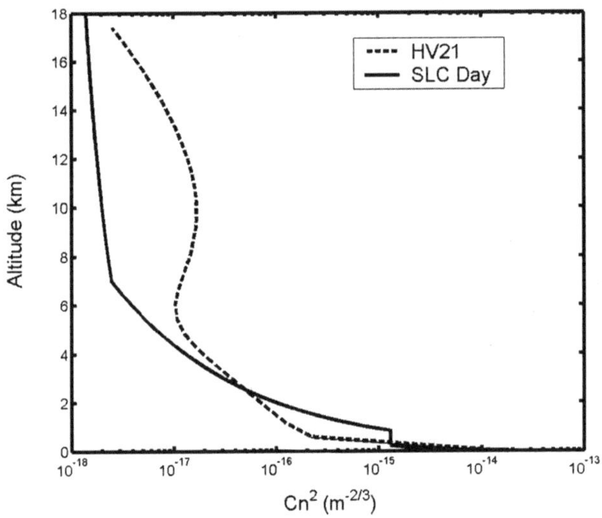

Fig. 5.1 Variations of $\mathcal{C}_n^2$ with altitude.

The numerical evaluation of the critical parameters requires the knowledge of the refractive index structure constant, $\mathcal{C}_n^2$, and wind profiles as a function of altitude. The behavior of such refractive index structure constant, $\mathcal{C}_n^2$, with height depends on both local conditions such as local terrain, the telescope dome etc., as well as on the planetary boundary layer. Since most of the above parameters are directly or indirectly related with $\mathcal{C}_n^2$, therefore for a particular optical path it needs to be modeled. The widely preferred two models are: (i) Submarine Laser Communication (SLC)-Day turbulence model and (ii) Hufnagel-Valley model (Hufnagel, 1974, Valley, 1980). The former is described as,

$$\begin{aligned} \mathcal{C}_n^2(h) &= 0 && : 0\ m < h < 19\ m, \\ &= 4.008 \times 10^{-13} h^{-1.504} && : 19\ m < h < 230\ m, \\ &= 1.300 \times 10^{-15} && : 230\ m < h < 850\ m, \\ &= 6.352 \times 10^{-7} h^{-2.966} && : 850\ m < h < 7{,}000\ m, \\ &= 6.209 \times 10^{-16} h^{-0.6229} && : 7000\ m < h < 20{,}000\ m, \end{aligned} \tag{5.58}$$

while the latter is described as:

$$\mathcal{C}_n^2(h) = 2.2 \times 10^{-23} h^{10} e^{-h} + 10^{-16} e^{-h/1.5} + A e^{-h/0.1}, \tag{5.59}$$

in which the parameter A is normally set to $1.7 \times 10^{-14}\mathrm{m}^{-2/3}$.

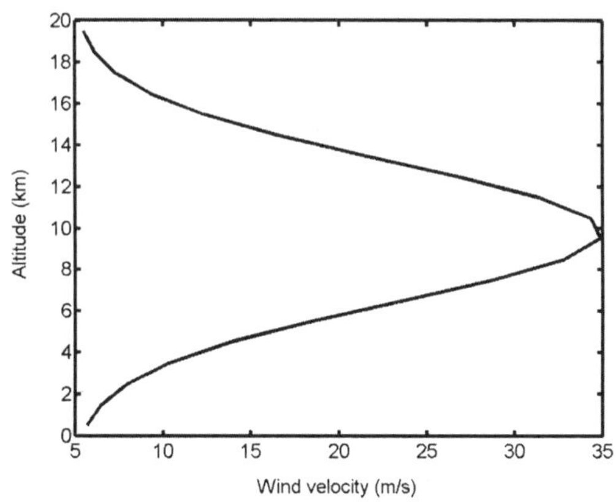

Fig. 5.2 Variations of wind velocity with altitude.

The profile of the variations of $\mathcal{C}_n^2$ with altitude is displayed in Figure (5.1). As the profile varies from site to site and from time to time, this model described in equation (5.59) may provide a rough estimate of the layer structure. The wind velocity profile most often applied to turbulence problems is the Burfton model,

$$v(h) = v_g + 30 e^{-[(h - 9400)/4800]^2}, \tag{5.60}$$

where v_g is the ground wind speed parameter, usually $v_g = 5\ m/sec$.

The function described by equation (5.60) is illustrated in Figure (5.2). The significant scale lengths, in the case of local terrain, depend on the local objects that introduce changes primarily in the inertial subrange, and temperature differentials. Such scale lengths in this zone depend on the nearby objects while the refractive index structure constant, $C_n^2 \propto h^{-2/3}$.

In real turbulent flows, turbulence is usually generated at solid boundaries. Near the boundaries, shear is the dominant source (Nelkin, 2000), where scale lengths are roughly constant. In an experiment, conducted by Cadot et al. (1997), it was found that Kolmogorov scaling is a good approximation for the energy dissipation, as well as for the torque due to viscous stress. They measured the energy dissipation and the torque for circular Couette flow with and without small vanes attached to the cylinders to break up the boundary layer. The theory of the turbulent flow in the neighborhood of a flat surface applies to the atmospheric surface layer (a few meters above the ground). Schematically there are three classes of situations in the atmosphere such as small-scale, medium-scale, and large-scale perturbations. These situations can be generalized as follows:

(1) The small-scale perturbations occur in the lowest part of the atmosphere, under surface boundary layer due to the ground convection, extending up to a few kilometer (km) height of the atmosphere, where shear is the dominant source of turbulence (scale lengths are roughly constant and the temperature structure constant, $C_T^2 \propto h^{-2/3}$). There are small turbulent cells ranging from 0.05 to 0.3 m in size that are produced by the temperature difference between the ground and the air and small-scale irregularities of the ground. During day-time, this layer turns out to be as high as 1 km, while during night-time it becomes as low as a few meters. In this region, the temperature increases up to a limit, known as inversion layer.

(2) The medium-scale perturbation zone is known to be at heights above inversion layer above which the temperature decreases, and convection dominates. The turbulence with dimensions ranging from a few tens of meters to several kilometers occur essentially between 1 km and 10 km in altitude. They are produced by ascending currents, convection or non-laminar winds fostered by the existence of large topographic features like mountains, hills or valleys. Wind plays a role in carrying them along and responsible for variations of refraction with periods ranging from a few seconds to a few tens of seconds. The free convection

layer associated with afore-mentioned orographic disturbances, where the scale lengths are height dependent, ($C_T^2 \propto h^{-4/3}$). The turbulence concentrates into a thin layer of 100-200 m thickness, where the value of C_n^2 increases by more than an order of magnitude over its background level. It reaches a minimum value of the order of 10^{-17} $m^{-2/3}$ around 6-9 km with slight increase to a secondary maximum near the tropopause and decreases further in the stratosphere (Roddier, 1981). Masciadri et al. (1999) have noticed that the value of C_n^2 increases about 11 km over Mt. Paranal, Chile. According to them the value of orographic disturbances also play an important role, while the behaviour of the former is independent of the location (Barletti et al., 1976).

(3) The large-scale perturbations originate in the tropopause around 10 to 15 km high where wind shears may develop and create systematic pressure gradients with strong turbulence as the temperature gradient vanishes slowly. The refractive index at such heights is smaller than 1.0001, hence the turbulence gives marginal effects on images. But a systematic horizontal pressure gradient modifies the refraction as a function of the direction from the observed on the ground. This provides an unmodelled contribution to refraction causing errors in the evaluation of the measured position of a star. The time evolution of turbulence at such heights can be followed either by employing radar sounding or by analyzing stellar scintillation.

The turbulence which reaches a minimum just after the sunrise and steeply increases until afternoon, is primarily due to the solar heating of the ground (Hess, 1959). It decreases to a secondary minimum after sunset and slightly increases during night. At 12 m above the ground the typical values of the refractive index constant, C_n^2, are found to be 10^{-13} $m^{-2/3}$, during day-time and 10^{-14} $m^{-2/3}$, during night time (Kallistratova and Timanovskiy, 1971). These values are height-dependent; (i) $z^{-4/3}$, under unstable daytime conditions, (ii) $z^{-2/3}$, under neutral conditions, and (iii) a slow decrease under stable condition, say during night time (Wyngaard et al., 1971).

5.3 Statistical properties of the propagated wave through turbulence

Propagation theory through a turbulent atmosphere is complex. Among other things, it includes a statistical description of the properties of the atmosphere and their effects on the statistics of amplitude and phase of the incident waves. At a first approximation, one assumes the turbulence of the atmosphere to be stationary, which may allow to derive some kind of statistical behavior, although the actual behavior may strongly deviate from that of mean atmosphere.

Spatial correlational properties of the turbulence-induced field perturbations, are evaluated by combining the basic turbulence theory with the stratification and phase screen approximations. The variance of the ray can be translated into a variance of the phase fluctuations. For calculating the same, Roddier (1981) used the correlation properties for propagation through a single (thin) turbulence layer and then extended the procedure to account for many such layers. The layer is non-absorbing and its statistical properties depend on the altitude. It is assumed that the refractive index fluctuations between the individual layers are statistically independent (Tatarski, 1961). Several investigators (Goodman, 1985, Troxel et al., 1994) have argued that individual layers can be treated as independent provided the separation of the layer centers is chosen large enough so that the fluctuations of the log amplitude and phase introduced by different layers are uncorrelated. The method set out by Roddier (1981) for the wave propagation through the atmosphere runs as follows.

Let a horizontal monochromatic plane wave of wavelength λ be propagating from a distant star at zenith towards a ground based observer. Each point of the atmosphere is designated by a horizontal coordinate vector $\vec{x}$ and an altitude, h, above the ground. The scalar vibration located at coordinates $(\vec{x}, h)$ is described by its complex disturbance, $U_h(\vec{x})$,

$$U_h(\vec{x}) = |U_h(\vec{x})|e^{i\psi_h(\vec{x})}, \tag{5.61}$$

At each altitude h, the phase fluctuation of the wavefront, $\psi_h(\vec{x})$, is referred to its average value so that for any h, $\langle\psi_h(\vec{x})\rangle = 0$. In addition, the unperturbed complex disturbance outside the atmosphere is assumed to be unity, so that $U_\infty(\vec{x}) = 1$.

5.3.1 *Contribution of a thin layer*

Let a layer of turbulent air of thickness be δh and height h above the ground. Here δh is chosen to be large compared to the scale of the turbulent eddies, but small enough for the phase screen approximation (diffraction effects to be negligible over the distance, δh). When this wave is allowed to pass through such a thin layer, the complex disturbance of the plane wavefront after passing through the layer is expressed as,

$$U_h(\vec{x}) = e^{i\psi_h(\vec{x})}. \tag{5.62}$$

The complex disturbance is assumed to be unity at the layer input. The phase-shift, $\psi(\vec{x})$, introduced by the refractive index fluctuations, $n(x, z)$ inside the layer is given by,

$$\psi_h(\vec{x}) = \kappa \int_h^{h+\delta h} n(\vec{x}, z) dz, \tag{5.63}$$

where $\kappa = 2\pi/\lambda$ and z is a variable defining length along the path of propagation.

If a point source is observed through a telescope, the turbulence limited point spread function (PSF) can be obtained by computing the Fourier integral of the coherence function in the telescope pupil. One of the main tasks of the turbulence theory is to connect the atmospheric properties to such coherence function, and thus to its Fourier transform, known as PSF, in the telescope focal plane. Considering the rest of the atmosphere calm and homogeneous, the second order moment of the complex random field at layer output, $U_h(\vec{x})$, is the coherence function,

$$\mathcal{B}_h\left(\vec{\xi}\right) = \left\langle U_h(\vec{x}) U_h^*\left(\vec{x} + \vec{\xi}\right)\right\rangle = \left\langle e^{i\left[\psi(\vec{x}) - \psi\left(\vec{x} + \vec{\xi}\right)\right]}\right\rangle. \tag{5.64}$$

The term, $\psi(\vec{x})$, is considered to be sum of the large number of independent variables (the refractive indices $n(\vec{x}, z)$ due to the equation 5.63). In the case of the thicker layers than the individual turbulent cells, many independent variables may contribute phase shift and according to the central-limit theorem, $\phi(\vec{x})$ follow Gaussian statistics, i.e.,

$$\left\langle e^{izv}\right\rangle = \int_{-\infty}^{\infty} e^{izx} \mathcal{P}_v(x) dx = e^{-\frac{1}{2}\left\langle v^2\right\rangle z^2}, \tag{5.65}$$

in which $\mathcal{P}_v(x)$ denotes the Gaussian (or normal) distribution (see Appendix B) of the random variable v.

Roddier (1981) pointed out the similarity of equation (5.64) to the Fourier transform of the probability density function, of the expression in square brackets at unit frequency. By considering v as the Gaussian distributed phase difference, $\psi(\vec{x}) - \psi(\vec{x} + \vec{\xi})$ and z equals unity, the expression for the coherence function can be recast into,

$$\mathcal{B}_h\left(\vec{\xi}\right) = e^{-\frac{1}{2}\left\langle\left|\psi(\vec{x}) - \psi\left(\vec{x}+\vec{\xi}\right)\right|^2\right\rangle}. \tag{5.66}$$

The quantity, $\psi(\vec{x}) - \psi(\vec{x}+\vec{\xi})$, has Gaussian statistics with zero mean. The expression in square brackets in the equation (5.64) is considered to be Gaussian as well. Fried (1966) had introduced the two dimensional (2-D) horizontal structure function, $\mathcal{D}_\psi\left(\vec{\xi}\right)$, of the phase, $\psi(\vec{x})$,

$$\mathcal{B}_h\left(\vec{\xi}\right) = e^{-\frac{1}{2}\mathcal{D}_\psi\left(\vec{\xi}\right)}. \tag{5.67}$$

Hence, the structure function, $\mathcal{D}_\psi\left(\vec{\xi}\right)$, for the phase fluctuations is defined as,

$$\mathcal{D}_\psi\left(\vec{\xi}\right) = \left\langle|\psi(\vec{x}) - \psi\left(\vec{x}+\vec{\xi}\right)|^2\right\rangle. \tag{5.68}$$

5.3.2 *Computation of phase structure function*

The coherence function, $\mathcal{B}_h(\vec{\xi})$, is seen to be a function of the phase structure function, which is dependent on the refractive index fluctuation. Let the covariance of phase, $\mathcal{B}_\psi(\vec{\xi})$, be defined as,

$$\mathcal{B}_\psi\left(\vec{\xi}\right) = \left\langle\psi\left(\vec{x}\right)\psi\left(\vec{x}+\vec{\xi}\right)\right\rangle \tag{5.69}$$

$$= \kappa^2\int_h^{h+\delta h} dz \int_h^{h+\delta h}\left\langle n(\vec{x},z)n\left(\vec{x}+\vec{\xi},z'\right)\right\rangle dz'. \tag{5.70}$$

The value of $\psi(\vec{x})$ is replaced from equation (5.63) into equation (5.69) which can be recast into,

$$\mathcal{B}_\psi\left(\vec{\xi}\right) = \kappa^2\int_h^{h+\delta h} dz \int_{h-z}^{h+\delta h-z} \mathcal{B}_n\left(\vec{\xi},\zeta\right) d\zeta, \tag{5.71}$$

in which $\zeta = z' - z$ and the 3-D refractive index covariance is,

$$\mathcal{B}_n\left(\vec{\xi},\zeta\right) = \left\langle n\left(\vec{x},z\right)n\left(\vec{x}+\vec{\xi},z'\right)\right\rangle. \tag{5.72}$$

Since the thickness of the layer, δh, is large compared to the correlation scale of the fluctuations, the integration over ζ can be extended from $-\infty$

to ∞, which leads to,

$$\mathcal{B}_\psi\left(\vec{\xi}\right) = \kappa^2 \delta h \int_{-\infty}^{\infty} \mathcal{B}_n\left(\vec{\xi}, \zeta\right) d\zeta. \tag{5.73}$$

The phase structure function, $\mathcal{D}_\psi(\vec{\xi})$, is related to the phase covariance, $\mathcal{B}_\psi(\vec{\xi})$, by,

$$\mathcal{D}_\psi\left(\vec{\xi}\right) = 2\left[\mathcal{B}_\psi(\vec{0}) - \mathcal{B}_\psi(\vec{\xi})\right]. \tag{5.74}$$

By taking equation (5.74) into account, one obtains

$$\begin{aligned}
\mathcal{D}_\psi\left(\vec{\xi}\right) &= 2\kappa^2 \delta h \int_{-\infty}^{\infty} \left[\mathcal{B}_n\left(\vec{0}, \zeta\right) - \mathcal{B}_n\left(\vec{\xi}, \zeta\right)\right] d\zeta \\
&= 2\kappa^2 \delta h \int_{-\infty}^{\infty} \left[\left(\mathcal{B}_n(\vec{0}, 0) - \mathcal{B}_n\left(\vec{\xi}, \zeta\right)\right) - \left(\mathcal{B}_n(\vec{0}, 0) - B_n\left(\vec{0}, \zeta\right)\right)\right] d\zeta \\
&= \kappa^2 \delta h \int_{-\infty}^{\infty} \left[\mathcal{D}_n\left(\vec{\xi}, \zeta\right) - \mathcal{D}_n\left(\vec{0}, \zeta\right)\right] d\zeta,
\end{aligned} \tag{5.75}$$

with

$$\mathcal{D}_n\left(\vec{\xi}, \zeta\right) = 2\left[\mathcal{B}_n\left(\vec{0}, 0\right) - \mathcal{B}_n\left(\vec{\xi}, \zeta\right)\right], \tag{5.76}$$

as the refractive index structure function.

The scale of the phase perturbations fails exactly in the range where $\mathcal{D}_n(r) = \mathcal{C}_n^2 r^{2/3}$ is valid. The refractive index structure function defined in equation (5.48) is evaluated as,

$$\mathcal{D}_n\left(\vec{\xi}, \zeta\right) = \mathcal{C}_n^2\left(\xi^2 + \zeta^2\right)^{1/3}. \tag{5.77}$$

With $\xi = |\vec{\xi}|$ and with the help of equation (5.74), one obtains,

$$\begin{aligned}
\mathcal{D}_\psi\left(\vec{\xi}\right) &= 2\kappa^2 \mathcal{C}_n^2 \delta h \int_{-\infty}^{\infty} \left[\left(\vec{\xi}^2 + \zeta^2\right)^{1/3} - \zeta^{2/3}\right] d\zeta \\
&= \frac{2\Gamma(1/2)\Gamma(1/6)}{5\Gamma(2/3)} \kappa^2 \mathcal{C}_n^2 \xi^{5/3} \delta h.
\end{aligned} \tag{5.78}$$

The structure function of phase fluctuations due to Kolmogorov turbulence in a layer of thickness, δh is thus obtained as,

$$\mathcal{D}_\psi\left(\vec{\xi}\right) = 2.914 \kappa^2 \mathcal{C}_n^2 \xi^{5/3} \delta h. \tag{5.79}$$

The covariance of the phase is deduced by substituting equation (5.79) in equation (5.67),

$$\mathcal{B}_h\left(\vec{\xi}\right) = e^{-\frac{1}{2}\left[2.914\kappa^2 C_n^2 \xi^{5/3}\delta h\right]}. \tag{5.80}$$

5.3.3 *Effect of Fresnel diffraction*

The covariance of the phase at the ground level due to thin layer of turbulence at some height off the ground can be derived by employing Fresnel diffraction since the optical wavelengths are much smaller than the scale of observed wavefront perturbations. At ground level, the complex field $U_0(\vec{x})$ is the field diffracted from the layer output. In terms of Fourier optics, the diffracted field is expressed as a 2-D convolution,

$$U_0\left(\vec{x}\right) = U_h\left(\vec{x}\right) \star \frac{e^{i\pi\vec{x}^2/\lambda h}}{i\lambda h}, \tag{5.81}$$

with respect to the variable $\vec{x}$. Here $\star$ denotes convolution parameter, which is the impulse response for Fresnel diffraction.

The Fourier transform of the convolution operator is the transfer function for Fresnel diffraction. The power spectrum as well as the coherence of the complex field are invariant under Fresnel diffraction. The coherence function, $\mathcal{B}_0(\vec{\xi})$, at the ground level is given by,

$$\mathcal{B}_0\left(\vec{\xi}\right) = \left\langle U_0\left(\vec{x}\right) U_0^*\left(\vec{x}+\vec{\xi}\right)\right\rangle. \tag{5.82}$$

By putting equation (5.81) into this equation (5.82), one may write,

$$\begin{aligned}\mathcal{B}_0\left(\vec{\xi}\right) &= \left\langle U_h\left(\vec{x}\right) U_h^*\left(\vec{x}+\vec{\xi}\right)\right\rangle \star \frac{e^{i\pi x^2/\lambda h}}{i\lambda h} \star -\frac{e^{-i\pi x^2/\lambda h}}{i\lambda h} \\ &= e^{-\frac{1}{2}\mathcal{D}_\psi\left(\vec{\xi}\right)} = \mathcal{B}_h\left(\vec{\xi}\right),\end{aligned} \tag{5.83}$$

where the Fourier transform of

$$\frac{e^{i\pi x^2/\lambda h}}{i\lambda h} \star -\frac{e^{-i\pi x^2/\lambda h}}{i\lambda h} = \delta\left(x\right), \tag{5.84}$$

is the Dirac delta function.

Thus equation (5.83) shows that the coherence of the complex field at ground level is the same as that of the complex field at the output of the turbulent layer. For high altitude layers, the complex field fluctuates both in phase and in amplitude (scintillation). Therefore the wave structure function, $\mathcal{D}_{\psi}(\vec{\xi})$, is not strictly true as at the ground level. The turbulence layer acts as a diffraction screen; however, correction in the case of astronomical observation remains small (Roddier, 1981).

5.3.4 *Contribution of multiple turbulent layers*

According to Roddier (1981), the fluctuations produced by several layers are the sum of the fluctuations that each layer would produce. Statistically they are independent, hence their power spectrum is the sum of the power spectra of the fluctuations as well. Let the atmosphere be divided into a series of layers of thickness, δ_j along the propagation path. The thickness is large enough so that, to a good approximation, the fluctuations of the log amplitude and phase introduced by different layers are not correlated. Let the turbulence be located in a number of thin layers between altitudes h_j and $h_j + \delta h_j$. The complex disturbance, U_{h_j}, at the output of the layer j is related to the complex disturbances, $U_{h_j+\delta h_j}$, at the input by,

$$U_{h_j}(\vec{x}) = U_{h_j+\delta h_j}(\vec{x})\, e^{i\psi_j(\vec{x})}, \tag{5.85}$$

where $\psi_j(\vec{x})$ is the phase fluctuation introduced by layer j.

Since ψ_j is statistically independent of $U_{h_j+\delta h_j}$, the field coherence at the output is related to the coherence at the input by,

$$\begin{aligned}\mathcal{B}_{h_j}(\vec{\xi}) &= \left\langle U_{h_j}(\vec{x})\, U^*_{h_j}\left(\vec{x}+\vec{\xi}\right)\right\rangle \\ &= \left\langle U_{h_j+\delta h_j}(\vec{x})\, U^*_{h_j+\delta h_j}\left(\vec{x}+\vec{\xi}\right)\right\rangle \left\langle e^{i\left[\psi_j(\vec{x}) - \psi_j\left(\vec{x}+\vec{\xi}\right)\right]}\right\rangle.\end{aligned} \tag{5.86}$$

From equations (5.64) and (5.80), one gets,

$$\left\langle e^{i\left[\psi_j(\vec{x}) - \psi_j\left(\vec{x}+\vec{\xi}\right)\right]}\right\rangle = e^{-\frac{1}{2}\left[2.914\kappa^2 C_n^2(h_j)\xi^{5/3}\delta h_j\right]}. \tag{5.87}$$

The coherence function is multiplied by this equation (5.87) through each layer. It remains unaffected by Fresnel diffraction between layers. The wave structure function after passing through N layers can be expressed as the sum of the N wave structure functions associated with the individual layer, i.e.,

$$\mathcal{D}\left(\vec{\xi}\right) = \sum_{j=1}^{N} \mathcal{D}_j\left(\vec{\xi}\right). \tag{5.88}$$

For each layer, the coherence function is multiplied by the term,

$$e^{-\frac{1}{2}\left[2.914\kappa^2 C_n^2(h_j)\xi^{5/3}\delta h_j\right]},$$

therefore, the coherence function at the ground level is written as,

$$\begin{aligned}\mathcal{B}_0\left(\vec{\xi}\right) &= \prod_{j=1}^{n} e^{-\frac{1}{2}\left[2.914\kappa^2 C_n^2(h_j)\xi^{5/3}\delta h_j\right]} \\ &= e^{-\frac{1}{2}\left[2.914\kappa^2\xi^{5/3}\sum_{j=1}^{n} C_n^2(h_j)\delta h_j\right]}.\end{aligned} \tag{5.89}$$

in which the term, $2.914\kappa^2 \int_{-\infty}^{\infty} C_n^2(z)dz$ is a constant determined by the path of propagation, the wavelength and the particular environmental conditions.

It is noted here that between the layers the coherence function remains unaffected. The expression (equation 5.89) may be generalized for the case of a continuous distribution of turbulence by taking the integral extended all over the turbulent atmosphere,

$$\mathcal{B}_0\left(\vec{\xi}\right) = e^{-\frac{1}{2}\left[2.914\kappa^2\xi^{5/3}\int_{-\infty}^{\infty} C_n^2(h)\delta h\right]}, \tag{5.90}$$

thus the phase structure function is given by,

$$\mathcal{D}_\psi\left(\vec{\xi}\right) = 2.914\kappa^2\xi^{5/3}\int_{-\infty}^{\infty} C_n^2(h)\delta h. \tag{5.91}$$

When a star at a zenith distance[4], γ, is viewed through all of the turbulence atmosphere, the thickness δh of each layer is multiplied by $\sec\gamma$ and to a good approximation, the coherence function $\mathcal{B}_0\left(\vec{\xi}\right)$ on the telescope

[4]Zenith distance is known to be the angular distance of the object from the zenith at the time of observation.

aperture plane is expressed as,

$$\mathcal{B}_0\left(\vec{\xi}\right) = e^{-\frac{1}{2}\left[2.914\kappa^2\xi^{5/3}\sec\gamma\int_0^\infty \mathcal{C}_n^2(h)\delta h\right]}. \tag{5.92}$$

The integral of equation (5.92) is computed along the vertical whose lower bound is the height of the observatory and the upper bound may be considered as the height at which $\mathcal{C}_n^2$ turns out to be insignificant, say at 10 to 15 km.

Thus using the relationship between the coherence function and phase structure function (equation 5.80), the phase structure function, $\mathcal{D}_\psi\left(\vec{\xi}\right)$, at the ground level is deduced as,

$$\mathcal{D}_\psi\left(\vec{\xi}\right) = 2.914\kappa^2\xi^{5/3}\sec\gamma\int_0^\infty \mathcal{C}_n^2(h)\delta h. \tag{5.93}$$

If the phase is provided in the dimension of meter, it describes the physical shape of the turbulent wavefront which is independent of wavelength. A wavefront sensor can be used in the optical band to determine the shape of the wavefront.

5.4 Imaging in randomly inhomogeneous media

Unlike ideal conditions (see section (4.3.1), in which the achievable resolution in an imaging experiment is limited only by the imperfections in the optical system, the image resolution gets severely affected when light from an object traverse through an irregular atmospheric layer before reaching the telescope. Consider the propagation of light through the iso-planatic patch[5], it experiences the same wavefront error. The impulse response is constant due to time invariance condition and is referred to freeze the seeing.

Consider that the complex disturbances of the image, $U(\vec{\alpha})$, in which $\vec{\alpha} = \mathbf{x}/f$ is the 2-D position vector and f the focal length of the telescope, is diffracted in the telescope focal plane. The observed illumination at the focal plane of a telescope in presence of the turbulent atmosphere as a function of the direction $\vec{\alpha}$,

$$S(\vec{\alpha}) = \langle U(\vec{\alpha})U^*(\vec{\alpha})\rangle = \frac{1}{\mathtt{A}_p}\left|\mathcal{F}[\widehat{U}(\vec{u})\widehat{P}(\vec{u})]\right|^2. \tag{5.94}$$

[5]Iso-planatic patch is the area over which atmospheric point spread function is invariant over the entire field.

The term $S(\vec{\alpha})$ is known as the instant point spread function (PSF) produced by the telescope and the atmosphere and $\widehat{P}(\mathbf{u})$ the pupil transfer function. It is stated that the PSF is the square of the complex disturbances of the Fourier transform of the complex pupil function, and thus the instant transfer function of the telescope and the atmosphere takes the form,

$$\begin{aligned}\widehat{S}(\vec{f}) &= \int_{-\infty}^{\infty} U(\vec{\alpha})U^*(\vec{\alpha})e^{-i2\pi\vec{x}\cdot\vec{f}}d\vec{\alpha} \\ &= \int_{-\infty}^{\infty} S(\vec{\alpha})e^{-i2\pi\vec{\alpha}\cdot\vec{f}}d\vec{\alpha},\end{aligned} \tag{5.95}$$

where $\vec{f}$ is the spatial frequency vector expressed in radian^{-1} and $|\vec{f}|$ is its magnitude.

According to the autocorrelation theorem (see Appendix B), the Fourier transform of the squared modulus (equation 5.94) is the autocorrelation of $\widehat{U}(\vec{u})\widehat{P}(\vec{u})$, hence,

$$\widehat{S}(\vec{f}) = \frac{1}{\mathtt{A}_p}\int_{-\infty}^{\infty} \widehat{U}(\vec{u})\widehat{U}^*(\vec{u}+\vec{f})\widehat{P}(\vec{u})\widehat{P}^*(\vec{u}+\vec{f})d\vec{f}. \tag{5.96}$$

Equation (5.96) is describes the spatial frequency content $\widehat{S}(\vec{f})$ of images taken through the turbulent atmosphere. For a non-turbulent atmosphere, $\widehat{U}(\vec{u}) = 1$, the equation (5.96) shrinks to the telescope transfer function (see section 4.1.3).

5.4.1 *Seeing-limited images*

The term 'seeing' refers to the total effect of distortion in the path of starlight through different contributing layers of the atmosphere, such as, (i) the free atmosphere layer (above 1 km height), (ii) the boundary layer (less than 1-km), and (iii) the surface layer, up to the detector placed at the focus of the telescope. Let the modulation transfer function of the atmosphere and a simple lens based telescope in which the PSF is invariant to spatial shifts be described as in Figure (5.3). If a point-like object is observed through such a telescope, the turbulence induced PSF, known as seeing-disc, is derived by computing Fourier integral of the coherence function over the telescope aperture.

A point source at a point $\vec{x}'$ anywhere in the field of view produces a pattern $S(\vec{x} - \vec{x}')$ across the image. If the atmospheric degradations are iso-planatic all over the telescope field of view, the irradiance distribution

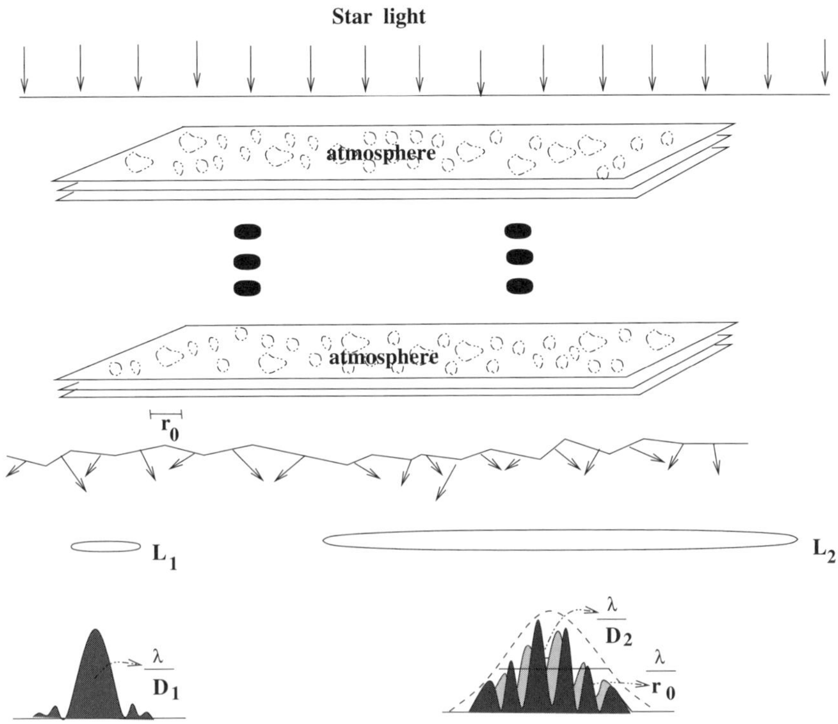

Fig. 5.3 Plane-wave propagation through the multiple turbulent layers. L_1 and L_2 represent the small and large telescopes with respective diameters D_1 and D_2.

from the object $O(x)$ is related to the instantaneous irradiance distribution $I(x)$ by a convolution relation,

$$I(\vec{x}) = \int_{-\infty}^{\infty} O(\vec{x}')S(\vec{x} - \vec{x}')d\vec{x}' = O(\vec{x}) \star S(\vec{x}), \tag{5.97}$$

where $\vec{x}(= x, y)$ is the 2-D position vector, $\vec{x}'$ the deviation of a stellar image from its mean position, $S(\vec{x})$ the instantaneous illumination (PSF of the telescope and the atmosphere) of a point source, and $\star$ denotes convolution,

The Fourier space relationship between the object and the image is,

$$\widehat{I}(\vec{u}) = \widehat{O}(\vec{u})\widehat{S}(\vec{u}), \tag{5.98}$$

where $\vec{u} = u, v$ is the 2-D spatial frequency vector, $\widehat{O}(\vec{u})$ the transform of the object, $\widehat{S}(\vec{u})$ the transfer function of the telescope-atmospheric combination, and

$$\widehat{I}(\vec{u}) = \int_{-\infty}^{\infty} I(\vec{x}) e^{-i2\pi\vec{u}\cdot\vec{x}} d\vec{x}; \qquad \widehat{O}(\vec{u}) = \int_{-\infty}^{\infty} O(\vec{x}) e^{-i2\pi\vec{u}\cdot\vec{x}} d\vec{x},$$
$$\widehat{S}(\vec{u}) = \int_{-\infty}^{\infty} S(\vec{x}) e^{-i2\pi\vec{u}\cdot\vec{x}} d\vec{x}. \tag{5.99}$$

The time-averaged PSF of both atmosphere and optical system is defined as the intensity distribution of the image of an incoherent object whose Gaussian image distribution is given by the probability density function. The seeing-limited or long-exposure[6] PSF is defined by the ensemble average, i.e., $\langle S(\vec{x})\rangle$. In practice, taking long-exposure means averaging over different realizations of the atmosphere, and thus provide long-exposure PSF. Conventional imaging in astronomy is associated with long-exposure integration, $\Delta t \gg \tau_0$, in which Δt is the exposure time, τ_0 the atmospheric coherence time. It is in general more than 20 msecs in visible wavelengths and for the infrared (IR) wavebands, it is on the order of more than 100 msecs. The effect of both telescope and atmosphere is usually considered as a random linear filtering. The relation between the average irradiance, $< I(\vec{x}) >$, and the radiance $O(\vec{x})$ of the resolved object is given by,

$$\langle I(\vec{x})\rangle = O(\vec{x}) \star \langle S(\vec{x})\rangle\,, \tag{5.100}$$

and its Fourier transform is,

$$\left\langle \widehat{I}(\vec{u})\right\rangle = \widehat{O}(\vec{u}) \left\langle \widehat{S}(\vec{u})\right\rangle, \tag{5.101}$$

where, $< \widehat{I}(\vec{u}) >$ is the long-exposure image spectrum, $\widehat{O}(\vec{u})$ is the object spectrum, and $< \widehat{S}(\vec{u}) >$ the transfer function for long-exposure images.

The aberrations of optical trains in the system can also reduce the image sharpness performance greatly. In large optics, small local deviations from perfect curvature are manifested as high spatial frequency aberrations. The intensity PSF of a point source $\langle S(\vec{x})\rangle$ is equivalent to evaluating Wiener spectrum of the scaled pupil function that is made up of two factors, such as aperture function, and random variations of the disturbance due to light arising from refractive index fluctuations of the atmosphere. In order to describe the spatial frequency content $< \widehat{S}(\vec{u}) >$ of images obtained through

[6]Long-exposure is the frame integration time that is greater than the freezing time of the atmosphere.

the turbulent atmosphere, equation (5.96) can be recast as,

$$\left\langle \widehat{S}(\vec{u}) \right\rangle = \int_{-\infty}^{\infty} \left\langle \widehat{U}(\vec{u})\widehat{U}^*(\vec{u}+\vec{u}) \right\rangle \widehat{P}(\vec{u})\widehat{P}^*(\vec{u}+\vec{u})d\vec{u}$$
$$= \widehat{\mathcal{B}}(\vec{u}).\widehat{\mathcal{T}}(\vec{u}), \tag{5.102}$$

with

$$\widehat{\mathcal{B}}(\vec{u}) = \left\langle \widehat{U}(\vec{u})\widehat{U}^*(\vec{u}+\vec{u}') \right\rangle, \tag{5.103}$$

as the wave transfer function, also known as atmosphere transfer function and $\widehat{\mathcal{T}}(\vec{u})$ the optical transfer function of the telescope.

Equation (5.102) contains the important result that the optical transfer function for long-exposure images, $< \widehat{S}(\vec{u}) >$, is the product of the telescope transfer function, $\widehat{\mathcal{T}}(\vec{u})$, and the atmosphere transfer function, $\widehat{\mathcal{B}}(\vec{u})$. Fried (1966) proposed to define the resolving power, $\mathcal{R}$, of a telescope as the integral of the optical transfer function for the combined effect of the atmosphere and telescope system,

$$\mathcal{R} = \int_{-\infty}^{\infty} \widehat{S}(\vec{u})d\vec{u} = \int_{-\infty}^{\infty} \widehat{\mathcal{B}}(\vec{u}).\widehat{\mathcal{T}}(\vec{u})d\vec{u}. \tag{5.104}$$

The diffraction-limited resolving power of a small telescope with an unobscured circular aperture of diameter, $D \ll r_0$, in which D is the diameter of the telescope and r_0 the coherence length of atmospheric turbulence, depends on its optical transfer function (see equation 4.80).

An atmospheric transfer function is defined in terms of its coherence length, r_0. This length, introduced by Fried (1966) known as Fried's parameter, which may be defined as the diameter of a circular pupil that would produce the same diffraction-limited full width at half maximum[7] (FWHM) of a point source image as the atmospheric turbulence would with an infinite-sized mirror. Such a parameter essentially determines the iso-planatic limit of turbulence. It is the measure of the distance over which the refractive index fluctuations are correlated. The RMS phase variation over an aperture of diameter r_0 is given by $\langle \sigma \rangle^2 = 1.03\ \mathrm{rad}^2$. These aspects elucidate the atmospheric turbulence by r_0-sized patches of constant phase and random phases between the individual patches. This causes blurring of the image which limits the performance of any terrestrial large telescope.

[7]Full width at half maximum (FWHM) is the width measured at half level between the continuum and the peak of the line.

With a critical diameter r_0 for a telescope, one finds,

$$\int_{-\infty}^{\infty} \widehat{\mathcal{B}}(\vec{u})d\vec{u} = \int_{-\infty}^{\infty} \widehat{\mathcal{T}}(\vec{u})d\vec{u}. \tag{5.105}$$

If r_0 is smaller than the telescope aperture, it produces a mean spot size larger than that possible with diffraction-limited image. Therefore, the eddies produce different atmospheric indices of reflection in different parts of light seen by a large telescope. Such a telescope is susceptible to atmospheric distortions in turbulent air, which limits its performance. The resolving power of a large telescope, $(D \gg r_0)$, is dominated by the effects of the atmospheric turbulence. In such a situation, the wave transfer function becomes narrow compared to the telescope transfer function, hence the resolution becomes,

$$\begin{aligned}\mathcal{R}_{atm} &= \int_{-\infty}^{\infty} \widehat{\mathcal{B}}(\vec{u})d\vec{u} = \int_{-\infty}^{\infty} e^{-3.44\,(\lambda f/r_0)^{5/3}} d\vec{u} \\ &= \int_{-\infty}^{\infty} e^{-\frac{1}{2}\mathcal{D}_{\psi}(\lambda\vec{u})} d\vec{u} = \frac{\pi}{4}\left(\frac{r_0}{\lambda}\right)^2. \end{aligned} \tag{5.106}$$

It may be reiterated that the atmospheric transfer function, $\widehat{\mathcal{B}}(\vec{u})$, decreases faster than the telescope transfer function, $\widehat{\mathcal{T}}(\vec{u})$, at visible wavelengths. The PSF of a turbulence degraded image is the Fourier transform of the former. The angular width of such a spread function is dictated by λ/r_0.

5.4.2 *Atmospheric coherence length*

Atmospheric coherence length measures the effect of atmospheric turbulence on optical propagation. A large value of coherence diameter implies that the turbulence is weak, while a small value means that these statistical correlations fall off precipitously. According to equation (5.92), the atmosphere transfer function, $\widehat{\mathcal{B}}(\vec{u})$, is given by,

$$\widehat{\mathcal{B}}(\vec{u}) = \mathcal{B}_0(\lambda\vec{u}) = e^{-\frac{1}{2}\left[2.914\kappa^2(\lambda u)^{5/3}\sec\gamma\int_0^{\infty} C_n^2(h)dh\right]}. \tag{5.107}$$

The resolving power of the large telescope can be recast into,

$$\mathcal{R} = \frac{6\pi}{5}\Gamma\left(\frac{6}{5}\right)\left(\frac{1}{2}\left[2.914\kappa^2\lambda^{5/3}\sec\gamma\int_0^{\infty} C_n^2(h)dh\right]\right)^{-6/5}. \tag{5.108}$$

As discussed in the preceding chapter (section 4.3.1), the resolving power, $\mathcal{R}$, of an optical system is given by the integral over the optical transfer function (see equation 4.79), and hence by placing $D = r_0$, equation (5.107) takes the form of,

$$\widehat{\mathcal{B}}(\vec{u}) = e^{-3.44\,(\lambda u/r_0)^{5/3}}, \qquad \text{or} \tag{5.109}$$

$$\mathcal{B}_0\left(\vec{\xi}\right) = e^{-3.44\,(\xi/r_0)^{5/3}}. \tag{5.110}$$

Equation (5.110) is the simplified expression for the coherence function in terms of Fried's parameter, r_0. By combining equation (5.110) with the expression (equation 5.83), the expression for the phase structure function, $\mathcal{D}_\psi\left(\vec{\xi}\right)$, for Kolmogorov turbulence is written in terms of the coherence length, r_0, introduced by Fried (1966) as,

$$\mathcal{D}_\psi\left(\vec{\xi}\right) = \left\langle \left|\psi\left(\vec{x}\right) - \psi\left(\vec{x} + \vec{\xi}\right)\right|^2 \right\rangle \simeq 6.88\left(\frac{\xi}{r_0}\right)^{5/3}. \tag{5.111}$$

The significance of the factor 6.88 is given by,

$$\begin{aligned} 2.914\kappa^2\lambda^{5/3}\sec\gamma\int_0^\infty C_n^2(h)dh &= 2\left[\left(\frac{24}{5}\right)\Gamma\left(\frac{6}{5}\right)\right]^{5/6}\left(\frac{r_0}{\lambda}\right)^{-5/3} \\ &= 6.88\left(\frac{r_0}{\lambda}\right)^{-5/3}, \end{aligned} \tag{5.112}$$

in which $2[(24/5)\Gamma(6/5)]^{5/6}$ denotes the effective propagation path length,

$$\mathcal{D}_\psi\left(\vec{\xi}\right) = \begin{cases} \mathcal{D}\left(\vec{\xi}\right) & \text{near field,} \\ \dfrac{1}{2}\mathcal{D}\left(\vec{\xi}\right) & \text{otherwise.} \end{cases} \tag{5.113}$$

It is pertinent to note that for long-exposure MTF, there is no definition between the near-field and far-field cases. As the value of $\vec{u}$ approaches the cut off frequency of $\widehat{\mathcal{S}}_0(\vec{u})$, the effect of using a short-exposure[8] becomes more and more important. The short-exposure time should be sufficiently short, $\Delta t \ll \tau_0$, to eliminate image wander as a blurring mechanism. Such an exposure time is of the order of a few milliseconds to a few tens of milliseconds in visible wavelengths, while for infrared wavebands, it is on the order of 100 msecs. By comparing equation (5.111) with equation (5.92), yields an expression for r_0 in terms of the angle away from the zenith and

[8]Short-exposure is an integration time, which is shorter than the evolution time of the atmospheric turbulence.

an integral over the refractive index structure constant of the atmospheric turbulence,

$$
\begin{aligned}
r_0 &\simeq \left(\frac{6.88}{\mathcal{D}}\right)^{3/5} \\
&= \left[0.423\kappa^2 \sec\gamma \int_0^\infty C_n^2(h)dh\right]^{-3/5}. \qquad (5.114)
\end{aligned}
$$

The resolving power, $\mathcal{R}$, of a telescope is essentially limited by its diameter smaller than the atmospheric coherence length, r_0. It is limited by the atmosphere when the telescope diameter is larger than r_0. The size of the seeing-disc is of the order of $1.22\lambda/r_0$. Larger r_0 values are associated with good seeing conditions. The dependence of r_0 on wavelength is given by $r_0 \propto \lambda^{6/5}$ and dependence on zenith angle by $r_0 \propto (\cos\gamma)^{3/5}$. With $r_0 \propto \lambda^{6/5}$, the seeing is $\lambda/r_0 \propto \lambda^{-1/5}$. This means that seeing decreases slowly with increasing wavelength.

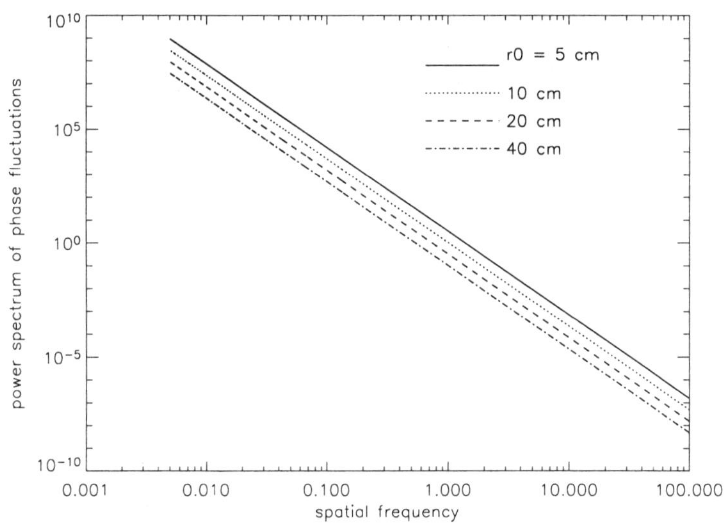

Fig. 5.4 Kolmogorov power spectrum of phase fluctuations against various spatial frequencies; four different values of atmospheric coherence length, r_0, are taken (Courtesy: R. Sridharan).

Since the structure function is related to the power spectrum, $\Phi_\psi(\vec{\kappa})$, (see section 5.4.1), one may write,

$$\mathcal{D}_\psi(\vec{\xi}) = 2\int_0^\infty \Phi_\psi(\vec{\kappa})\left[1 - \cos\left(2\pi\vec{\kappa}\cdot\vec{\xi}\right)\right] d\vec{\kappa}. \tag{5.115}$$

On using the integral,

$$\int_0^\infty x^{-p}[1 - J_0(ax)]dx = \frac{\pi b^{p-1}}{2^p[\Gamma(p+1)/2]^2 \sin[\pi(p-1)/2]},$$

in which J_0 is the zeroth order Bessel function, one derives the power spectrum of the phase fluctuations due to Kolmogorov turbulence as,

$$\Phi_\psi(\vec{\kappa}) = 0.023 r_0^{-5/3}\kappa^{-11/3}. \tag{5.116}$$

Figure (5.4) displays the Kolmogorov power spectrum of phase fluctuations. The Wiener spectrum of phase gradient after averaging with the telescope aperture, D, due to Kolmogorov turbulence is deduced as,

$$\Phi_\psi(\vec{\kappa}) = 0.023 r_0^{-5/3}\kappa^{-11/3}\left|\frac{2J_1(\pi D\kappa)}{\pi D\kappa}\right|^2, \tag{5.117}$$

where J_1 is the first order Bessel function describing Airy disc, the diffraction-limited PSF, which is the Fourier transform of the circular aperture.

5.4.3 *Atmospheric coherence time*

By employing Taylor hypothesis of frozen turbulence, according to which the variations of the turbulence caused by a single layer may be modeled by a frozen pattern that is moved across the telescope aperture by the wind in that layer. Assuming that a static layer of turbulence moves with constant speed $\vec{v}$ in front of the telescope aperture, the phase at a point $\vec{x}$ at time $t + \tau$ is expressed as,

$$\psi(\vec{x}, t+\tau) = \psi(\vec{x} - \vec{v}\tau, t), \tag{5.118}$$

and the temporal phase structure function is,

$$\mathcal{D}_\psi(\vec{v}\tau) = \left\langle |\psi(\vec{x}, t) - \psi(\vec{x} - \vec{v}\tau, t)|^2\right\rangle, \tag{5.119}$$

Equation (5.119) for temporal structure function of the wavefront phase provides the mean square error phase error associated with a time delay, τ. Such a structure function depends individually on the two coordinates parallel and perpendicular to the wind direction. Though time evolution is complicated in the case of multiple layers contributing to the total turbulence, the temporal behavior is characterized by atmospheric coherence time τ_0. In the direction of the wind speed an estimate of the correlation time yields the temporal coherence time, τ_0,

$$\tau_0 \simeq \frac{r_0}{|\vec{v}|}, \tag{5.120}$$

in which v is the wind velocity in the turbulent layer of air.

The parameter of atmospheric coherence time, τ_0 is of paramount importance for the design of optical interferometers, as well as of adaptive optics systems. The wavelength scaling of such a coherence time, τ_0 is the same of that of atmospheric coherence length, r_0, i.e., $\tau_0 \propto \lambda^{6/5}$. The time scale for the temporal changes is usually much longer than the time it takes the wind to blow the turbulence past the telescope aperture. A wind speed of $v = 10$ m/sec and Fried's parameter of $r_0 = 20$ cm provide a coherence time $\tau_0 = 20$ msec. The coherence time, τ_0, is a highly variable parameter depending on the effective wind velocity. Its value is in the range of a few milliseconds (msec) in the visible during normal seeing conditions and can be as high as to ~ 0.1 sec (s) in excellent seeing condition. Variations in r_0 from 5% to 50% are common within a few seconds; they can reach up to 100% sometimes. Davis and Tango (1996) have measured the value of atmospheric coherence time that varied between ~1 and ~7 msec with the Sydney University stellar interferometer (SUSI).

5.4.4 *Aniso-planatism*

Aniso-planatism is a well known problem in compensating seeing, the effects of which distort image of any celestial object, both for post-processing imaging technique like speckle interferometry (see section 6.3), as well as for adaptive optics (see chapter 7) systems. The angle over which such distortions are correlated is known as iso-planatic patch, θ. It is is the area of sky enclosed by a circle of radius equal to the iso-planatic angle. The radius of this patch increases with wavelength. For a visible wavelength it is about a few arcseconds, while for 2.2 μm, it is 20-30 arcseconds. The lack of iso-planaticity of the instantaneous PSF is the most important limitation in the compensated field-of-view (FOV) related to the height of turbulence layers. This effect, called aniso-planaticity, occurs when the volume of turbulence experienced by the reference object differs from that experienced by the target object of interest, and therefore experience different phase variations. Departure from iso-planaticity yields in non-linear degradations. The more the angular distance, θ, more the degradation of the image quality takes place. The angular aniso-planatic aberration can be evaluated from Kolmogorov spectrum. By invoking the equation (5.93), the wavefront variance for two beams, $\langle\sigma_\theta\rangle^2$, is deduced as,

$$\begin{aligned}\langle\sigma_\theta\rangle^2 &= 2.914\kappa^2 \sec\gamma \int_{-\infty}^{\infty} C_n^2(h)(\theta h \sec\gamma)^{5/3}\delta h \\ &= 2.914\kappa^2 (\sec\gamma)^{8/3}\theta^{5/3} \int_{-\infty}^{\infty} C_n^2(h) h^{5/3}\delta h \\ &= \left(\frac{\theta}{\theta_0}\right)^{5/3}, \qquad \mathrm{rad}^2, \end{aligned} \tag{5.121}$$

where θ_0 is defined as the iso-planatic angle,

$$\theta_0 \simeq \left[2.914\kappa^2(\sec\gamma)^{8/3}\int_{-\infty}^{\infty} C_n^2(h)h^{5/3}\delta h\right]^{-3/5}, \tag{5.122}$$

and $h\sec\gamma$ the distance at zenith angle γ with respect to the height h.

A comparison of the equation (5.122) with that of the expression for the Fried's parameter, r_0 (equation 5.114) provides the following relationship:

$$\theta_0 = (6.88)^{-3/5}\frac{r_0}{L\sec\gamma} = 0.314\frac{r_0}{L\sec\gamma}, \tag{5.123}$$

where L is the mean effective height of the turbulence.

Like the atmospheric coherence length, r_0, the iso-planatic angle, θ_0, also increases as the (6/5) power of the wavelength, but it decreases as the (-8/5) power of the airmass.

5.5 Image motion

Motion of an image takes place if the scale of the wavefront perturbation is larger than the diameter of the telescope aperture. Perturbations that are smaller than the telescope aperture yields blurring of the image. Blurring is defined as the average width of the instantaneous PSF. The image motion and blurring are produced by unrelated parts of the turbulence spectrum, and are thus statistically independent.

In order to analyze the imaging process at the focal plane of a telescope, certain assumptions are to be made about the phase distribution in the telescope. Neglecting the effects of the Fresnel diffraction, let the turbulent atmosphere be represented by a single thin layer in the telescope aperture. The average gradient of the phase distribution in the said aperture determines the position of the image at the focal point of the telescope. The power spectrum, $W_\psi(\vec{f})$ of the phase, ψ can be expressed as,

$$W_\psi(\vec{f}) = \kappa^2 W_n(\vec{f}, 0)\delta h, \tag{5.124}$$

with

$$W_n(f_x, f_y, f_z) = (2\pi)^3 \Phi_n(2\pi f_x, 2\pi f_y, 2\pi f_z), \tag{5.125}$$

is the 3-D power spectrum of angular fluctuations.

Assuming Kolmogorov's law (equation 5.51) to be valid,

$$W(\vec{f}, 0) = (2\pi)^3 \times 0.033 C_n^2(h)\delta h = 9.7 \times 10^{-3} f^{-11/3} C_n^2(h), \tag{5.126}$$

so that

$$\begin{aligned} W_\psi(\vec{f}) &= 9.7 \times 10^{-3} \kappa^2 f^{-11/3} C_n^2(h)\delta h \\ &= 0.38\lambda^{-2} f^{-11/3} C_n^2(h)\delta h. \end{aligned} \tag{5.127}$$

With $\kappa = 2\pi/\lambda$. By integrating equation (5.127), $W_\psi(\vec{f})$, in the near-field approximation is deduced as,

$$W_\psi(\vec{f}) = 0.38\lambda^{-2} f^{-11/3} \int_{-\infty}^{\infty} C_n^2(h)dh. \tag{5.128}$$

5.5.1 *Variance due to angle of arrival*

The statistical properties of the gradient $\vec{\alpha}$ of the wavefront without averaging over the telescope aperture are described as follows. Since the optical rays are normal to the wavefront surface, the fluctuations of their angle are related to the fluctuations of the wavefront slope. Let two components α_x and α_y be the independent Gaussian random variables, hence, according to Roddier (1981),

$$\alpha_x(\vec{x}) = -\frac{\lambda}{2\pi}\frac{\partial \psi_0(\vec{x})}{\partial x}, \qquad \alpha_y(\vec{x}) = -\frac{\lambda}{2\pi}\frac{\partial \psi_0(\vec{x})}{\partial y}. \tag{5.129}$$

Here the two components $\alpha_{x,y}$ are considered as a function of the horizontal coordinates, $\vec{x}$.

The power spectra of these variables are related to the phase, ψ,

$$W_{\alpha_{x,y}}(\vec{f}) = \lambda^2 f_{x,y}^2 W_\psi(\vec{f}), \tag{5.130}$$

in which f_x and f_y are the respective x and y components of the frequency vector, $\vec{f}$.

The variance of these variables are derived as,

$$\langle \alpha_{x,y} \rangle^2 = \lambda^2 \int_{-\infty}^{\infty} f_{x,y}^2 W_\psi(\vec{f}) d\vec{f}. \tag{5.131}$$

The standard deviation (see Appendix B), $\langle \sigma \rangle$, of the angle of arrival is deduced as,

$$\langle \sigma \rangle^2 = \lambda^2 \int_{-\infty}^{\infty} \vec{f}^2 W_\psi(\vec{f}) d\vec{f}. \tag{5.132}$$

On replacing the value of $W_\psi(\vec{f})$ for near-field approximations (equation 5.128) into equation (5.132), gives after integration over all directions in the frequency domain,

$$\langle \sigma \rangle^2 \propto \int_0^{\infty} f^{-2/3} df \int_0^{\infty} C_n^2(h) dh. \tag{5.133}$$

By neglecting the central obscuration, the variance is derived in terms of outer scale of turbulence, L_0 and the high frequency cut-off produced by averaging over the aperture diameter, D, i.e.,

$$\begin{aligned} \langle \sigma \rangle^2 &\propto \int_{1/L_0}^{1/D} f^{-2/3} df \int_0^{\infty} C_n^2(h) dh \\ &\propto \left[D^{-1/3} - L_0^{-1/3} \right] \int_0^{\infty} C_n^2(h) dh. \end{aligned} \tag{5.134}$$

For a small aperture, it is written,

$$\langle \sigma \rangle^2 \propto D^{-1/3} \int_0^{\infty} C_n^2(h) dh. \tag{5.135}$$

Due to Fried (1966) and Tatarski (1961), equation (5.135) can be recast into,

$$\langle \sigma \rangle^2 = 0.364 \left(\frac{\lambda}{r_0} \right)^2 \left(\frac{D}{r_0} \right)^{-1/3} \qquad \text{arcsec}^2, \tag{5.136}$$

in which the value of Fried's parameter, r_0 is taken from equation (5.114), λ/r_0 the seeing-disc in arcsec, and the quotient D/r_0 describes the imaging process in the telescope which relates the size of the seeing-disc to the FWHM of the Airy pattern λ/D.

Equation (5.136) states that the dependence of the variance on $D^{-1/3}$, that is with decrease of the size of the telescope aperture, the variance of the image motion increases.

5.5.2 *Scintillation*

Small-scale perturbations play by far the most important role in deteriorating an astronomical image. The temporal variation of higher order aberrations due to the movement of small cell causes dynamic intensity fluctuations. If the propagation length is of the order of $\simeq r_0^2/\lambda$ or longer, the rays diffracted at the turbulence cells interfere with each other, which in turn, causes intensity fluctuations in addition to the phase fluctuations. This interference phenomenon is highly chromatic, known as scintillation. The statistical properties of this phenomenon have been experimentally investigated (Roddier, 1981 and references therein).

Scintillation is one of the most disturbing phenomena, which either focus light or disperse it, causing apparent enhancement or dimming of light intensity. The motion of small cells produces a motion of the enhanced or dimmed images, known as agitation of the image. Although the scintillation is week for application of adaptive optics and interferometry, it is important to take into consideration for high performance adaptive optics (AO) system that is designed for the direct detection of exo-solar planets. In such programmes, it is necessary to correct the wavefront errors so well that intensity fluctuations become important.

The intensity variations are usually expressed as the fluctuation of the log of the amplitude, known as log-amplitude fluctuations. According to the Kolmogorov spectrum, such fluctuations are produced by eddies with sizes of the order of $\sqrt{\lambda L}$ in which,

$$L \simeq \left(\frac{\int_{-\infty}^{\infty} C_n^2(h) h^{5/3} \delta h}{\int_{-\infty}^{\infty} C_n^2(h) \delta h} \right)^{3/5}, \tag{5.137}$$

is the mean effective height of the turbulence.

The amount of scintillation, known as scintillation index, σ_I^2, is defined as the variance of the relative intensity fluctuations. By determining the relative intensity fluctuations $\delta I/I$, the effects of scintillation can be quantified. The scintillation index is related to the variance σ_χ^2 of the relative amplitude fluctuations, χ, by $\sigma_I^2 = 4\sigma_\chi^2$, in which,

$$\sigma_\chi^2 = \int_{-\infty}^{\infty} W_\chi(\vec{f})d\vec{f}, \tag{5.138}$$

where the power spectra $W_\chi(\vec{f})$ is given by,

$$W_\chi(\vec{f}) = 0.38\lambda^{-2} f^{-11/3} \int_0^\infty \left(\sin \pi\lambda h f^2\right)^2 C_n^2(h)dh.$$

The variance of the log-intensity fluctuations, σ_I^2, is given by,

$$\sigma_I^2 = 19.2\lambda^{-7/6} (\sec\gamma)^{11/6} \int_0^\infty C_n^2(h)h^{5/6}dh. \tag{5.139}$$

Equation (5.139) is valid for small apertures with diameter $D \ll \sqrt{\lambda L}$.

Scintillation is reduced for larger apertures since it averages over multiple independent sub-apertures. This changes the amplitude of the intensity fluctuations; it changes the functional dependence on zenith angle, wavelength, and turbulence height as well. Considering the telescope filtering function, $|\widehat{P}_0(\vec{f})|^2$, and with aperture frequency cut-off $f_c \sim D^{-1}$ sufficiently small, so that, $\pi\lambda h f_c^2 \ll 1$, equation (5.139) translates into,

$$\sigma_I^2 \propto D^{-7/3} (\sec\gamma)^3 \int_0^\infty C_n^2(h)h^2dh, \tag{5.140}$$

which is valid for $D \gg \sqrt{\lambda L}$ and $\gamma \leq 60°$ shows the decrease of the scintillation amplitude with aperture size (Tatarski, 1961).

5.5.3 *Temporal evolution of image motion*

One may also estimate the effect of moving turbulence by invoking Taylor hypothesis of frozen turbulence in which the atmospheric density perturbations are assumed to be constant over the time it takes wind to blow them across a given aperture. Since the time scales of eddy motion are smaller than the frequency of interest, it is necessary to use temporal mode by assuming that a frozen piece of turbulent air is being transported through the wavefront by a wind with a component of velocity, v, perpendicular to the direction of propagation. The parallel component of the same is assumed not to affect the temporal statistical properties of the wavefront entering the aperture. With multiple layers contributing to the total turbulence, the time evolution becomes more complicated. The temporal power spectrum of the phase fluctuations is derived from the spatial power spectrum $\Phi(\vec{\kappa})$.

With $\vec{v}$ being parallel to the x-axis, $\kappa_x = f/v$ and an integration over κ_y is performed to obtain the temporal power spectrum, $\Phi_\tau(f)$,

$$\begin{aligned}\Phi_\tau(f) &= \frac{1}{v}\int_{-\infty}^{\infty} \Phi\left(f/v, \kappa_y\right) d\kappa_y \\ &= 0.077 r_0^{-5/3} \frac{1}{v}\left(\frac{f}{v}\right)^{-8/3}.\end{aligned} \tag{5.141}$$

Since solution of this integral is improbable, Taylor (1994) provided an approximation for the power spectrum at low and high frequencies that may be simplified by assuming single dominant layer with wind speed $\vec{v}$. The power spectral density of the centroid motion for the low frequency (LF) and high frequency (HF) are derived respectively as,

$$W(\vec{f})_{LF} = 0.097\left(\frac{r_0}{\vec{v}}\right)^{1/3}\left(\frac{\lambda}{r_0}\right)^2 f^{-2/3} \qquad \text{arcsec}^2/\text{Hz}, \tag{5.142}$$

$$W(\vec{f})_{HF} = 0.0013\left(\frac{D}{\vec{v}}\right)^{-8/3}\left(\frac{\lambda}{r_0}\right)^2\left(\frac{D}{r_0}\right)^{-1/3} f^{-11/3}. \tag{5.143}$$

The power spectrum decreases with $f^{-2/3}$ in the low frequency region and is independent of the size of the aperture, while in the high frequency region the spectrum is proportional to $f^{-11/3}$ decreasing with D^{-3}.

5.5.4 *Image blurring*

Astronomers use the measurements of motion and blurring in order to estimate the degradation of the image due to the atmospheric turbulence. Since the image motion does not degrade short-exposures, its effect on the long-exposure can be removed by employing a fast automatic guider or by adding post-detected short-exposure properly centered images. The remaining degradation is known as blurring. Let $\vec{\alpha}'$ be the deviation of a stellar image from its average position. For image motion described by statistically independent Gaussian random processes of zero mean, the probability density function, $\mathcal{P}(\vec{\alpha}')$, may be described as,

$$\mathcal{P}(\vec{\alpha}') = \frac{1}{\pi\langle\sigma\rangle^2} e^{-|\vec{\alpha}'|^2/\langle\sigma\rangle^2}. \tag{5.144}$$

The time-averaged intensity distribution of the image, formed by a telescope, undergoing random motion is given by the convolution of its motion-free distribution and the probability density function describing its motion.

Considering that $S_0(\vec{\alpha}') = S(\vec{\alpha} + \vec{\alpha}')$ is an instantaneous illumination in the central image of a point source, the conventional long-exposure image $\langle S(\vec{\alpha})\rangle$ is given by,

$$\begin{aligned}\langle S(\vec{\alpha})\rangle &= \left\langle \int_{-\infty}^{\infty} S_0(\vec{\alpha} - \vec{\alpha}')\mathcal{P}(\vec{\alpha}')d\vec{\alpha}' \right\rangle \\ &= \langle S_0(\vec{\alpha})\rangle\, \mathcal{P}(\vec{\alpha}). \end{aligned} \tag{5.145}$$

The total optical transfer function (OTF) is the product of OTF of the aperture, as well as OTF of the turbulence and is expressed as,

$$\left\langle \widehat{S}(\vec{f})\right\rangle = \left\langle \widehat{S}_0(\vec{f})\right\rangle \widehat{\mathcal{P}}(\vec{f}), \tag{5.146}$$

in which the Fourier transform of $\mathcal{P}(\vec{x})$ is,

$$\widehat{\mathcal{P}}(\vec{f}) = e^{-\pi^2 \langle\sigma\rangle^2 f^2}. \tag{5.147}$$

By substituting the value of the variance from equation (5.136) into this equation (5.147) one gets,

$$\widehat{\mathcal{P}}(\vec{f}) = e^{-3.44(\lambda f/D)^{1/3}(\lambda f/r_0)^{5/3}}, \tag{5.148}$$

and therefore, the transfer function, $\left\langle \widehat{S}_0(\vec{f})\right\rangle$, associated with Kolmogorov turbulence, assuming the telescope is located in the near-field as well as in the far-field of the turbulence is derived as (Fried, 1966),

$$\left\langle \widehat{S}_0(\vec{f})\right\rangle = \begin{cases} \widehat{\mathcal{T}}(\vec{f})e^{-3.44(\lambda f/r_0)^{5/3}\left[1-(\lambda f/D)^{1/3}\right]}, & \text{for near field,} \\ \widehat{\mathcal{T}}(\vec{f})e^{-3.44(\lambda f/r_0)^{5/3}\left[1-\frac{1}{2}(\lambda f/D)^{1/3}\right]}, & \text{otherwise.} \end{cases} \tag{5.149}$$

in which

$$\left\langle \widehat{S}(\vec{f})\right\rangle = \widehat{\mathcal{T}}(\vec{f})e^{-3.44(\lambda f/r_0)^{5/3}}, \tag{5.150}$$

and the telescope transfer function is similar to the OTF of a circular aperture (Goodman, 1968),

$$\widehat{\mathcal{T}}(\vec{f}) = \begin{cases} \dfrac{2}{\pi}\left[\cos^{-1}\left(\dfrac{\lambda f}{D}\right) - \dfrac{\lambda f}{D}\sqrt{1-\left(\dfrac{\lambda f}{D}\right)^2}\right] & \text{for } f < \dfrac{D}{\lambda} \\ 0 & \text{otherwise,} \end{cases}$$

For the near case, this assumes that the phase structure function at the telescope is equal to that at the turbulence layer. The discrepancy between the two phase structure function is small for the conditions of astronomical observations (Roddier, 1981). Equation (5.149) describes the transfer function for short-exposure images that are re-centered and added (Fried, 1966). The PSF of a short-exposure image is the ensemble average of such re-centered images and can be found by taking the Fourier transform of the OTF numerically. The transfer functions for both the low and high frequency part are given by,

$$\left\langle \left|\widehat{S}_0(\vec{f})\right|^2 \right\rangle_{LF} = \widehat{\mathcal{T}}^2(\vec{f}) e^{-3.44(\lambda f/r_0)^{5/3}\left[1-(\lambda f/D)^{1/3}\right]} \qquad f < \frac{r_0}{\lambda}, \tag{5.151}$$

$$\left\langle \left|\widehat{S}_0(\vec{f})\right|^2 \right\rangle_{HF} = 0.342 \widehat{\mathcal{T}}(\vec{f}) \left(\frac{r_0}{D}\right)^2 \qquad f \gg \frac{r_0}{\lambda}. \tag{5.152}$$

The increasing variance of the image motion with smaller apertures is attributed to the increase of the power spectrum in the high frequency region. If the aperture of the telescope is larger than the outer scale of turbulence, L_0, the image motion is reduced below the values predicted by Kolmogorov statistics.

5.5.5 *Measurement of r_0*

Measurement of r_0 is of paramount importance to estimate the seeing at any astronomical site. Systematic studies of this parameter would help in understanding the various causes of the local seeing, such as thermal inhomogeneities associated with the building. Degradation in image quality may takes place because of opto-mechanical aberrations of the telescope as well.

Stellar image profiles provide a mean to estimate the atmosphere transfer function. But a detector with high dynamic range is required to obtain such profiles. Moreover it is sensitive to telescope aberrations, misalignment, and focusing errors.

A qualitative method to measure r_0 is based on the short-exposure images using speckle interferometric technique (Labeyrie, 1970). The averaged autocorrelation of these images contains both the autocorrelations of the seeing-disc together with the autocorrelation of mean speckle cell. Figure (5.5) displays the autocorrelation of α And observed at the Cassegrain fo-

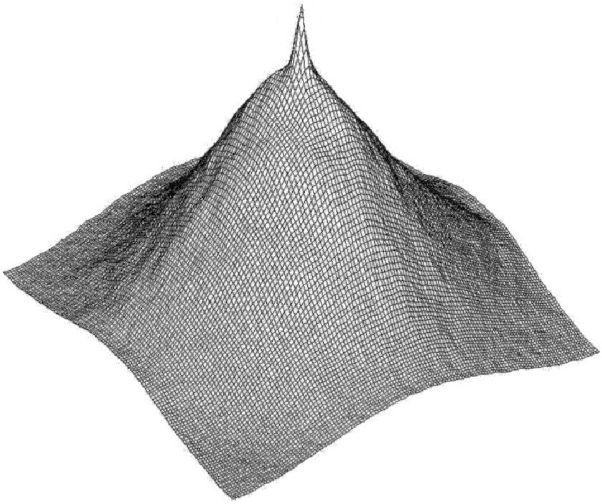

Fig. 5.5 Autocorrelation of seeing disc derived from a specklegram (speckle patterns in a picture) of α And that was recorded at the 2.34 meter VBT, Kavalur, India (Saha et al. 1999a) together with the autocorrelation of the speckle component.

cus of 2.34 meter Vainu Bappu Telescope (VBT), Vainu Bappu Observatory (VBO), India, which consists of the width of the seeing-disc, as well as of the width of the speckle component (the sharp peak). It is the width of the speckle component of the autocorrelation that provides the information on the size of the object being observed (Saha and Chinnappan, 1999, Saha et al., 1999a). The form of transfer function, $< |\widehat{S}(\vec{u})|^2 >$, can be obtained by calculating Wiener spectrum of the instantaneous intensity distribution from a point source.

The seeing fluctuates on all time scales down to minutes and seconds. Figure (5.6) displays the microfluctuations of r_0 at a step of $\sim$150 msec observed at the 2.34 meter VBT, Kavalur, India, on a different night, 28 February, 1997. At a given site, r_0 varies dramatically night to night. It can be a factor 2 better than the median or vice versa. The scaling of r_0 with zenith angle, as well as with wavelength (see equation 5.114) has practical consequences.

5.5.6 *Seeing at the telescope site*

Image of an astronomical source deteriorates due to temperature variations within the telescope building, known as dome seeing, and difference in temperature of the primary mirror surface with its surroundings as well.

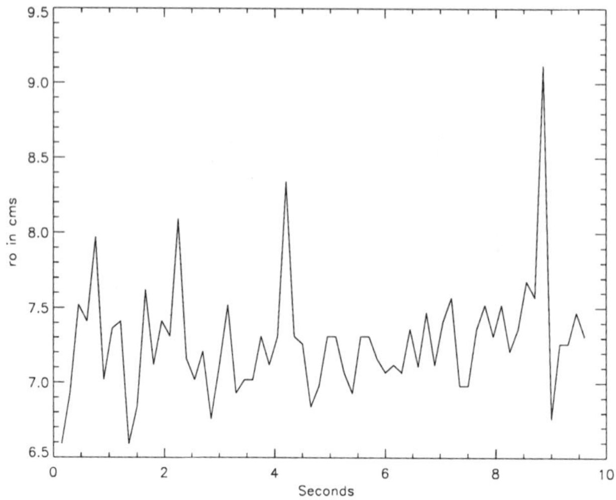

Fig. 5.6 Microfluctuations of r_0 as observed at the 2.34 meter VBT, Kavalur, India on 28 February, 1997 (Saha and Yeswanth, 2004).

The seeing at a particular place depends on the topography of the place, composition of Earth, height, dust level in the atmosphere, temperature gradients, atmospheric turbulence, wind velocity etc. The seeing angle or seeing-disc, θ_s, is a parameter that determines the image quality. It is defined as the FWHM of a Gaussian function fitted to a histogram of image position in arcsec. The quality of seeing is characterized by Fried's parameter, i.e.,

$$\theta_s = 0.976\frac{\lambda}{r_0}, \tag{5.153}$$

in which λ is the wavelength of observation.

Though the effect of the different layer turbulence has been receiving attention to identify the best site, the major sources of image degradation predominantly come from the thermal and aero-dynamic disturbances in the atmosphere surrounding the telescope and its enclosure, namely (i) thermal distortion of primary and secondary mirrors when they get heated up, (ii) dissipation of heat by the secondary mirror (Zago, 1995), (iii) rise in temperature at the primary cell, and (iv) at the focal point causing temperature gradient close to the detector etc. In what follows the empirical results that are obtained from various observations are described in brief.

5.5.6.1 *Wind shears*

Wind shears produce eddy currents of various sizes in the atmosphere. The turbulent phenomena associated with heat flow and winds in the atmosphere occur predominantly due to the winds at various heights, convection in and around the building and the dome, that shields the telescope from the effects of wind buffeting, cooled during the day. any obstructed location near the ground, off the surface of the telescope structure, inside the primary mirror cell, etc. The cells of differing sizes and refractive indices produced by this phenomena, move rapidly across the path of light, causing the distortion on the shape of the wave-front and variations on the intensity and phase. Both wind speed and wind direction appear to affect degradation in image quality. Cromwell et al. (1988) have found θ_s increases with increasing wind speed in the observed range of 0-18 m/sec; θ_s was found to be sensitive to the wind direction as well.

5.5.6.2 *Dome seeing*

The performance of an accurately fabricated telescope may deteriorate due to its enclosure which has a few degrees temperature variations with its surroundings. The contribution from dome and the mirror should be kept minimum following good thermal engineering principles.

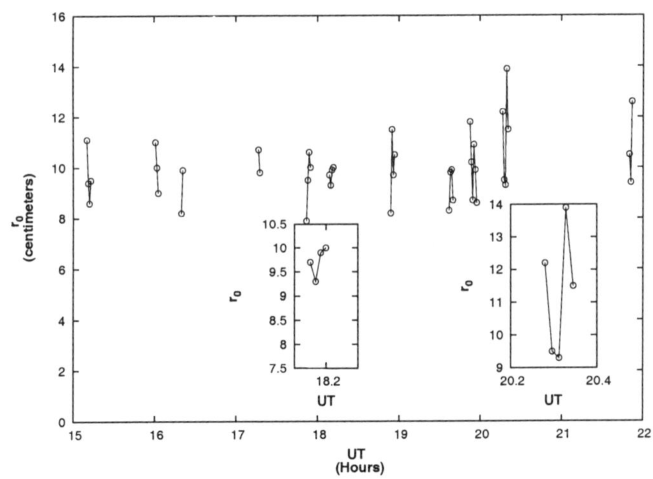

Fig. 5.7 A set of r_0 values from 12 different stars acquired on 28-29 March, 1991, at VBT, Kavalur, India (Saha and Yeswanth, 2004).

Studies pertaining to the correlation of observed PSF with the difference between the dome temperature and ambient have found to be a much weaker trend that for the mirror. In fact, a $4-5^\circ$ C difference in temperature between the outside and inside of the dome causes a seeing degradation amounting to 0.5" only (Racine et al., 1991). The reported improvement of seeing at the 3.6 m CFHT, is largely due to the implementation of the floor chilling system to damp the natural convection, which essentially keeps the temperature of the primary mirror closer to the air volume (Zago, 1995).

Figure (5.7) displays the plot of Fried's parameters (r_0), as obtained at the Cassegrain focus of the 2.34 m VBT, Kavalur, India, using the speckle camera[9]. These values are calculated from several sets of specklegrams of twelve different stars acquired on 28/29th March 1991 (Saha and Yeswanth, 2004). The parameter r_0 reaches its maximum value of 0.139 m at 20.329 UT, which corresponds to a seeing of $0.98''$. A poor seeing of $1.7''$ occurs at 17.867 UT. In Figure (5.7), a few sets of plots of r_0 (shown insets) depict points at which the value of r_0 changes not more than 1-2 cm during an interval of 1 min., while another set shows a variation of as high as 5 cm.

Various corrective measures have been proposed to improve the seeing at the telescope site. These are:

- insulating the surface of the floors and walls,
- introducing active cooling system to eliminate local temperature effects, heat dissipation from motor and electronic equipment on the telescope during the night and elsewhere in the dome,
- installing ventilator to generate a sucking effect through the slit to counteract the upward action of the bubbles (Racine, 1984, Ryan and Wood, 1995), and
- maintaining a uniform temperature in and around the primary mirror of the telescope (Saha and Chinnappan, 1999)

In order to remove the difference in temperature between inside and outside the dome, several ventilating fans were installed at Anglo-Australian telescope (AAT). Ventilation may generate dome flow velocities of the same order as natural convection in the presence of temperature gradients (Zago, 1995). Operation of such methods during observing may cause degradation in image quality. Iye et al. (1992) pointed out that flushing wind through the dome can be used if the turbulence generated inside the dome is dominant. If the turbulence brought from outside into the dome is larger than

[9] A camera records an event by taking in light signals and turning that into an image.

the turbulence generated inside, flushing by natural wind may turn out to be detrimental.

5.5.6.3 *Mirror seeing*

Mirror seeing is the most dominant contributer for bad seeing at ground level and has the longest time-constant, of the order of several hours depending on the size and thickness of the mirror, for equilibrating with the ambient temperature. Further, owing to reflection of optical beam by the mirror, the wavefront degrades twice by the turbulence near the mirror. The spread amounts to 0.5" for a 1° difference in temperature. The production of mirror seeing takes place very close to its surface, $\sim$ 0.02 m above (Zago, 1995 and references therein). The free convection above the mirror depends on the excess temperature of its surface above the ambient temperature with an exponent of 1.2. In what follows, some of the studies that point to mirror seeing as the dominant factor in degradation of image quality are outlined.

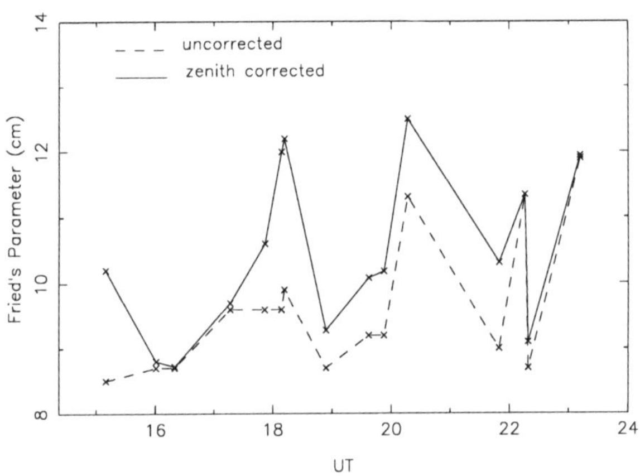

Fig. 5.8 Nighttime variations of r_0 at the 2.34 meter VBT site, Kavalur, India, on 28-29 March, 1991(Saha and Chinnappan, 1999).

Saha and Chinnappan (1999) have measured the night-time variation of Fried's parameter, as obtained at the Cassegrain focus of the 2.34 m VBT, Kavalur, India, using the speckle camera. Figure (5.8) displays the nighttime variations of r_0 on 28-29 March 1991 at the 2.34 meter VBT site. The solid line curve of the figure is for the zenith distance corrected value, while the dotted curve is for the uncorrected value. It is found that average observed r_0 is higher during the later part of the night than the earlier part, implying that the seeing improves gradually in the later part of the night at an interval of several minutes. This might indicate that the slowly cooling mirror creates thermal instabilities that decreases slowly over the night (Saha and Chinnappan, 1999); the best seeing condition may last only for a few minutes.

It may be necessary to maintain a uniform temperature in and around the primary mirror of the telescope to avoid the degradation of the seeing. Iye et al. (1991) have made extensive measurements of the mirror seeing effect with a Shack-Hartmann wavefront analyzer and opined that a temperature difference of $< 1° \ C$ should be maintained between the mirror and its ambient. The mirror seeing becomes weak and negligible if the mirror can be kept at a 1° lower temperature than the surrounding air (Iye et al., 1992). While, Racine et al., (1991) found after analyzing 2000 frames of CCD data that are obtained with high resolution camera at the 3.6 m Canada-France-Hawaii Telescope (CFHT), Mauna-Kea, Hawaii, that mirror seeing sets in as soon as the mirror is measurably warmer than the ambient air and is quite significant if it is warmer by 1°. Gillingham (1984) reported that the ventilation of the primary mirror of the 3.9 m Anglo-Australian telescope (AAT) was found to improve the seeing when the mirror is warmer than the ambient dome air, and degrade the seeing when mirror is cooler than the latter (Barr et al., 1990).

Chapter 6

Speckle imaging

6.1 Speckle phenomena

When a fairly coherent source is reflected from a surface in which the surface variation is greater than or equal to the wavelength of incident radiation, the optical wave resulting at any moderately distant point consists of many coherent wavelets, each coming from a different microscopic element of the surface. Each point on the surface absorbs and emits light. These microscopic elements produce a diffracted wave. The scattering medium introduces random path fluctuations on reflection or transmission. The large and rapid variations of the phases are the product of these fluctuations with wave vector, $\vec{\kappa}$. A change in frequency of the light changes the scale of the phase fluctuations. The intensity at a point in the far-field of the scattering medium reveals violent local fluctuations. Waves leave the scattering medium with uniform amplitudes, albeit become non-uniform rapidly as the waves propagate. The interference among numerous randomly dephased waves from scattering centers on the surface results in the granular structures of intensity. These structures containing dark and bright spots, called 'speckle' were first noticed while reconstructing an image from a hologram[1] and has been considered to be a kind of noise and was a bane of holographers. The speckle grains may be identified with the

[1] A hologram records the intensity and directional information of an optical wavefront (Gabor, 1948). If a coherent beam from a laser is reflected from an object and combined with light from a reference beam at the photographic film, a series of interference fringes are produced. These fringes form a type of diffraction pattern on the photographic film. Illuminating such a hologram with the same reference beam, diffraction from the fringe pattern on the hologram reconstructs the original object beam in both intensity and phase. With the advent of modern CCD cameras with sufficient dynamic range, the imaging of digital holograms is made possible, which can be used to optically store, retrieve, and process information.

coherence domains of the Bose-Einstein statistics[2].

A speckle pattern carries information on the position and surface structure of a rough object or on a star that is viewed through the atmosphere. Such patterns are non-symmetric, greatly distorted, or have numerous pockets of high or low intensity. The appearance of the speckle pattern is almost independent of the characteristics of the surface. The scale of the granularity depends on the grain size of diffusing surface and the distance at which the pattern is observed. The capacity of speckle patterns to carry information was utilized for interferometric measurements on real objects. From the speckle patterns, information can be derived about the non-uniformity of surfaces, which was found to be useful for metrological applications, and for stellar observations, by the technique called speckle interferometry. Such an interferometry is performed at the image plane.

Stellar speckle interferometry is a technique for obtaining spatial information on celestial objects at the diffraction-limited resolution of a telescope, despite the presence of atmospheric turbulence. In the work related to this interferometry (Labeyrie, 1970), speckle patterns in partially spatial coherent light have been extensively studied. The structure of speckles in astronomical images is a consequence of constructive and destructive bi-dimensional interference between rays coming from different zones of incident wave. One may observe the turbulence induced boiling light granules visually in a star image at the focus of a large telescope using a strong eyepiece. A typical speckle of an extended object, which is larger than the seeing-disc size, has as much angular extent as the object. With the increase in bandwidth of the light, for astronomical speckles in particular, more and more streakiness appears.

The wave passing through the atmospheric turbulence cannot be focused to a diffraction-limited image, instead it gets divided into a number of speckles fluctuating rapidly as the refracting index distribution changes with a typical correlation time of a few milliseconds (msecs). A plane wavefront passing through refractive index inhomogeneities suffers phase

[2]Bose-Einstein statistics determines the statistical distribution of identical indistinguishable bosons (particles with integer-spin) over the energy states in thermal equilibrium. This was introduced initially for photons by Bose and generalized later by Einstein. The average number of particles, n_i, in i^{th} energy level is given by,

$$\langle n_i \rangle = \frac{1}{e^{(E_i - u)/k_B T} - 1},$$

where E_i is the energy of a particle in i-state, u the chemical potential, $k_B (= 1.38 \times 10^{-23} \mathrm{JK}^{-1})$ the Boltzmann constant, and T the absolute temperature.

fluctuations; the plane wavefront no longer remains in a single plane. When such aberrated wavefronts are focused onto the focal plane of a telescope, the image is blurred.

6.1.1 *Statistical properties of speckle pattern*

Depending on the randomness of the source, spatial or temporal, speckles tend to appear. Spatial speckles may be observed when all parts of the source vibrate at the same constant frequency but with different amplitude and phase, while temporal speckles are produced if all parts of it have uniform amplitude and phase. With a heterochromatic vibration spectrum, in the case of random sources of light, spatio-temporal speckles are produced.

The formation of speckles stems from the summation of coherent vibrations having different random characteristics. The statistical properties of speckle pattern depend both on the coherence of the incident light and the random properties of medium. The complex amplitude at any point in the far field of a diffusing surface illuminated by a laser is obtained by summing the complex amplitudes of the diffracted waves from individual elements on the surface. Adding an infinite number of such sine functions would result in a function with 100% constructed oscillations (Labeyrie, 1985).

Let a speckle pattern be produced by illuminating a diffuse object with a linearly polarized monochromatic wave. The complex amplitude of an electric field, $U(\vec{r}, t) = a(\vec{r})e^{i2\pi\nu t}$ in which ν is the frequency of the wave and $\vec{r} = x, y, z$ the position vector at an observing point, consists of a multitude of de-phased contributions from different scattering regions of the uneven surface. Thus The phasor amplitude, $a(\vec{r})$, is represented as a sum of many elementary phasor contributions,

$$\begin{aligned} a(\vec{r})e^{i\psi} &= \frac{1}{\sqrt{N}}\sum_{k=1}^{N} A_k(\vec{r}) \\ &= \frac{1}{\sqrt{N}}\sum_{k=1}^{N} |a_k(\vec{r})|\, e^{i\psi_k}, \end{aligned} \tag{6.1}$$

where $A(\vec{r}) = |a(\vec{r})|\, e^{i\psi(\vec{r})}$, is the resultant complex amplitude (see equation 2.31), a_k and ψ_k are the amplitude and the phase respectively of the wave from the kth scatterer, and N the total number of scatterers.

Let the moduli of the individual complex amplitudes be equal, while their phases, after subtracting integral multiples of 2π are uniformly distributed between $-\pi$ and π. The probability density function for the phase ψ of the equation (2.31) is given by,

$$\mathcal{P}(\psi) = \frac{1}{2\pi}, \qquad \text{for } -\pi \leq \psi < \pi. \tag{6.2}$$

This reduces to random-walk problem. The probability density function of the real and imaginary parts of the complex amplitude is given by (Goodman, 1975),

$$\mathcal{P}_{r,i}\left(a_{r,i}\right) = \frac{1}{2\pi \left\langle\sigma\right\rangle^2} e^{-\left(a_r^2 + a_i^2\right)/2\left\langle\sigma\right\rangle^2}, \tag{6.3}$$

in which

$$\left\langle\sigma\right\rangle^2 = \lim_{N\to\infty} \frac{1}{N} \sum_{k=1}^{N} \frac{\left\langle\left|a_k\right|^2\right\rangle}{2}, \tag{6.4}$$

is a constant, and $\langle\,\rangle$ stands for ensemble average.

The most common value of the modulus is zero, and the phase has a uniform circular distribution. For such a speckle pattern, the complex amplitude of the resultant, $A(\vec{r})$, obeys the Gaussian statistics (Goodman, 1975). The probability density function of the intensity,

$$I(\vec{r}) = \lim_{T\to\infty} \frac{1}{2T} \int_{-T}^{T} \left|U(\vec{r},t)\right|^2 dt = \left|a(\vec{r})\right|^2, \tag{6.5}$$

of the wave obey negative exponential distribution,

$$\mathcal{P}(I) = \begin{cases} \dfrac{1}{\langle I\rangle} e^{-I/\langle I\rangle}, & I \geq 0, \\ 0, & I < 0, \end{cases} \tag{6.6}$$

where $\langle I\rangle = 2\left\langle\sigma\right\rangle^2$ is the intensity of the speckle pattern averaged over many points in the scattered field, which is associated with that polarization component.

Equation (6.6) implies that the fluctuations about the mean are pronounced. A measure of the contrast, $\mathcal{V}$, in the speckle pattern is the ratio of its standard deviation, $\langle\sigma\rangle$ to its mean, i.e.,

$$\mathcal{V} = \frac{\langle\sigma\rangle}{\langle I\rangle}. \tag{6.7}$$

For the polarized wave, the contrast is equal to unity. Due to the high contrast, speckle is extremely disturbing for the observer. Its presence yields a significant loss of effective resolution of an image. The appearance of such a speckle is independent of the nature of the surface, but its size increases with the viewing distance and the f-number of the imaging optics. The statistics presented in equation (6.6) do not reveal the variations in amplitude, intensity and phase in the speckle pattern. The second-order probability density function is required to determine the intensity distribution in the speckle pattern. By summing the intensities in two such images with mean intensities of $\langle I/2 \rangle$, the intensity is given by,

$$\mathcal{P}(I) = \frac{4I}{\langle I \rangle^2} e^{-2I/\langle I \rangle}, \tag{6.8}$$

in which

$$\langle I \rangle^2 = \langle \sigma \rangle^2 = \left\langle I^2 \right\rangle - \langle I \rangle^2, \tag{6.9}$$

is the variance of the intensity and $\left\langle I^2 \right\rangle = 2\langle I \rangle^2$ the second moment of the mean intensity.

Thus the standard deviation, $\langle \sigma \rangle$, of the probability density distribution, $\mathcal{P}(I)$, in polarized speckle patterns equals the mean intensity.

6.1.2 *Superposition of speckle patterns*

The addition of polarized speckle patterns is of practical importance. The intensity measured in many experiments at a single point in space is considered as resulting from a sum of two or more polarized speckle patterns. This can be added either on an amplitude basis or on an intensity basis. An example of addition of speckle patterns on the basis of an amplitude is in speckle shear interferometry where two such patterns are shifted followed by superposition. The complex amplitude of scattered light at any point is given by the superposition principle.

In speckle interferometry that is performed in the laboratory, the resultant speckle pattern arises when a speckled reference beam is used. This phenomena is due to the coherent superposition of two speckle patterns. Let the complex field, $A(\vec{r})$, be yielded from the addition of N different speckle patterns on an amplitude basis, i.e.,

$$A(\vec{r}) = \sum_{k=1}^{N} A_k(\vec{r}), \tag{6.10}$$

where A_k is the individual component field.

The individual fields A_k being speckles are zero-mean circular complex Gaussian random variables, the correlation that exists between the kth and l^{th} field is described by the ensemble average $\langle A_k A_l^* \rangle$. It is to be noted that the real or imaginary part of A, which is a correlated sum, in general, of Gaussian random variables, is Gaussian. Hence the real and imaginary parts of the total field, A, are also Gaussian random variables. The total intensity, $I = |A|^2$, obeys negative exponential statistics as the intensities of the component speckle patterns do. The statistics of intensity of the speckle pattern in the case of the addition of speckle patterns on an amplitude basis remain unchanged, aside from the scaling constant.

When the speckle patterns are added on the basis of the intensity, that is, if the two speckle patterns are recorded on the same photographic plate, the speckle statistics is modified and is governed by the correlation coefficient. Let the total intensity, I be composed of a sum of N speckle patterns, i.e.,

$$I(\vec{r}) = \sum_{k=1}^{N} I_k(\vec{r}), \tag{6.11}$$

in which $I = |A|^2$ and $I_k = |A_k|^2$.

The correlation coefficient of two random variables X and Y (Jones and Wykes, 1983) is given by,

$$\mathcal{C}(X,Y) = \frac{\langle XY \rangle - \langle X \rangle \langle Y \rangle}{\langle \sigma_X \rangle \langle \sigma_Y \rangle}, \tag{6.12}$$

in which $\langle \sigma_X \rangle = \sqrt{\langle X^2 \rangle - \langle X \rangle^2}$, and $\langle \sigma_Y \rangle = \sqrt{\langle Y^2 \rangle - \langle Y \rangle^2}$.

The correlation coefficient turns out to be zero if X and Y are independent, i.e, $\langle XY \rangle = \langle X \rangle \langle Y \rangle$. The correlation existing between the N intensity components is written in terms of correlation coefficients,

$$\mathcal{C}_{kl} = \frac{\langle I_k I_l \rangle - \langle I_k \rangle \langle I_l \rangle}{\sqrt{\langle I_k^2 \rangle - \langle I_k \rangle^2}\sqrt{\langle I_l^2 \rangle - \langle I_l \rangle^2}}. \tag{6.13}$$

The correlation coefficient equals unity when the ratio of the intensities is one and the two speckle patterns are in phase.

6.1.3 *Power-spectral density*

Figure (6.1) shows that the distances travelled by various rays differ when an imaginary coherent optical system is thought of as yielding an intensity distribution resulting from interference between diffracted components. Interference of the coherent wavelets yields the speckle pattern. Speckles of this kind are referred to as objective speckles, since they are present without further imaging. Here, at any point Q the amplitude is given by the sum of a set of amplitude vectors of random phase which when added together gives a random resultant amplitude. As the point is varied, say, to P, the resultant amplitude and hence, intensity, will have a different random value. It is this random intensity variation that is known as the speckle effect (Jones and Wykes, 1989).

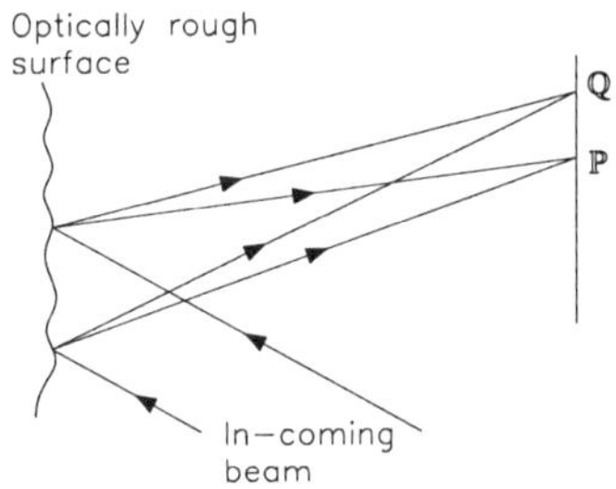

Fig. 6.1 Objective speckle formation due to an optically rough surface.

Consider that a monochromatic light is incident on an uneven surface in free space. The complex field, $U(\vec{x})$, representing the speckle field, is observed without any intervening optical element (see Figure 6.1) at a distance, s, across a plane parallel with the uneven plane, $\vec{\xi}$, in which $\vec{x} = x, y$ and $\vec{\xi} = \xi, \eta$ are the 2-D positional vectors. The autocorrelation (see Appendix B) of the intensity distribution, $I(\vec{x}) = |U(\vec{x})|^2$ in the plane $\vec{x}$, is given by

$$\mathcal{A}_I(\vec{x}_1, \vec{x}_2) = \langle I(\vec{x}_1) I(\vec{x}_2) \rangle , \tag{6.14}$$

where $\langle\,\rangle$ stands for the average over an ensemble of the uneven surface.

For circular complex Gaussian fields, the equation (6.14) is expressed as,

$$\mathcal{A}_I(\vec{x}_1, \vec{x}_2) = \langle I(\vec{x}_1) \rangle \langle I(\vec{x}_2) \rangle + |J_U(\vec{x}_1, \vec{x}_2)|^2 , \tag{6.15}$$

where $J_U(\vec{x}_1, \vec{x}_2) = \langle U(\vec{x}_1) U^*(\vec{x}_2) \rangle$ is the mutual intensity of the field.

With the help of the van Cittert-Zernike theorem (see section 3.5) of coherence theory, the mutual intensity of the observed fields is derived by the Fourier transform of the intensity distribution incident on the scattering spot $|P(\vec{\xi})|^2$, in which $P(\vec{\xi})$ represents the amplitude of the field incident on the scattering spot. The mutual coherence factor is defined as,

$$\mu_U(\vec{x}_1, \vec{x}_2) = \frac{J_U(\vec{x}_1, \vec{x}_2)}{\sqrt{J_U(\vec{x}_1, \vec{x}_1)}\sqrt{J_U(\vec{x}_2, \vec{x}_2)}}. \tag{6.16}$$

The complex coherent factor is derived as,

$$\mu_U(\vec{p}) = \frac{\int_{-\infty}^{\infty} \left|P(\vec{\xi})\right|^2 e^{-i\kappa\vec{p}\cdot\vec{\xi}} d\vec{\xi}}{\int_{-\infty}^{\infty} \left|P(\vec{\xi})\right|^2 d\vec{\xi}}, \tag{6.17}$$

in which $\vec{p} = p, q$ is the 2-D position vector. and the mutual intensity, J_U, across a plane at distance s from the source is,

$$J_U(\vec{x}_1, \vec{x}_2) = \left(\frac{\kappa}{2\pi s}\right)^2 k \int_{-\infty}^{\infty} \left|P(\vec{\xi})\right|^2 e^{-i\kappa\vec{\xi}_1 \cdot [\vec{x}_1 - \vec{x}_2]} d\vec{\xi}_1, \tag{6.18}$$

where k is the proportionality constant.

Thus the autocorrelation function of the speckle intensity takes the form,

$$\mathcal{A}_I = \langle I \rangle^2 \left[1 + |\mu_U(\vec{p})|^2\right]. \tag{6.19}$$

By using Wiener-Khintchine theorem, the power spectral density, $\widehat{\Gamma}_I(\vec{u})$, of the speckle intensity distribution, $I(\vec{x})$, is derived (Goldfisher, 1965). Hence, after applying Fourier transform, the equation (6.19) can be recast as,

$$\widehat{\Gamma}_I(\vec{u}) = \langle I \rangle^2 \left\{ \delta(\vec{u}) + \frac{\int_{-\infty}^{\infty} \left|P(\vec{\xi})\right|^2 \left|P(\vec{\xi} - \lambda\vec{u})\right|^2 d\vec{\xi}}{\left[\int_{-\infty}^{\infty} \left|P(\vec{\xi})\right|^2 d\vec{\xi}\right]^2} \right\}, \tag{6.20}$$

where $\vec{u} = \vec{x}/\lambda$ is the 2-D spatial frequency vector (u, v) and δ the Dirac delta function.

Equation (6.20) states that the power spectral density of the speckle pattern consists of delta function component with zero spatial frequency, $\vec{u} = 0$, plus a component extended over frequency. Let a rough surface, illuminated by a laser light, be imaged on the recording plane by placing a lens in the speckle field of the object (see Figure 6.2). The image appears a random intensity variations as in the case of objective speckles, but in this

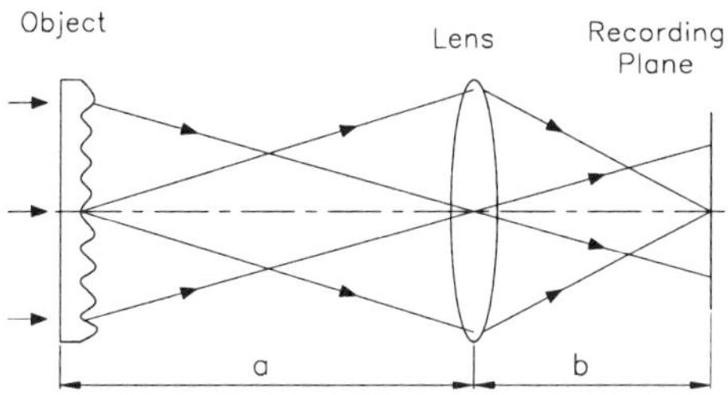

Fig. 6.2 Schematic diagram for illustrating the formation of subjective speckle pattern.

case the speckle is called subjective. Indeed, every imaging system alters the coherent superposition of waves and hence produces its own speckle size. Due to interference of waves from several scattering centers in the aperture the randomly dephased impulse response functions are added, yielding in a speckle. The size of this subjective speckle is governed by the Airy disc (see section 3.6.3.2). The fringe separation, θ, is given by, $\theta = 1.22\lambda b/D$, in which b is the distance between the lens and the image plane, and D the diameter of the lens. In terms of aperture ratio, $F\#$, of the lens, and magnification, $\mathrm{M}(= b/a)$, with a as the object distance, the speckle size is written as, $\theta = 1.22(1+\mathrm{M})\lambda F\#$. The average speckle size decreases as the aperture of the imaging system increases, although the aberrations of the system do not alter the speckle size. The control of speckle size by $F\#$ is used in speckle metrology to match the speckle size with the pixel size of a CCD array detector.

Since the disturbance of the image, $U(\vec{x})$, at a point $\vec{x}$ is the convolution of the object, $O(\vec{x})$ and the point spread function of the optical system, $K(\vec{x})$, mathematically one writes,

$$U(\vec{x}) = O(\vec{x}) \star K(\vec{x}). \tag{6.21}$$

in which $\star$ stands for convolution.

The spectral correlation function becomes,

$$\begin{aligned}\widehat{\Gamma}(\vec{x}_1,\nu_1;\vec{x}_2,\nu_2) &= \left\langle \widehat{U}(\vec{x}_1,\nu_1)\widehat{U}^*(\vec{x}_2,\nu_2)\right\rangle \\ &= \left\langle \widehat{O}(\vec{x}_1,\nu_1)\star\widehat{O}^*(\vec{x}_2,\nu_2)\right\rangle \star \left[\widehat{K}(\vec{x}_1,\nu_1)\widehat{K}(\vec{x}_2,\nu_2)\right].\end{aligned} \tag{6.22}$$

The first term of the RHS of the equation (6.22) has the properties of the surface, while the second term describes the properties of the imaging system.

6.2 Speckle pattern interferometry with rough surface

The speckle patterns in quasi-monochromatic light may arise either when the light is reflected from the scattering surface or when it passes through the turbulent atmosphere. The properties of these patterns can be determined by using an analysis that are used in polychromatic light. Since the intensity in a speckle pattern is the sum of the intensities from all point of the source, the variables ν_1 and ν_2 are replaced by the co-ordinates of two point sources $\mathtt{Q}_1(\vec{r}_1')$, and $\mathtt{Q}_2(\vec{r}_2')$. The function that governs the statistical properties of the pattern is the correlation of the disturbance in the speckle pattern at a point $\mathtt{P}_1(\vec{r}_1)$ produced by a source at $\mathtt{Q}_1(\vec{r}_1')$, with that at $\mathtt{P}_2(\vec{r}_2)$ produced by a source point at $\mathtt{Q}_2(\vec{r}_2')$. The angular correlation function is represented by,

$$\Gamma_U(\vec{r}_1, \vec{r}_2; \vec{r}_1', \vec{r}_2') = \langle U(\vec{r}_1, \vec{r}_1')U^*(\vec{r}_2, \vec{r}_2')\rangle\,. \tag{6.23}$$

6.2.1 *Principle of speckle correlation fringe formation*

Consider that a plane wavefront is split into two components of equal intensities by a beamsplitter similar to that of Michelson classical interferometer (see Figure 6.3). Of these two wavefronts, one arises from the object and is speckled, while the other could be a specular by reflected or a diffused reference wave, reflected off the optically rough surfaces. They interfere on recombination and are recorded in the image plane of the lens-aperture combination. The second exposure is recorded on the same plate after the object undergoes any physical deformation with the camera position remaining unchanged. This record on development is known as double-exposure specklegram. Fringes that are contours of the constant out-of-plane displacement, are more prominently the higher order terms observed on spatial filtering.

Let $U_1(\vec{r}, t) = A_1(\vec{r}, t)e^{i\psi_1(\vec{r},t)}$ and $U_2(\vec{r}, t) = A_2(\vec{r}, t)e^{i\psi_2(\vec{r},t)}$, in which $A_{j=1,2}(\vec{r}, t)$ and $\psi_{j=1,2}(\vec{r}, t)$, correspond respectively to the randomly varying amplitude and phase of the individual image plane speckles. The resulting intensity distribution is expressed as,

$$\begin{aligned}\mathcal{I}_1(\vec{r},t) &= U_1^2(\vec{r},t) + U_2^2(\vec{r},t) + 2U_1(\vec{r},t)U_2(\vec{r},t)\cos[\psi_1(\vec{r},t) - \psi_2(\vec{r},t)] \\ &= I_1(\vec{r},t) + I_2(\vec{r},t) + 2\sqrt{I_1(\vec{r},t)I_2(\vec{r},t)}\cos\psi(\vec{r},t),\end{aligned} \tag{6.24}$$

where $I_1(\vec{r},t) = |U_1(\vec{r},t)|^2$, $I_2(\vec{r},t) = |U_2(\vec{r},t)|^2$ are the intensities of the two beams and $\psi(\vec{r},t) = \psi_1(\vec{r},t) - \psi_2(\vec{r},t)$ the random phase.

Unlike in classical interferometry where the resultant intensity distribution represents a fringe pattern, this distribution represents a speckle pattern. In order to generate a fringe pattern, the object wave has to carry an additional phase that may arise due to deformation of the object. With the introduction of phase difference, δ, the intensity distribution is written as,

$$\mathcal{I}_2(\vec{r},t) = I_1(\vec{r},t) + I_2(\vec{r},t) + 2\sqrt{I_1(\vec{r},t)I_2(\vec{r},t)}\cos[\psi(\vec{r},t) + \delta]. \tag{6.25}$$

In a double-exposure specklegram, the two speckle patterns derived from the object in its two states are superimposed. Two shifted speckle patterns are simultaneously recorded with the space variant shift. Simultaneous superimposition of two such patterns are not required if the time between the two illuminations is less than the persistence time (~100 msec). The resulting intensity distribution, $\mathcal{I}(\vec{r},t)$, is represented by,

$$\begin{aligned}\mathcal{I}(\vec{r},t) &= \mathcal{I}_1(\vec{r},t) + \mathcal{I}_2(\vec{r},t) \\ &= 2\left[I_1(\vec{r},t) + I_2(\vec{r},t) + 2\sqrt{I_1(\vec{r},t)I_2(\vec{r},t)}\cos\left(\psi + \frac{\delta}{2}\right)\cos\left(\frac{\delta}{2}\right)\right].\end{aligned} \tag{6.26}$$

The first two terms of the RHS of the equation (6.26) represent a random intensity distribution and as such are due to the superposition of two speckle patterns. The third term is the intermodulation term in which $\cos(\psi + \delta/2)$ provides random values and $\cos(\delta/2)$ is the deterministic variable.

Following equation (6.12), the correlation coefficient of $I_1(\vec{r},t)$ and $I_2(\vec{r},t)$ defined in equations (6.24 and 6.25) can be deduced as,

$$\mathcal{C}(\delta) = \frac{\langle \mathcal{I}_1(\vec{r},t)\mathcal{I}_2(\vec{r},t)\rangle - \langle \mathcal{I}_1(\vec{r},t)\rangle\langle \mathcal{I}_2(\vec{r},t)\rangle}{\sqrt{\langle \mathcal{I}_1^2(\vec{r},t)\rangle - \langle \mathcal{I}_1(\vec{r},t)\rangle^2}\sqrt{\langle \mathcal{I}_2^2(\vec{r},t)\rangle - \langle \mathcal{I}_2(\vec{r},t)\rangle^2}}. \tag{6.27}$$

By considering $I_1(\vec{r},t)$, $I_2(\vec{r},t)$, and $\psi(\vec{r},t)$ as independent variables which can be averaged separately and $\langle\cos\psi(\vec{r},t)\rangle = \langle\cos(\psi(\vec{r},t) + \delta)\rangle = 0$, the equation (6.27) can be evaluated. From the equation (6.9), it can be observed $\langle I^2 \rangle = 2\langle I \rangle^2$ and assuming $\langle I_1(\vec{r},t)\rangle = \langle I_2(\vec{r},t)\rangle = \langle I(\vec{r},t)\rangle$, one may write,

$$\mathcal{C}(\delta) = \frac{1}{2}(1+\cos\delta)\,. \tag{6.28}$$

Thus the correlation turns out to be zero or unity whenever δ becomes,

$$\delta = \begin{cases} (2m+1)\pi & m = 0,1,2,3,\cdots, \\ 2m\pi & . \end{cases} \tag{6.29}$$

As δ varies over the object surface, the intensity variation is seen on a gross scale. Such a variation is termed as fringe pattern. These fringes are highly speckled.

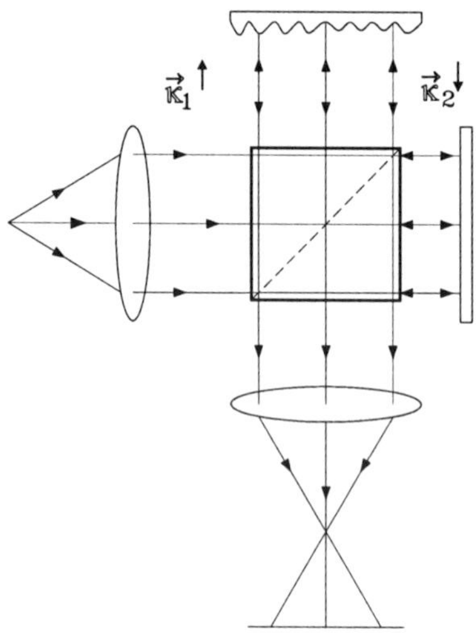

Fig. 6.3 The Michelson interferometer arrangement for out-of-plane displacement sensitive speckle interferometer.

If $\langle I_2(\vec{r},t)\rangle = \rho\langle I_1(\vec{r},t)\rangle$, in which ρ is the ratio of averaged intensities of the two speckle patterns, the correlation coefficient can be recast into,

$$\mathcal{C}(\delta) = \frac{1+\rho^2+2\rho\cos\delta}{(1+\rho)^2}, \tag{6.30}$$

which has a maximum value of unity when $\delta = 2m\pi$ and a minimum value of $[(1-\rho)/(1+\rho)]^2$ when $\delta = (2m+1)\pi$.

Following Figure (6.3), the phase difference due to deformation is deduced as,

$$\delta = (\vec{\kappa}_2 - \vec{\kappa}_1) \cdot d(\vec{r}) = \frac{4\pi d_z}{\lambda}, \tag{6.31}$$

in which $\vec{\kappa}_1$, $\vec{\kappa}_2$ are the propagation vectors in the direction of illumination and observation respectively and $d(\vec{r}) = d_x, d_y, d_z$ the displacement vector at a point on the object.

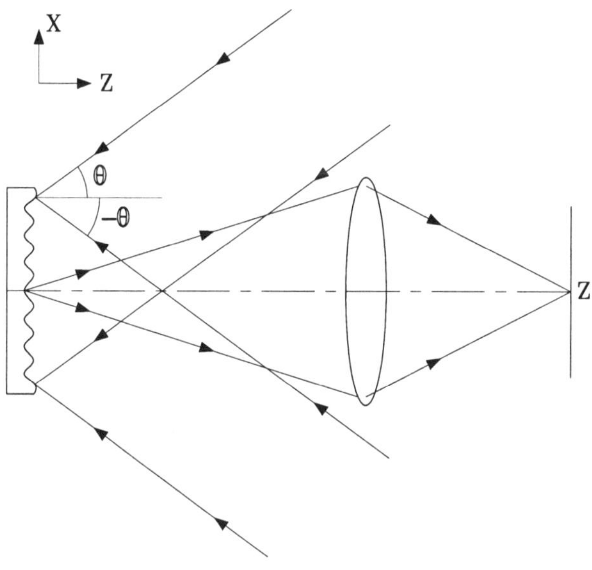

Fig. 6.4 Schematic diagram for in-plane displacement measurement.

It is noted here that the phase difference depends on the out-of-plane displacement component, d_z. Bright fringes occurs along the lines where,

$$d_z = \frac{1}{2} m\lambda. \tag{6.32}$$

When the object is illuminated by two plane wavefronts, U_0 and U_0', inclined at equal and opposite angles, θ, to object normal (see Figure 6.4) and observation is made along the optical axis, the observation generates fringes which are sensitive to in-plane displacement (Leendertz, 1970). If the phase difference introduced by deformation is governed by the equation (6.31), the relative phase change of the two beams is derived as,

$$\begin{aligned}\delta &= \delta_2 - \delta_1 \\ &= (\vec{\kappa}_2 - \vec{\kappa}_1) \cdot d(\vec{r}) - (\vec{\kappa}_2 - \vec{\kappa}_1') \cdot d(\vec{r}) \\ &= (\vec{\kappa}_1' - \vec{\kappa}_1) \cdot d(\vec{r}) \\ &= \frac{4\pi}{\lambda} d_x \sin\theta, \end{aligned} \tag{6.33}$$

in which δ_2 and δ_1 are the phases acquired due to two wavefronts, U_0 and U_0' and $\vec{\kappa}_1'$ and $\vec{\kappa}_1$ the propagation vectors of the said wavefronts.

Bright fringes are discernible when,

$$\frac{4\pi}{\lambda} d_x \sin\theta = 2m\pi. \tag{6.34}$$

Thus,

$$d_x = \frac{m\lambda}{2\sin\theta}. \tag{6.35}$$

It may be noted here that the arrangement is sensitive only to the x-component of the in-plane displacement.

6.2.2 *Speckle correlation fringes by addition*

In order to produce speckle correlation fringes either addition or subtraction method based on the electronic addition or subtraction of the signals corresponding to the deformed and initial states of the object may be employed. In the case of the former, the CCD output in terms of voltage is proportional to the added intensities. Invoking equation (6.26), one may write,

$$V = V_1 + V_2 \propto \mathcal{I}_1 + \mathcal{I}_2. \tag{6.36}$$

This technique is employed for the observation of time-averaged fringes. A dual pulsed laser can be used for such method. The contrast of the fringes is defined as the standard deviation of the intensity, i.e.,

$$\langle\sigma\rangle = 2\left[\langle\sigma_1\rangle^2 + \langle\sigma_2\rangle^2 + 8\langle I_1\rangle\langle I_2\rangle\cos^2\frac{\delta}{2}\right]^{1/2}, \tag{6.37}$$

in which $\langle\sigma_1\rangle$ $\langle\sigma_2\rangle$ are the standard deviations of the respective intensities I_1, I_2.

The standard deviation (see Appendix B), $\langle\sigma\rangle$, varies between maximum and minimum values when,

$$\langle\sigma\rangle = \begin{cases} 2\left[\langle\sigma_1\rangle^2 + \langle\sigma_2\rangle^2 + 2I_1I_2\right]^{1/2}, & \delta = 2m\pi, \\ 2\left[\langle\sigma_1\rangle^2 + \langle\sigma_2\rangle^2\right]^{1/2}, & \delta = (2m+1)\pi, \end{cases} \tag{6.38}$$

The contrast of these fringes can be increased considerably after removing the DC speckle component, $2[\langle I_1\rangle + \langle I_2\rangle]$, by filtering. Bright and dark fringes can be envisaged when two speckle patterns are correlated and decorrelated respectively. The resulting brightness, B_r, of the video monitor is given by,

$$B_r = k\left[\langle\sigma_1\rangle^2 + \langle\sigma_2\rangle^2 + 2\langle I_1\rangle\langle I_2\rangle\cos^2\frac{\delta}{2}\right]^{1/2}, \tag{6.39}$$

in which k is a constant of proportionality.

On averaging the intensity over time, τ, the equation (6.24) can be written as,

$$\mathcal{I}(\tau) = I_1 + I_2 + \frac{2}{\tau}\sqrt{I_1I_2}\int_0^\tau \cos\left[\psi + 2\kappa a(t)\right]dt, \tag{6.40}$$

with $\kappa = 2\pi/\lambda$ and $a(t) = a_0\sin\omega t$ in which a_0 is the amplitude of the vibration across the object surface.

Equation (6.40) is deduced as,

$$\mathcal{I}(\tau) = I_1 + I_2 + 2\sqrt{I_1I_2}J_0(2\kappa a_0)\cos\psi, \tag{6.41}$$

where J_0 is the zero order Bessel function.

A variation in the contrast of the speckle pattern can be observed on the monitor. With high pass filtration followed by rectification, the brightness, B_r,

$$B_r = k\left[\langle\sigma_1\rangle^2 + \langle\sigma_2\rangle^2 + 2\langle I_1\rangle\langle I_2\rangle J_0^2(2\kappa a_0)\right]^{1/2}. \tag{6.42}$$

The maxima and minima of the fringe corresponds to the maxima and minima of J_0^2.

6.2.3 *Speckle correlation fringes by subtraction*

Since the typical size of the speckles ranges between 5 to 100 μm, either a standard television camera or a CCD camera is necessary to record their pattern. This process, known as electronic speckle pattern interferometry (ESPI), enables real-time correlation fringes to be displayed directly upon a video monitor. The present day CCD camera is able to process digitally by using appropriate software. Let a CCD camera be placed in the image plane of a speckle interferometer. The analogue video signal V_1 from the sensor is sent to an analogue-to-digital (A/D) converter for sampling the signal at the video rate and records it as a digital frame in the memory of a PC for processing. An additional phase δ is introduced soon after the loading of the object, therefore the intensity distributions of these records in the detector are given respectively by,

$$\mathcal{I}_1(\vec{x}) = |U_1(\vec{x})|^2 + |U_2(\vec{x})|^2 + 2U_1(\vec{x})U_2(\vec{x})\cos\psi(\vec{x}), \tag{6.43}$$

$$\mathcal{I}_2(\vec{x}) = |U_1(\vec{x})|^2 + |U_2(\vec{x})|^2 + 2U_1(\vec{x})U_2(\vec{x})\cos[\psi(\vec{x}) + \delta], \tag{6.44}$$

in which $\vec{x} = x, y$ is the 2-D positional vector.

The subtracted signal is written as,

$$\begin{aligned} V &= V_2 - V_1 \propto \mathcal{I}_2(\vec{x}) - \mathcal{I}_1(\vec{x}) \\ &= 4U_1(\vec{x})U_2(\vec{x}) \sin\left(\psi + \frac{\delta}{2}\right) \sin\left(\frac{\delta}{2}\right), \end{aligned} \tag{6.45}$$

with V_1, V_2 as the output camera signals, which are proportional to the input intensities.

The signal has positive and negative values. In order to avoid this loss of signal, a negative DC bias is added before being fed to the monitor. The brightness of the monitor is given by,

$$B_r' = 4k\left|\sqrt{I_1 I_2}\sin\left(\psi + \frac{\delta}{2}\right)\sin\left(\frac{\delta}{2}\right)\right|, \tag{6.46}$$

The term $\sin(\psi + \delta/2)$ in equation (6.46) represents the high frequency speckle noise and the interference term $\sin(\delta/2)$ modulates the speckle term. Due to subtraction procedure, the DC speckle terms are eliminated. The bright and dark fringes would be discernible wherever δ turns out to be, $(2m + 1)\pi$ and $2m\pi$ respectively in which $m = 0, 1, 2, \cdots$. The brightness, B_r, varies between maximum and minimum values and is given by,

$$B_r' = \begin{cases} 2k\sqrt{I_1 I_2}, & \delta = (2m+1)\pi, \\ \\ 0, & \delta = 2m\pi. \end{cases} \tag{6.47}$$

In contrast with speckle pattern interferometry, the bright and dark fringes due to the subtraction method are envisaged when two speckle patterns are decorrelated and correlated respectively.

6.3 Stellar speckle interferometry

In short-exposure images, the movement of the atmosphere is too sluggish to have any effect. The speckles recorded in the image are a snapshot of the atmospheric seeing at that instant. Labeyrie (1970) showed that information about the high resolution structure of a stellar object could be inferred from speckle patterns using Fourier analysis. This technique is is used widely to decipher diffraction-limited spatial Fourier spectrum and image features of stellar objects.

Let a point source be imaged through the telescope by using the pupil function consisting of two small sub-pupils (θ_1, θ_2), separated by a distance, d, corresponding to the two seeing cells separated by a vector $\lambda\vec{u}$. Each sub-pupil diffracts the incoming light and one obtains linear interference fringes with narrow spatial frequency bandwidth. Such a pupil is small enough for the field to be coherent over its extent. Any stellar object is too small to be resolved through a single sub-pupil. Atmospheric turbulence causes random phase fluctuations of the incoming optical wavefront, so the random variation of phase difference between the two sub-pupils leads to the random motion of the amplitude and phase of the sinusoidal fringe move within a broad PSF envelope, which are determined by the amplitude and phase of the mutual intensity transmitted by the exit pupil.

If the phase shift between the sub-pupils is equal or greater than the fringe spacing λ/d, fringes will disappear in a long exposure, hence one may follow their motion by recording a sequence of short exposures to freeze the instantaneous fringe pattern. The turbulence does not affect the instantaneous contrast of the fringes produced, but for a long exposure, the phase and the random perturbation of the logarithm of the amplitude vary through a reasonable portion of the ensemble average of all the values. The phase delays introduced by the atmospheric turbulence shift the fringe pattern randomly and smear the fringe pattern during long-exposure.

The introduction of a third sub-pupil, which is not collinear with the former two sub-pupils provides three non-redundant pairs of sub-pupils and yields the appearance of three intersecting patterns of moving fringes. Covering the telescope aperture with r_0-sized sub-pupils synthesizes a filled aperture interferometer. In the presence of many such pair of sub-pupils, the interfering fringes produce enhanced bright speckles of width, $\sim \lambda/D$, in which λ is the wavelength of interest and D the diameter of the telescope aperture.

According to the diffraction theory (Born and Wolf, 1984), the total amplitude and phase of the spectral component of image intensity, I, obtained in the focal plane of the system is the result of addition of all such fringes with frequencies that take proper account of both amplitudes and their spatial phases. The major component, $\widehat{I}(\vec{u})$, at the frequency, $\vec{u}$, is produced by contributions from all pairs of points with separations $\lambda\vec{u}$, with one point in each aperture. With increasing distance of the baseline between two subapertures, the fringes move with an increasingly larger amplitude. No such shift is observed on long-exposure images, which implies the loss of high frequency components of the image. In the presence of severe atmospheric turbulence with a short-exposure, the interference fringes are preserved but their phases are randomly distorted. This produces speckles arising from the interference of light from many coherent patches distributed over the full aperture of a telescope.

Constructive interference of the fringes would show an enhanced bright speckle. One speckle with unusually high intensity, resulting in an image Strehl ratio of 3 to 4 times greater than the median Strehl ratio for short-exposure specklegrams. A vast majority of the light is distributed in a large number of fainter speckles. In the case of a complex object, each of these fainter speckles contributes noise to the image. This gives rise to poor image quality.

High-frequency angular information is contained in a specklegram composed of such numerous short-lived speckles. The number of speckles, n_s, per image is defined by the ratio of the area occupied by the seeing disc, $1.22\lambda/r_0$ to the area of a single speckle, which is of the same order of magnitude as the Airy disc of the aperture (see equation 4.78),

$$n_s = \left(\frac{\lambda}{r_0}\right)^2 : \left(\frac{\lambda}{D}\right)^2 = \left(\frac{D}{r_0}\right)^2. \tag{6.48}$$

The number of photons, n_p, per speckle is independent of its diameter. The structure of the speckle pattern changes randomly over short intervals of time. The speckle lifetime, τ_s (milliseconds) is defined by the velocity dispersion in the turbulent atmosphere,

$$\tau_s \sim \frac{r_0}{\Delta\nu}, \tag{6.49}$$

where $\Delta\nu$ is the velocity dispersion in the turbulent seeing layers across the line of sight.

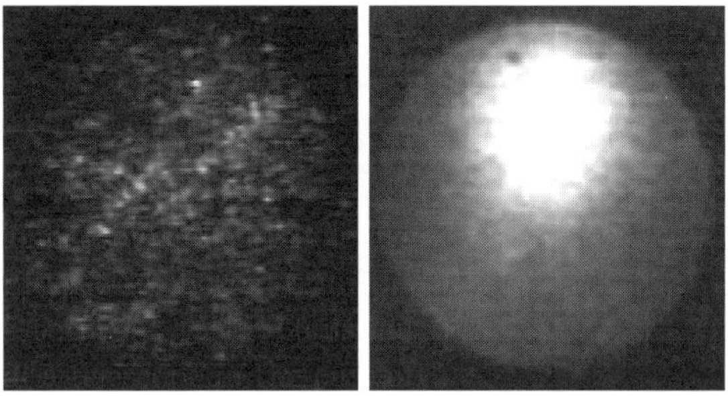

(a) (b)

Fig. 6.5 (a) Instantaneous specklegram of a close binary star HR4689 taken at the Cassegrain focus of 2.34 meter VBT, situated at Kavalur, India on 21st February, 1997 (Saha and Maitra, 2001), and (b) Result of summing 128 speckle pictures of the same, which tends to a uniform spot.

Key to the stellar speckle interferometry is to take very fast images in which the atmosphere is effectively frozen in place. Under typical atmospheric conditions, speckle boiling can be frozen with exposures in the range between 0.02 s and 0.002 s or shorter for visible wavelength, while they are between 0.1 s and 0.03 s for infrared wavelength. If the speckle pattern is not frozen enough due to the long integration time, intermediate spatial frequencies vanish rapidly. The exposure time, Δt, is to be selected accordingly to maximize the S/N ratio. Figure (6.5a) illustrates speckles of a binary star HR4689, recorded at the Cassegrain focus of 2.34 m Vainu Bappu Telescope (VBT), Vainu Bappu Observatory (VBO), Kavalur, India, (Saha and Maitra, 2001).

6.3.1 *Outline of the theory of speckle interferometry*

Speckle interferometry estimates the modulus square of the Fourier transform of the irradiance from the specklegrams of the object of interest. This is averaged over the duration of the short, narrow bandpass exposure. It is pertinent to note that the bandpass should be narrow enough to provide temporal coherence over the whole image plane. It is determined by 100 $nm/(D\theta_s)$, in which θ_s is the seeing disc and D the size of the telescope diameter in meters. An ensemble of such specklegrams, $I_j(\vec{x}), j = t_1, t_2, t_3, \ldots, t_N$, constitute an astronomical speckle observation. By obtaining a short-exposure at each frame the atmospheric turbulence is frozen in space at each frame interval.

In an imaging system, the equation (5.97) should be modified by the addition of a noise component, $N(\vec{x})$. The variability of the corrugated wavefront yields 'speckle boiling' and is the source of speckle noise that arises from difference in registration between the evolving speckle pattern and the boundary of the PSF area in the focal-plane. In general, the specklegrams have additive noise contamination, $N_j(\vec{x})$, which includes all additive measurement of uncertainties. This may be in the form of:

(1) photon statistics noise and
(2) all distortions from the idealized iso-planatic model represented by the convolution of $O(\vec{x})$ with $S(\vec{x})$, which includes non-linear geometrical distortions.

A specklegram represents the resultant of diffraction-limited incoherent imaging of the object irradiance convolved with the intensity PSF. The quasi-monochromatic incoherent imaging equation applies,

$$I(\vec{x},t) = O(\vec{x}) \star S(\vec{x},t) + N(\vec{x},t), \tag{6.50}$$

where $I(\vec{x},t)$ is the intensity of the degraded image, $S(\vec{x})$ the space-invariant blur impulse response, $\star$ denotes convolution, $\vec{x} = x, y$ is the 2-D position vector, and $O(\vec{x})$ the intensity of an object at a point anywhere in the field of view, and $N(\vec{x},t)$ is the noise.

The recorded speckle image of the speckle pattern is related to the object irradiance by,

$$\frac{1}{\Delta t}\int_0^{\Delta t} I(\vec{x},t)dt = O(\vec{x}) \star \frac{1}{\Delta t}\int_0^{\Delta t} S(\vec{x},t)dt + \frac{1}{\Delta t}\int_0^{\Delta t} N(\vec{x},t)dt, \tag{6.51}$$

in which Δt is the exposure time.

Denoting the noise spectrum as $\widehat{N}(\vec{u},t)$, the Fourier spectra of the degraded images are,

$$\frac{1}{\Delta t}\int_0^{\Delta t} \widehat{I}(\vec{u},t)dt = \widehat{O}(\vec{u}).\frac{1}{\Delta t}\int_0^{\Delta t} \widehat{S}(\vec{u},t)dt + \frac{1}{\Delta t}\int_0^{\Delta t} \widehat{N}(\vec{u},t)dt. \tag{6.52}$$

where $\vec{u} = u, v$ is the 2-D spatial frequency vector, $\widehat{O}(\vec{u},t)$ the object spectrum, and $\widehat{S}(\vec{u},t)$ the blur transfer function.

In measuring the average transfer function for some image frequency $\vec{u}$, the amplitudes of the components are added. If each speckle picture is squared, so that at all frequencies the picture transforms are real and positive, the sum of such transforms retains the information at high frequency, and therefore, yields greatly improved signal-to-noise (S/N) ratio. By integrating autocorrelation function of the successive short-exposure records rather than adding the images themselves, the diffraction-limited information can be obtained. Indeed, autocorrelation of a speckle image preserves some of the information in the way which is not degraded by the co-adding procedure. The analysis of data may be carried out in two equivalent ways. In the spatial domain, the ensemble average space autocorrelation is found giving the resultant imaging equation. The ensemble average of the Wiener spectrum is found by writing,

$$\left\langle \left|\widehat{I}(\vec{u})\right|^2 \right\rangle = \left|\widehat{O}(\vec{u})\right|^2 \cdot \left\langle \left|\widehat{S}(\vec{u})\right|^2 \right\rangle + \left\langle \left|\widehat{N}(\vec{u})\right|^2 \right\rangle. \tag{6.53}$$

The term $|\widehat{S}(\vec{u})|^2$ describes how the spectral components of the image are transmitted by the atmosphere and the telescope. Such a function is unpredictable at every moment, albeit its time-averaged value can be determined if the seeing conditions are unchanged. Since $|\widehat{S}(\vec{u})|^2$ is a random function in which the detail is continuously changing, its ensemble average becomes smoother. This form of transfer function can be obtained by calculating Wiener spectrum of the instantaneous intensity distribution from the reference star (unresolved star) for which all observing conditions are required to be identical to those for the target star. However, the short-exposure PSFs for two stars (target and reference) separated by more than θ_0 are different. Such a comparison is likely to introduce deviation in the statistics of speckles from the expected model based on the physics of the atmosphere. This, in turn, would result either in the suppression or in the enhancement of intermediate spatial frequencies, which unheeded could lead to the discovery of rings or discs around some poor unassuming star! One must be cautious in interpreting high resolution data. It is essential to choose the point source calibrator as close as possible, preferably within 1° of the programme star; the object and calibrator observations should be interleaved to calibrate for changing seeing condition by shifting the telescope back and forth during the observing run to equalize seeing distributions for both target and reference.

The quality of the image degrades due to (i) variations of airmass between the object and the reference or of its time average, (ii) deformation of mirrors or to a slight misalignment while changing its pointing direction, (iii) bad focusing, and (iv) thermal effect from the telescope as well. Another problem in speckle reconstruction technique arises from the photon noise. It causes the bias that may change the result of reconstruction. The photon bias has to be compensated during the reconstruction procedure using the average photon profile determined from the acquisition of photon fields.

6.3.2 *Benefit of short-exposure images*

It is reiterated that the seeing-limited PSF is defined by the ensemble average (equation 5.100), where the distribution of speckles becomes uncorrelated in time and the statistics of irradiance turn out to be Gaussian over the seeing disc. The obtainable resolution of a telescope in the case of long-exposure (see section 5.4.1) is governed by the form of average transfer function, $<\widehat{S}(\vec{u})>$, while in the case of short-exposure it is governed by the energy transfer function, $<|\widehat{S}(\vec{u})|^2>$. The result of summing several specklegrams from a point source can result in a uniform patch of light a few arcseconds ($''$) wide, destroying the finer details of an image (see Figure 6.5b). The Fourier transform of such a composite picture is the same as the sum of the individual picture transforms because the transformation is a linear process. From the analytical expression for the power spectrum of short-exposure images, proposed by Korff (1973), it is found that the asymptotic behavior extends up to the diffraction cut-off limit of the telescope, which means that the typical size of a speckle is of the order of the Airy disc of a given aperture. The short-exposure transfer function includes the high spatial frequency component, while the low-frequency part of $<|\widehat{S}(u)|^2>$ corresponds to a long-exposure image with the wavefront tilts compensated.

Image gets smeared during the long-exposure by its random variations of the tilt, which becomes larger than what is determined by the stationary mean atmospheric seeing angle. The image sharpness and the MTF are affected by wavefront tilt. A random factor associated with the tilt is extracted from the MTF before being taken the average. In a long-exposure image, no shift can be visualized in the fringe movement. When the major component, $\widehat{I}(\vec{u})$, at the frequency, $\vec{u}$, is averaged over many frames, the resultant for frequencies greater than r_0/λ, tends to zero. The phase-difference between the two sub-pupils is distributed uniformly between $\pm\pi$. The Fourier component performs a random walk in the complex plane and averages to zero:

$$\left\langle \widehat{I}(\vec{u}) \right\rangle = 0, \qquad u > r_0/\lambda, \tag{6.54}$$

and the argument of which (from equation 5.101), is given by,

$$arg\left|\widehat{I}(\vec{u})\right| = \psi(\vec{u}) + \theta_1 - \theta_2, \tag{6.55}$$

in which $\psi(\vec{u})$ is the Fourier phase at $\vec{u}$, $arg|\,|$ stands for "the phase of", and $\theta_{j=1,2}$ represent the apertures corresponding to the seeing cells.

In the case of autocorrelation technique, the major Fourier component $\widehat{I}(\vec{u})$ of the fringe pattern is averaged as a product with its complex conjugate and so the atmospheric phase contribution is eliminated and the averaged signal is non-zero. Therefore, the resulting representation in the Fourier space is,

$$\left\langle \widehat{I}^{\mathcal{A}_c}(\vec{u})\right\rangle = \left\langle \widehat{I}(\vec{u})\widehat{I}^*(\vec{u})\right\rangle = \left\langle \left|\widehat{I}(\vec{u})\right|^2\right\rangle \neq 0. \tag{6.56}$$

However, in such a technique, the complete symmetry of the operation prevents the preservation of Fourier phase information; for an arbitrary shape object the information cannot be recovered. The argument of this equation (6.56) is always zero,

$$\begin{aligned} arg\left|\widehat{I}(\vec{u})\widehat{I}(-\vec{u})\right| &= \theta^{\mathcal{A}_c}(\vec{u}) \\ &= \psi(\vec{u}) + \theta_1 - \theta_2 + \psi(-\vec{u}) - \theta_1 + \theta_2 = 0, \end{aligned} \tag{6.57}$$

where $\psi(\vec{u})$ is the phase of $\widehat{I}(\vec{u})$ and $arg|\,|$ defines the phase of the complex number $mod\,2\pi$.

6.3.3 *Data processing*

If the transfer function, $< |\widehat{S}(\vec{u})|^2 >$, is known, the object transfer function, $|\widehat{O}(\vec{u})|^2$, can be estimated. In practice, the PSF is determined from the short-exposure images taken from an unresolved star close to the target star of interest (within the field of view). Usually specklegrams of a brightest possible reference star are recorded to ensure that the S/N ratio of reference star is much higher than the S/N ratio of the programme star. In the absence of such data from a brighter one, one should take larger samples than the target star to map the PSF due to the telescope and the atmosphere accurately. The size of the data sets is constrained by the consideration of the S/N ratio. The Fourier transform of a point source (delta function) is a constant, C_δ. For a point source, the equation (6.53) reduces to,

$$\left\langle \left|\widehat{I}_s(\vec{u})\right|^2 \right\rangle = C_\delta^2 . \left\langle \left|\widehat{S}(\vec{u})\right|^2 \right\rangle . \tag{6.58}$$

To find C_δ^2, one has to find the boundary condition. At the origin of the Fourier plane, $\vec{S}(\vec{u} = 0)$ is unity. This is true for an incoherent source. Hence, C_δ^2 is given by,

$$C_\delta^2 = \left\langle \left|\widehat{I}_s(\vec{0})\right|^2 \right\rangle / \left\langle \left|\widehat{S}(\vec{0})\right|^2 \right\rangle \simeq \left\langle \left|\widehat{I}_s(\vec{0})\right|^2 \right\rangle . \tag{6.59}$$

Using equation (6.59) in equation (6.58) gives,

$$\left\langle \left|\widehat{S}(\vec{u})\right|^2 \right\rangle = \left\langle \left|\widehat{I}_s(\vec{u})\right|^2 \right\rangle / \left\langle \left|\widehat{I}_s(\vec{0})\right|^2 \right\rangle . \tag{6.60}$$

Thus from the equations (6.60) and (6.53), the power spectrum of the object is recast as,

$$\left|\widehat{O}(\vec{u})\right|^2 = \frac{\left\langle \left|\widehat{I}(\vec{u})\right|^2 \right\rangle}{\left[\left\langle \left|\widehat{I}_s(\vec{u})\right|^2 \right\rangle / \left\langle \left|\widehat{I}_s(\vec{0})\right|^2 \right\rangle\right]} . \tag{6.61}$$

Hence, the power spectrum of the object is the ratio of the average power spectrum of the image to the normalized average power spectrum of the point source. By Wiener-Khintchine theorem, the inverse Fourier transform of equation (6.61) yields the spatial domain estimate,

$$\mathcal{A}_c\left[O(\vec{x})\right] = \mathcal{F}^-\left[\left|\widehat{O}(\vec{u})\right|^2\right] , \tag{6.62}$$

where, $\mathcal{A}_c$ and $\mathcal{F}^-$ stand for autocorrelation and inverse Fourier transform respectively.

In the case of an equal magnitudes binary star with an angular separation between the components less than the seeing disc of $\sim \lambda/r_0$ size, the image represents a superposition of two identical speckle patterns. The vectors connecting individual speckles from the two components, are equal to the projected position of the stars on the sky. The binary system can be studied from its Fourier transform pattern or from its averaged autocorrelation. However, the inherent property of the autocorrelation method is to produce double images with 180° ambiguities of a binary source (see Figure 6.6). Nevertheless, the reconstruction of object autocorrelation in

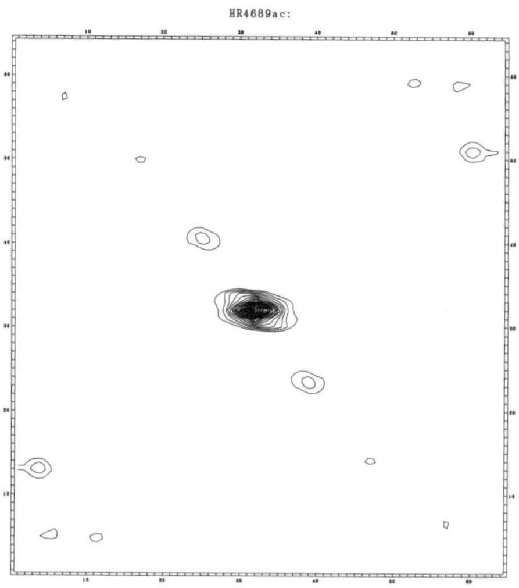

Fig. 6.6 Autocorrelation of a binary star HR4689. The axes of the figure are the pixel values; each pixel value is 0.015 arc-seconds. The central contours represent the primary component. One of the two contours on either side of the central one displays the secondary component. The contours at the corners are the artifacts.

case of the components in a group of stars retrieves the separation, position angle, and the relative magnitude difference at low light levels. Saha and Venkatakrishnan (1997), found the usefulness of the autocorrelation technique in obtaining the prior information on the object for certain applications of the image processing algorithms.

6.3.4 *Noise reduction using Wiener filter*

Disadvantage with a division as in equation (6.61), is that the zeros in the denominator corrupts the ratio and spurious high frequency components are created in the reconstructed image. Moreover, a certain amount of noise is inherent in any kind of observation. In this case, noise is primarily due to thermal electrons in the CCD interfering with the signal. Most of this noise is in the high spatial frequency regime. In order to get rid of such a high frequency noise as much as possible, Saha and Maitra (2001) had developed an algorithm, where a Wiener parameter is added to PSF power spectrum in order to avoid zeros in the PSF power spectrum. The notable advantage of such an algorithm is that the object can be reconstructed with a few frames. Often, it may not be possible to gather sufficient number of images within the time interval over which the statistics of the atmospheric turbulence remains stationary.

In this process, a Wiener filter is employed in the frequency domain. The technique of Wiener filtering[3] damps the high frequencies and minimizes the mean square error between each estimate and the true spectrum. Applying such a filter is a process of convolving the noise degraded image, $I(\vec{x})$, with the Wiener filter. The original image spectrum, $\widehat{I}(\vec{u})$, is estimated from the degraded image spectrum, $\widehat{I}'(\vec{u})$, by multiplying the image spectrum with the Wiener filter, $\widehat{W}(\vec{u})$, i.e.,

$$\widehat{I}(\vec{u}) = \widehat{I}'(\vec{u}).\widehat{W}(\vec{u}). \tag{6.63}$$

However, this would reduce the resolution in the reconstructed image. The advantage is the elimination of spurious higher frequency contributions. The Wiener filter, in the frequency domain, has the following functional form:

$$\widehat{W}(\vec{u}) = \frac{\widehat{H}^*(\vec{u})}{\left|\widehat{H}(\vec{u})\right|^2 + \dfrac{\left|\widehat{P}_n(\vec{u})\right|^2}{\left|\widehat{P}_s(\vec{u})\right|^2}}. \tag{6.64}$$

in which $\widehat{H}(\vec{u})$ is the Fourier transform of the point spread function (PSF), $|\widehat{P}_s(\vec{u})|^2$ and $|\widehat{P}_n(\vec{u})|^2$ the power spectra of signal and noise respectively.

The term $|\widehat{P}_n(\vec{u})|^2/|\widehat{P}_s(\vec{u})|^2$ can be interpreted as reciprocal of S/N ratio. In this case, the noise is due to the CCD. In the algorithm developed by Saha and Maitra (2001), the Wiener parameter is chosen according to the S/N ratio of the image spectrum. In practice, this term is usually just a constant, a 'noise control parameter' whose scale is estimated from the noise power spectrum. In this case, it assumes that the noise is white and that one can estimate its scale in regions of the power spectrum where the signal is zero (outside the diffraction-limit for an imaging system). The expression for the Wiener filter simplifies to:

$$\widehat{I}(\vec{u}) = \frac{\left|\widehat{P}_{ref}(\vec{u})\right|^2}{\left|\widehat{P}_{ref}(\vec{u})\right|^2 + w}, \tag{6.65}$$

[3]The classic Wiener filter that came out of the electronic information theory where diffraction-limits do not mean much, is meant to deal with signal dependent 'colored' noise.

where, w is the noise-variance and is termed as Wiener filter parameter in the program.

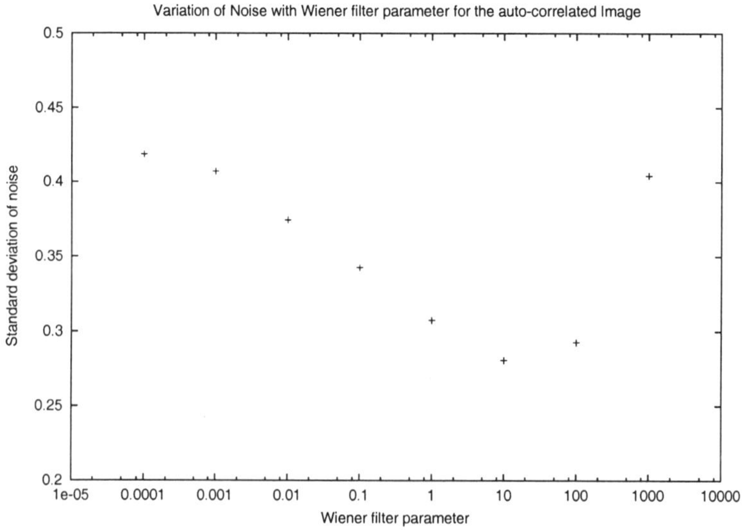

Fig. 6.7 $\langle\sigma\rangle$ vs. Wiener filter parameter (WFP) plot.

In order to obtain an optimally autocorrelated image, a judicious choice of the Wiener filter parameter is made by Saha and Maitra (2001). They have reconstructed autocorrelated images for a very wide range of Wiener filter parameter (WFP) values. A small portion (16×16 pixels) of each image, far from the centre, is sampled to find-out the standard deviation in the intensity values of the pixels, $\langle\sigma\rangle$. The plot of standard deviation thus obtained against the Wiener filter parameter (see Figure 6.7) shows a minimum. The abscissa corresponding to this minimum gives the optimum Wiener filter parameter value. The nature of $\langle\sigma\rangle$ vs. WFP plot is understood as follows:

The noise in the data is primarily in the high frequency region whereas the signal is at a comparatively low frequency. With zero WFP, there is no filtering and hence there is ample noise. As the WFP value is gradually increased from zero, more and more high frequency noise is cut off and the $\langle\sigma\rangle$ goes down. It attains a minimum when WFP value is just enough to retain the signal and discard the higher frequency noise. However, at higher WFP region, blurring of the image starts to occur. This leads to the sharp increase in $\langle\sigma\rangle$. The 'ringing' effect due to sharp cutoff comes into play.

Computer simulations were carried out by convolving ideal star images having Gaussian profile with a random PSF to generate speckle pattern. The plot of $\langle\sigma\rangle$ vs. WFP also provides similar results though some of the major sources of noise, e.g., thermal noise due to electron motion in the CCD, effects due to cosmic rays on the frame, etc. were not incorporated in the simulation.

6.3.5 *Simulations to generate speckles*

Computer simulations of the intensity distribution in the focal point of a telescope for the Fried's parameter (r_0) were attempted (Venkatakrishnan et al. 1989) in order to demonstrate the destructions of the finer details of an image of a star by the atmospheric turbulence. The simulated smallest contours should have the size of the Airy disc of the telescope. A power spectral density of the form,

$$\mathcal{E}(\vec{\kappa}) \propto \frac{L_0^{11/3}}{(1+\vec{\kappa}^2 L_0^2)^{11/6}}, \tag{6.66}$$

where $\vec{\kappa} = \kappa_x, \kappa_y$, L_0 is the outer scale of turbulence, was multiplied with a random phase factor, $e^{i\psi}$, one for each value of $(\vec{\kappa})$, with ψ uniformly distributed between $-\pi$ and π.

The resulting 2-D pattern in $\vec{\kappa}$ space was Fourier transformed to obtain single realization of the wavefront, $W(\vec{x})$. The Fraunhofer diffraction pattern of a piece of this wavefront with the diameter of the entrance pupil provides angular distribution of amplitudes, while the squared modulus of this field gives the intensity distribution in the focal plane of the telescope. The sum of several such distributions would show similar concentric circles of equal intensity.

Laboratory simulation is also an important aspect for the accurate evaluation of the performance of speckle imaging system. Systematic use of simulated image is required to validate the image processing algorithms in retrieving the diffraction limited information. Atmospheric seeing can be simulated at the laboratory by introducing disturbances in the form of a glass plate with silicone oil (Labeyrie, 1970). Saha (1999a) had introduced various static dielectric cells (SDC) of various sizes etched in glass plate with hydrofluoric acid. Several glass plates with random distribution of SDCs of known sizes were made and used in the experiment. The phase-differences due to etching lie between 0.2λ and 0.7λ. In order to obtain the light beam from a point source, similar to the star in the sky, an artificial star image was developed by placing a pair of condensing lenses along with micron-sized pin-hole in front of the source. The beam was collimated with a Nikkon lens; the wavefronts from this artificial star enter a simulated telescope whose focal ratio is 1:3.25. The image was slowed down to $f/100$ in order to discern the individual speckles with a high power microscope objective. The magnifying optics are required in front of the camera to rescale the image so that there are at least 2 or more pixels for each telescope diffraction limit. The speckles were recorded through an interference filter of 10 nm bandwidth centered on 557.7 nm.

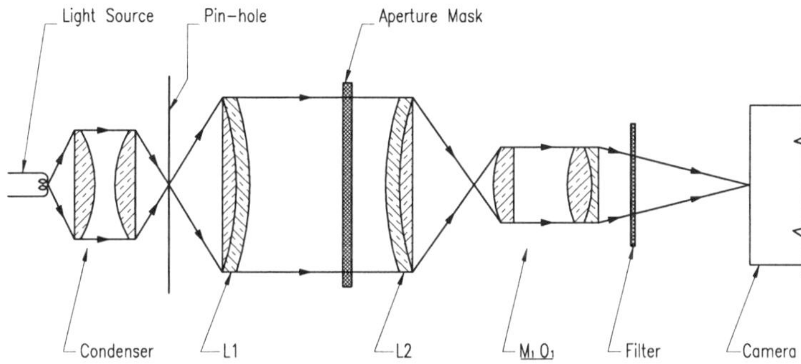

(a)

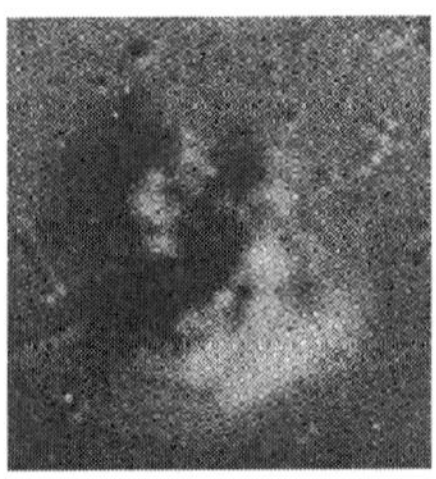

(b)

Fig. 6.8 (a) Schematic of laboratory set up to simulate speckles; (b) fringes from an artificial star, and generated speckles.

Figure (6.8) depicts (a) the laboratory set up to simulate speckles from an artificial star, (b) speckles obtained in the laboratory through the aforementioned narrow band filter. The image was digitized with the photometric data system (PDS) 1010M micro-densitometer and subsequently were processed by the COMTAL image processing system of the VAX 11/780 and associated software. A method called clipping[4] technique was used to

[4]Clipping is done by taking differences in grey levels along any direction. But usu-

enhance the contrast in grey levels. The laboratory set up was sensitive enough to detect aberrations produced by the objective lenses, as well as micro-fluctuations in the speckle pattern caused by vibrations.

6.3.6 *Speckle interferometer*

A speckle interferometer is a high quality diffraction limited camera where magnified (~f/100) short exposure images can be recorded. However, atmospherically induced chromatic dispersion at zenith angle larger than a few degrees causes the individual speckles to become elongated by a significant amount. This elongation is approximated through the following equation,

$$\Delta\theta \approx 5 \times 10^{-3} \tan\theta \qquad \text{arcsec nm}^{-1}, \tag{6.67}$$

in which θ is the angle from the zenith.

For example, with $\theta = 45°$ and a bandwidth $\Delta\lambda$ of 20 nm, the elongation would be approximately 0.1 arcsec that is equal to the diffraction-limit of 1 meter telescope. For a large telescope, this amounts to the diameter of several speckles. In order to compensate for such a dispersion either a computer controlled counter-rotating dispersion correcting Risley prisms or a narrow-band filter are necessary to be incorporated in the speckle camera system. The expression of the spectral bandwidth that can be used is given by,

$$\frac{\Delta\lambda}{\lambda} = 0.45 \left(\frac{r_0}{D}\right)^{5/6}. \tag{6.68}$$

The admitted bandwidth is a function of λ^2, so a large bandwidth can be used for an infrared (IR) speckle interferometry, while a very narrow band filter is necessary to be used in optical band. Inability to correct the dispersion precisely at large angle results in a rotationally asymmetrical transfer function and the presence of any real asymmetry of the object may be obscured. In the following, the salient features of a few speckle interferometers are enumerated.

(1) Saha et al. (1999a) had built the camera system with extreme care so as to avoid flexure problems which might affect high precision measurements of close binary star systems etc., in an unfavorable manner.

ally the direction of maximum variations in grey levels is taken. Clipping enhances sharp features; therefore, they do not represent actual grey levels. The clipped image is superposed on the histogram-equalized original image.

The mechanical design of this instrument was made in such a way that flexures at different telescope orientations do not cause image motion. The design analysis has been carried out with the modern finite element method[5] (Zienkiewicz, 1967) and computer aided machines were used in manufacturing to get dimensional and geometrical accuracies.

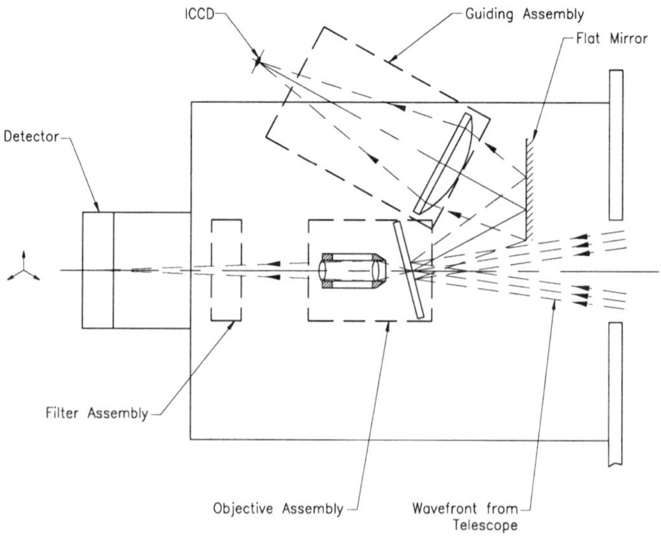

Fig. 6.9 Optical layout of the speckle interferometer (Saha et al. 1999a)

A high precision hole of aperture, called diaphragm, with a diameter of ~350 μm, at an angle of 15° on the surface of a low expansion glass, allows the image of the target to pass on to the microscope objective. The function of such a diaphragm is to exclude all light except that coming from a small area of the sky surrounding the star under study. This aperture is equivalent to a field of ~9 arcsecs at the Prime focus

[5]Finite element analysis requires the structure to be subdivided into a number of basic elements like beams, quadrilateral and solid prismatic elements etc. A complete structure is built up by the connection of such finite elements to one another at a definite number of boundary points called nodes and then inputting appropriate boundary constraints, material properties and external forces. The relationship between the required deformations of the structure and the known external forces is $[K]\{\mathbf{d}\} = \{\mathbf{F}\}$, where, $[K]$ is the stiffness matrix of the structure, $\{\mathbf{d}\}$ is the unknown displacement vector and $\{\mathbf{F}\}$ is the known force vector. All the geometry and topology of the structure, material properties and boundary conditions go into computation of $[K]$. The single reason for universal application of finite element method is the ease with which the matrix, $[K]$ is formulated for any given structure.

and to a field of ~2.25 arcsecs at the Cassegrain focus of the VBT. The field covered by this aperture of the flat at Prime focus of the VBT is sufficient to observe both the object and the reference star simultaneously, if the latter is located within the iso-planatic domain around the object. The portion of the light beam outside this aperture is reflected and constitutes the guiding path (see Figure 6.9) and re-imaged on an intensified CCD for guiding. Such a guiding system helps to monitor the star-field during the observation, which provides an assurance about the quality of the data that need to be collected. Instrumental errors such as (i) improper tracking of telescope and (ii) obstruction in the light path of the main detector can also be noticed, hence the corrective measures may be taken immediately. If there is a drift in the star position across the detector due to inaccurate tracking rate or an unbalanced telescope and shift in the star position due to disturbances to the telescope like a gust of wind has also required to be corrected in order to obtain useful data.

(2) Another system developed at Observatoire de la Calern (formerly CERGA) by Labeyrie (1974) had a concave grating, later replaced by a holographic concave grating, instead of interchangeable filters. Such a grating provides the necessary filtering in a tunable way. In addition, the holographic concave gratings have low stray light levels than those of classically ruled gratings and posses no ghosts in the spectral image. These gratings are recorded on spherically concave substrates with equidistant and parallel grooves. Their optical properties are same as the ruled gratings. The object's spectrum can be displayed by the sensor while adjusting wavelength and bandwidth selection decker. Thus, spectral features of interest, such as Balmer emission lines, may be visualized and isolated down to 1 nm bandwidth. The sensor used for this specklegraph was a photon-counting camera system.

A digital correlator (Blazit, 1976) has been used for real time data reduction. During each frame scan period, the correlator discriminates photon events, computes their position in the digital window covering the central 256×256 pixels of the monitor field, then computes all possible vector differences between photon positions in the frame considered (up to 10^4 differences per frame), and finally integrates in memory a histogram of these difference vectors. The memory content is considered as the autocorrelation function of short-exposure images. Such an approach is found to be better suited to a digital processor than the equivalent Fourier treatment used when the computation was achieved by an optical analog method. Subsequently, another specklegraph had been developed by the same group in order to remove the effect of centreur hole. In this system a pupil splitter was installed to create 2×2 channels. Duplicate images are two different sortings of the same photon distribution: information up to the telescope cut-off frequency may be saved by cross-correlating them.

6.3.7 *Speckle spectroscopy*

The application of speckle interferometric technique to speckle spectroscopic observations enables to obtain spectral resolution with high spatial resolution of astronomical objects simultaneously. Information is concentrated in narrow spectral intervals in astrophysics and can be obtained from narrow band stellar observations. There are various types of speckle spectrograph, such as:

(1) objective speckle spectrograph (slitless) that yields objective prism spectra with the bandwidth spanning from 400 nm – 800 nm (Kuwamura et al., 1992),
(2) wide-band projection speckle spectrograph that yields spectrally dispersed 1-dimensional (1-D) projection of the object (Grieger et al., 1988), and
(3) slit speckle spectrograph, where the width of the slit is comparable to the size of the speckle (Beckers et al., 1983).

A prism or a grism (grating on a prism) can be used to disperse 1-d specklegrams. In the case of projection spectrograph, the projection of 2-d specklegrams is carried out by a pair of cylindrical lenses and the spectral dispersion is done by a spectrograph. Baba et al., (1994) have developed an imaging spectrometer where a reflection grating acts as disperser. Two synchronized detectors record the dispersed speckle pattern and the specklegrams of the object simultaneously. They have obtained stellar spectra of a few stars with the diffraction-limited spatial resolution of a telescope, by referring to the specklegrams. Figure (6.10) depicts the concept of a speckle spectroscopic camera.

Mathematically, the intensity distribution, $W(\vec{x})$, of an instantaneous objective prism speckle spectrogram can be derived as,

$$W(\vec{x}) = \sum_m O_m(\vec{x} - \vec{x}_m) \star \mathcal{G}_m(\vec{x}) \star S(\mathbf{x}), \tag{6.69}$$

where, $O_m(\vec{x}-\vec{x}_m)$ denotes the mth object pixel and $\mathcal{G}_m(\vec{x})$ is the spectrum of the object pixel.

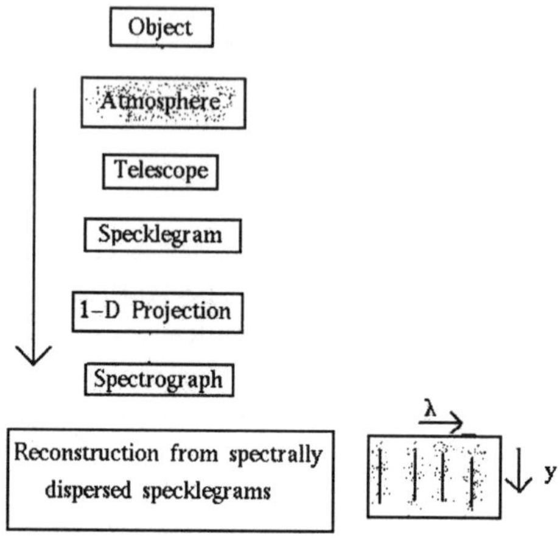

Fig. 6.10 Concept of a speckle spectrograph.

In the narrow wavelength bands (<30 nm), the PSF, $S(\vec{x})$, is wavelength independent. The objective prism spectrum, $\sum_m O_m(\vec{x}-\vec{x}_m)\star\mathcal{G}_m(\vec{x})$, can be reconstructed from the speckle spectrograms.

6.3.8 *Speckle polarimetry*

Polarized light carries valuable information about the origin of light, and the various physical parameters which have been responsible for its generation. The importance of such observations in astronomy is to obtain information such as the size and shapes of dust envelopes around stars, the size and shape of the dust grains, and magnetic fields. Among other astronomical objectives worth investigating are:

- the wavelength dependence of the degree of polarization and the rotation of the position angle in stars with extended atmospheres, and
- the wavelength dependence of the degree of polarization and position angle of light emitted by stars present in very young ($\leq 2\times10^6$ years) clusters and associations.

Polarization measurements with conventional technique have provided information in astronomy in such diverse areas as star forming region, galactic magnetic fields, comets, planetary surfaces and binary systems. Most of the celestial objects in the night sky posses perfect spherical symmetry and their radiation follow Planckian law of radiation. Thus the received light is totally unpolarized. There is, however many situations where asymmetry is introduced by, for example, the presence of circumstellar dust around young stars, patchy distribution of dust in star forming regions, etc. The light reaching from such systems are polarized and carry informations about their structures. If optically-thick obscuring matter blocks the line-of-sight to the nucleus of an AGN like NGC 1068, the broad permitted lines are invisible. A small fraction of radiation may be scattered into the direction of an observer if other sight-lines are not completely obscured. The absorbing matter is thus not sky-covering as seen by the active galactic nuclei (AGN), and is generically imagined to be in a torus geometry. The polarization of the radiation may reveal that the reflector consists of electrons in a plasma, presumably ionized by direct AGN radiation. The narrow-line region in all AGN lies above the torus and is irradiated by the photoionizing continuum.

An imaging polarimeter may be employed at the telescope in order to study the extended astrophysical object like reflection nebulae, structures near star forming regions, dusty active galaxies etc. These objects are typically a few arcminutes in size and often fainter than 20th magnitude per square arcsecond. The complexity of structure present in these objects often demands observations with an angular resolution of a few arcseconds. Dichroic extinction of background starlight by magnetically aligned dust particles at the periphery of dark molecular clouds in the galaxy, introduces polarization in the starlight and hence imaging polarimeter can be used to obtain magnetic field maps of these regions.

The effect on the statistics of a speckle pattern is the degree of depolarization caused by the scattering at the surface. If the light is depolarized, the resulting speckle field is considered to be the sum of two component speckle fields produced by scattered light polarized in two orthogonal directions. The intensity at any point is the sum of the intensities of the component speckle patterns (Goodman, 1975). These patterns are partially correlated; therefore, a polarizer that transmits one of the component speckle patterns is used in the speckle camera system.

The advantage of using speckle camera over conventional imaging polarimeter is the capability of monitoring the short-time variability of the atmospheric transmission. Another advantage of high resolution polarimetry is that of using it as a tool to get insight on the binary star mechanism. Speckle polarimetry has been used to obtain polarimetric information with sub-arc-second resolution of astronomical objects. Falcke et al., (1996) developed a speckle imaging polarimeter consisting of a rotatable, achromatic $\lambda/2$ - retardation mica plate in front of a fixed polarization filter. These elements were installed on a single mount and inserted into the optical axis in front of the telescope focus of their speckle camera.

6.4 Pupil-plane interferometry

In the pupil-plane interferometry, the interference of the beam takes place in the pupil-plane. This type of interferometer is typically made using wire grids or dielectric beam splitters. A pair of plane parallel plates or wedges or a biprism or a grating or use of beam separation by polarization as well. Such a technique was developed to overcome some limitations of speckle interferometry, one major shortcoming of which is the random attenuation of the spectral components introduced by the atmospheric turbulence.

A Fizeau-Michelson interferometry is free from speckle noise, albeit most of the light is thrown away by the mask on the telescope pupil. For a point object that is smaller than the seeing disc size, wavefront shearing interferometers provide a means of using light more efficiently without introducing speckle noise. Like speckle interferometry, the true object visibility is recorded by employing the short-exposure. The major advantage of this technique over speckle interferometry is to obtain better S/N ratio on bright sources (Roddier and Roddier, 1988). It also makes a better use of the detector dynamic range to detect faint extended structure like stellar envelopes.

6.4.1 *Estimation of object modulus*

A two-dimensional Fourier transform of an interferogram produces an array of complex numbers with a central peak and a pair of side lobes. A window is generally set to isolate one of the side lobes, following which the reduced data set is Fourier transformed back yielding a map of the amplitude and phase of the fringes. In the case of stellar pupil-plane interferometry, the phases are distorted by atmospheric turbulence, hence amplitudes may be considered. The square of the amplitudes provides an estimate of the object energy spectrum. Similar windows may be set up at the same distance from the central peak, but outside of the side lobes and Fourier transformed back would provide an estimate of noise amplitude. Because of the non-uniformity of the irradiance in the interferogram the DC level or continuum should also be mapped by setting a window on the central peak and an inverse Fourier transform of which is necessary to be taken. At each point on the pupil, the object amplitudes can be corrected in this manner for the continuum non-uniformity. A generalized fringe pattern of an interferogram is expressed as,

$$I(\vec{r}) = A(\vec{r}) + \Re B(\vec{r}) e^{i2\pi \vec{f_0} \cdot \vec{r} - \psi(\vec{r})}, \tag{6.70}$$

$$= A(\vec{r}) + \frac{1}{2} B(\vec{r}) e^{i[\psi(\vec{r}) + 2\pi \vec{f_0} \cdot \vec{r}]} + \frac{1}{2} B(\vec{r}) e^{-i[\psi(\vec{r}) + 2\pi \vec{f_0} \cdot \vec{r}]}, \tag{6.71}$$

where $\Re$ stands for the real term, $\vec{r}(= x, y, z)$ is the position vector, $\vec{f_0}$ the spatial fringe frequency vector, $A(\vec{r})$, $B(\vec{r})$ define the intensity modulation arising from the imperfections in the optical system. Taking,

$$C(\vec{r}) = \frac{1}{2} B(\vec{r}) e^{i\psi(\vec{r})}, \tag{6.72}$$

and $C^*(\vec{r})$ its complex conjugate, the equation (6.71) is recast as,

$$I(\vec{r}) = A(\vec{r}) + C(\vec{r}) e^{i2\pi \vec{f_0} \cdot \vec{r}} + C^*(\vec{r}) e^{-i2\pi \vec{f_0} \cdot \vec{r}}. \tag{6.73}$$

The local phase information in the interferogram relates to the local wave-front slope. Applying the Fourier transform on the intensity distribution one finds,

$$\widehat{I}(f, y) = \widehat{A}(f, y) + \widehat{C}(f - f_0, y) + \widehat{C}^*(f + f_0, y), \tag{6.74}$$

in which $\widehat{I}$ denotes the Fourier transform of intensity distribution, $\widehat{A}$ the Fourier transform the function $A(\vec{r})$, $\widehat{C}$ the Fourier transform of the function $C(\vec{r})$, and f the spatial frequency in x-axis.

In order to estimate the phase of the object Fourier transform from the pupil-plane interferograms, several methods have been developed. One such method is called phase-unwrapping (see section 9.3.1), where it reconstructs a continuous wavefront from the phase principal value, the process of which is the most complicated part of fringe analysis (see section 9.3.1). Using a maximum entropy algorithm (see section 9.2.4), one may also reconstruct the atmospherically degraded star image (Roddier and Roddier, 1988).

The main drawback of pupil-plane interferometry is its limitation to small objects since the coherence area on the pupil, according to the Zernike theorem, varies as the inverse of the object area. For an extended object, the coherence area is limited by the object area rather than seeing disc area. For a bright object the S/N ratio varies as the square root of the number of photons per coherence area, it decreases as the inverse square root of the object area. One can observe such phenomena in incoherent holography. Another drawback comes from the detector elements, which is required to record the interference fringes. It is generally twice the number needed for a speckle experiment. There are several pupil-plane interferometry techniques. A brief account of some of them is discussed in the following sections.

6.4.2 *Shear interferometry*

Shear or (shift) interferometry, a technique that measures wavefront phase through intensity measurements, was introduced by Bates (1947). The basic approach is to produce an interference pattern between the wavefront and a sheared or displaced replica of itself. In this way, the phase variations, which cannot be measured directly because of its high temporal frequency, are converted into intensity variations (fringe pattern) in the pupil-plane. The techniques that measure the displacements[6] components have been discussed in section (6.2.1).

Shearing interferometry is a direct method based on the principle of self-referencing, and does not require a coherent reference source; it measures the wavefront slope. It combines the wavefront with a shifted version of itself to form interferences, yielding the derivatives of the displacement directly. An interferogram can be produced by comparing a wavefront with a sheared image of itself by constructing an interferometer, in which the interfering beams follow the same path with a small separation. The advantage of such an arrangement is that it is less affected by outside disturbances. In shear interferometry, either a point on the object is imaged as two points or two points on the object are imaged as a single point. These object-plane shear and image-plane shear are related through the magnification of the imaging lens.

Bates' shear is known as lateral or linear shear. The other noted shears such as radial, reversal, and rotational shears are also used in shear inter-

[6]A stress analyst is involved in the strains rather than displacements. The strain is obtained from the displacement data by differentiation, which may lead to large errors.

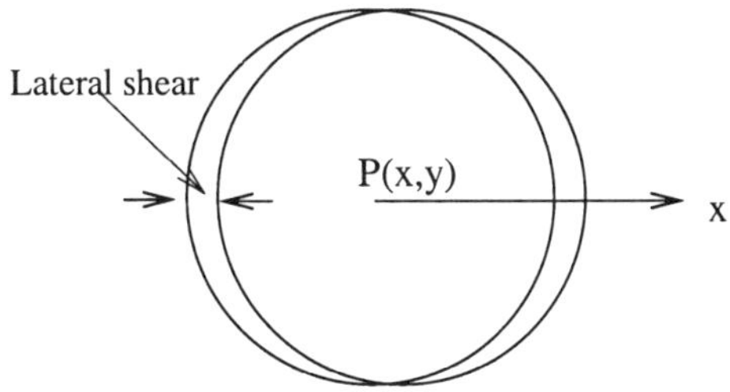

Fig. 6.11 Example of shear in lateral shearing interferometer.

ferometry.

(1) Linear shear interferometer (LSI;Cognet, 1973): There are many arrangements to obtain lateral shear. One of the arrangements is to divide the amplitude of the incident wavefront, but do not change the shape of the wavefront (see Figure 6.11) below shows the schematic of the lateral shear.
The wavefront error may be expressed as, $W(x,y)$, in which x,y are the coordinates of the point P. When this wavefront is sheared in the x-direction by an amount, s, the error at the same point for the sheared wavefront is $W(x-s,y)$. The resulting path difference, ΔW at P between the original and sheared wavefront is,

$$\Delta W \simeq W(x,y) - W(x-s,y). \tag{6.75}$$

Thus information in the lateral shearing interferometry is $\partial W/\partial x$ in angular measure. The intensity at a point in the interferogram can be written as,

$$I = I_1 + \cos\left[\frac{2\pi}{\lambda}\Delta W(x,y)\right]. \tag{6.76}$$

Hence, the interference fringe is the representation of ΔW. The original wavefront is deduced by fitting the interference fringes to Zernike polynomials (see section 7.2).
There are several other types of interferometers to obtain lateral shear, in which other optical principles are used, namely diffraction, polarization etc. Figure (6.12a) displays the (a) computer simulated, as well as (b) polarization shear interferograms using a polarization shearing interferometer (see Figure 6.12b) that has been developed using birefringent prisms (Lancelot, 2006). In this system, the cone of light passing

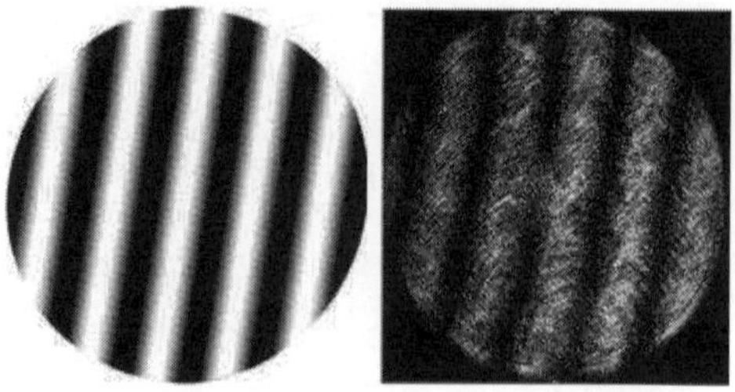

(a) (b)

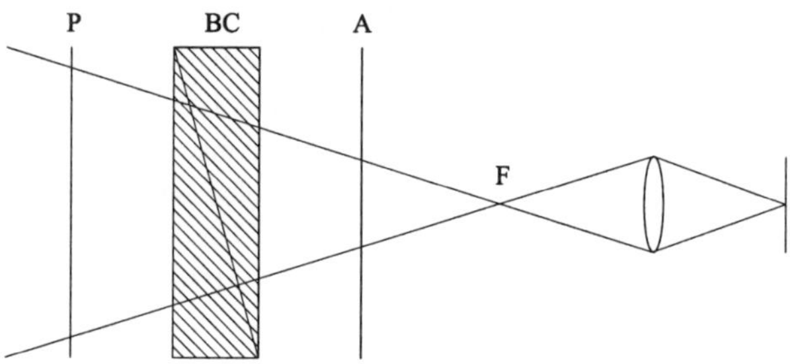

Fig. 6.12 Top: (a) 2-D view of computer simulated polarization shear interference patterns; (b) 2-D view of polarization shear interferometric fringes obtained in the laboratory (Courtesy: J. P. Lancelot). Bottom: Schematic diagram of a polarization shearing interferometer using Babinet compensator (BC): P - polarizer, A - analyzer, F - focus. The shear is given by $2(n_e - n_o)\tan\theta f$, in which n_o and n_e are the refractive indices of ordinary and extra-ordinary rays respectively, θ the compensator wedge angle, and f the focal length of the system.

through a Babinet compensator[7] (BC) would produce fringe pattern due to the different phase change introduced between the extraordinary and the ordinary vibrations at different points during the oblique passage of the ray. The lateral displacement between the rays is thus different at different distances from the optical axis of the BC. The interference fringes are produced by two partially or totally superimposed pupil images created by introducing a beam splitter. At each point, interference occurs from the combination of only two points on

[7] A Babinet compensator is a birefringent quartz crystal consisting of two thin prisms cemented together to form a thin parallel plate.

the wavefronts at a given baseline, and therefore, behaves as an array of Fizeau-Michelson interferometers. A two dimensional, i.e., shearing in x- and y-directions simultaneously, is achieved using two crossed prisms placed at either side of the focus in a single interferogram.

It may be necessary to make two orthogonal measurements to assess the full wavefront tilt. To be precise, at the entrance of the sensor, the wavefront is usually split into two similar channels, one with a x-shear device and the other y-shear device. Each channel is equipped with a detector array to measure a map of the wavefront gradient; each detector pixel corresponds to the sub-aperture area in the telescope pupil. The detector planes are divided into contiguous sub-apertures for maximum efficiency, hence the detector array determines the spatial sampling of the wavefront. The area of one detector provides a spatial filtering of the phase gradients. The measurement represents the average slope of the optical path difference in the shear direction. Before detection the beams are split again and laterally shifted (sheared) with respect to each other. All the baselines are identical and a single object Fourier component is measured at a time.

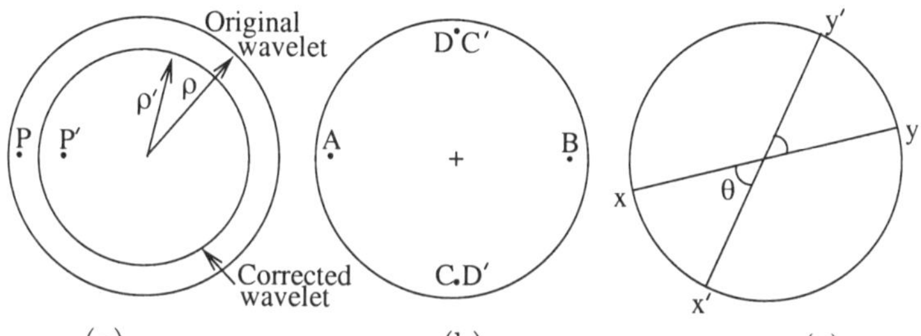

Fig. 6.13 Three methods to obtain shear between two wavefronts such as: (a) radial shear, (b) reversal shear, and (c) rotational shear interferometers.

An important property of lateral shearing interferometers (LSI) is their ability to work with partially coherent light, which offers better S/N ratio on bright sources, and is insensitive to calibration errors due to seeing fluctuations and telescope aberrations. Such interferometers may be used in adaptive optics (AO) system as wavefront sensor.

(2) Radial shear interferometer: This interferometer produces two interfering wavefronts with identical deformation, but one of the wavefronts is contracted or expanded with respect to each other so that correspond-

ing points can be regarded as sheared apart radially (see Figure 6.13a). If ρ' and ρ'' are the ratio of the radial distance to the maximum radius of the contracted and expanded wavefront, the wavefront that interfere at two points separated by a distance, $d(= \rho' - \rho'')$, on the interference pattern can be given by,

$$\Delta W = W(\rho', \theta) - W(\rho'', \theta). \tag{6.77}$$

One of the radial shear interferometer designed by Hariharan and Sen (1961) uses two lenses of differing focal length to produce radial shear, $R(= f_1/f_2)$.

(3) Reversal shear interferometer: This interferometer produces two wavefronts in which deformations on one wavefront are symmetrical with respect to those on the other wavefront (see Figure 13b). The reversal of the wavefront about an arbitrary axis is equivalent to reversion about the x-axis followed by a lateral shear s in the y-direction. In this case,

$$\Delta W = W(\rho, \theta) - W(\rho', \theta). \tag{6.78}$$

The reversal shear interferometer has no sensitivity to symmetric aberration. Such interferometers are constructed using Koéstar prisms. The shortcomings of this system comes from the loss of light as it is reflected back to the sky and produces low contrast fringes owing to mismatch of orthogonal polarizations. This interferometer produces 1-d Fourier transform, providing information on all the object Fourier components in one direction. Sequential measurements are required to map the object 2-D Fourier transform.

(4) Rotation shear interferometer: The rotational shear interferometer rotates one pupil image, usually about the optical axis, by a small angle with respect to the other (see Figure 13c). If the rotation axis coincides with the centre of the pupil, the two images overlap. Each baseline measures a different spatial frequency, mapping the whole frequency plane up to a cut-off frequency. This interferometer provides an interferogram defined by,

$$\Delta W = W\left(\rho, \theta - \frac{\phi}{2}\right) - W\left(\rho, \theta + \frac{\phi}{2}\right), \tag{6.79}$$

where ϕ is the rotation of one wavefront with respect to the other. A rotational shear of π produces a 2-D map of the object Fourier transform up to the cut-off frequency, $\vec{f}_c$, of the telescope. A rotational angle,

β, produces a 2-D map up to the spatial frequency,

$$\vec{f} = \vec{f_c} \sin\left(\frac{\beta}{2}\right). \tag{6.80}$$

This effect allows to match the Fourier plane coverage to the extent of the object Fourier transform. All the object Fourier components within telescope diffraction cutoff frequency are measured simultaneously.
An ordinary Twyman-Green interferometer can be converted to a rotation shear interferometer by means of several method. A 180° rotational shear is achieved using two prisms. Another interferometer is based on the Sagnac or cyclic interferometer, in which the rotational shearing angle, ϕ is produced by a Dove prism of $\phi/4$ within the closed loop of the interferometer. Variable rotational shear was used to develop a similar interferometer by Roddier and Roddier, (1983), where one roof prism is allowed to rotate around the optical axis. They have introduced a phase plate between the beam splitter and the roof prism to correct the mismatch of polarization. Roddier and Roddier (1988) applied the principle of Michelson's interferometry in rotation-shear interferometer in order to measure the spatial coherence.

6.5 Aperture synthesis with single telescope

Two promising methods, viz., (i) speckle masking (see chapter 9.3) technique (Lohmann et al. 1983) and (ii) non-redundant aperture masking (NRM) technique are based on the principle of phase-closure method (Jennison, 1958). It is reiterated that according to the van Cittert-Zernike theorem the image of an object is the Fourier transform of the measured visibility or the cross-correlation between the various sub-apertures. To recover the Fourier phases of the source brightness distribution from the observations, it is necessary to detect fringes on a large baselines, therefore, enables one to reconstruct images. The concept of using three antennae arranged in a triangle was first introduced in radio astronomy by Jennison (1958) in late fifties.

6.5.1 *Phase-closure method*

Closure-phases are insensitive to the atmospherically induced random phase errors, as well as to the permanent phase errors introduced by the telescope aberrations in optics. Since any linear phase term in the object cancels out, this method is insensitive to the position of the object but sensitive to any object phase non-linearity.

Equation (6.56) represents the simplest form of the phase closure technique but unfortunately phase information is not preserved. The phase-closure requirement can be redefined in terms of spatial frequency vectors forming a closed loop. The phase-closure is achieved (equation 6.56) in the power spectrum by taking pairs of vectors of opposite sign, $-\mathbf{u}$ and $+\mathbf{u}$, to form a closed loop. In three aperture interferometry, the observed phases, ψ_{ij}, on the different baselines contain the phases of the source Fourier components $\psi_{0,ij}$ and also the error terms, θ_j, θ_i, introduced by errors at the individual antennas and by the atmospheric variations at each antenna. The observed fringes are represented by the following equations,

$$\psi_{12} = \psi_{0,12} + \theta_2 - \theta_1. \tag{6.81}$$

$$\psi_{23} = \psi_{0,23} + \theta_3 - \theta_2. \tag{6.82}$$

$$\psi_{31} = \psi_{0,31} + \theta_1 - \theta_3. \tag{6.83}$$

where the subscripts refer to the antennae at each end of a particular baseline. The closure phase, β_{123} is the sum of phases of the source Fourier components and is derived as,

$$\beta_{123} = \psi_{12} + \psi_{23} + \psi_{31}, \tag{6.84}$$

$$= \psi_{0,12} + \psi_{0,23} + \psi_{0,31}. \tag{6.85}$$

Equation (6.85) implies cancellations of the antennae phase errors. Using the measured closure phases and amplitudes as observables, the object phases are determined (mostly by least square techniques, viz., singular value decomposition, conjugate gradient method). From the estimated object phases and the calibrated amplitudes, the degraded image is reconstructed.

Figure (6.14) displays the laboratory simulation of fringes produced by three apertures arranged in a circular mode at a regular interval. The simulated interference fringes produced by several apertures are also made by Saha et al. (1988). In their simulations, the diffraction effects were neglected. The shape of the fringes change with increasing number of baselines between the apertures. The number of baselines is defined by $N(N-1)/2$, in which N is the number of apertures. If the number of apertures exceeds 9, the shapes of the fringes were becoming circular and the side lobes are discernible at the outer periphery. Saha et al. (1988) had found that the same trend being repeated in their experimental simulations. It is seen

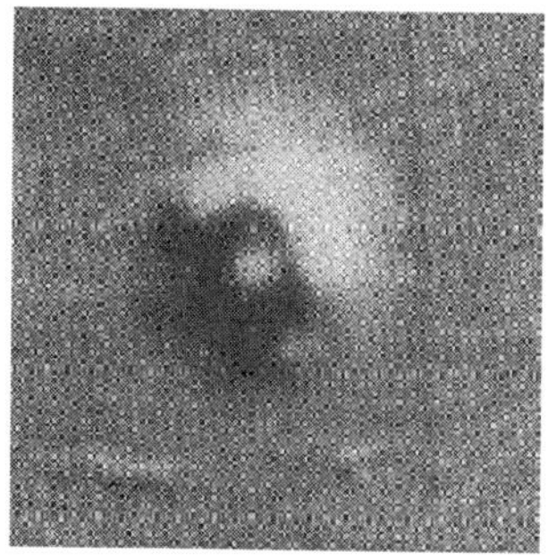

Fig. 6.14 3-D view of fringe patterns at the laboratory through 3-hole aperture mask (Saha, 1999a).

that the image is $2\pi/N$ fold degenerate if N is even while it is π/N fold degenerate if N is odd. A similarity of the experimentally obtained shape and computer simulations was observed.

Baldwin et al. (1986) reported the measurements of the closure-phases obtained at a high light level with a three hole aperture mask set in the pupil-plane of the telescope; interference patterns of the star were recorded using CCD as sensor. Saha et al. (1988) had conducted similar experiment around the same time by placing an aperture mask of 3-holes, 10 cm in diameter, arranged in a triangle, over 1 meter telescope at VBO, Kavalur and tried to record the interference pattern with a 16 mm movie camera. The advantage of placing aperture mask over the telescope, in lieu of pupil mask is to avoid additional optics. But a curious modulation of intensity in the fringe pattern was noticed, therefore, unable to proceed further. The modulation could result from a time-independent aberration in the optical system of the telescope (Saha et al. 1988).

6.5.2 *Aperture masking method*

Aperture masking method is a hybrid techniques (speckle interferometry with non-redundant pupils), allowing diffraction-limited imaging from ground based telescopes. In this method, a pattern of holes with sizes $\leq r_0$ in diameter is cut in a plate and placed in the pupil plane of a large telescope. Such a mask introduces a series of overlapping two-hole interference patterns on the image plane. The Fourier amplitude and the phase of each baseline can be recovered from the interference fringes. The interference patterns, in a series of short-exposure, contain information about structure of the object at the spatial frequencies from which an image of the same can be reconstructed by measuring the visibility amplitudes and closure phases. This system has several advantages. These are:

- an improvement of signal-to-noise (S/N) ratios for the individual visibility and closure-phase measurements,
- attainment of the maximum possible angular resolution by using the longest baselines,
- built-in delay to observe objects at low declinations, and
- produces the optical aperture synthesis maps of high dynamic range.

However, this system is restricted to high light levels, because the instantaneous coverage of spatial frequencies is sparse and most of the available light is discarded, albeit the potential of this in the optical domain was demonstrated by the spectacular images produced with aperture-masking of a single telescope (Tuthill et al. 2000). Equation (4.10) is recast as,

$$I(\vec{x}_1) = \iint_{-\infty}^{\infty} K(\vec{x}_1 - \vec{x}_0')K^*(\vec{x}_1 - \vec{x}_0'')J_P(\vec{x}_0'; \vec{x}_0'')d\vec{x}_0'd\vec{x}_0'', \tag{6.86}$$

where $\vec{x}_0'(= x_0', y_0')$, and $\vec{x}_0''(= x_0'', y_0'')$ are the position vectors of the fields.

Equation (6.86) illustrates that in the general case, the system is linear with respect to the mutual intensities. This system has a 4-D transfer function and the result depends on the respective width of the pupil function of the system, P and the coherence area, $\mathtt{A}_c$, of the incoming wave. If the coherence area is larger than the pupil function, i.e., $\mathtt{A}_c \gg P$, the system is known to be coherent regime, or in the reverse condition, i.e., $\mathtt{A}_c \ll P$, it is known as incoherent regime.

In a general case, the image formation with a given $\mathtt{A}_c$ and P, one looks into the problem in a different way. Let the pupil, P, be subdivided into a set of sub-apertures, p_j, of diameter, d, that is small enough for the field to be coherent over its extent, i.e., $d \ll \mathtt{A}_c$. By doing so, one ensures that the object is not resolved through a single sub-aperture and that the field over each sub-aperture is considered as a small portion of a plane wave with negligible distortions.

In the presence of these pair of sub-apertures, the total amplitude and phase of the spectral component of image intensity, I, obtained in the focal plane of the system is the result of addition of all such fringes with frequencies, taking proper account of both amplitudes and their spatial phases. The associated intensity, according to the diffraction theory (Born and Wolf, 1984), is determined by the expression,

$$I = \sum_{n,m} \langle U_n U_m^* \rangle$$
$$= |\sum_{n,m} U_n|^2 = \sum_{n,m} |U_n|^2 + \sum_{n \neq m} \sum_{m} U_n U_m^*. \tag{6.87}$$

The term, $U_n U_m^*$, is multiplied by $e^{i\psi}$, where, ψ is the random instantaneous shift in the fringe pattern. Each sub-aperture is small enough for the field to be coherent over its extent. The first term on the RHS of the equation (6.87) is the sum of the irradiance produced by each sub-aperture, which does not contain high resolution information, while the second term that describes the interference through the cross product, contains high resolution information. The average values of the latter measures the coherence of the wavefront. Equation (6.87) implies that an image can be reconstructed from sequential measurements of all cross products using pairs of sub-apertures. For n such apertures, there are $n(n-1)$ independent baselines, with $n-1$ unknown phase errors. This technique, known as aperture synthesis, implies that by using many telescopes in an interferometric array, most of the phase informations can be retrieved.

At a point source $O(\vec{\alpha}) = \delta(\vec{\alpha} - \vec{\alpha}_0)$, the mutual intensity of the light, $\mu(\vec{u}) = 1$ for all spatial frequencies and the visibility of the fringe pattern is maximum for all baselines, $\mathcal{V} = 1$. The coherent addition of all the fringes provides a sharp concentration at the image position $\vec{\alpha}_0$. In a circular pupil, Airy pattern is obtained.

Since a pair of sub-apertures produces a fringe pattern, the image produced by the pupil, P, is the sum of fringe patterns, which is the Fourier expansion. Each coherence term is a Fourier component of the expansion. This relationship is known as van Cittert-Zernike theorem. The spacing of the fringes depends on the distance between the sub-apertures; the smaller the distance, larger the fringe spacing and vice versa. If the pupil is larger than the coherence area, for baseline vector $\vec{B}$, the modulus of the mutual intensity of the light, $|\mu(\vec{u})| \leq 1$. In the limit $\vec{B} \gg \sqrt{\mathtt{A}_c}$, the visibility of these fringes is null; the image lacks high spatial frequencies, implying that the optical system has spatially resolved the object.

6.5.3 *Non-redundant masking interferometer*

The masks can be categorized either as non-redundant or partially redundant. The former method consists of arrays of small holes, in which no two pairs of holes have the same separation vector. Its basic idea is the same as that of multi-element interferometry that is implemented at radio wavelengths. The latter technique is designed to provide a compromise between minimizing the redundancy of spacings and maximizing both the throughput and the range of spatial frequencies investigated. Several groups have obtained the fringe patterns using both non-redundant and partially redundant aperture mask of N-holes at large or moderate telescope (Haniff et al., 1987, Monnier et al., 1999). The salient features of a few of these instruments run as follows.

(1) The instrument developed for the Hale 5 meter telescope (Nakajima et al. (1989) used a f/2.8 (ϕ 85 mm) Nikkon camera lens to collimate f/3.3 primary beam of the telescope. The lens formed an image of the primary mirror at a distance of about 85 mm where a mask was placed on a stepper-motor-driven rotary stage controlled by a PC. Another identical lens forms a second focus (scale is 12 arcsec/mm). This image was expanded by microscope objective (X80), enabling to sample 0.15 arcsec/mm on the detector. A narrow band interference filter (630 nm, FWHM 3 nm) was placed between the microscope objective and the detector. The resistive anode position sensing photon counting detector was used to record the interference patterns.
(2) The University of Sydney (Bedding et al. 1992) had developed a Coudé based masked aperture-plane interference telescope (MAPPIT) for the 3.9 meter Anglo-Australian telescope to investigate interferometry with non-redundant masks. A field lens re-images the telescope pupil down to diameter of 25 mm and the aperture mask is placed where the pupil image is formed. Dove prism is used to rotate the field, allowing coverage of all position angles on the sky. Dispersed fringes are produced using a combination of image and pupil plane imaging. The camera lens and microscope objective produce an image of a star in one direction. In the orthogonal direction, the detector receives the dispersed pupil image. The mask holes play the role of the spectrograph slit. A cylindrical lens is used as the spectrograph's camera lens. Image photon counting system is used in this experiment to record the dispersed fringe pattern.

Chapter 7

Adaptive optics

7.1 Basic principles

Thus far the deleterious effects of the atmosphere on the image due to density inhomogeneities in the path of optical wavefront have been discussed in chapter 5, and the retrieval of image by means of off-line processing is highlighted in chapter 6. This chapter enumerates the recent developments in technology that have enabled the correction of perturbations in the wavefronts in real time by incorporating a controllable counter wavefront distortion which both spatially and temporally follows that of the atmosphere. For long-exposure and high resolution imaging, it is essential to compensate for atmospheric turbulence in real time. Such a technique, referred to adaptive optics (AO) system, refers to optical systems which adapt to correct the optical effects introduced by the medium between the object and its image. It performs two main functions such as, (i) sensing the wavefront perturbations, as well as (ii) compensating the disturbances in real time.

Unlike active optics, where it corrects the wavefront distortions once in a few tens of seconds, adaptive optics correction, depending on wavelength, may vary from a few milliseconds to a few tens of milliseconds. The advantages of the AO system over the post-detection image restoration techniques are the immediate enhancement in the sharpness of the image. For unresolved sources, such a system attempts to gather as many photons in as small an image area as possible, thus enhancing the image contrast against the sky background, thereby improving the resolution. For resolved sources, the improved resolution extends imaging to fainter and more complex objects. Apart from the astronomical applications, the AO systems can be applied in the fields of (i) long range surveillance, with high resolution imaging and tracking, (ii) free space or under water line-of-sight optical communications, (iii) opthalmology, (iv) metrology, and (v) three dimensional (3-D) imaging.

Development of AO system has evolved over the years due to the contribution of numerous scientists and engineers. At the initial stages of its development, the suggestions made by Babcock (1953) have been researched thoroughly by U.S. military. Consistent efforts of scientists, spread over the last two decades, have led to the development of the AO technology that is being employed in astronomy in particular. The requirements of AO system for the military applications appear to be quite different from the astronomical AO systems, which require a small number of modes, a wide optical bandwidth and a small time bandwidth. The defence AO systems require a large number of modes, a narrow optical bandwidth mostly monochromatic and a high time bandwidth.

In the field of observational astronomy, AO system is able to reach deep space enabling to unravel the secrets of the universe. One of its most successful applications has been in imaging of Neptune's ring arcs. AO system is useful for spectroscopic observations, low light level imaging with very large telescopes, and ground-based long baseline optical interferometers (LBOI; Labeyrie, 1975, Saha, 2002 and references therein). Unlike optical phase conjugation method[1] where it is limited to compensate the atmosphere induced aberrations, AO is a multi-disciplinary subject. Among the host of current technologies, its use and practice have become well known in the defence, bio-medical, and astronomical communities. Liang et al. (1997) have constructed a camera equipped with adaptive-optics, which allows one to image a microscopic size of single cell in the living human retina. They have shown that a human eye with adaptive-optics correction can resolve fine gratings that are invisible to the unaided eye.

7.1.1 *Greenwood frequency*

Unlike angular anisoplanatism (see section 5.4.4), the temporal anisoplanatism occur due to a time delay between the propagation of two beams coupled to such wind. In order to construct an adaptive optics system the dynamic behavior of atmospheric turbulence is necessary to be corrected.

[1] Optical phase conjugation method, known also as wavefront reversal or time reversal solution, is a process that involves the use of nonlinear optical processes to exactly reverse the propagation direction and phase variation of a beam of light, thereby causing the return beam to exactly retrace the path of the incident beam. The reverse beam is called conjugate beam.

Two parameters such as the gain[2] and the bandwidth are adjusted according to the number of Zernike modes. Greenwood (1977) determined the required bandwidth, f_G, which is referred also as Greenwood frequency, for full correction by assuming a system in which the static case corrects the wavefront and the remaining aberrations were due to finite bandwidth of the control system. He derived the mean square residual wavefront error as a function of servo-loop bandwidth for a first order controller, which is given by,

$$\langle \sigma_{cl} \rangle^2 = \left(\frac{f_G}{f_c} \right)^{5/3} \quad rad^2, \tag{7.1}$$

where f_c is the frequency at which the variance of the residual wavefront error is half the variance of the input wavefront, known as $3db$ closed-loop bandwidth of the wavefront compensator, and f_G the required bandwidth.

In order to determine afore-mentioned bandwidth, f_G, Greenwood (1977) employed the power spectrum of the phase fluctuations of the wavefront and applied the optical transfer function. This frequency essentially computes the necessary frequency at which aberrations must be applied to account for the temporal nature of the atmospheric turbulence. For a single turbulent layer the Greenwood frequency is defined by the relation,

$$f_G = \frac{0.426 v}{r_0}, \tag{7.2}$$

with v as the velocity of the wind in m s^{-1} and r_0 the atmospheric coherence length, the value of which is an essential parameter in designing proper AO system for a telescope.

According to the equation (7.2), if the effective wind speed is 15 m/s and the size of the Fried's parameter is 60 cm, the required bandwidth for full correction is 11 Hz (Glindemann et al. 2000). For imaging in near-IR to ultraviolet, the AO system bandwidths need to have a response time of the order of a few hundreds to thousand hertz (Hz). It is easier to achieve diffraction-limited information using AO systems at longer wavelengths since the effects of turbulence in the infrared region are weaker, therefore a fewer corrections are required for IR observations.

It is to reiterate that in the long-exposure, image is spread by its random variations of the wavefront tilt. The variance of such a tilt over an aperture

[2]Gain is stated as the power amplification in dB, which is the ratio of output to input power explained in dB($= 10 \log P_o/P_i$), with P_o as the output power and P_i the input power.

of diameter, D, is described by equation (5.136). By defining seeing disc, θ_s, as the FWHM of a Gaussian function fitted to a histogram of image position in arcsec,

$$r_0 = 1.009D\left(\frac{\lambda}{\theta_s D}\right)^{6/5}. \tag{7.3}$$

This equation (7.3) states that the atmospheric coherence length, r_0, depends upon the wavelength to the power of 6/5. For an aperture of the order of 1 m the equation (7.3) is approximated to be,

$$r_0 \sim \left(\frac{\lambda}{\theta_s}\right)^{6/5}. \tag{7.4}$$

The above equation (7.4) implies that the width of seeing limited images, $1.22\lambda/r_\circ \propto \lambda^{-1/5}$ varies with λ.

The number of degrees of freedom, i.e., the number of actuators on the deformable mirror (DM) and the number of sub-aperture in the wavefront sensor, in an AO system should be determined by the following equation,

$$\left(\frac{D}{r_0}\right)^2 \propto \lambda^{-12/5}. \tag{7.5}$$

7.1.2 *Thermal blooming*

Apart from the critical atmospheric parameters that are discussed in chapter 5, knowledge of a parameter, called thermal blooming, may also be essential to any system designer and needs to be characterized accurately (Tyson, 1991). It occurs whenever the atmosphere absorbs sufficient energy from a beam to change the local refractive index. The resultant self-induced distortion is known as thermal blooming. The expansion of the beam size (the blooming) is caused by the resonant absorption of the high power laser energy by atmospheric molecules, resulting in non-linear response of the atmosphere. There is a critical power, P_{cr} which can be transmitted through the atmosphere for which this effect is non-existent.

For a zero velocity wind case, thermal blooming appears, since the lowest index of refraction occurs near the center of the beam. The beam intensity at the center of the beam is the highest. This atmospheric negative lens causes the beam to defocus; this phenomena is also known as thermal defocusing. An important point to be noted is that when the wind or an artificial disturbance in the wind due to the wind slewing, causes the beam to take on a characteristic crescent shaped pattern. In the case of a high power laser beam propagating through the uniformly absorbing atmosphere, the intensity changes according to,

$$I(\vec{r}, L) = I(\vec{r}, 0)e^{-\alpha L - N_b \exp(-r^2/a^2)}, \tag{7.6}$$

in which N_b is the blooming distortion number, α the linear absorption coefficient, and L the propagation path.

The distortion becomes asymmetric in the presence of wind, thus blooming strength turns out to be,

$$N_b = \frac{-2.94 P_b \alpha L^2}{\pi a^3 n \rho v C_P}\left(\frac{dn}{dT}\right), \tag{7.7}$$

where a is the radius of the beam, P_b the beam power, C_P the specific heat at constant pressure, ρ the density of air, dn/dT the change of refractive index with temperature, n the refractive index of the medium, and for a focused beam, the effective radius over a propagation path turns out to be, $a_{eff} = (La\lambda/\pi)^{(1/3)}$.

From the equation (7.7), it is observed that the thermal blooming strength is a function of beam radius a and wind velocity. The thermal blooming effect represented by the blooming distortion number, N_b, on the propagation of the beam spreads the energy physically and thus reduces it on-axis. This reduction takes the form,

$$\mathcal{S}_r = \frac{1}{1 + k' N_b^m}, \tag{7.8}$$

where $\mathcal{S}_r$ is the Strehl ratio, $k' = 0.0626$ and m =2 for an infinite Gaussian beam, while for a uniform beam the respective parameters, k' and m may be 0.01 and 1.2.

The equation (7.8) takes the form in case of modification by the Strehl ratio for thermal blooming (Tyson, 1991),

$$I_{bl} \propto \frac{Pa^2 \mathcal{S}_r}{\lambda^2 L^2} \simeq \frac{P}{1 + kP^m}, \tag{7.9}$$

in which the constant $k = k'(N_b/P)^m$ absorbs the parameter that affects the blooming strength.

This equation (7.9) states that the intensity of the bloomed beam is non-linear in power, P. The thermal blooming effect essentially needs to be compensated during the propagation of the high power laser beams through the atmosphere. Although by increasing slew rate of wind, the blooming strength can be reduced; it is improbable since day-to-day variation of wind is not predictable.

7.2 Wavefront analysis using Zernike polynomials

In optics, the aberrations are often represented by a normalized set of orthogonal polynomials defined on a unit circle, known as Zernike polynomials (Born and Wolf, 1985). Zernike (1934) deduced them from the Jacobi polynomials (also known as hypergeometric polynomials). Zernike polynomials form a system for recording refractive errors, including sphere, cylinder, and higher aberrations. The advantage of such polynomials is that the low order terms are related to the classical aberrations like astigmatism, coma, and spherical aberrations. However, the non-rotationally symmetric aberrations like coma and astigmatism are decomposed into two components, one along the x-axis and the other along y-axis. These may be combined in a single aberration with a certain orientation, which depends on the magnitude of the two components.

Aberrations in the optics of an AO system must be smaller than residual atmospheric aberrations (after correction by the AO system); care should be taken in order to reduce off-axis aberrations. Since the telescope aperture is circular and in the presence of the atmospheric random aberrations, it is useful to express the wavefront in terms of Zernike polynomials. For an annular pupil like a conventional Cassegrain telescope, in which the central hole in the primary mirror allows reflected light from the secondary to pass, these polynomials are required to be suitably modified to take this into account. Since the annulus preserves the azimuthal symmetry, the radial functions of the same azimuthal order are orthogonalized with respect to each other (Mahajan, 1998).

A normalized set of Zernike polynomials was used by Noll (1976) for the application to Kolmogorov turbulence. In this the RMS value of each polynomial over the circle is set equal to one. The phase across the aperture is expanded in terms of these polynomials following which the temporal behavior of the expansion coefficients is studied.

7.2.1 *Definition of Zernike polynomial and its properties*

Let the circle polynomials of Zernike be $Z_n^m(\vec{\xi})$, in which $\vec{\xi} = \xi, \eta$ is the two-dimensional (2-D) position vector. A Zernike polynomial is defined in polar coordinates, i.e., $\xi = \rho \sin\theta$, and $\eta = \rho\cos\theta$, on a circle of unit radius ($R < 1$), hence,

$$Z_n^m(\rho\sin\theta, \rho\cos\theta) = R_n^m(\rho)e^{im\theta}, \tag{7.10}$$

where n indicates the radial degree of the Zernike polynomials and m the azimuthal frequency.

Both these constants, n and m, are always integers and satisfy $m \leq n$ and $n - m =$ even. Consequently, only polynomials with certain combinations of n and m exist. The Zernike polynomials follow the orthogonality relation and thus the orthogonality and normalizing properties are expressed as,

$$\pi^{-1}\int_{|\vec{\xi}^2|\leq 1} V_n^{m*}(\vec{\xi})Z_{n'}^{m'}(\vec{\xi})d\vec{\xi} = \frac{\delta_{mm'}\delta_{nn'}}{n+1}, \tag{7.11}$$

in which δ_{ij} is the Kronecker symbol and * denotes the complex conjugate.

The radial functions, $R_n^m(\rho)$, are polynomials in ρ containing the powers $\rho^n, \rho^{n-2}, \cdots \rho^m$. The radial polynomials satisfy the relation,

$$\int_0^1 R_n^m(\rho)R_n^{m*}(\rho)\rho d\rho = \frac{\delta_{nn'}}{2(n+1)}. \tag{7.12}$$

The arbitrary wavefront, $\psi(\vec{\rho})$, in which $\vec{\rho} = \rho, \theta$ over the unit circle may be represented as an infinite sum of Zernike polynomials,

$$\psi(\vec{\rho}) = \sum_{j=1}^{\infty} a_j Z_j(\vec{\rho}), \tag{7.13}$$

where $Z_j(\vec{\rho})$ is the Zernike polynomial of order j, and a_j the coefficients of expansion that is given by,

$$a_j = \int_{\text{aperture}} \psi(\vec{\rho})Z_j(\vec{\rho})W(\rho)d\vec{\rho}. \tag{7.14}$$

The aperture weight function, $W(\rho)$,

$$W(\rho) = \begin{cases} 1/\pi & \rho \leq 1, \\ 0 & \text{otherwise}, \end{cases} \tag{7.15}$$

is added so that the integral can be taken over all space. The Zernike polynomials are given by the formulae,

$$Z_j(\rho,\theta) = \sqrt{n+1}R_n^m(\rho) \begin{cases} \sqrt{2}\cos(m\theta), & m \neq 0, \quad \text{for j even,} \\ \sqrt{2}\sin(m\theta), & m \neq 0, \quad \text{for j odd,} \\ 1, & m = 0, \end{cases} \tag{7.16}$$

with

$$\begin{aligned} R_n^m(\rho) &= \frac{1}{(n-m)/2!\rho^m}\left\{\frac{d}{d(\rho^2)}\right\}^{(n-m)/2} \left\{(\rho^2)^{(n+m)/2}(\rho^2-1)^{(n-m)/2}\right\} \\ &= \sum_{s=0}^{(n-m)/2} \frac{(-1)^s(n-s)!}{s!\,[(n+m)/2-s]!\,[(n-m)/2-s]!}\rho^{n-2s}. \end{aligned} \tag{7.17}$$

The index j is used to order the modes, the first of which are illustrated in Table III (see Appendix A) and s is a mode-ordering number. Each mode has a coefficient, which is a number with +ve or -ve sign. The Zernike coefficient (in micron) specifies how much of that mode (aberration) is present. Zernike modes are grouped into Zernike orders.

Low order Zernike polynomials have a direct correspondence to familiar optical aberrations such as defocus, tilt, astigmatism, coma etc. The second order include spherical defocus and astigmatism. Figure (7.1) depicts Zernike polynomials in terms of such aberrations. The normalization can be chosen such that for all permissible values of n and m becomes,

$$R_n^m(1) = 1. \tag{7.18}$$

The other property of Zernike polynomials that requires for an analysis of atmospheric turbulence is their Fourier transform, $\widehat{Z}_j(\kappa,\psi)$, i.e.,

$$W(\rho)Z_j(\rho,\theta) = \int_{-\infty}^{\infty} \widehat{Z}_j(\kappa,\psi)e^{-i2\pi\vec{\kappa}\cdot\vec{\rho}}d\vec{\kappa}. \tag{7.19}$$

and thus one writes from the equation (7.16) as,

$$\begin{aligned} \widehat{Z}_j(\kappa,\psi) &= \sqrt{n+1}\frac{J_{n+1}(2\pi\kappa)}{\pi\kappa} \\ &\times \begin{cases} (-1)^{(n-m)/2}i^m\sqrt{2}\cos(m\theta), & m \neq 0,\ j \text{ even}, \\ (-1)^{(n-m)/2}i^m\sqrt{2}\sin(m\theta), & m \neq 0,\ j \text{ odd}, \\ (-1)^{n/2}, & m = 0. \end{cases} \end{aligned} \tag{7.20}$$

in which κ and θ are the modulus and argument of $\vec{\kappa}$, and $J_n(x)$ the Bessel function of order n.

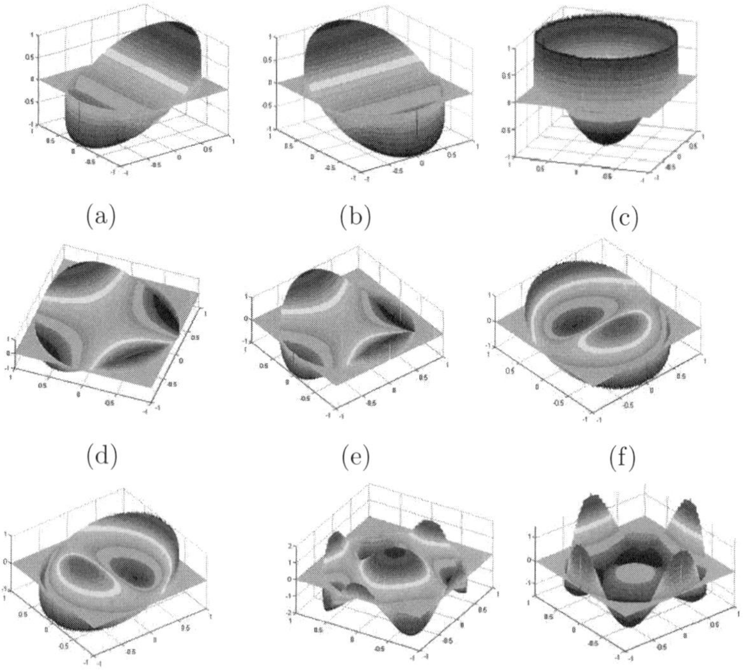

(a) (b) (c)
(d) (e) (f)
(g) (h) (i)

Fig. 7.1 3-D view of optical aberrations for the Zernike coefficients: (a) tilt about y-axis, (b) tilt about x-axis, (c) defocus, (d) astigmatism at 0°, (e) astigmatism at 45°, (f) y-axis coma, (g) x-axis coma, (h) 3rd order spherical, and (i) 3rd order defocus and spherical (combined); these aberrations are generated at the laboratory (Courtesy: V. Chinnappan).

By substituting this equation (7.20) into the equation (7.19), an integral representation for the radial function, R_n^m is derived,

$$R_n^m(\rho) = 2\pi(-1)^{(n-m)/2} \int_0^\infty J_{n+1}(2\pi\kappa) J_m(2\pi\kappa\rho) d\kappa. \tag{7.21}$$

7.2.2 *Variance of wavefront distortions*

The variance of any Zernike coefficient is determined by the total power in its spectrum,

$$\langle \sigma_{a_j} \rangle^2 = \int_0^\infty \mathcal{A}_c(f) df. \tag{7.22}$$

The general expression for the covariance of the expansion coefficients derived by Noll (1976) for n, $n' \neq 0$, as

$$\langle a_j a_{j'} \rangle = \begin{cases} \dfrac{0.046}{\pi}\left(\dfrac{R}{r_0}\right)^{5/3}\sqrt{(n+1)(n'+1)} \\ \quad \times(-1)^{(n+n'-2n)/2}\delta_{mm'} \\ \times\displaystyle\int_0^\infty \frac{J_{n+1}(2\pi\kappa)J_{n'+1}(2\pi\kappa)}{\kappa^2}\kappa^{-8/3}d\kappa, & (j-j')\ \text{even}, \\ 0 & (j-j')\ \text{odd}, \end{cases} \tag{7.23}$$

where n and n' the radial degrees and the integral term turns out to be,

$$\int_0^\infty \frac{J_{n+1}(2\pi\kappa)J_{n'+1}(2\pi\kappa)}{\kappa^2}\kappa^{-8/3}d\kappa = \frac{\Gamma(14/3)\Gamma\left[(n+n'-5/3)/2\right]}{\Gamma\left[(n-n'+17/3)/2\right]\Gamma\left[(n'-n+17/3)/2\right]\Gamma\left[(n+n'+23/3)/2\right]}, \tag{7.24}$$

in which Γ is Euler's Gamma function.

When $j = j'$, the equation (7.23) reduces to,

$$\left\langle \sigma_{a_j} \right\rangle^2 = \frac{0.046}{\pi}\left(\frac{R}{r_0}\right)^{5/3}(n+1)\int_0^\infty \frac{J_{n+1}^2(2\pi\kappa)}{\kappa}\kappa^{-8/3}d\kappa. \tag{7.25}$$

The mean square residual aberration is expressed as the variance of the difference between the uncorrected phase and of the removed modes. Let the aberration due to first J Zernike polynomials be,

$$\psi_J(\vec{\rho}) = \sum_{j=1}^{J} a_j Z_j(\vec{\rho}). \tag{7.26}$$

The mean square residual error is expressed as,

$$\Delta = \int_{\text{aperture}} \left\langle [\psi(\vec{\rho}) - \psi_c(\vec{\rho})]^2 \right\rangle W(\vec{\rho})d\vec{\rho}. \tag{7.27}$$

By inserting equation (7.25) into equation (7.27) and $\langle a_j \rangle = 0$,

$$\Delta_J = \langle \psi^2(\vec{\rho}) \rangle - \sum_{j=1}^{J} \left\langle |a_j|^2 \right\rangle, \tag{7.28}$$

in which $< \psi^2(\vec{\rho}) >$ is the variance of the phase fluctuations, which is infinite for the Kolmogorov spectrum.

Removing the piston term provides a finite value for the variance of the residual aberrations. The first few values of Δ_J are shown in Table IV (see Appendix A). The numbers in the table show that, for a telescope of diameter r_0 or less, the first three modes need to be corrected in order to achieve a large improvement in the optical quality. Of course to gain better improvement, many modes need need to be corrected for. For the removal of higher orders, i.e., $J > 10$, Noll (1976) provided an approximation for the phase variance, as

$$\Delta_J = 0.2944 J^{-\sqrt{3}/2} \left(\frac{D}{r_0}\right)^{5/3} \qquad \text{rad}^2. \tag{7.29}$$

It may be reiterated that the quality of an imaging system is measured by the Strehl ratio, $\mathcal{S}_r$, (see section 4.1.4). If the RMS wavefront error is smaller than about $\pi^2/4$, it is approximated to the equation (4.45). For a large telescope, i.e., $D > r_0$, the Strehl ratio steeply decreases with telescope diameter. Since $r_0 \propto \lambda^{6/5}$, the Strehl ratio, $\mathcal{S}_r$, also decreases sharply with decreasing wavelength.

7.2.3 *Statistics of atmospheric Zernike coefficients*

If the phase obeys Kolmogorov statistics, one can determine the covariance of the Zernike coefficients corresponding to the atmospheric phase aberrations. Noll (1976) had used a normalized set of Zernike polynomials for the application of Kolmogorov statistics. The convenience of the Zernike polynomials is that one derives individually the power in each mode such as, tilt, astigmatism or coma. This helps in calculating the residual aberration after correcting a specified number of modes with an adaptive optics system.

In order to specify the bandwidth requirements of AO systems, the temporal evolution of Zernike mode should be deduced (Noll, 1976, Roddier et al. 1993). A Zernike representation of the Wiener spectrum of the phase fluctuations due to Kolmogorov turbulence (see equation 5.116) can be obtained by evaluating the covariance of the expansion coefficients in equation (7.13). Combining the definition of the expansion coefficients a_j from equation (7.12) and adding the time dependence of the phase across the aperture, the temporal covariance can be defined as,

$$\begin{aligned} C_{a_j}(\tau) &= \langle a_j(t) a_j(t+\tau) \rangle \\ &= \iint_{aperture} Z_j(\vec{\rho}) W(\vec{\rho}) C_\psi(\vec{\rho}, \vec{\rho}\,', \tau) Z_j(\vec{\rho}\,') W(\vec{\rho}\,') d\vec{\rho} d\vec{\rho}\,', \end{aligned} \tag{7.30}$$

in which the integral contains the covariance of phase,

$$C_\psi(\vec{\rho}, \vec{\rho}', \tau) = \langle \psi(\vec{\rho}, t)\psi(\vec{\rho}', t+\tau) \rangle \,. \tag{7.31}$$

By using the power law on both the variables, $\vec{\rho}$ and $\vec{\rho}'$, the equation (7.30) can be expressed in Fourier space,

$$C_{a_j}(\tau) = \iint\limits_{-\infty}^{\infty} \widehat{Z}_j^*(\vec{\kappa})\Phi(\vec{\kappa}, \vec{\kappa}', \tau)\widehat{Z}_j(\vec{\kappa}')d\vec{\kappa}d\vec{\kappa}', \tag{7.32}$$

with $\Phi(\vec{\kappa}, \vec{\kappa}', \tau)$ as the spatial Fourier transform of $C_\psi(\vec{\rho}, \vec{\rho}', \tau)$ with respect to both $\vec{\rho}$ and $\vec{\rho}'$, and the spatial wave numbers are represented by, $\vec{\kappa}$, $\vec{\kappa}'$.

Following Noll (1976), one finds,

$$\Phi_\psi(\vec{\kappa})\delta(\vec{\kappa} - \vec{\kappa}') = 0.023 r_0^{-5/3} \kappa^{-11/3} \delta(\vec{\kappa} - \vec{\kappa}'). \tag{7.33}$$

This equation (7.33) is a direct consequence of the equation (5.116), and the auto-correlation theorem,

$$\mathcal{A}_c(f) = \int_{-\infty}^{\infty} C_{a_j}(\tau) e^{-i2\pi f\tau} d\tau, \tag{7.34}$$

With $\vec{\kappa} = \vec{\kappa}'$, the term, $\Phi_\psi(\vec{\kappa})\delta(\vec{\kappa} - \vec{\kappa}')$, denotes the spatial autocorrelation function of the phase across the aperture.

The Fourier transform of such a function is the spatial power spectrum, $\Phi_\psi(\vec{\kappa})$ (see equation 5.116), while in the case of $\vec{\kappa} \neq \vec{\kappa}'$, turbulence theory does not provide any information about the term, $\Phi_\psi(\vec{\kappa})\delta(\vec{\kappa} - \vec{\kappa}')$, and therefore the delta function is introduced in the equation (7.33). By invoking similarity theorem, we may write

$$\Phi_\psi(\vec{\kappa})\delta(\vec{\kappa} - \vec{\kappa}') = 0.023 \left(\frac{R}{r_0}\right)^{5/3} \kappa^{-11/3} \delta(\vec{\kappa} - \vec{\kappa}'), \tag{7.35}$$

By substituting the equation (7.18) into the equation (7.32), one obtains

$$C_{a_j}(\tau) = \frac{0.046}{\pi} \left(\frac{R}{r_0}\right)^{5/3} \int_0^\infty \frac{J_{n+1}^2(2\pi\kappa)}{\kappa^2} e^{(-i2\pi\widehat{v}\tau\kappa/R)} \kappa^{-8/3} d\kappa, \tag{7.36}$$

where $\widehat{v}$ is the perpendicular velocity of the wind.

The equation (7.36) is a Zernike matrix representation of the Kolmogorov phase spectrum. It is noted here that the effect of the Taylor hypothesis is to introduce a periodic envelope function to the transform. Its frequency dependence on the radius of the aperture, R, average wind velocity, $\widehat{v}$ and the time, τ, and is given by $\widehat{v}\tau/R$. The power spectrum should be real and is undefined for negative frequencies, hence

$$\int_0^\infty e^{-i2\pi\widehat{v}\tau\kappa/R}e^{-i2\pi f\tau}df = \delta\left(f - \frac{\widehat{v}\kappa}{R}\right). \tag{7.37}$$

The resulting power spectra shows dependence on the radial degree of the Zernike polynomial at low frequencies and a high frequency behavior proportional to $f^{-17/3}$ that is independent of Zernike mode. In the low frequency domain, the Zernike tip and tilt spectra decreases with $f^{-2/3}$. The transient frequency between the high and low frequency regions is given by,

$$f_t^n \approx 0.3(n+1)\hat{v}/D, \tag{7.38}$$

which is approximately equal to the bandwidth required to correct for the Zernike mode in an AO system.

7.3 Elements of adaptive optics systems

Normally a Cassegrain type telescope is used in the adaptive optics imaging system which transmits the beacon as well as receives the optical signal for the wavefront sensor (WFS). The other required components for implementing an AO system are:

- image stabilization devices that are the combination of deformable reflecting surfaces, i.e., flexible mirrors such as tip-tilt mirrors, deformable mirrors (DM); these mirrors are, in general, continuous surface mirrors with a mechanical means of deformation to match the desired conjugate wavefront,
- a device that measures the distortions in the incoming wavefront of starlight, called wavefront sensor,
- wavefront phase error computation (Roggemann et al. 1997 and references therein), and
- post-detection image restoration (Roddier, 1999).

In addition, a laser guide star (beacon) may also be needed to improve the signal-to-noise (S/N) ratio for the wavefront signal since the natural guide stars are not always available within iso-planatic patch. A typical Adaptive Optics Imaging System is illustrated in Figure (7.2).

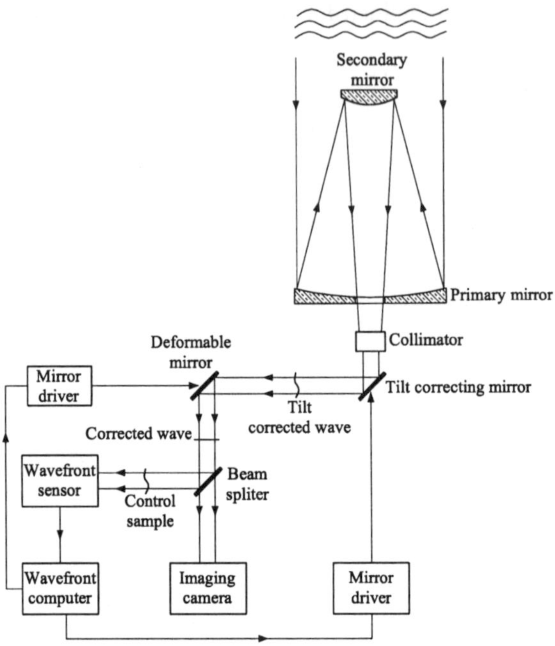

Fig. 7.2 Schematic of the adaptive optics imaging system.

Beam from a telescope is collimated and fed to a tip/tilt mirror to remove low frequency tilt errors. After traveling further, it reflects off of a deformable mirror that eliminates high frequency wavefront errors. A beam-splitter splits the beam into two parts; one is directed to the wavefront sensor and the other is focused to form an image. The former measures the residual error in the wavefront and provides information to the actuator control computer to compute the deformable mirror actuator voltages. This process should be at speeds commensurate with the rate of change of the corrugated wavefront phase errors.

Performance of such an AO system close to the diffraction limit of a telescope can be achieved in the limit of when

- the angular separation between the turbulence probe and the object of interest is smaller than the iso-planatic angle,
- the spacing between the control elements on the DM is well matched

to the turbulence coherence length, and

- a sufficiently high update rate is maintained, i.e., less than inverse of the coherence time.

7.3.1 *Steering/tip-tilt mirrors*

A tip-tilt mirror in an AO system is used for correction of atmospheric turbulence. Such a mirror compensates a tracking error of the telescope automatically as well. It corrects the tilts of the wavefront in two dimensions; a rapidly moving tip-tilt mirror makes small rotations around two of its axes. Implementation of the dynamically controlled active optical components lead-zirconate-titanate[3] (PZT) consisting of a tip-tilt mirror system in conjunction with closed-loop control electronics has several advantages: (i) conceptually the system is simple, and (ii) field of view is wider (Glindemann, 1997). A steering mirror[4] is mounted to a flexure support system that may be tilted fast about its axis of the spring/mass system in order to direct a image in x, y-plane. Such a mirror is used for low frequency wavefront corrections of the turbulence-induced image motions, as well as thermal and mechanical vibrations of optical components. Effectively it is in use for various dynamic applications of active and adaptive optics including precision scanning, tracking, pointing and laser beam or image stabilization.

Tip-tilt mirrors are generally designed to cater for the dynamic application in mind with appropriate dynamic range, tilt resolution and frequency bandwidth. In an AO system, a tip-tilt corrector (see Figure 7.3) is required as one of the two main phase correctors along with a deformable mirror for beam or image stabilization by correcting beam jitter and wander[5].

Tip-tilt corrections require the largest stroke, which may be produced by flat steering mirrors. The amount of energy required to control the tilt is

---

[3]Lead-zirconate-titanate (PZT)s typically consist of laminated stacks of piezoelectric material encased in a steel cylinder. A modulated high-voltage signal is applied to the PZT. This gives rise to small increments of motion. PZT actuators may produce large force in a smaller package at much greater frequency response.

[4]A steering mirror is a glass or metal mirror mounted to a flexure support system, which may be moved independent of the natural frequency of the spring/mass system to direct a light source. It can be used to perform a variety of emerging optical scanning, tracking, beam stabilization and alignment. Such devices have become key components in diverse applications such as industrial instrumentation, astronomy, laser communications, and imaging systems.

[5]Beam wander is the first order wavefront aberration that limits the beam stabilization and pointing accuracy onto the distant targets.

Fig. 7.3 Steering mirror.

related to the stroke (amplitude) as well as to the bandwidth requirements for such mirrors (Tyson, 1991). The amplitude and bandwidth considerations of the disturbance may drive the requirements for the tilt mirror. The inertia of the scanning flat plate mirror for a constant diameter-thickness ratio is proportional to D^5, in which D is the diameter of the mirror. The subsequent force that is needed to move the mirror is proportional to τ_{max}/D, where τ_{max} is the maximum required torque. The steering mirrors with high bandwidth operation can be electronically controlled to tilt around two orthogonal axes (tip-tilt movements) independently. The tip-tilt mirror has three piezoelectric actuators kept in a circle separated by 120°. Hence a two-axis to three-axis conversion is to be carried out. The three piezoelectric actuators expand or contract when DC voltage is applied across them. These actuators can be applied with voltage range of 250 V and -25 V DC. They are essentially capacitor load.

Steering mirror systems are limited to two Zernike modes (x and y-tilt). However the two-axis tilt mirror suffers from the thermal instabilities and cross-talk between the tilting axes at high frequencies. A higher order system compensating many Zernike mode is required to remove high frequency errors. Glindemann (1997) discussed the analytic formulae for the aberrations of the tip-tilt corrected wavefront as a function of the tracking algorithm and of the tracking frequency. A tip-tilt tertiary mirror system has been developed for the Calar Alto 3.5 m telescope, Spain, that corrects the rapid image motion (Glindemann et al., 1997).

7.3.2 *Deformable mirrors*

The incoming wavefront error is both in amplitude and phase variations; the latter is the predominant one. After measuring the phase fluctuations, they can be corrected by implementing an optical phase shift, ψ, by producing optical path difference,

$$\psi = \frac{2\pi}{\lambda}\delta, \tag{7.39}$$

in which $\delta = \Delta(ne)$, is the variance of the optical path, n the refractive index, e the geometrical path spatial distribution of the corrector.

The phase of the wavefront can be controlled by changing either the propagation velocity or the optical path length. Geometrical path difference, Δe, can be introduced by deforming the surface of a mirror, and index spatial difference, Δn, can be produced by birefringent electro-optical materials. The surface of such mirrors is electronically controlled in real time to create a conjugate surface enabling compensation of the wavefront distortion such that perturbations of the turbulence induced incident wavefronts are canceled as the optical field reflects from their surface.

The characteristics of the DMs are dictated by spatial and temporal properties of the phase fluctuations and required degree of corrections. The primary parameters of deformable mirror based AO system are the number of actuators, the control bandwidth and the maximum actuator stroke. For astronomical AO systems, the DMs are suited for controlling the phase of the wavefront. The required actuators are proportional to $(D/r_0)^2$, in which D is the telescope diameter and r_0 the Fried's parameter. Depending on the wavelength of the observations, the desired Strehl ratio, and the brightness of the wavefront reference source, the number of actuators varies from two (tip-tilt) to several hundred. The required stroke is proportional to $\lambda(D/r_0)^2$, and the required optical quality, i.e., RMS surface error, varies in proportion with the observed wavelength. The response time of the actuator is proportional to the ratio r_0/v_w. The typical actuator response time is about a few milliseconds. With the decrease of corrections, it increases.

At the initial stages of AO development, Babcock (1953) suggested to use Eidophor system, a mirror in a vacuum chamber is covered with a thin layer upon which a modulated beam from an electron gun is deposited in a rastered pattern. The transient changes in the slope of the oil film is formed by the induced local forces of surface repulsion. The wavefront is locally tilted by refraction in traversing the film. However, the technological development at that time did not permit to proceed further.

7.3.2.1 *Segmented mirrors*

A variety of deformable mirrors (DM) are available for the applications of (i) high energy laser focusing, (ii) laser cavity control, (iii) compensated imagery through atmospheric turbulence etc. These mirrors can be either segmented mirror or continuous face-plate mirrors that has single continuous surface. There are two varieties of segmented mirrors: (i) piston and (ii) piston and tilt. In the former category, the actuators are normally push-pull type in which each segment can be pushed or pulled in the direction perpendicular to the mirror plane. The latter category mirrors have elements that can be tilted as well. The advantages of segmented mirrors are that

- they can be combined in rectangular arrays to form larger mirrors and
- each element can be controlled independently of the others as there are no interaction between elements.

But the disadvantages of such a system include problems with diffraction effects from the individual elements and interelement alignment. The gap between the elements may be the source of radiation in infrared wave-band, which deteriorates the image quality. In order to deform such a mirror, a wide variety of effects, viz., magnetostrictive, electromagnetic, hydraulic effects, have been used. Refractive index varying devices such as smectic liquid crystals (SLC) and other ferroelectric or electro-optic crystal devices have been used with limited success to implement phase control. Frequency response and amplitude limitations have been the limiting factors for the crystal devices. Reflective surface modifying devices such as segmented mirrors and continuous surface DMs are very successful in several high end applications.

7.3.2.2 *Ferroelectric actuators*

Since the number of actuators are large, there is a need for controlling all the actuators almost simultaneously; the frequency of control is about 1 KHz. In the deformable mirrors, two kinds of piezo actuators are used namely, stacked and bimorph actuators. The present generation piezoelectric actuators are no longer discrete, but ferroelectric wafers are bounded together and treated to isolate the different actuators. In the operational deformable mirrors, the actuators use the ferroelectric effect in the piezoelectric or electrostrictive form. A piezoelectric effect occurs when an electric field is applied to a permanently polarized piezoelectric ceramic, it induces a deformation of the crystal lattice and produces a strain proportional to the electric field. For a disc shaped actuator, the effect of a longitudinal electric field, E is to proportional to the change in the relative thickness, $\Delta e/e$.

The typical values of the longitudinal piezoelectric coefficient vary from 0.3 to 0.8 μm/kV. In order to obtain a stroke of several microns with voltages of a few hundred volts, which is compatible with solid state electronics, several discs are stacked and are electrically connected in parallel. The applied maximum electric field, E_{max}, for a given voltage is limited to a lower value of hysteresis. The minimum thickness, e, turns out to be V/E_{max}, in which V is the voltage. The maximum displacement, Δe, produced by stacked actuators of height, h, is expressed as,

$$\Delta e \propto hE_{max}. \tag{7.40}$$

The piezoelectric materials generally exhibit hysteresis, a cycle characterizing the behavior of polarization and strain with respect to the electric field, which increases as the applied electric field approaches the depolarization field. The hysteresis cycle is characterized by the response stroke with respect to alternating applied voltage and the stroke for the the zero voltage during the cycle. The relative hysteresis, H_{rel}, is given by,

$$H_{rel} = \frac{\Delta S}{S_{max} - S_{min}}, \tag{7.41}$$

in which ΔS is the stroke difference for the zero voltage and S_{max} and S_{min} are the respective maximum and minimum strokes.

Thus the phase delay, $\Delta\psi$ is derived as,

$$\Delta\psi = \sin\left(\frac{\Delta S}{S_{max} - S_{min}}\right), \tag{7.42}$$

While an electrostrictive effect generates a relative deformation, $\Delta e/e$, which is proportional to the square of the applied electric field, E (Uchino et al. 1980), i.e.,

$$\frac{\Delta e}{e} \propto E^2. \tag{7.43}$$

In the electrostrictive materials like lead-magnesium-niobate (PMN), the change in thickness is thickness dependent. In piezoelectric ceramics, the deformation induced by an electric field is due to the superposition of both the piezoelectric and electrostrictive effects. The value of the relative hysteresis depends on the temperature for electrostrictive materials.

7.3.2.3 *Deformable mirrors with discrete actuators*

Deformable mirrors using discrete actuators are used in astronomical AO systems at various observatories (Shelton and Baliunas, 1993, Wizinovitch et al., 1994). This type of deformable mirror (DM) contains a thin deformable face-sheet mirror on a two-dimensional array of electrostrictive stacked actuators supported by rugged baseplate as shown in Figure (7.4). In some cases, actuators are not produced individually, but rather a multi-layer wafer of piezo-ceramic is separated into individual actuators. When some voltage, V_i, is applied to the ith actuator, the shape of DM is described by the influence function[6] $D_i(\vec{x})$, in which $\vec{x}(= x, y)$ is the 2-D position vector, multiplied by V_i. When all actuators are driven, assuming the linearity of the responses of all the actuators, the surface of the mirror, $S(\vec{x}, t)$, can be modeled as,

$$S(\vec{x}, t) = \sum_{i=1}^{N} V_i(t) D_i(\vec{x}), \tag{7.44}$$

where $V_i(t)$ is the control signal applied at time, t and the influence function of the ith actuator at position $\vec{x}$ at the mirror sheet,

$$D_i(\vec{x}) = e^{-\left[(\vec{x} - \vec{x}_i)/d_s\right]^2}, \tag{7.45}$$

in which $\vec{x}_i$ is the location of the ith actuator and d_s the inter-actuator spacing.

The influence function for the ith actuator may be modeled by the Gaussian function that is often used to model PZT or micro-machined deformable mirror (MMDM). The problems may arise from the complexity of the algorithm to control the mirror surface as the actuators are not allowed to move independently of each other. Assuming that each actuator acts independently on a plate that is unconstrained at the edge, a kind of form of the influence function can be found. The fundamental resonant frequency of the mirror is provided by the lowest resonant frequency of the plate and of the actuators. The dynamic equation of the deformation, W,

[6] If one of the actuators is energized, not only the surface in front of this actuator is being pulled, but because of the continuous nature of the deformable mirror, the surface against nearby non-energized actuator also changes. This property is called mirror influence function. It resembles a bell-shaped (or Gaussian) function for DMs with continuous face-sheet (there is some cross-talk between the actuators, typically 15%).

of a plate is given by,

$$s_p \nabla^2 W(\vec{x}) = \rho_p t'_p \left(\frac{\nu_p}{2\pi}\right)^2 W(\vec{x}), \tag{7.46}$$

where

$$\nabla^2 = \frac{\partial^2}{\partial x^2} + \frac{\partial^2}{\partial y^2},$$

denotes two-dimensional Lapacian operator, $s_p[= E_p t_p'^3/12r^2(1-\sigma_p^2)]$, is the stiffness of the clamped plate of radius r, E_p and σ_p respectively the Young's modulus, and the Poisson coefficient of the plate, and t'_p, ρ_p and ν_p respectively the thickness, the mass density, and the characteristic frequency of the plate.

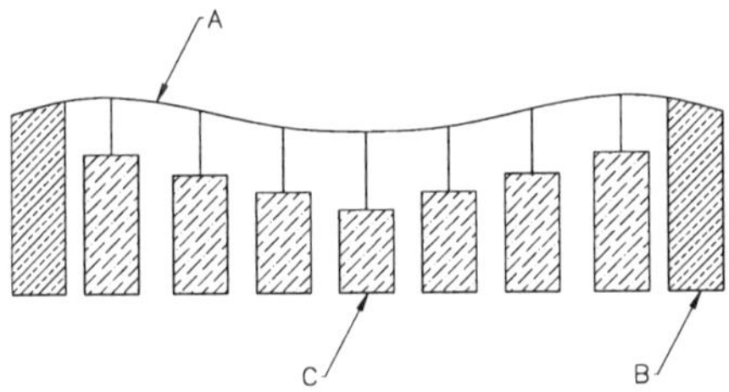

Fig. 7.4 Electrostrictive stacked actuators mounted on a baseplate. A stands for Glass facesheet, B for Mirror support collar, and C for Electro-distortive actuator stack.

The stiffness of the actuators, s_a, depends on the surface, S, of a section, Young's modulus, E_a, and the actuator's height, h, i.e., $s_a = E_a \mathrm{S}/h$. Following these points, the resonant frequency, ν_p, for part of the plate clamped to the actuator spacing distance, d_s, as well as the lowest compression resonant frequency for a clamped-free actuators are deduced as,

$$\nu_p = \frac{Ct'_p}{d_s^2}\sqrt{\frac{E_p}{\rho_p\left(1-\sigma_p^2\right)}},$$
$$\nu_a = \sqrt{\frac{s_a}{2m}}, \tag{7.47}$$

in which $C \simeq 1.6$ is a constant and m is the mass of the actuator.

Ealey (1991) states that the ratio between these two ratios, i.e., ν_p/ν_a turns out to be $4t'_p h/d_s^2$. If the height of the actuator, h, is large, the lowest resonant frequency is that of the actuators, but if h decreases, the frequency increases to that of the face-plate. If the stiffness of the actuator is larger than the stiffness of the face-plate, there is little coupling. The deformation of the plate may be 20-30% smaller than the deformation of the actuator. This is due to the high mechanical coupling. A multi-channel high voltage amplifier must have a short response time, despite a high capacitive load of DM electrodes. For high bandwidth applications such DMs are preferred and further it could be easily cooled.

7.3.2.4 *Bimorph deformable mirror (BDM)*

The name bimorph mirror came from the structure that controls its shape. It is made from two thin layers of materials bonded together. Piezoelectric bimorph plates consist of either a metal plate and a piezoelectric plate, such as PZT or of two piezoelectric plates which are bonded together. The former is known as unimorph, while the latter is called bimorph. A piezoelectric bimorph operates in a manner similar to the bimetal strip in a thermostat. One layer is a piezo-electric material such as PZT, which acts as an active layer and the other is the optical surface, known as passive layer, made from glass, Mo or Si or both pieces may be PZT material. This passive layer is glued to the active layer and coated with a reflective material. The bottom side of the piezoelectric disc is attached with many electrodes; the outer surface between the two layers acting as a common electrode. The PZT electrodes need not be contiguous.

When a voltage is applied to an electrode, one layer contracts and the opposite layer expands, which produces a local bending. The local curvature being proportional to voltage, these DMs are called curvature mirrors. Let the relative change in length induced on an electrode of size l be Vd_{31}/t', in which d_{31} is the transverse piezoelectric coefficient, t' the thickness of the wafer, and V the voltage. Neglecting the stiffness of the layer, and three-dimensional effects, the local radius of curvature, r, turns out to be,

$$r = \frac{t'^2}{2Vd_{31}}. \tag{7.48}$$

The sensitivity of the bimorph, S_b, for a spherical deformation over the diameter, D, is expressed as,

$$S_b = \frac{D^2}{8rV} = \frac{D^2 d_{31}}{4t'^2}. \tag{7.49}$$

The geometry of electrodes in BDM as shown in Figure (7.5) is radial-circular, to match the telescope aperture with central obscuration. For a given number of electrodes (i.e. a given number of controlled parameters) BDMs reach the highest degree of turbulence compensation, better than segmented DMs. BDM very well suits with the curvature type wavefront sensor. Modal wavefront reconstructor is preferred with BDM control. However, such mirrors cannot reproduce all the Zernike polynomials without the application of a gradient at the edges.

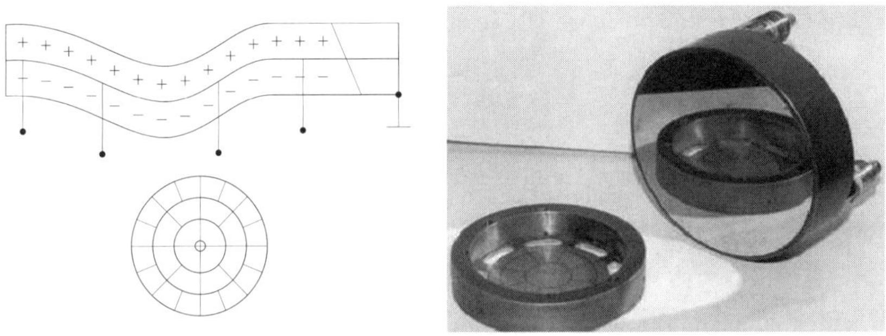

(a) (b)

Fig. 7.5 (a) Geometry of electrode in bimorph deformable mirror and (b) typical BDM.

There is no such simple thing as influence functions for bimorph DMs. The surface shape as a function of applied voltages must be found from a solution of the Poisson equation which describes deformation of a thin plate under a force applied to it. The boundary conditions must be specified as well to solve the equation (7.49). In fact, these DMs are made larger than the beam size, and an outer ring of electrodes is used to define the boundary conditions - slopes at the beam periphery. The mechanical mounting of a bimorph DM is delicate: on one hand, it must be left to deform, on the other hand it must be fixed in the optical system. Typically, 3 V-shaped grooves at the edges are used.

7.3.2.5 *Membrane deformable mirrors*

A membrane mirror consists of a thin flexible reflective membrane, stretched over an array of electrostatic actuators. These mirrors are being manufactured for use in AO system. An integrated electrostatically controlled adaptive mirror has the advantage of integrated circuit compatibility with high optical quality, thus exhibiting no hysteresis.

Flexible mirrors such as MMDM in silicon can be deformed by means of electrostatic forces. The membrane remains flat if voltage differential is not applied to the actuators. When a voltage is applied, the electrostatic attraction between electrodes, individual responses superimpose to form the necessary optical figure. The local curvature of the surface is represented by (Tyson, 1991),

$$\nabla^2 W(\vec{x}) = -\frac{P(\vec{x})}{T(\vec{x})}, \tag{7.50}$$

where the external pressure (force/area) at position $\vec{x}$ is,

$$P(\vec{x}) = \epsilon_a \frac{|V(\vec{x})|^2}{d^2(\vec{x}, P)}, \tag{7.51}$$

and the membrane stress/length ratio,

$$T(\vec{x}) = \frac{E_p\, t_m\, \Delta^2}{2(1 - \sigma_P)}, \tag{7.52}$$

in which ϵ_a is the dielectric constant of air, $V(\vec{x})$ the potential distribution on the actuator, $d(\vec{x})$ is distance between actuator and membrane, E_p the Young's modulus, t_m the thickness of membrane, σ_P the Poisson ratio of membrane material, and Δ the in-plane membrane elongation due to stretching.

The mirror consists of two parts: (i) the die with the flexible mirror membrane and (ii) the actuator structure. A low stressed nitride membrane forms the active part of MMDM. In order to make the membrane reflective and conductive, the etched side is coated with a thin layer of evaporated metal, usually aluminum or gold. Reflective membranes, fabricated with this technology have a good optical quality. Assembly of the reflective membrane with the actuator structure should ensure a good uniformity of the air gap so that no additional stress or deformations are transmitted onto the mirror chip. All components of a MMDM except the reflective membrane can be implemented using PCB technology. Hexagonal actuators are connected to conducting tracks on the back side of the PCB by means of vias (metalized holes). These holes reduce the air damping, extending the linear range of the frequency response of a micro-machined mirror to at least 1 kHz, which is much better than similar devices.

The influence function is primarily determined by the relative stiffness of actuators and face-sheet. Stiffer actuator structures may reduce inter-actuator coupling but require high central voltages. A more practical approach is to reduce the stiffness of the face-sheet material by reducing its

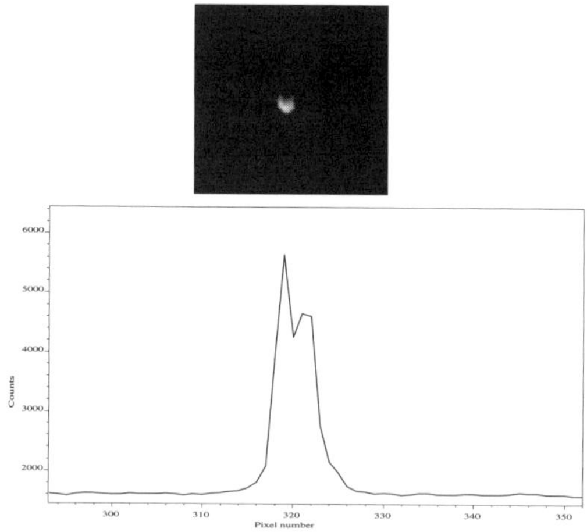

(a)

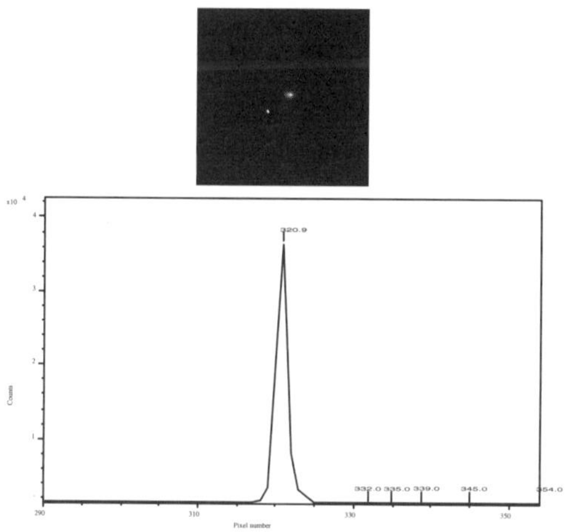

(b)

Fig. 7.6 (a) Un-corrected image (top) of a point source taken with a Cassegrain telescope and its cross section (bottom), (b) corrected image (top) of the said source with a tip-tilt mirror for tilt error correction and other high frequency errors with a MMDM, and its cross section (bottom); images are twice magnified for better visibility (Courtesy: V. Chinnappan).

thickness and or elastic modulus and by increasing the inter-actuator spacing. Figure (7.6) displays the images captured by ANDOR Peltier cooled electron multiplying CCD camera with 10 msec exposure time in the laboratory set up using the MMDM. It is found that an aberrated image having

6.4 pixel FWHM can be sharpened to have 3.5 pixel and the peak intensity has increased from 5,610 counts to 36,500 counts (Chinnappan, 2006).

7.3.2.6 *Liquid crystal DM*

A different class of wavefront actuation, represented by the liquid crystal half-wave phase shifter, is suitable for narrow band applications (Love et al. 1995). Wavefront correction in AO is generally achieved by keeping the refractive index constant by tuning the actual path length with a mirror. An optical equivalent is to fix the actual path length and to tune the refractive index. This could be achieved using many different optical materials; a particularly convenient class of which is liquid crystals[7] (LC) because they can be made into closely packed arrays of pixels which may be controlled with low voltages. When an electric field is applied, the molecular structure is changed, producing a change in refractive index, Δn. This produces a change in the optical path, according to,

$$\Delta W = t\Delta n, \tag{7.53}$$

in which t is the cell thickness.

Electrically addressed nematic liquid crystals (NLC) are generally used for the wavefront correction in conventional AO system, whereas optically addressed the SLCs are also being used to develop an unconventional AO with all optical correction schemes. These crystals differ in their electrical behavior. Ferroelectricity is the most interesting phenomenon for a variety of SLCs. NLCs provide continuous index control, compared with the binary modulation given by ferroelectric liquid crystals (FLC). They are having lower frame rates so it is not the best device for the atmospheric compensation under strong turbulent conditions.

The FLCs are optically addressed in which the wave plates whose retardance is fixed but optical axis can be electrically switched between two states. Phase only modulation with a retarder whose axis is switchable is more complicated than with one whose retardance can be varied. The simplest method involves sandwiching a FLC whose retardance is half a wave in between two fixed quarter wave plates. FLCs have the advantage that they can be switched at KHz frame rates, but the obvious disadvantage is that they are bistable. The use of binary algorithm in wavefront correction

[7] Liquid crystal refers to a state of matter intermediate between solid and liquid and are classified in nematic and smectic crystals. The fundamental optical property of the LCs is their birefringence. They are suitable for high spatial resolution compensation of slowly evolving wavefronts such as instrument aberrations in the active optics systems.

is the simplest approach to develop closed-loop control. The basic wavefront correction algorithm is: whenever the wavefront error is greater that $\lambda/2$ then correction of $\lambda/2$ is applied.

7.3.3 *Deformable mirror driver electronics*

Electronics for the actuator system are the most complex, and by far the most expensive part of the system itself, typically accounting for 2/3rd of its cost. In an extreme example, the first 2000 channel mirror built had approximately 125 electronic component per control channel just for the driver. These drivers are incredibly safe but so complex as to be unreliable. The power supply delivers analog high voltage output signals to the actuators from the input digital low voltage signals supplied by the control computer.

The main component of the single channel driver electronics is high voltage operational amplifier. But it is required to have a feed back loop which limits the available current and shuts the driver down in case of the actuator failure or short circuit. This prevents damaging the mirror by power dissipation in the actuator.

Apart from the high voltage amplifiers, a power supply comprising of a stabilized high voltage generator is required. Such a generator is characterized by the maximum delivered current, which depends on the spectral characteristics of the required correction. A voltage driver, frequently with an analog-to-digital (A/D) converter on the output provides the information to the main system computer on the status of each corrector channel. Today analog inputs are generally insufficient since most wavefront controllers are digital, so each channel has its own digital-to-analog (D/A) converters for the input. The actuator load is a low loss capacitor which must be charged and discharged at the operating rate, typically upto 1 KHz. The required current, i, to control a piezoelectric actuator is given by,

$$i = C_t \frac{dV}{dt}, \tag{7.54}$$

where C_t is the capacitance of the actuator and its connection wire, and V the control voltage that is proportional to the stroke, i.e., the optical path difference.

The peak power consumption can be written as,

$$P_{peak} = \sqrt{2} V_{max} i_{peak}. \tag{7.55}$$

Thus each driver is a linear power amplifier with peak rating of 1-10 W per channel. Certainly every channel is not operating at its full rating all the time. Though the power dissipation is low, the capacitive load gives rise to a high instantaneous current at high frequencies which, with the high voltage, produces large reactive power. The capacitance for the free actuator, $C_a(= \epsilon\epsilon_0 \mathrm{S}/e)$, in which ϵ, ϵ_0 are respectively the relative and vacuum permittivity, and e the capacitance, is required to be considered, since the capacitance for the connection wire is negligible.

Since the temporal Fourier spectrum of the current, i, is proportional to the product of temporal frequency, ν, and the temporal Fourier spectrum of δ, the spectral current density, $\widehat{\Phi}_i$, is given by,

$$\widehat{\Phi}_i \propto \nu^2 \widehat{\Phi}_\delta, \tag{7.56}$$

in which $\widehat{\Phi}_\delta$ is the spectral density of optical path difference, δ.

Thus, the current fluctuations variance required for the actuator, $\langle\sigma_i\rangle^2$, can be written as,

$$\langle\sigma_i\rangle^2 = \frac{C_t^2}{K^2} \int \nu^2 \widehat{\Phi}_\delta(\nu) d\nu, \tag{7.57}$$

where K is the sensitivity of the actuator and is defined as stroke/voltage, which lies between a few μm/kV to a few tens of μm/kV.

7.3.4 *Wavefront sensors*

It is to reiterate that the wavefront is defined as a surface of a constant optical path difference (OPD). The instrument that measures the OPD function is referred to as wavefront sensor. It is a key subsystem of an AO system, which consists of front end optics module and a processor module equipped with a data acquisition, storage, and sophisticated wavefront analysis programs. It estimates the overall shape of the phase-front from a finite number of discrete measurements that are, in general, made at uniform spatial intervals.

Wavefront sensors, that are capable of operating with incoherent (and sometimes extended) sources using broad-band light or white light, are useful for the application in astronomy. These sensors should, in principle, be fast and linear over the full range of atmospheric distortions. The phase of the wavefront does not interact with in any measurable way, hence the wavefront is deduced from intensity measurements at one or more planes.

The algorithms to unwrap the phase and to remove this ambiguity are also slow.

Two paradigms for wavefront sensing such as interferometric and geometric wavefront sensors are employed. The problem of measuring wavefront distortions is common to optics, particularly in the fabrication and control of telescope mirrors, and typically solved with the help of interferometers. These interferometers consider the interference between different parts of the wavefront, for example lateral shear interferometer (see section 6.6.2). Since the interferometric fringes are chromatic in nature and also faint stars (even laser guide stars are not coherent enough to work), are used for such measurements the starlight is not filtered. These sensors should be capable of utilizing the photons very efficiently.

Geometric wavefront sensors such as SH, curvature, and pyramid wavefront sensors rely on the light rays travelling perpendicular to the wavefront. With wavefront sensors measurements are made on:

- the intensity distribution of the image produced by the entire wavefront,
- a reference wavefront of the same or slightly different wavelength combined with the wavefront to produce interference fringes, and
- the wavefront slope, i.e., the first derivative of small zones of the wavefront.

A realization of the first approach the multi-dither technique which requires very bright sources, and thus is applicable only for the compensation of high power laser beams. The second approach is also difficult to implement for astronomical application because of the nature of the astronomical light sources. The third approach can be made by using either a shearing interferometer or a SH sensor and is required to be employed for astronomical applications. These sensors measure phase differences over small baselines on which the wavefront is coherent. They are generally sensitive to wavefront local slopes. In the following sections some of the most commonly used wavefront sensors in astronomical telescope system are enumerated.

7.3.4.1 *Shack Hartmann (SH) wavefront sensor*

Hartmann (1900) developed a test, known as Hartmann's screen test, to evaluate the optical quality of telescope's primary mirror when it was being fabricated. A mask comprising of an array of holes is placed over the aperture of the telescope, and an array of images are formed by the mirror for a parallel beam. In the presence of any surface errors, distorted image spots can be noticed. As the location of the mirror possessing error is known, it may be worked upon further to reduce the error.

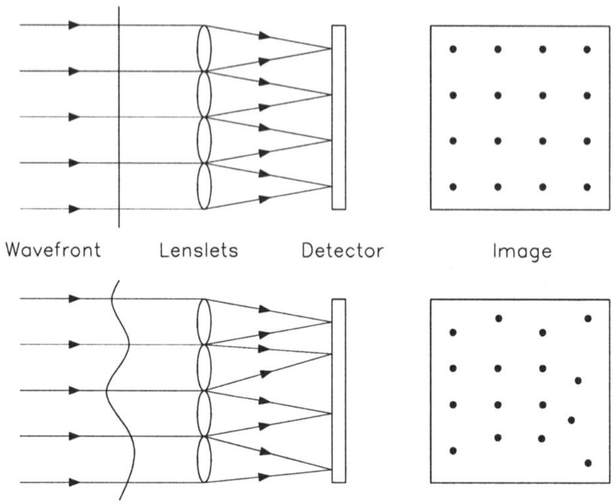

Fig. 7.7 Schematic diagram of Shack Hartmann wavefront sensor.

Hartmann test was the front runner of the modern Shack-Hartmann (SH) wavefront sensor (Shack and Hopkins, 1977), which was the first design permitting to measure the wavefront error, and was developed in the late 1960s to improve the images of satellites taken from the Earth. Such a sensor divides the pupil into sub-pupils (see Figure 7.7) and measures a vector field, i.e., the local wavefront tilts (a first derivative) along two orthogonal directions.

The beam at the focal plane of the telescope is transmitted through a field lens to a collimating doublet objective and imaged the exit pupil of the former on to a lens-let array. Each lens-let defines a sub-aperture in the telescope pupil and is of typically 300 to 500 μm in size. These lenses are arranged in the form of a square grid and accurately positioned from one another. The lens-let array is placed at the conjugate pupil plane in order to sample the incoming wavefront. If the wavefront is plane, each lenslet forms an image of the source at the focus, while the disturbed wavefront, to a first approximation, each lenslet receives a tilted wavefront and forms an off-axis image in the focal plane. The measurement of image position provides a direct estimate of the angle of arrival of wave over each lenslet.

Dimensions of the lens-lets are often taken to correspond approximately to r_0, though Tallon and Foy (1990) suggested that depending on the number of turbulent layers, the size of the sub-pupils can be made significantly larger than the latter. The value of r_0 varies over the duration of observation, therefore, a minimal number of lenslet array for a given aperture size is required. The test consists of recording the ray impacts in a plane slightly before the focal plane. If optics are perfect, the recorded spots would be exactly distributed as the position of lenslets but on a smaller scale.

Shack-Hartmann wavefront sensor requires a reference plane wave generated from a reference source in the instrument, in order to calibrate precisely the focus positions of the lenslet array. Due to aberrations, light rays are deviated from their ideal position, producing spot displacements. The centroid (center of gravity) displacement of each of these subimages provides an an estimate of the average wavefront gradient over the subaperture. A basic problem in this case is the pixellation in the detectors like CCDs on the estimator. If the detector consists of an array of finite-sized pixels, the centroids or the first order moments, C_x, C_y, of the image intensity with respect to x- and y-axes are given by,

$$C_x = \frac{\sum_{j,j} x_{i,j} I_{i,j}}{\sum_{i,j} I_{i,j}}, \qquad C_y = \frac{\sum_{j,j} y_{i,j} I_{i,j}}{\sum_{i,j} I_{i,j}}, \tag{7.58}$$

in which $I_{i,j}$ are the image intensities on the detector pixels and $x_{i,j}$, $y_{i,j}$ the coordinates of the positions of the CCD pixels, (i, j).

Because of the normalization by $\sum_{i,j} I_{i,j}$, the SH sensor is insensitive to scintillation. The equation (7.58) determines the average wavefront slope over the subaperture of area A_{sa}, in which sa stands for subaperture. Thus the first order moment, C_x can be recast as,

$$\begin{aligned} C_x &= \frac{\iint_{im} I(u,v)\, u\, du dv}{\iint_{im} I(u,v) du dv} \\ &= \frac{f}{\kappa}\int_{sa} \frac{\partial \psi}{\partial x} dx dy = \frac{f}{\kappa}\int_0^{d/2}\int_0^{2\pi} \frac{\partial \psi}{\partial x} \rho d\rho d\theta, \end{aligned} \tag{7.59}$$

where $\kappa = 2\pi/\lambda$ is the wave number, f the focal length of the lenslets, and ψ the wavefront phase.

By integrating these measurements over the beam aperture, the wavefront or phase distribution of the beam can be determined. In particular the space-beam width product can be obtained in single measurement. The intensity and phase information can be used in concert with information about other elements in the optical train to predict the beam size, shape, phase and other characteristics anywhere in the optical train. Moreover, it also provides the magnitude of various Zernike coefficients to quantify the different wavefront aberrations prevailing in the wavefront. The variance for the angle of arrival, $\alpha_x = C_x/fM$, in which M is the magnification between the lenslet plane and the telescope entrance plane, is given by,

$$\langle\sigma_x\rangle^2 = 0.17\left(\frac{\lambda}{r_0}\right)^2\left(\frac{D_{sa}}{r_0}\right)^{-1/3} \qquad \text{arcsec}^2, \tag{7.60}$$

where D_{sa} is the diameter of the circular subaperture; this equation can be written for the y-direction as well.

In order to minimize the read noise effect, a small number of pixels per sub-aperture is used. The smallest detector size per such sub-aperture is 2×2 array, called a quad-cell[8]. Let I_{11}, I_{12}, I_{21}, and I_{22} be the intensities measured by the four quadrants,

$$\begin{aligned} C_x &= \frac{\theta_b}{2}\frac{I_{11} + I_{21} - I_{12} - I_{22}}{I_{11} + I_{12} + I_{21} + I_{22}}, \\ C_y &= \frac{\theta_b}{2}\frac{I_{21} + I_{12} - I_{11} - I_{22}}{I_{11} + I_{12} + I_{21} + I_{22}}, \end{aligned} \tag{7.61}$$

in which θ_b is the angular extent of the image.

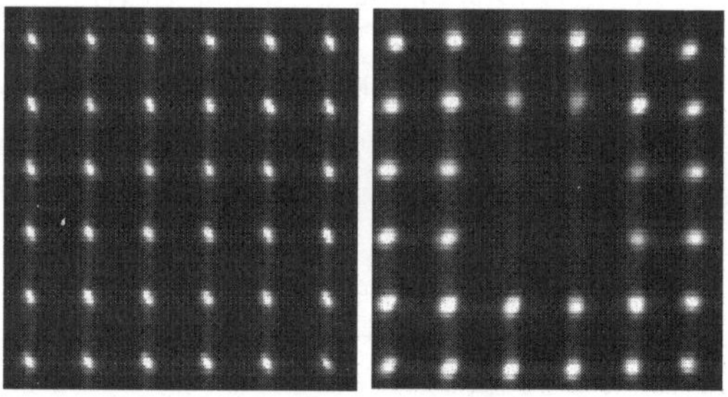

(a) (b)

Fig. 7.8 Intensity distribution at the focal plane of a 6×6 lenslet array captured by the EMCCD camera (a) for an ideal case at the laboratory and (b) an aberrated wavefront taken through a Cassegrain telescope. (Courtesy: V. Chinnappan).

For a diffraction-limited image, $\theta_b = \lambda/d$, in which d is the size of the lenslet, while under atmospheric turbulent conditions, $\theta_d \approx \lambda/r_0$. Figure

[8]Quad-cell sensors have a non-linear response and have a limited dynamic range.

(7.8a) depicts the distribution of intensity at the focal plane of a 6×6 lenslet array illuminated by the test beam, while Figure (7.8b) shows the aberrated intensity wavefront taken through a Cassegrain telescope. Careful observation of the lenslet spots would reveal deviations in the spot position. It is to be noted that four missing spots in the middle is due to the central hole in the primary mirror of a Cassegrain telescope.

The major advantage of Shack-Hartmann sensor is its high optical efficiency. The other notable advantages are that it measures directly the angles of arrival, and therefore works well with incoherent white light extended sources (Rousset, 1999 and references therein); it is able to operate with continuous or pulsed light sources. It has a high spatial and temporal resolution, large dynamic range and no 2π ambiguities. This type of sensors have already been used in AO systems (Fugate et al., 1991, Primmerman et al., 1991).

7.3.4.2 *Curvature sensing*

The curvature sensor (CS) has been developed by Roddier (1988c, 1990) to make wavefront curvature measurements instead of wavefront slope measurements. It measures a signal proportional to the second derivative of the wavefront phase. The Lapacian of the wavefront, together with wavefront radial tilts at the aperture edge, are measured, providing data to reconstruct the wavefront by solving the Poisson equation with Neumann boundary conditions[9]. Such a sensor works well with incoherent white light (Rousset, 1999) as well. The advantages of such an approach are:

- since the wavefront curvature is a scalar, it requires one measurement per sample point,
- the power spectrum of the curvature is almost flat, which implies that curvature measurements are more effective than tilt measurements, and
- flexible mirrors like a membrane or a bimorph can be employed directly to solve the differential equation, because of their mechanical behavior, apriori removing any matrix multiplication in the feedback loop; they can be driven automatically from the CS (Roddier, 1988).

This technique is a differential Hartmann technique in which the spot displacement can be inverted. The principle of the CS is depicted in Figure (7.9), in which the telescope of focal length f images the light source in

[9]Neumann boundary conditions specify the normal derivative of the function on a surface.

its focal plane. The CS consists of two detector arrays placed on either side of focus. The first and second detector arrays record the intensity distributions in an intra-focal plane $P_1(\vec{x})$ and in an extra-focal plane $P_2(\vec{x})$ respectively. A local wavefront curvature in the pupil produces an excess of illumination in one plane and a lack of illumination in other. A field lens is used for symmetry in order to re-image the pupil. A pair of out-of-focus images are taken in these planes. Hence, by comparing spot displacement on each side of the focal plane, one can double the test sensitivity.

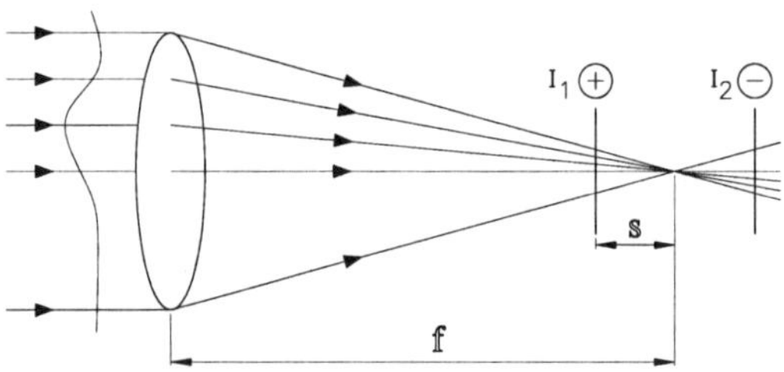

Fig. 7.9 Curvature wavefront sensor.

The difference between two plane intensity distribution, $I_1(\vec{x})$ and $I_2(\vec{x})$, is a measurement of the local wavefront curvature inside the beam and of the wavefront radial first derivative at the edge of the beam. It is a measure of wavefront slope independent of the mask irregularities. The computed sensor signals are multiplied by a control matrix to convert wavefront slopes to actuator control signals, the output of which are the increments to be applied to the control voltages on the DM. Subsequently, the Poisson equation is solved numerically and the first estimate of the aberrations is obtained by least squares fitting Zernike polynomials to the reconstructed wavefront. A conjugate shape is created using this data by controlling a deformable mirror, which typically compose of many actuators in a square or hexagonal array.

As the normalized difference, C_n, is used for the comparison, and $I_1(\vec{x})$ and $I_2(\vec{x})$ are measured simultaneously, the sensor is not susceptible to the non-uniform illumination due to scintillation. The normalized intensity difference is written as,

$$\begin{aligned} C_n &= \frac{I_1(\vec{x}) - I_2(\vec{x})}{I_1(\vec{x}) + I_2(\vec{x})} \\ &= \frac{f(f-s)}{s}\left[\frac{\partial W(\vec{x})}{\partial n}\delta_c - P(\vec{x})\nabla^2 W(\vec{x})\right], \end{aligned} \tag{7.62}$$

where the quantity $\partial W(\vec{x})/\partial n$ is the radial first wavefront derivative in the outward direction perpendicular to pupil edge, $\vec{x} = x, y$ the 2-D position vector, $P(\vec{x})$ the transmission function of the pupil, f the focal length of the telescope, s the distance between the focal point and the intra/extra-focal plane, and δ_c the Dirac distribution around the pupil edge.

Both the local wavefront slope and local wavefront curvature can be mapped with the same optical setup, doubling the number of reconstructed points on the wavefront. A high resolution detector with almost zero read-out noise is required for such a sensor. The first astronomical images obtained from a low-order adaptive optical imaging system using a curvature sensor was reported by Roddier (1994). The CFHT adaptive optics bonnette (AOB), PUEO (Arsenault et al., 1994), is based on the variable curvature mirror (Roddier et al., 1991) and has a 19-zone bimorph mirror (Rigaut et al., 1998). In order to drive a flexible membrane mirror, Roddier et al. (1991) employed sound pressure from a loudspeaker that is placed behind the said mirror. They could provide a feedback loop that adjusts the power to the loudspeaker to maintain a constant RMS tip-tilt signal error.

7.3.4.3 *Pyramid WFS*

Another wavefront sensor based on a novel concept, called pyramid wavefront sensor has been developed by Ragazzoni (1996) and evaluated the limiting magnitude for it to operate in an adaptive optics system (Esposito and Riccardi, 2001). This sensor (see Figure 7.10) is able to change the continuous gain and sampling, thus enabling a better match of the system performances with the actual conditions on the sky.

Pyramid sensor consists of a four-faces optical glass pyramidal prism that behaves like an image splitter and is placed with its vertex at the focal point. When the tip of the pyramid is placed in the focal plane of the telescope and a reference star is directed on its tip, the beam of light is split into four parts. Using a relay lens located behind the pyramid, these four beams are then re-imaged onto a high resolution detector, obtaining

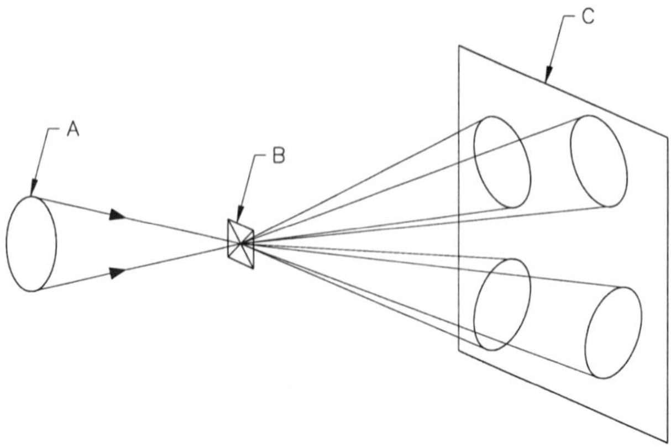

Fig. 7.10 Pyramid wavefront sensor. A stands for light beam coming from telescope, B for Pyramid, and C for detector.

four images of the telescope pupil. The intensity distributions in the $j(=1,2,3,4)$th pupil are represented by $I_1(\vec{x})$, $I_2(\vec{x})$, $I_3(\vec{x})$, and $I_4(\vec{x})$, in which $\vec{x} = x, y$ is the 2D position vector. Since the four edges of the pyramid act like a knife-edge (or Foucault) test system, these images contain essential information about the optical aberrations introduced in the beam from the atmosphere. These parameters can be used to correct the astronomical images.

The phase can be retrieved by using phase diversity technique (see section 6.3.10). The notable advantages of the pyramid sensor are:

- the sub-apertures are defined by the detector pixels since there is no lenslet array; the number of sub-apertures for faint object can be reduced by binning, and
- the amplitude of the star wobble can be adjusted as a trade-off between the smaller wobble (sensitivity) and the larger wobble (linearity); at small amplitudes the sensitivity of such a sensor can be higher than SH sensor (Esposito and Riccardi, 2001).

However the pyramid sensor introduces two aberrations:

- at the pupil plane, there is a rotating plane mirror that displaces the apex of the pyramid with respect to the image at the focal plane, and
- it divides the light at the focal plane in the same fashion as the lenslets

of the SH sensor divide the light at the pupil plane.

7.3.5 *Wavefront reconstruction*

As stated earlier in the preceding section (7.3.4), the wavefront sensor measures the local wavefront tilt or curvature yielding local wavefront tilt or curvature as a function of transverse ray aberrations defined at specific pupil locations. Since the wavefront is continuous, such local measurements are stitched together so that a continuous wavefront profile is generated. Such a process is known as wavefront reconstruction; it generates the OPD function.

Wavefront reconstructor converts the signals into phase aberrations and measures any remaining deviations of the wavefront from ideal and sends the corresponding commands to the DM. The small imperfections of DM like hysteresis or static aberrations are corrected automatically, together with atmospheric aberrations. The real-time computation of the wavefront error, as well as correction of wavefront distortions, involves digital manipulation of data in the wavefront sensor processor, the reconstructor and the low-pass filter; the output is converted to analog drive signals for the deformable mirror actuators. The functions are to compute (i) sub-aperture gradients, (ii) phases at the corners of each sub-aperture, (iii) low-pass filter phases, as well as to provide (iv) actuator offsets to compensate the fixed optical system errors and real-time actuator commands for wavefront corrections.

A direct method of retrieving the wavefront is to use the derivations of the Zernike polynomials expressed as a linear combination of Zernike polynomials (Noll 1976). Let the measurements of wavefront sensor data be represented by a vector, $\vec{S}$, whose length is twice the number of sub-apertures, N, for a SH sensor because of measurement of slopes in two directions and is equal to N for curvature wavefront sensor. The unknowns (wavefront), a vector, $\vec{\psi}$, specified as phase values on a grid, or more frequently, as Zernike coefficients is given by,

$$\vec{\psi} = \vec{B} \cdot \vec{S}, \tag{7.63}$$

where $\vec{B}$ is the reconstruction or command matrix, $\vec{S}$ the error signal, and $\vec{\psi}$ the increment of commands which modifies slightly previous actuator state, known as closed-loop operation.

7.3.5.1 *Zonal and modal approaches*

In order to apply a phase correction, the information of the wavefront derived from the measured data are employed to close the loop. The phase reconstruction method finds the relationship between the measurements and the unknown values of the wavefronts and can be categorized as being either zonal or modal, depending on whether the estimate is either a phase value in a local zone or a coefficient of an aperture function (Rousset, 1999). In these methods, the optical beam aperture is divided into sub-apertures and the wavefront phase slope values are computed in each of these sub-apertures using difference in centroids from reference image and an aberrated image; the wavefront is constructed from these slope values.

Approach due to the former deals with the wavefront expressed in terms of the OPD over a small spatial area or zone, while the latter approach is known when the wavefront is expressed in terms of coefficients of the modes of a polynomial expression over the whole-aperture. If the low order systematic optical aberrations such as tilt, defocus, astigmatism etc. are dominant, the modal analysis and corrections are generally used, while in the presence of high order aberrations, the zonal approach is employed. In the first approximation, the relation between the measurements and unknown is assumed to be linear. The matrix equation between $\vec{S}$ and $\vec{\psi}$ is read as,

$$\vec{S} = \vec{A}\vec{\psi}, \tag{7.64}$$

in which $\vec{A}$ is called the interaction matrix and is determined experimentally in an AO system.

The Zernike polynomials are applied to a DM and the reaction of wavefront sensor to these signals is recorded. The reconstructor matrix, $\vec{B}$, (see equation 7.63) performs the inverse matrix and retrieves wavefront vector from the measurements. A least-square solution that consists of the minimization of the measurement error, σ_s,

$$\sigma_s = \left|\left|\vec{S} - \vec{A}\vec{\psi}\right|\right|^2, \tag{7.65}$$

in which $||\quad||$ is the norm of a vector, is useful since the number of measurements is more than the number of unknowns.

The least-square solution is generally employed where the wavefront phase, $\vec{\psi}$, is estimated so that it minimizes the error, σ_s. The resulting reconstructor is recast as,

$$\vec{B} = \left(\vec{A}^t A\right)^{-1} \vec{A}^t, \tag{7.66}$$

with $\vec{A}^t$ as the transpose of $\vec{A}$.

The matrix $\vec{A}^t \vec{A}$ is singular, therefore some parameters or combinations of parameters are not constrained by the data. The phase is determined up to a constant by its derivatives. The wavefront sensor is insensitive to wavefront constant over the aperture (piston mode). In order to solve the matrix inversion, singular value decomposition algorithm is being employed. By using a priori information, i.e, the statistics of wavefront perturbations (covariance of Zernike modes) alongwith the wavefront sensor noise on the signal properties, another reconstructor matrix similar to a Wiener filter (see section 5.3.2) may be achieved. This technique, known as iterative method (will be discussed in chapter 9), looks for a solution that provides the minimum expected residual phase variance, which in turns gives rise to the maximum Strehl's ratio.

The shape of an optical wavefront is represented by a set of orthogonal entire pupil modal functions. One possible approach is to apply Zernike polynomials as spatially dependent functions. Let the phase be represented by the coefficient of expansion in a set of functions, Z_i, called modes. The reconstruction calculates a vector of coefficients, $\vec{\psi} = \{\psi_i\}$, using a relation similar to the equation (7.11). The computed phase anywhere in the aperture (Rousset, 1999),

$$\psi(\vec{r}) = \sum_i \psi_i Z_i(\vec{r}), \tag{7.67}$$

in which $i = 1, 2, \cdots n$ is the mode, with n the number of modes in the expansion.

The interaction matrix, $\vec{A}$, is calculated using the analytic expression of the modes, $Z_i(\vec{r})$. The two elements of $\vec{A}$ for a Shack-Hartmann sensor are represented by,

$$\begin{aligned} A^x_{i,j} &= \frac{1}{A_{sa}} \int_j \frac{\partial Z_i(\vec{r})}{\partial x} d\vec{r}, \\ A^y_{i,j} &= \frac{1}{A_{sa}} \int_j \frac{\partial Z_i(\vec{r})}{\partial y} d\vec{r}, \end{aligned} \tag{7.68}$$

where j stands for the sub-aperture and A_{sa} for the area of the sub-aperture.

In an AO system, the noise is propagated from the measurements to the commands in the reconstruction process. The expression for the minimal variance reconstructor involves the interaction matrix and the covariance matrices of noise, $\vec{C}_n$ and atmospheric perturbations. The maximum a posteriori probability approach can also be used. The noise of the reconstructed phase, $\langle\sigma\rangle^2$, for any reconstructor, $\vec{B}$, is given by,

$$\langle\sigma\rangle^2 = \frac{1}{n}\sum_i Var\left(\psi_i\right) = \frac{1}{n}T_r\left(\vec{B}\vec{C}_n\vec{B}^t\right), \tag{7.69}$$

in which $\vec{B}\vec{C}_n\vec{B}^t$ is the noise covariance matrix of $\vec{\psi}$, $\vec{C}_n$ the covariance matrix of measurements (a diagonal matrix with elements $\langle\sigma_{ph}\rangle^2$ in case of uncorrelated noise), and T_r the sum of diagonal matrix elements.

The equation (7.69) allows to compute the noise propagation coefficient relating the wavefront measurement error to the error of the reconstructed phases.

7.3.5.2 *Servo control*

Control systems are often called as either process (regulator) control or servomechanisms. In the case of the former, the controlled variable or output is held to a constant or desired value, like a human body temperature, while the latter vary the output as the input varies. These systems are known as closed-loop control systems[10], in which they respond to information from somewhere else in the system.

In a temporal control of the closed-loop, the control system is generally a specialized computer, which calculates aberrations from the wavefront-sensor measurements, the commands sent to the actuators of the deformable mirror. In order to estimate the bandwidth requirements for the control system, one needs to know how fast the Zernike coefficients change with time. The calculation must be done fast (depending on the seeing), otherwise the state of the atmosphere may have changed rendering the wavefront correction inaccurate. The required computing power needed can exceed several hundred million operations for each set of commands sent to a 250-actuator deformable mirror. The measured error signal by wavefront sensor as shown in the Figure (7.11) is given by,

$$e(t) = x(t) - y(t), \tag{7.70}$$

[10] An open-loop control system does not use feedback. It has application in optics, for example, when a telescope points at a star following the rotation of the Earth.

in which $x(t)$ is the input signal (e.g. a coefficient of some Zernike mode) and $y(t)$ the signal applied to the DM.

The error signal should be filtered before applying it to DM, or else the servo system would be unstable. In the frequency domain this filter is,

$$\widehat{y}(f) = \widehat{e}(f)\widehat{G}(f), \tag{7.71}$$

in which $\widehat{G}(f)$ is the Laplace transfer function (see Appendix B) of the control system, called open-loop transfer function.

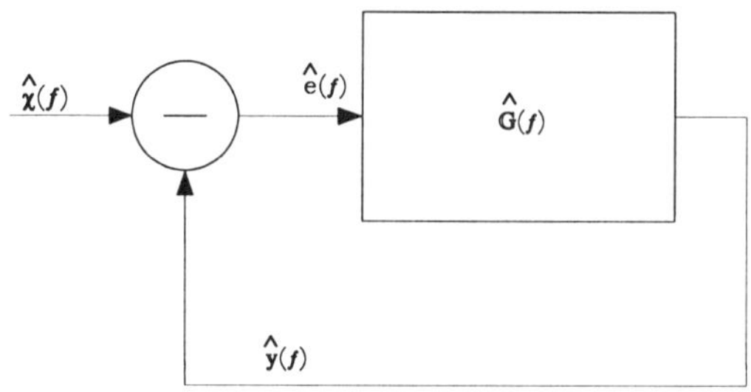

Fig. 7.11 Schematic diagram of the control system (Roddier, 1999).

The equation (7.71) can be recast as,

$$\begin{aligned}\widehat{e}(f) &= \widehat{x}(f) - \widehat{y}(f) \\ &= \widehat{x}(f) - \widehat{e}(f)\widehat{G}(f),\end{aligned} \tag{7.72}$$

where $\widehat{x}(f)$, $\widehat{y}(f)$, $\widehat{e}(f)$, are the Laplace transform of the control system input, $x(t)$, output, $y(t)$, and the residual error, $e(t)$.

Thus the transfer functions for the closed-loop error, χ_c, and for the closed-loop output, χ_o are deduced respectively as,

$$\chi_c = \frac{\widehat{e}(f)}{\widehat{x}(f)} = \frac{1}{1+\widehat{G}(f)}, \qquad \text{and} \tag{7.73}$$

$$\chi_o = \frac{\widehat{y}(f)}{\widehat{x}(f)} = \frac{\widehat{G}(f)}{1+\widehat{G}(f)}. \tag{7.74}$$

By replacing open-loop transfer function, $\widehat{G}(f)$ with g/f, in which $g = 2\pi\nu_c$ is the loop gain, ν_c the $3db$ closed-loop bandwidth of the control system, and $f = 2i\pi\nu$, the closed-loop error transfer function of the time frequency, ν, so that,

$$\widehat{E}(\nu) = \frac{i\nu}{\nu_c + i\nu}. \tag{7.75}$$

Thus the power spectrum of the residual error, $e(t)$ is derived as,

$$|\widehat{E}(\nu)|^2 = \frac{\nu^2}{\nu_c^2 + \nu^2}. \tag{7.76}$$

The response time to measure the wavefront signal by the wavefront sensor is represented by, $\widehat{G}(f)e^{-2i\pi\tau\nu}$, for a delay of time, τ. A certain time in computing the control signal is also required, since the response of the DM is not instantaneous due to its resonance and hysteresis. The transfer function, $\widehat{G}(f)$ accumulates additional phase delays with increasing frequency, hence the delay turns out to be larger than π. This implies that the servo system amplifies the errors. Such a system becomes unstable when the modulus of the closed-loop transfer function exceeds 1. It should be noted here that the closed-loop bandwidth is about 1/10 of the lowest DM response frequency.

7.3.6 *Accuracy of the correction*

The error signal measured by the wavefront sensor accompanies noise. The optimum bandwidth ensuring best performance of an adaptive optics (AO) system depends on the (i) brightness of the guide star, (ii) atmospheric time constant, and (iii) correction order. The main sources for errors in such a system are mean square deformable mirror fitting errors, $\langle\sigma_F\rangle^2$, the mean square detection error, $\langle\sigma_D\rangle^2$, the mean square prediction error, $\langle\sigma_P\rangle^2$, and the mean square aniso-planatic error, $\langle\sigma_{\theta_0}\rangle^2$, (Roddier, 1999). The overall mean square residual error in wavefront phase, $\langle\sigma_R\rangle^2$ is given by,

$$\langle\sigma_R\rangle^2 = \langle\sigma_F\rangle^2 + \langle\sigma_D\rangle^2 + \langle\sigma_P\rangle^2 + \langle\sigma_{\theta_0}\rangle^2. \tag{7.77}$$

The capability to fit a wavefront with a finite actuator spacing is limited, hence it leads to the fitting error. The fitting error phase variance, $\langle\sigma_F\rangle^2$, is described by,

$$\langle\sigma_F\rangle^2 = k\left(\frac{d_s}{r_0}\right)^{5/3}, \tag{7.78}$$

where the spatial error is a function of the coherence length r_0, the size of the interactuator center-to-center spacing d_s of the deformable mirror, and the coefficient, k, that depends on influence functions of DM and on the geometry of the actuator.

The equation (7.78) relates that the variance of the wavefront fitting error decreases as the 5/3rd power of the mean actuator spacing. Such an error depends on how closely the wavefront corrector matches the detection error. The detection error is the reciprocal to the signal-to-noise (S/N) ratio of the wavefront sensor output which can be expressed as,

$$\langle\sigma_D\rangle^2 = \left[\chi\eta\frac{2\pi d_s}{\lambda}\frac{d}{S/N}\right]^2, \tag{7.79}$$

in which d is the spot size in radians, η the reconstructor noise propagator, and χ the closed-loop transfer function.

If a plane wave is fitted to the wavefront over a circular area of diameter, d, and its phase is subtracted from the wavefront phase (tip-tilt removal), the mean square phase distortion reduces to,

$$\langle\sigma_{\psi'}\rangle^2 = 0.134\left(\frac{d}{r_0}\right)^{5/3}. \tag{7.80}$$

The prediction error is due to the time delay between the measurement of the wavefront disturbances and their correction. By replacing ξ in equation (5.111) with a mean propagation velocity with modulus, $\bar{v}$ (an instantaneous spatial average), the temporal structure function of the wavefront phase, $\mathcal{D}_\psi(\tau)$, is determined as,

$$\mathcal{D}_\psi(\tau) = 6.88\left(\frac{\bar{v}}{r_0}\right)^{5/3}. \tag{7.81}$$

The time delay error, $\langle\sigma_\tau\rangle^2$, can be expressed as,

$$\langle\sigma_\tau\rangle^2 = 6.88\left(\frac{\bar{v}\tau}{r_0}\right)^{5/3}. \tag{7.82}$$

This equation (7.82) shows that the time delay error, $\langle\sigma_\tau\rangle^2$ depends on two parameters, viz., $\bar{v}$ and r_0 which vary with time independent of each other. The acceptable time delay, τ_0, known as Greenwood time delay (Fried, 1993) for the control loop is given by,

$$\tau_0 = (6.88)^{-3/5}\frac{r_0}{\bar{v}} = 0.314\frac{r_0}{\bar{v}}. \tag{7.83}$$

It is noted here that the delay should be less than τ_0 for the mean square phase error to be less than 1 radian. The equation (7.82) can be recast into,

$$\langle\sigma_\tau\rangle^2 = \left(\frac{\tau}{\tau_0}\right)^{5/3}. \tag{7.84}$$

From the equation (5.83), the atmospheric transfer function, $\widehat{\mathcal{B}}(\vec{f})$, is recast as,

$$\widehat{\mathcal{B}}(\vec{f}) = e^{-\frac{1}{2}\mathcal{D}_\psi(\lambda\vec{f})}. \tag{7.85}$$

With the AO system, the large-scale wavefront distortions having largest amplitude are compensated. The effect of smoothing off the structure function at level $2\langle\sigma\rangle^2$, in which $\langle\sigma\rangle^2$ is the variance of remaining uncorrelated small-scale wavefront distortions. The $\langle\sigma\rangle^2$ turns out to be smaller with better corrections. the atmospheric transfer function, $\widehat{\mathcal{B}}(\vec{f})$, at low frequency decreases, but converges to a constant,

$$\widehat{\mathcal{B}}(\infty) = e^{-\langle\sigma\rangle^2}. \tag{7.86}$$

The image quality degrades exponentially with the variance of the wavefront distortion. To a good approximation, the Strehl ratio, $\mathcal{S}_r$, can be written as,

$$\mathcal{S}_r \approx e^{-\langle\sigma\rangle^2}. \tag{7.87}$$

By inserting the equation (7.84) into the equation (7.87), one may derive the decrease of Strehl ratio as a function of the time delay in the servo loop,

$$\mathcal{S}_r \approx e^{-\left(\frac{\tau}{\tau_0}\right)}. \tag{7.88}$$

It may be reiterated that limitation due to lack of iso-planaticity of the instantaneous PSF occurs since the differences between the wavefronts coming from different directions. For a single turbulent layer at a distance $h\sec\gamma$, the mean square error, $\langle\sigma_{\theta_0}\rangle^2$, on the wavefront is obtained by replacing ξ with $\theta h\sec\gamma$. As stated in chapter 5, several layers contribute to image degradation in reality. The mean square error due to aniso-planaticity, $\langle\sigma_{\theta_0}\rangle^2$, is expressed as,

$$\langle\sigma_{\theta_0}\rangle^2 = 6.88\left(\frac{\theta L \sec\gamma}{r_0}\right)^{5/3}, \tag{7.89}$$

in which L is the mean effective height of the turbulence.

The equation (7.89) shows that the mean square aniso-planaticity error, $\langle\sigma_{\theta_0}\rangle^2$, depends on two independent parameters, viz., the weighted average of the layer altitude, L and the atmospheric coherence length, r_0. Recalling the equation (5.121) for the iso-planatic angle, θ_0, for a given distance, θ, between the target of interest and the guide star, the residual wavefront error due to aniso-planatism is estimated as,

$$\langle\sigma_{\theta_0}\rangle^2 = \left(\frac{\theta}{\theta_0}\right)^{5/3}. \tag{7.90}$$

The mean square calibration error, $\langle\sigma_{cal}\rangle^2$, may also add to the misery. The determining factor of such an error arises from deformable mirror flattening and non-common path errors.

Another limitation of measurement errors comes from the detector noise as well. This may be in the form of photon noise as well as read noise, which can deteriorate the performance of correction system for low light level. An ideal detector array senses each photon impact and measures its position precisely. The fundamental nature of noise of such a detector is produced by the quantum nature of photoelectron. A single photon event provides the centroid location with a mean square error equal to the variance of the intensity distribution. In a system that consists of a segmented mirror controlled by a Shack-Hartmann sensor, let θ'' be the width of a subimage. The mean square angular error, $\langle\sigma_{\theta'}\rangle^2$, on local slopes is of the order of θ'^2, and therefore, the mean square angular error, $\langle\sigma_{\theta'}\rangle^2$, is given by,

$$\langle\sigma_{\theta'}\rangle^2 = \frac{\theta''^2}{p_n}, \tag{7.91}$$

where $p_n = n_p d^2$ is the independent photon events provided by the guide star, n_p the number of photons, and d the size of the sub-aperture over which an error, θ', on the slope angle produces an error, $\delta = \theta' d$, on the optical path with variance,

$$\langle\sigma_\delta\rangle^2 = \frac{\theta''^2}{n_p}. \tag{7.92}$$

Assume that each sub-aperture is larger than the atmospheric coherence length, r_0, each sub-image is blurred with angular size, $\theta'' \simeq \lambda/r_0$, hence the variance can be derived as,

$$\langle\sigma_\delta\rangle^2 = \frac{\lambda^2}{n_p r_0^2}. \tag{7.93}$$

Taking the help of the equation (7.80), the fitting error for a segmented mirror in terms of optical path fluctuations, one may express,

$$\langle\sigma_F\rangle^2 = 0.134\frac{1}{\kappa^2}\left(\frac{d}{r_0}\right)^{5/3}, \tag{7.94}$$

in which $\kappa = 2\pi/\lambda$ is the wavenumber and d the spot size.

A variance contribution from the non-linear effects of thermal blooming, $\langle\sigma_{bl}\rangle^2$, that is a function of the blooming strength, N_b, and the number of modes corrected, N_{mod}, should be taken into account as well. The approximation is is given by,

$$\langle\sigma_{bl}\rangle^2 = \frac{\sqrt{2}}{5\pi^4}\frac{N_b^2}{N_{mod}^{2.5}}. \tag{7.95}$$

7.3.7 *Reference source*

Implementation of adaptive optics system depends on the need for bright unresolved reference source for the detection of wavefront phase distortions and the size of iso-planatic angle. Observations of such a source within iso-planatic patch help to measure the wavefront errors by means of a wavefront sensor, as well as to map the phase on the entrance pupil. The most probable solution to such a problem is to make use of artificial laser guide stars (Foy and Labeyrie, 1985), though the best results are still obtained with natural guide stars (NGS), they are too faint in most of the cases; their light is not sufficient for the correction. The number of detected photons, n_p per cm^2, for a star of visual magnitude m (see section 10.2.2.1) striking the Earth's surface is,

$$n_p = 8 \times 10^{(3-0.4m)}\Delta\tau\eta_{tr}\int \eta_d(\lambda)d\lambda \qquad \text{cm}^{-2}, \tag{7.96}$$

with $\Delta\tau$ as the integration time (seconds), η_{tr} the transmission coefficient of the system, $\eta_d(\lambda)$ the quantum efficiency of the detector and the integral is over the detector bandwidth expressed in nanometers.

The integral in the equation (7.96) is taken over the detector bandwidth and is expressed in nanometers. The number of stars brighter than 12 m_v is, $1.45e^{0.96m_v}$ stars rad^{-2}. According to which there are 150,000 stars rad^{-2} brighter than 12 m_v are available. Since the number of iso-planatic patches in the sky is about 10^9, these stars are insufficient to provide one in each iso-planatic patch (Tyson, 2000).

With a poor beam divergence quality laser, the telescope's primary mirror can be used as an element of the laser projection system, while with a diffraction-limited laser, projection system can be side-mounted and bore-sighted to the telescope (Tyson, 2000). The beam is focused onto a narrow spot at high-altitude in the atmosphere in the direction of the scientific target. Light is scattered back to telescope from high altitude atmospheric turbulence, which is observed and used to estimate wavefront phase distortions. In principle, the LGS should descend through the same patch of turbulence as the target object. A laser may produce light from three reflections (Foy and Labeyrie, 1985):

(1) Resonance scattering: Existence of layer in the Earth's mesosphere containing alkali metals such as sodium (10^3 - 10^4 atoms cm^{-3}), potassium, calcium, at an altitude of 90 km to 105 km, permits to create laser guide stars.
(2) Rayleigh scattering: This kind of scattering refers to the scattering of light by air molecules (mainly nitrogen molecules, N_2) and tiny particles between 10 and 20 km altitude. It is more effective at short wavelengths. The degree of scattering varies as a function of the ratio of the particle diameter to the wavelength of the radiation, along with other factors including polarization, angle, and coherence. This scattering is considered as elastic scattering that involves no loss or gain of energy by the radiation[11].
(3) Mie scattering: This scattering arises from dust, predominantly for particles larger than the Rayleigh range. This scattering is not strongly wavelength dependent, but produces the white glare around the Sun in presence of particulate material in the air. It produces a pattern like lobe, with a sharper and intense forward lobe for larger particles.

Unlike Mie scattering by aerosol or cirrus clouds, which may be impor-

[11]Scattering in which the scattered photons have either a higher or lower photon energy is called Raman scattering. The incident photons interacting with the molecules in a fashion that energy is gained or lost so that the scattered photons are shifted in frequency. Both Rayleigh and Raman scattering depend on polarizability of the molecules.

tant at lower altitudes, but are usually variable and transient, scattering of the upward propagating laser beam is due to Rayleigh scattering; its strength depends on the atmospheric density. Since the density decreases with altitude, this limits the strength of the backscatter at high altitudes. The main drawback is the inadequate sampling of the atmospheric turbulence due to the lower height of the backscatter.

In order to produce backscatter light from Na atoms in the mesosphere, a laser is tuned to a specific atomic transition. Sodium atoms scatter beacon strongly from NaD_2 resonance line at 589.2 nm and NaD_1 resonance line at 589.6 nm. The sodium atom absorbs a photon, making the electrons jump to the first energy orbital above ground state, which is followed by the return of the atom to its ground state, accompanied by the emission of single photon. The probability of transition in the former line is higher than that of the latter line. The high altitude of the process makes it suitable for astronomical AO systems since it is closer to sampling the same atmospheric turbulence that a starlight from infinity comes through. However, the laser beacons from either of the Rayleigh scattering or of the sodium layer return to telescope are spherical wave, unlike the natural light where it is plane wave, hence some of the turbulence around the edges of the pupil is not sampled well.

Concerning the flux backscattered by a laser shot, Thompson and Gardner (1988) stressed the importance of investigating two basic problems: (i) the angular aniso-planatic effects and (ii) the cone effect. The problem arises due to the former if the natural guide stars are used to estimate the wavefront errors. The mean square residual wavefront error, $\langle\sigma_{\theta_0}\rangle^2$, due to aniso-planatism is provided by the equation (7.90). The iso-planatic angle is only a few arcseconds in the optical wavelengths and it is often improbable to locate a bright reference star within this angle of a target star. It is worthwhile to note that the size of the iso-planatic angle increase linearly with wavelength, even in the infrared only 1% of the sky contains bright enough reference star.

Since LGS is at finite altitude, H, above the telescope, while the astronomical objects are at infinity, the latter effect arises due to the parallax between these sources; the path between the LGS and the aperture is conical rather than cylindrical. The laser beacons, Rayleigh beacon in particular, suffer from this effect since it samples a cone of the atmosphere instead of a full cylinder of atmosphere, which results in annulus between the cone and the cylinder. A turbulent layer at altitude, h, is sampled differently by the laser and starlight. Due to this cone effect, the stellar wavefront may have a residual phase error while compensating the laser beacon by the AO system, which is given by (Foy, 2000),

$$\langle \sigma_c \rangle^2 = \frac{\kappa^2 \sec \gamma}{H} \int_0^H \mathcal{C}_n^2(h) h^2 (h^{1/3} - 1) dh, \tag{7.97}$$

where $\mathcal{C}_n^2$ is the refractive index structure constant.

In terms of telescope diameter, D (Fried and Belsher, (1994), one may write,

$$\langle \sigma_c \rangle^2 = \left(\frac{D}{d_0} \right)^{5/3}, \tag{7.98}$$

with $d_0 \approx 2.91\theta_0 H$ as the parameter characterizing the cone effect, which depends on the vertical distribution of the turbulence, the wavelength, the zenith angle, and the backscatter altitude.

In view of the discussions above, laser beacon from the resonance scattering from the mesospheric Na atom seemed to be more promising. Both pulsed and continuous wave laser are used to cause a bright compact glow in sodium laser guide star. Over the years, several observatories have developed the laser guide artificial star in order to palliate the limitations of low sky coverage[12]. The major advantages of an artificial laser guide star system are (i) it can be put anywhere and (ii) is bright enough to measure the wavefront errors. However, the notable drawbacks of using laser guide star are:

- Although, rays from the LGS and the astronomical source pass through the same area of the pupil, the path of the back scattered light of the laser guide star does not cross exactly the same layers of turbulence as the star beacon since the artificial light is created at a relatively lower height. This introduces a phase estimation error, the correction of which requires multiple laser guide stars surrounding the object of interest.
- The path of the artificial star light is the same as the path of the back scattered light, so the effects of the atmosphere on the wavefront tilt are cancelled out.
- Laser beacon is spread out by turbulence on the way up; it has finite spot size (typically 0.5 - 2 arcseconds).

[12]The fraction of the sky that is within range of a suitable reference star is termed as sky coverage. The sky coverage is relatively small, which limits the applicability of high resolution techniques in scientific observations

- It increases measurement error of wavefront sensor and cannot be useful for the spectroscopic sky conditions.
- It is difficult to develop artificial star with high powered laser. The powerful laser beacons directed to the sky are dangerous for aircraft pilots and satellites and enhance light pollution at the observatory as well.

The sky coverage remains low at short wavelengths as well, owing to the tip/tilt problem. It is improbable to employ laser guide star for the basic correction, since it moves around due to atmospheric refraction on the upward path of the laser beam. A different system must be developed to augment the AO system for tip/tilt.

7.3.8 *Adaptive secondary mirror*

Another way to correct the wavefront disturbance in real time is the usage of a adaptive (deformable) secondary mirror (ASM). Such a system has several advantages over the conventional system such as it (i) makes relay optics obsolete which are required to conjugate a deformable mirror at a reimaged pupil, (ii) minimizes thermal emission (Bruns et al. 1997), (iii) enhances photon throughput that measures the proportion of light which is transmitted through an optical set-up, (iv) introduces negligible extra infrared emissivity, (v) causes no extra polarization, and (vi) non-addition of reflective losses (Lee et al. 2000).

Fig. 7.12 Deformable secondary mirror at the 6.5 m MMT, Arizona (Courtesy: L. Close).

The 6.5 meter Multi Mirror Telescope (MMT), Mt. Hopkins Observatory, Arizona, USA has a 64-cm diameter ultra-thin (1.7 mm thick) secondary mirror with 336 active elements or actuators operating at 550 Hz (see Figure 7.12). Due to the interactuator spacing, the resonant frequency of such a mirror may be lower than the AO bandwidth. The actuators are basically like acoustical voice coils used in stereo systems. There is a 50 micron air space between each actuator and a magnet that is glued to the back surface of the ultra-thin secondary mirror. The viscosity of the air is sufficient to damp out any unwanted secondary harmonics or other vibrations (Wehinger, 2002).

The ASM system employs a Shack-Hartmann (SH) sensor with an array of small lenslets, which adds two extra refractive surfaces to the wavefront sensor optical beam (Lloyd-Hart, 2000). Such a system is used at the $f/15$ AO Cassegrain focus of the MMT. The corrected beam is relayed directly to the infrared science instrument, with a dichroic beam splitter passing light beyond 1 μm waveband and reflecting visible light back into the wavefront sensing and acquisition cameras. Owing to very low emissivity of the system, the design of such a system is optimized for imaging and spectroscopic observations in the 3-5 μm band. It is planned to install a similar mirror with 1000 actuators that has a diameter of 870 mm and a thickness of 2 mm, whose shape can be controlled by voice coil, at the Large Binocular Telescope (LBT).

7.3.9 *Multi-conjugate adaptive optics*

Due to severe isoplanatic patch limitations, a conventional AO system fails to correct the larger field-of-view (FOV). Such a correction may be achieved by employing multi-conjugate adaptive optics (MCAO) system. In this technique, the atmospheric turbulence is measured at various elevations and is corrected in three-dimensions with a number of altitude-conjugate DMs, generally conjugate to the most offending layers. Each DM is conjugated optically to a certain distance from the telescope. Such a system enables near-uniform compensation for the atmospheric turbulence over considerably wider FOV.

However, its performances depends on the quality of the wavefront sensing of the individual layers. A multitude of methods have been proposed. Apart from solving the atmospheric tomography, a key issue, it is apparent that a diversity of sources, sensors, and correcting elements are required to tackle the problem. The equation (7.104) reveals that the cone effect becomes more significant with increasing diameter of telescope. A solution may be envisaged to mitigate such a problem in employing a web of guide stars that allows for 3-d tomography of the atmospheric turbulence. Combining the wavefront sensing data from these guide stars, one can enable to reconstruct the three-dimensional (3-D) structure of the atmosphere and eliminate the problem of aniso-planatism (Tallon and Foy, 1990).

Ragazzoni et al. (2000) have demonstrated this type of tomography. This new technique pushes the detection limit by $\sim$1.7 mag on unresolved objects with respect to seeing limited images; it also minimizes the cone effect. This technique will be useful for the extremely large telescopes of 100 m class, e.g., the OverWhelmingly Large (OWL) telescope (Diericks and Gilmozzi, 1999). However, the limitations are mainly related to the finite number of actuators in a DM, wavefront sensors, and guide stars.

Chapter 8

High resolution detectors

8.1 Photo-electric effect

Photo-electric emission is the property possessed by certain substances that emit electrons, generally in vacuum, when they receive light or photons. This effect was first observed by Heinrich Hertz in 1887. Around 1900, P. Lenard (1862-1947) had observed that when ultraviolet radiation falls on a metal surface, it gets positively charged, which was interpreted later as being due to the emission of electrons from the metal surface when light of suitable frequency falls on it. It was felt that one cannot explain the radiation emitted from a heated body, strictly speaking a black body, on the basis of the laws of classical physics.

Following the introduction of the quantum nature of electromagnetic radiation by Planck (1901), Einstein (1905) brought back the idea that light might be made of discrete quanta[1] and postulated that the electromagnetic wave is composed of elementary particles called *lichtquanten*, which gave way to a more recent term 'photon'. He pointed out that the usual view that the energy of light is continuously distributed over the space through which it travels faces great difficulties when one tries to explain photoelectric phenomena as expounded by Lenard. He conceived the light quantum to transfer its entire energy to a single electron. This photon concept helped him to obtain his famous photo-electric equation and led to the conclusion that the number of electrons released would be proportional to the intensity of the incident light. The energy of the photon is expended in liberating the electron from the metal and imparting a velocity to it. If the energy, $h\nu$, is sufficient to release the electrons from the substance, the collected electrons

[1]Quanta means singular quantum and the word 'quantum' means a specified quantity or portion.

produce an electric current. This phenomenon is known as 'photo-electric effect'.

While explaining such an effect, Einstein (1905) mentioned that when exposed to light certain materials emit electrons, called 'photo-electrons'. The energy of the system obeys Fermi-Dirac distribution function[2]. At absolute zero, the kinetic energy of the electrons cannot exceed a definite energy, called the Fermi energy, E_F, a characteristic of the metal. Now the electrons inside the metal need certain amount of energy to come out of the metal and this deficiency is called surface potential barrier. When a photon with an energy, $h\nu$, larger than the binding energy of an electron hits an atom, it is absorbed. The electron is emitted with a Fermi level energy, E_F, equal to,

$$E_F = h\nu - E_k, \tag{8.1}$$

where E_k represents the kinetic energy of the ejected photo-electron.

Photo-electric effect cannot be explained on the basis of classical theory of physics, according to which the energy of radiation depends upon the intensity of the wave. If such an intensity is very low, it requires a considerable amount of time for an electron to acquire sufficient energy to come out of the metal surface. But with the proper frequency, in photo-electric effect, irrespective of the intensity, photo-emission commences immediately after the radiation is incident on the metal surface. The energy of emitted electrons depend on the frequency of the incident radiation; higher the frequency higher is the energy of the emitted electrons. The number of emitted electrons per unit time increased with increasing intensity of incident radiation.

8.1.1 *Detecting light*

The 'photo-detector' is a device that produces a sole electrical signal when a photon of the visible spectrum has been detected within its field-of-view, regardless of its angle of incidence. For an ideal photo-detector, the spectral

[2]Fermi-Dirac distribution applies to Fermions, possessing an intrinsic angular momentum of $\hbar/2$, where $\hbar$ is the Planck's constant, h, divided by 2π, obeying Pauli exclusion principle. The probability that a particle possesses energy, E, is

$$n(E) = \frac{1}{Ae^{(E - E_F)/k_BT} + 1},$$

where E_F is the Fermi energy.

responsivity[3], $R(\lambda)$, in amps(A)/watts(W) is given by,

$$R(\lambda) = \begin{cases} \dfrac{\lambda}{\lambda_P} R_P & \lambda \leq \lambda_P, \\ 0 & \text{elsewhere}, \end{cases} \tag{8.2}$$

where R_P is the peak responsivity, λ_P the wavelength at which R_P occurs.

The quantum efficiency (QE), η_d, is defined as the ratio of signal electrons generated and collected to the number of incoming photons. It determines the sensitivity of the detector. An ideal detector would have a QE of 100% and would be sensitive to all colors of light. The responsivity in terms of the QE, is determined by,

$$R = \frac{e\lambda}{hc}\eta_d, \tag{8.3}$$

in which $e(= 1.6 \times 10^{-19}\text{coulomb(C)})$ is the electron charge, h the Planck's constant, and c the velocity of light.

Several techniques have been developed to turn single electron into a measurable pulse of multiple electrons. A wide range of modern photodetectors, such as photo-multipliers, charge-coupled devices (CCD), and television cameras, have been developed. These sensors produce a electric current that is proportional to the intensity of the light.

The most desired properties for detectors used in high resolution imaging are low noise[4] and high readout speed. It is improbable to obtain these two properties at the same time. During the initial phases of the development in speckle imaging, a problem in the early 1970s was the data processing. Computers were not powerful enough for real-time processing and video recorders were expensive. One of the first cameras for such purpose, built by Gezari et al. (1972), was an intensified film movie[5] camera. Subsequently, a few observers used photographic films with an intensifier attached to it for recording speckles of astronomical objects (Breckinridge et al. 1979). Saha et al. (1987) used a bare movie camera, which could record the fringes and specklegrams of a few bright stars; a water-cooled

[3]Spectral responsivity of an optical detector is a measure of its response to radiation at a specified wavelength (monochromatic) of interest. If the entire beam falls within the active area (aperture) of the detector, the responsivity is equal to the ratio of detector response to beam radiant power, while in the case of a detector being placed in a radiation field which over-fills its aperture, it is equal to the ratio of detector response to the irradiance of the field.

[4]Noise describes the unwanted electronic signals, sometimes random or systematic contaminating the weak signal from a source.

[5]Movie is known to be a film running at 16 or more frames per second.

bare CCD was also used for certain interferometric observations by Saha et al. (1997c). However, the main drawback of a CCD is its serial readout architecture, limiting the speed of operation.

Owing to low quantum efficiency of the photographic emulsion[6], it is essential to use a high quality sensor for the high angular resolution interferometry, which enables one to obtain snap shots with a very high time resolution of the order of (i) frame integration of 50 Hz, or (ii) photon recording rates of several MHz. Such interferometry also requires to know the time of occurrence of each photo-event within less than 20 msecs. The performance of high resolution imaging relies on the detector characteristics, such as, (i) the spectral bandwidth, (ii) the quantum efficiency, (iii) the time lag due to the read-out of the detector, and (iv) the array size and the spatial resolution.

Ever since the successful development of a photon-counting detector system (Boksenburg, 1975), detectors for visible light interferometry have made incredible advances and operating near their fundamental limit in most wavelength regions. Photon-counting cameras for low-energy photons are the result of parallel progress in different fields of research, for example, gamma-ray imaging, night vision, and photometry. However, these cameras did not aim directly attempt at low-energy photon-counting imaging, and therefore separately brought the elements like micro-channel plates, image intensifiers, position sensitive anodes. Such elements are being employed in present day photon-counting cameras.

After long years of struggle to develop detectors like CP40 (Blazit, 1986), precision analogue photon address (PAPA; Papaliolios and Mertz 1982), and multi anode micro-channel array (MAMA; Timothy, 1983), commercial devices are produced for real time applications like adaptive optics. These cameras have high quantum efficiencies, high frame rates, and read noise of a few electrons.

8.1.2 *Photo-detector elements*

Photo-electric effect can occur if the interaction of light with materials results in the absorption of photons and the creation of electron-hole pairs. Such pairs change the conductance of the material. A metal should contain a very large number of free electrons, of the order of 10^{22} per cm^3, which

[6]Photo-sensitive emulsion absorbs light. An individual absorption process may lead to the chemical change of an entire grain in the emulsion, and thereby create a dark dot in the plate.

may be chosen for a given frequency of the light. The semi-transparent photo-surfaces (photo-cathodes) are generally coated with photo-electric material on the inner side. The main characteristics of such a photo-cathode is its quantum output that is to say that photo-electric effect manifests itself when a photon of a particular frequency strikes the photo-cathode. This cathode is generally maintained at a lower potential with respect to the anode[7]

For a photo-electric emission, the energies of photons impinging on the material surface provide the extra energy for electrons to overcome the energy barrier. The energies of these electrons obey the Fermi-Dirac statistics. At absolute zero, the kinetic energies cannot exceed a definite energy, which is characteristic of the metal. When the temperature is raised, a small fraction of these electrons can acquire additional energy. At low temperatures these electrons are unable to leave the metal spontaneously, their energy being insufficient to overcome the surface potential barrier or work function, ϕ_0. The work function of a metal is defined to be the minimum energy required to release an electrons from atomic binding. This energy is supplied by the photon of frequency, ν_0, such that $\phi_0 = h\nu_0$, in which ν_0 is the photoelectric threshold frequency and is a constant for the material. Below this frequency there is no emission and above it, there would be emission, even with faintest of radiation. The remaining energy would appear as kinetic energy of the released electron. For $\nu > \nu_0$, the emitted electrons have some extra energy, characterized by velocity, v and is given by the energy equation,

$$h\nu = \frac{1}{2}mv^2 + \phi_0. \tag{8.4}$$

As temperature is raised some of the electrons acquire extra energy and eventually they may come out of the metal as thermoionic emission, which obeys Richardson's law. This law states that the emitted current density, $\vec{J}$ is related to temperature, T by the equation,

$$\vec{J} = A_R T^2 e^{-\phi_0/k_B T}, \tag{8.5}$$

where $A_R(= 4\pi\, m\, e\, k_B^2/h^3)$ is the proportionality constant, known as Richardson's constant, m and e the mass and charge of an electron respectively, $k_B(= 1.38 \times 10^{-23}\mathrm{JK}^{-1})$ the Boltzmann constant, and T the

[7]An anode are generally is a positively charged terminal of a vacuum tube where electrons from the cathode travel across the tube toward it, and in an electroplating cell negative ions are deposited at the anode.

temperature of the device.

Since the photo-electric effect demonstrates the quantum nature of light, this effect can actually detect a single photon and hence is the most sensitive detector of radiation. The quantum outputs measured for each wavelength in a given space provide the spectral response of the photo-cathode. The sensitivity of a photo-cathode corresponds to the illumination of a black body of 2856 K. The flux of the emitted electrons is, in general, expressed in microamperes per lumen (lm; 1 lm $\approx l.47 \times 10$ 3W). Another way of expressing sensitivity is to use the radiance, corresponding to illumination by a monochromatic source of specific wavelength. The unit used in this case is often the milliampère per watt.

Photo-electric surfaces with high efficiency are not generally obtainable with metals, but rather with semiconductors. The metals have the same work function value determined by thermoionic emission and photo-electric emission, while the semiconductors do not have this property. Their thermoionic work function may differ considerably from their photo-electric work function though semiconductors are effective photo-emitting surfaces. For visible wavelength, the photo-electric effect occurs when the Fermi energy is of the order of 1.5 eV, what corresponds to that of alkaline metals such as sodium (Na), potassium (K), and cesium (Cs). Semiconductors solids, such as germanium (Ge), silicon (Si), and indium-gallium-arsenide (InGaAs) are suited as well for this purpose. These metals, therefore can be broadly used, in general in association with some antimony, in the manufacture of photo-cathode. It is to be noted that the quantum output of photo-cathodes based on alkaline metals is often too weak for expulsions with longer wave length.

In a photo-cathode, light enters from one side and the electrons are emitted from the other side into the vacuum. They are further amplified by a chain of cathodes. These cathodes simplify the collection of electrons and the concentrations of them on the 'dynode', which posses the property of emitting many more electrons than they receive under electron bombardment. This phenomenon is called 'secondary electrons' emission. The number of secondary electrons emitted depends on the nature of the surface, on the energy of the primary, as well as on the incident angle of the primary electrons. The ratio of the average number of secondary electrons emitted by a target to the primary electrons bombarding the dynode is characterized by the 'secondary emission ratio', δ. Among the emitted electrons three groups of electrons are distinguishable:

(1) primary electrons that are elastically reflected without of loss of energy (high energy species),
(2) primary electrons that are back diffused, low energy species in comparison, having continuous distribution of energy, and
(3) true secondary electrons (low energy species).

If the primary energy is sufficiently large, the energy distribution become independent of primary energy. Because, for highly energetic primaries a large number of secondaries are produced deep inside the material and hence, most of them are unable to escape, resulting in a dip of the number of secondaries as primary energy goes very high. The oxides of alkali metals prove to be very good dynode materials. In order to avoid field emission, the dynode surfaces must have some conducting property, i.e., the surface oxide layer must have some metal in excess.

The possibility of detecting photo-electrons individually was envisaged from 1916 by Elster and Geitel (1916), based on the design of the α particles counter by Rutherford and Geiger (1908). It consisted a bulb filled with gas containing a photo-cathode and an anode. The voltage between the electrodes was regulated so that the presence of a photo-electron could cause, by ionization of gas, a discharge (measured by an electrometer) and therefore detected.

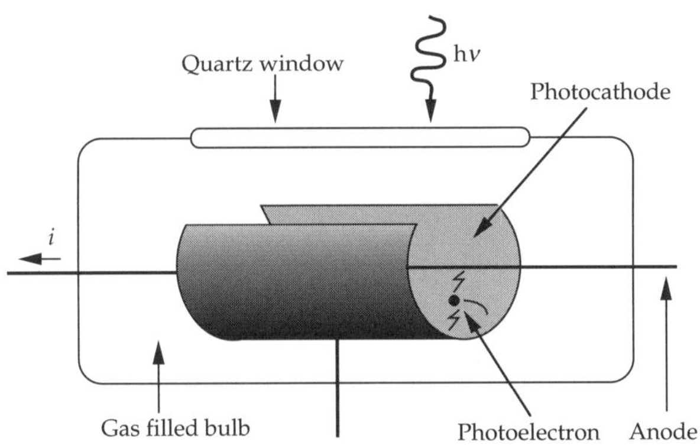

Fig. 8.1 Principle of a Geiger-Müller gas detector.

The counting of particles issued by radioactivity with the system of Geiger-Müller (1928), based on the system represented before, was adapted later by Locher (1932) to count visible photons. It accomplished several counters composed of a gas bulb, in which a photo-cathode had a three-quarters of cylinder shape and the anode was a simple wire aligned on the cylinder axis (see Figure 8.1). Several alkaline materials, associated with hydrogen, were tried for the photo-cathode. The most sensitive were potassium, but was more noisy; the dark noise[8] was 122/min, against only 5.8/min for cesium, and 4.7/min for sodium. At any rate, the flux maximum rate, limited by the reading electronics, did not exceed 300 counts/s. Because of their weak precision, gas detectors as stellar photometry sensors were abandoned in the 1940s.

8.1.3 *Detection of photo-electrons*

It is reiterated that an incoming photon with energy greater than the band-gap energy is absorbed within the semiconductor material, elevating a bound electron into the conduction band from the valance band. The remaining net positive charge behaves as a positively charged particle, known as a hole. Thus an electron-hole pair is created in the detector material, which carries a electric current, called 'photo-current'. The probability of converting light into electron-hole pairs depends on the material, the wavelength of the light, and the geometry of the detector. The efficiency of the detector, η_d is independent of the intensity and the detection frequency. It is the product of the probabilities that

(1) a photon incident on the front surface of the detector reaches the photon-sensitive semiconductor layer,
(2) a photon generates an electron-hole pair within the semiconductor, and
(3) an electron-hole pair is detected by the readout noise circuitry.

The performance of a detector depends on the dark current, which is measured as the signal generated in the absence of the external light, of the device and is due to the generation of electron-hole pairs by the effect of temperature, as well as by the arrival of photons. The dark current poses an inherent limitation on the performance of the device. It is substantially reduced at lower temperatures. The generation of this current can be described as a poison process. Thus the dark current noise is proportional to the square root of the dark current. This current depends on the material

[8]Dark noise is created by false pulses resulting from thermally generated electrons (so called dark signal). The noise arising from the dark charge is given by Poisson statistics as the square root of the charge arising form the thermal effects.

used and the manufacturing process and it is given by the relation,

$$i_d = \alpha e^{-\beta/k_B T}, \tag{8.6}$$

where α and β are the constants, k_B the Boltzmann constant, and T the temperature of device.

Photo-currents are sensed by measuring the voltage across a series resistor when no current passes through it. However, this voltage is sensitive to changes in the external resistance or in the intrinsic resistance of the semiconductor. Two methods are generally employed to record light such as:

(1) The electronic signal from the photo-detector is integrated for a time interval, Δt and record a photo-current which is proportional to the power of the light, $P(t)$, on the detector. The rate at which photons reach the detector is given by,

$$i(t) = \frac{n_e(t)e}{\Delta t} = \frac{P(t)e}{h\nu\eta_d}, \tag{8.7}$$

in which e is the charge of a single electron and $n_e(t)$ the number of electrons generated.

(2) Existence of photons means that, for a given collecting area, there exists a physical limit on the minimum light intensity for any observed phenomenon. The ability to detect individual each photons (or 'photo-event') in an image plan, thus giving the maximum possible signal-to-noise (S/N) ratio is called 'photon-counting'. The photons can be detected individually by a true photon-counting system. All the output signals above a threshold are, generally, counted as photon events provided the incoming photon flux is of a sufficiently low intensity that no more than one electron is generated in any pixel[9] during the integration period, and the dark noise is zero, and gain can be set at suitable level with respect to the amplifier readout noise. The readout noise is a component of the noise on the signal from a single pixel which is independent of the signal level. It occurs due to two components such as: (i) if the conversion from an analog-to-digital[10] (A/D) number is not

[9] For each point in an image, there is a memory location, called a picture element or pixel.

[10] Analog to digital conversion, also referred to as digitization, is the process by which charge from a detector is translated in to a binary form used by the computer. The term binary refers to the base 2 number system used. A 12-bit camera system will output 2 raised to the 12th power or 4096 levels. For applications requiring higher speed and less

perfectly repeatable and (ii) if the electronics of the camera introduces spurious electrons into the process with unwanted random fluctuations in the output. A value for gain, G, is given by,

$$G = \frac{c_m}{\langle\sigma\rangle^2 - \sqrt{\langle i_{ro}\rangle^2}}, \tag{8.8}$$

in which c_m is the mean counts, $\langle\sigma\rangle^2$ the variance, and $\langle i_{ro}\rangle^2$ the read-out noise.

The probability of obtaining a photon at a given location is proportional to the intensity at that location. Thus the probability distribution of photon impacts is the intensity distribution in the image. One single electron can be turned into an electronic pulse using processes for secondary electron emission. The pulses correspond to individual detection events (individual photons), which can be counted and processed. The count rate, $R(t)$ is proportional to optical power of the light incident on the detector, i.e.,

$$R(t) = \frac{n_e(t)}{\Delta t} = \frac{P(t)}{h\nu\eta_d}. \tag{8.9}$$

In order to characterize a photon-counting device, the statistics of the intensity of the output signal (expressed in electrons) triggered by a photo-event, known as 'pulse height distribution' (PHD), is measured. Figure (8.2) represents PHD by a curve number-of-counts vs. output signal intensity. A PHD curve displays a peak which may be characterized by its normalized peak-to-valley (PV) ratio, and its normalized full-width at half-maximum (NFWHM). Ideally, PV should tend to infinity and NFWHM to zero. Of course, the intensity corresponding to the peak has to be much larger than the maximum level of readout noise of the sensor (anode, CCD chip, etc.) whose output signal is the photon information carrier and that terminates the chain in the photon-counting device. In these conditions, the ideal false detection rate (or electron noise) FD$\rightarrow$ 0.

The marked advantage of a photon-counting technique is that of reading the signal a posteriori to optimize the correlation time of short exposures in order to overcome the loss of fringe visibility due to the speckle lifetime; the typical values for an object of $m_v = 12$ over a field of $2.5''$ are < 50 photons/msec with the narrow band filter. The other notable features of such a technique are:

---

dynamic range, 8 to 16-bit digitization is common. The higher the digital resolution, the slower the system throughout.

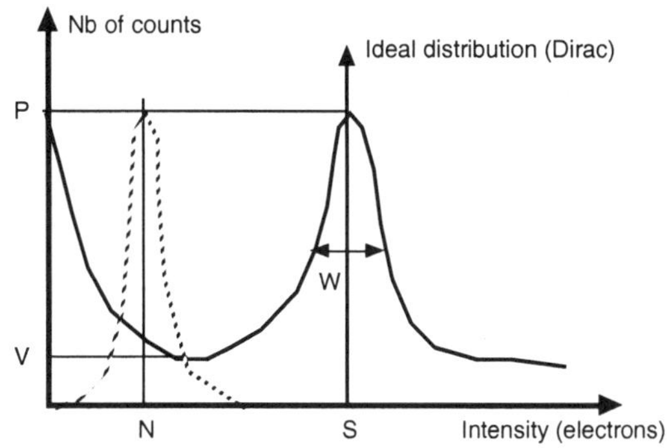

Fig. 8.2 Example of a usual PHD, in which the dashed curve stands for the noise statistics of the detector readout; the ideal PHD is a Dirac's peak (Morel and Saha, 2005).

(1) capability of determining the position of a detected photon to 10 μm to 10 cm,
(2) providing spatial event information by means of the position sensitive readout set-up; the encoding systems identify each event's location,
(3) ability to register individual photons with equal statistical weight and produces signal pulse (with dead time of ns), and
(4) possessing low dark noise, which is typically of the order of 0.2 counts $cm^{-2}s^{-1}$.

For high resolution imaging, a photon-counting system of very high temporal resolution of the order of several MHz is necessary to tune the integration time according to the value of r_0. Photon-counting hole is a problem of such systems connected with their limited dynamic range[11].

With photon-counting an important consideration is the level of clock induced charges, which do not arise from photons but are due to either spurious charges created by clocking charge over the surface of the CCD, or charges that are thermally generated. Since such a sensor and the processing electronics have a dead time, two or more photons arriving at the sensor at time intervals less than such a dead time can generate single electronic pulse. In order to overcome the shortcoming due to the former, the

[11]Dynamic range is defined as the peak (maximum) possible signal, the saturation level, to the readout noise. This ratio also gives an indication of the number of digitization levels that might be appropriate for a given sensor. For example, a system with a well depth of 85000 electrons and a readout noise of 12 electrons would have a dynamic range $= 20\log(85000/12)$ or 77 decibels (db).

count rate should be larger than the dark noise, while the latter effect can be looked into by adjusting input intensity so that average count rate is smaller in the dead time, else yields saturation.

A photo-electric substance gives a saturation current i_0 upon being illuminated by a luminous flux, Φ_0. The current being weak needs support of multipliers or amplifier. Such an amplifier may have wide bandwidth, and large gain. However there are different noises in the output. These are:

(1) Shot noise (Schottky effect): A current source in which the passage of each charge carrier is a statistically independent event delivers a current that fluctuates about an average value. If the illumination is constant and the rate of photo-electric events is large, the resulting current is a superposition of many such waveforms initiated at random times with a long time average rate. This gives rise to a fluctuating current. The fluctuating component of such a current is called shot noise; all incoming photons have an inherent noise, which is referred to as an photon shot noise. The magnitude of fluctuations depends on the magnitude of the charges on the individual carriers. Shot noise is due to discrete nature of electricity, which arises even at the initial state of the photo-cathode even in the presence of constant luminous flux. For a frequency band of $\Delta\nu$, the noise is estimated as,

$$\langle i_s \rangle^2 = 2ei_0\Delta\nu, \tag{8.10}$$

where i_0 is the photo-cathode current.

(2) Thermal (Johnson) noise: It is the random voltage noise produced in the resister (external) either at the input of an amplifier or at the output of the multiplier. In this all frequency components are present in equal intensity. Within the frequency band, $\Delta\nu$, the noise estimate is,

$$\langle i_{th} \rangle^2 = \frac{4k_BT\Delta\nu}{R}, \tag{8.11}$$

in which T the ambient temperature, and R the impedance of the circuit.

(3) Amplifier noise: This noise is described as,

$$\langle i_{amp} \rangle^2 \approx 2eG\Delta\nu F, \tag{8.12}$$

in which G is the voltage gain and F the excess noise factor.

The comparison between these two noises (equations 8.10 and 8.11) are given by,

$$\gamma = \frac{\langle i_s \rangle^2}{\langle i_{th} \rangle^2} = \frac{i_0 R}{2(k_B T/e)} = \frac{i_0 R}{5 \times 10^{-2}}. \tag{8.13}$$

It is important to note that the actual information that is the optical signal, $\langle i_{op} \rangle$, created by the light must be more than the total noise (addition of the variances of all these noises such as shot-noise, thermal-noise, and amplifier noise),

$$\langle i_{op} \rangle^2 \gg \langle i_s \rangle^2 + \langle i_{th} \rangle^2 + \langle i_{amp} \rangle^2 . \tag{8.14}$$

For successful determination of the signal, both shot-noise as well as thermal noise should be low. The voltage drop at the input of the amplifier (in case of photo-electric cell with an amplifier) should be at least 0.05 volts and hence, the flux sensitivity of the set-up would be very poor. To do away with it, multiplication is used so that how feeble the photo-cathode current may be, the output current is always large enough to produce 0.05 volts at the output resistance.

8.1.4 *Photo-multiplier tube*

Photomultiplier tube (PMT) is a very high sensitive detector and is useful in low intensity applications fluorescence spectroscopy. It is a combination of a photo-emitter followed by a current amplifier in one structural unit, which makes possible a very large amplification of the electric current (a photo-current) by the photosensitive layer from a faint light source. In an ordinary photo-electric cell, photon impinges on the photo-cathode made of photo-emitting substance and the electrons are directed to the anode, as a result a current is registered in the circuit. But in a PMT, photons impinging on photo-cathode liberate electrons, which are directed to dynode. The secondary electron emitted by the first dynode can be directed onto the second dynode which functions in the same manner as the first. This process may be repeated many times. If a multiplier has n such dynodes, each with same amplification factor, δ, the total gain or amplification factor for the PMT is δ^n.

Ejected by a photon, an electron from the photo-cathode creates a snow balling electron cloud on through the dynode path. These electrons hit the anode whose output current is large. These electrons are guided by a strong electrostatic field of several kilovolts. The system of dynodes should satisfy the conditions (Hiltner, 1962), viz., (i) each dynode surface receives the largest possible fraction of secondaries emitted by the preceding dynode, (ii) secondary electrons are emitted into an accelerating electric field, (iii) system is insensitive to perturbing fields, such as the Earth's magnetic field, (iv) ionic feedback is eliminated, and electron cold field emission is avoided.

The history of the photo-multiplier begins with the discovery of the secondary emission, the first implement of which to use was 'Dynatron', a system with negative resistance used for oscillators invented by Hull (1918). Later Zworykin (1936) conceived in a multiplier of electrons the 12 elements of which in secondary emission were made of a mixture of silver, zirconium and cesium. Subsequently, the Société Radioelectrique company perfected a tube called photo-multiplier 'MS-10' (Coutancier, 1940). It featured 10 dynodes of composition Ag-Cs2-O-Cs, providing a gain from 4,000 to 12,000. This tube was characterized by the use of a magnetic field from 10 - 20 milliteslas (mT), in order to apply to electrons a Lorentz force which, associated with the electrical field, making them bounce from an element to the other one (see Figure 8.3a).

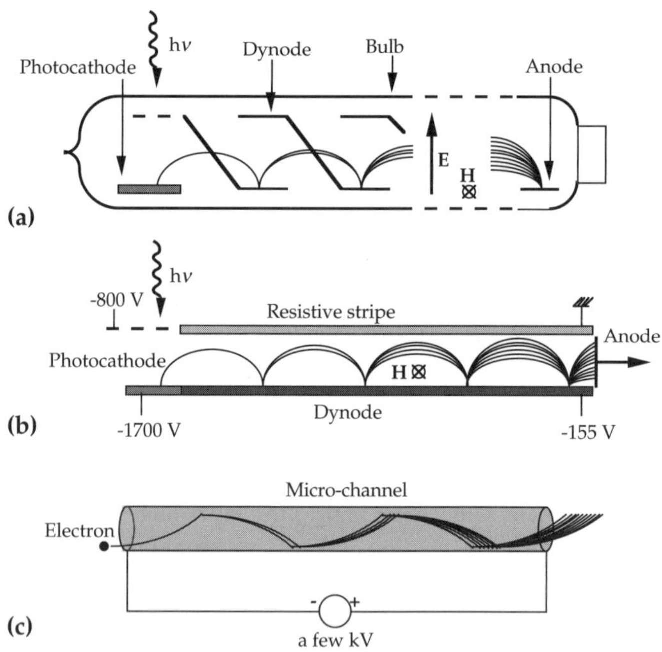

Fig. 8.3 Schematic diagram of a (a) photo-multiplier tube, (b) continuous dynode photo-multiplier, and (c) micro-channel.

Applications in the imagery of this detector regarded mechanical scan television (using the Nipkow's disc), which required only a detector 'mono-pixel'. However, the magnetic system of MS-10, even if it is left for photo-multipliers today, will have been the first stage the miniaturization of the multipliers of electrons. Heroux and Hinterreger (1960) resumed the idea of the magnetic photo-multiplier, but simplified it by using only one dynode, which consisted of a coating on a plate. The electrons bounced and were multiplied on this plate during their travel from the cathode to the anode (see Figure 8.3b). Goodrich and Wiley (1961) fabricated an identical system, providing a 10^7 electron gain, a few millimeter thick. The process of miniaturization was therefore already commenced. If a photo-multiplier, containing n dynodes, each of same amplification factor, δ, gives an overall amplification of δ^n, therefore we find,

$$i = i_0\delta^n, \tag{8.15}$$

and assuming after each stage the noise is likely to be the same Schottky form ($2ei_0\Delta\nu$) and gets multiplied by the same way as current. so,

$$\langle i_s\rangle^2 = 2e\delta i_0\Delta\nu + \delta^2(2e\delta i_0\Delta\nu). \tag{8.16}$$

After first stage, $i_1 = \delta i_0$, hence after n^{th} stage, one gets,

$$\begin{aligned}\langle i_n\rangle^2 &= 2ei_0\Delta\nu\delta^n\left[1+\delta+\delta^2+\cdots+\delta^n\right]\\ &= 2ei_0\Delta\nu\delta^n\left[\frac{1-\delta^{n+1}}{1-\delta}\right].\end{aligned} \tag{8.17}$$

The S/N ratio at the input (ip) divided by the S/N ratio at the output (op) can be specified as,

$$\frac{(\mathrm{S/N})_{\mathrm{ip}}}{(\mathrm{S/N})_{\mathrm{op}}} = \left[\frac{1-\delta^{n+1}}{1-\delta}\right]^{1/2} = \mathcal{A}. \tag{8.18}$$

Thus $\mathcal{A}$ is small when δ is greater, while it is small if n is small. But for all practical purposes, the noise introduced by increasing n is negligible when $\delta > 2$.

The gain of a photo-multiplier is given by,

$$G = \frac{\phi_\nu}{\phi_m}, \tag{8.19}$$

in which ϕ_m is the minimum luminous flux detectable with an ideal photo-multiplier and ϕ_ν the minimum flux detectable by the photo-cathode that is without multiplication (directly coupled with an amplifier).

Since a photo-cathode yields a current, i_a, that contains thermoionic current and current due to background light flux (sky background light) alongwith the true signal, i_m, one defines a modulation factor, Γ,

$$\Gamma = \sqrt{2}\frac{i_m}{i_a}. \tag{8.20}$$

Now one obtains,

$$G = \frac{i_\nu}{i_m} = \frac{\Gamma}{C\rho}\sqrt{\pi k_B T C}, \tag{8.21}$$

where ρ is the S/N ratio defined by i_m/i, C the capacitance, and i_ν can be expressed by,

$$i_\nu = \rho\sqrt{\left(\frac{4k_B T\Delta\nu}{R}\right)}. \tag{8.22}$$

Thus for $C = 20p\text{F}\,(10^{-12}$ farad or picofarad) and $\rho = 2$, for room temperature, $G = 1600\Gamma$. For a 100% modulated signal (in an ideal case, PMT should be cooled and it should have no background flux), $\Gamma = 1$ and $G = 1600$, i.e., the multiplier can detect 1600 times fainter flux than the amplifier. But for a non-ideal situation, efficiency of multiplier rapidly goes to the amplifiers. Now when an impedance transfer mechanism or electronic amplifier is used at the output (op) of the multiplier, in order that the shot noise is greater than the Johnson noise,

$$\text{M}^2 \geq \frac{5\times 10^{-2}}{i_0 R}. \tag{8.23}$$

So that high gain is required for cooled (small i_0) multiplier. It is to be noted that there is upper limit to the last dynode current, $\sim\ 10^{-7}\ A$. Thus a multiplier with higher number of stages reach the highest cut-off flux situation more quickly than the one with smaller stages. The latter can also be used to measure very weak flux using high amplifier input resistance.

The S/N ratio of the multiplier is given by,

$$\text{S/N} = \frac{\text{M}i_\phi}{\text{M}\left[2e(i_e + i_\phi)\Delta\nu\right]^{1/2}} = \frac{i_\phi}{\left[2e(i_e + i_\phi)\Delta\nu\right]^{1/2}}, \tag{8.24}$$

where i_ϕ and i_e are respectively the signal and extraneous components of the photo-cathode current.

Let i_Φ and i'_e respectively be the signal component of the multiplier current and the extraneous component of the output current, we may write,

$$i_\Phi = \mathrm{M} i_\phi; \qquad i'_e = \mathrm{M} i_e. \tag{8.25}$$

This equation (8.25) suggests that the S/N ratio is independent of the amplification, M, and increases with when

- i_ϕ is large, i.e., for a given light-flux, the quantum efficiency of the photo-cathode is large, thus the measurement of weakest light flux requires only photo-cathode of higher quantum efficiency (QE),
- i_e is smallest; it tends to very less thermionic current and background flux, and
- $\Delta\nu$ is small, i.e., the S/N ratio of the instrument can be improved by the time constant of the measurement.

Goodrich and Wiley (1962) achieved a major breakthrough by inventing the 'micro-channel', a simple glass tube in which the inner side is coated by a secondary emission semiconductor (see Figure 8.3c). A potential difference of some kilovolts (kV) is applied to ends of the tube in order to cause the multiplication of electrons. They pointed out that the electron gain of such a tube does not depend on the diameter, but on the ratio (length/diameter), in a proportional manner. With such dimensions, the parallel assembling of micro-channels in arrays, with an intention of making the enhancement of image became realistic. However, the gain of micro-channels is limited by the positive charges left by the secondary electron cascade, which goes against the electric field that is applied at the ends of the micro-channel; the gain can even decrease if a increases. The maximum electron gain of a micro-channel is a few 10,000.

8.1.5 *Image intensifiers*

Image intensifier refers to a series of imaging tubes that have been designed to amplify the number of photons internally so that a dim-lit scene may be viewed by a camera. It has the capability of imaging faint objects with relatively short-exposures. The flux of a zero magnitude star with spectral type AO at $\lambda = 0.63$ μm, the value at which silicon detectors have maximum efficiency, is 2.5×10^{-12} W/cm^2 per micron bandwidth (Johnson, 1966) and the photon energy, hc/λ, in which c the velocity of light, and λ the wavelength of light, is calculated as 3.5×10^{-19} joules. It is to be noted that the faintest stars visible by the naked eye are 6th magnitude and for a zero magnitude star, the flux, F_0, with spectral type AO is given by,

$$m - m_0 = -2.5 \log \frac{F}{F_0}, \tag{8.26}$$

where m and m_0 are the apparent magnitudes of two stars of fluxes F and F_0 respectively.

The term, F/F_0, in the equation (8.26) is the ratio of the observed stellar flux over the flux given by a zero magnitude AO star. By dividing F_0 with the photon energy, the corresponding photon flux turns out to be, $F_0 \approx 8 \times 10^6$ photons $s^{-1}cm^{-2}$. This value can be placed in equation (8.26), in order to derive the photon flux of a given star, i.e.,

$$F = 8 \times 10^6 \times 10^{(3 - 0.4m)}. \tag{8.27}$$

The number of detected photons, n_p per cm^2 is dictated by equation (7.96). The photo-cathode being subjected to an electrical field when the intensifier works, the equation (8.4) becomes,

$$\vec{J} = A_R T^2 e^{-(\phi_0 - \Delta\phi_0)/k_B T}, \tag{8.28}$$

in which $\Delta\phi_0(= \sqrt{e^3 \vec{E}/(4\pi\epsilon_0)}$, e represents the elementary charge, $\vec{E}$ the electrical field in the photo-cathode, and $\epsilon_0(= 8.8541 \times 10^{-12}(\mathrm{F})m^{-1})$, the permittivity at vacuum.

In order to bring the image intensity above the dark background level of the photographic plate, Lallemand (1936) introduced an archetype new imaging device, using a monitoring screen, generally known as 'phosphor', onto which the energy of each accelerated electron from the photo-cathode was converted into a burst of photons (spot). Such a device consists of a 35 cm glass tube, with a potassium photo-cathode at one end, and 8 cm diameter zinc sulfide monitoring screen at the other end (which may be replaced with a photographic plate for recording the image). The focusing of electrons was performed by an electrostatic lens made by an inner silver coating on the tube and by a magnetic lens consisting of a 10 cm coil fed with a 0.5 A current. The accelerating voltage inside the tube was 6 kV, providing intensified images by the collision of accelerated photo-electrons onto the screen or the plate. Figure (8.4) depicts the schematic diagram of a Lallemand tube used for astronomy.

The description of the first operational Lallemand tube on a telescope dates from 1951 (Lallemand and Duchesne, 1951). Similar tubes have been

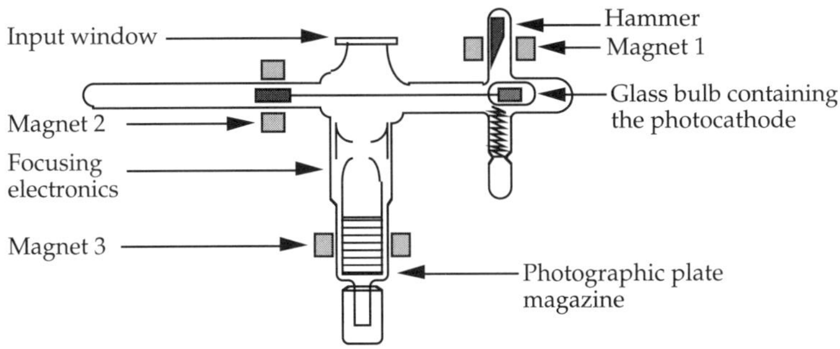

Fig. 8.4 The Lallemand tube. The ring magnet 1 is employed to move the hammer and break the glass bulb containing the photo-cathode. The ring magnet 2 is used to bring the photo-cathode behind the input window, while the ring magnet 3 is used to change the photographic plate.

employed until the beginning of the 1970s for faint object imaging. However, the sensitivity of Lallemand tubes did not reach the quantum limit because of the dark threshold of the photo plates. Moreover, such tubes are made of glass, and hence very fragile and inconvenient for operation on telescopes.

A similar version was also developed by Holst et al. (1934), where they used proximity focusing[12] without electronic or magnetic lenses, in which the photo-cathode and the phosphor were separated by a few millimeters. They are completely free of geometric distortion and feature high resolution over the photo-cathode's useful area. The image magnification is 1:1. The other advantages include: (i) their immunity against electrical and electromagnetical stray-fields, and (ii) ability to function as fast electronic shutters in the nanosecond range. This tube, inspite of its poor resolution owed to its structure, was constructed in numbers during the second world war for observation in infrared (Pratt, 1947). The industrial production of first generation (Gen I) image intensifiers began from the 1950s; they were developed in most cases for nocturnal vision. The tubes in this category feature high image resolution, a wide dynamic range, and low noise.

A common type of detector is based on the television (TV) camera. A photo-electron accelerates under 15 kV about 900 photons by striking a phosphorus of type P-20, where they form an image in the form of an electric charge distribution. Following exposure, the charge at different points of the electrode is read by scanning its surface with electron beam,

[12]The proximity focus intensifiers of new generation are of compact mechanical construction with their length being smaller than their diameter.

row by row. this produces a video signal that can be transformed into a visible signal on a TV tube. The information can be stored in digital form. In order to overcome the problem of photon gain limitation of these intensifiers, cascades of Gen I intensifiers were used for high sensitivity cameras.

Further development on how to build arrays of micro-channels, known today as 'micro-channel plates' (MCP), were carried out in the 1960s. The operational MCPs, known as 'second generation' (Gen II) image intensifiers were ready to be mounted at the telescopes in 1969 (Manley et al., 1969). The photo-electrons are accelerated into a channel of the MCP releasing secondaries and producing an output charge cloud of about $10^3 - 10^4$ electrons with 5 - 10 kilovolt (KV) potential. With further applied potential of $\sim$ 5 - 7 KV, these electrons are accelerated to impact a phosphor, thus producing an output pulse of $\sim 10^5$ photons.

However the channels of these MCPs had a 40-micron diameter. Such MCPs offer a larger gain compared to Gen-I image intensifiers, but a smaller quantum efficiency, due to the fact that some electrons, ejected from the photo-cathode by a photon, do not enter any micro-channel. Several electronic readout techniques have been developed to detect the charge cloud from a high gain MCP. However, the short-comings of the MCPs are notably due to its local dead-time which essentially restricts the conditions for use of these detectors for high spatial resolution applications. These constraints are also related with the luminous intensity and the pixel size.

Third generation (Gen III) image intensifiers are similar in design to Gen II intensifiers, with a GaAs photo-cathode that offers a larger QE ($\sim$0.3) than multi-alkali photo-cathodes (Rouaux et al. 1985). Such tubes employs proximity focus and have a luminous sensitivity of approximately 1.200 μA/lm. The main advantage is in the red and near infrared; they are not appropriate for ultraviolet. However, the high infrared sensitivity makes these tubes more susceptible to high thermal noise.

Of course, an alternative to the MCP is the microsphere plate (Tremsin et al. 1996) comprising of a cluster of glass beads whose diameter is about 50 μm each. These beads have a secondary emission property. Electrons are, therefore multiplied when they cross a microsphere plate. Compared to MCPs, microsphere plates require a less drastic vacuum (10^{-2} Pa), a reduction of ion return and a faster response time (about 100 ps). The drawback is a poor spectral resolution (2.5 lp/mm). Hence, they can be used for PMTs, but not for image intensifiers.

8.2 Charge-coupled device (CCD)

Boyle and Smith (1970) introduced charge-coupled devices (CCD) to the imaging world at Bell Laboratories. CCD refers to a semiconductor chip consisting of bi-dimensional array of sensors, called pixels separated by insulating fixed walls and having no electronic connection; each pixel size is about a few μm and the number of electrons per pixels is nearly 1×10^5 to 5×10^5. The concept of such a device was initially developed as an electronic analogue to the magnetic bubble memories. The architecture of a CCD has three functions (Holst, 1996), such as (i) charge generation and collection (magnetic bubble formation), (ii) charge transfer by manipulating the electrode gate voltages (bubble propagation by an external rotating magnetic field), and (iii) the conversion of charge into a measurable voltage (bubble detection as either true or false); these are adopted from the magnetic bubble memory technology.

In order to cover large areas in the sky, several CCDs can be formed into a mosaic. Usages of the modern CCDs camera system became a major tool in the fields of astronomy because of its low-noise characteristics. It was introduced for observational purposes in late seventies of the last century (Monet, 1988). To-day it is the most commonly used imaging device in other scientific fields such as biomedical science and in commercial applications like digital cameras as well.

The operating principle of a CCD is based on the photoelectric effect. Unlike a photo-multiplier where the photoelectrons leave the substratum in order to produce an electric current, CCD allows them to remain where they are released, thus creating an electronic image, analogous to the chemical image formed in a photographic plate. The CCD is made up of semiconductor plate (usually p-type silicon).

Silicon has a valency of four and the electrons in the outermost shell of an atom pair with the electrons of the neighbouring atoms to form covalent bonds. A minimum of 1.1 eV (at 300° K) is required to break one covalent bond and generate a hole-electron pair. This energy could be supplied by the thermal energy in the silicon or by the incoming photons. Photons of energy 1.1 eV to 4 eV generate a single electron-hole pair, while photons of higher energy generate multiple pairs. The charge pattern in the silicon lattice reflects that of the incident light. However, these generated electrons, if left untrapped, would recombine into the valence band within 100 μsecs. If a positive potential is applied to the gate, the generated electrons could be collected under this electrode forming a region of holes. The holes would diffuse into the substrate and be lost. Thus the electrons generated by the incoming photons could be collected in the respective pixels. These electrons should be counted to reproduce the pattern of the incident light which is termed as the image.

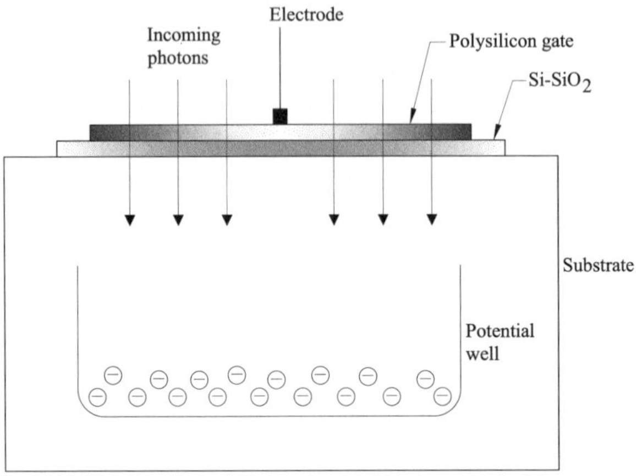

Fig. 8.5 Typical potential well.

The basic structure of a CCD is an array of metal oxide semiconductor (MOS capacitors), which can accumulate and store charge due to their capacitance. The chip is a thin wafer of silicon consisting of millions of photo sites each corresponding to a MOS capacitor. The charge generation and collection can be easily understood in terms of a simple parallel plate MOS capacitor which holds the electrical charge. The MOS structure is formed by applying a metal electrode on top of a epitaxial p-type silicon material separated by a thin layer of insulation, usually silicon-silicon dioxide (Si-SiO_2). When a positive potential is applied to the electrode, the holes are repelled from the region beneath the Si-SiO_2 layer and a depletion region is formed. This depletion region is an electrostatic potential well (see Figure 8.5) whose depth is proportional to the applied voltage. The free electrons are generated by the incident photons, as well as by the thermal energy, and are attracted by the electrode and thus get collected in the potential well. The holes, that are generated, are repelled out of the depletion region and are lost in the substrate. The electrons and the holes that are generated outside the depletion region, recombine before they could be attracted towards the depletion region.

The number of MOS capacitor structures in a single pixel is determined by the number of phases (Φ) in operation. In a three-phase CCD, three such capacitors are placed very close to each other in a pixel. The center electrode is biased more positively than the other two and the signal electrons are collected under this phase, which is called the collecting phase. The other two phases are called barrier phases.

The whole CCD array can be conceived as shift registers arranged in the form of columns close to each other. The electrons, that are collected, should be shifted along the columns. Highly doped p-regions, called channel stops, are deposited between these columns so that the charges do not move across the columns. Every third electrode in these shift registers are connected to the same potential. The electrodes of each pixel in a column are connected to the corresponding electrodes in other columns also. By manipulating the voltages on these electrodes, the charges can be shifted along the columns. This array is the imaging area and is referred as the parallel register.

A similar kind of CCD shift register is arranged at right angle to the imaging area, which is called output register or serial register. The charges should be shifted horizontally from pixel to pixel onto an on chip output amplifier, where the collected charge would be converted into a working voltage.

The CCD is exposed to the incident light for a specified time, known as the exposure time. During such time the central electrode in each pixel is kept at a more positive potential than the other two (3-phase CCD). The charge collected under this electrode should be measured. First the charges should be shifted vertically along the parallel register (the columns) onto the output register. After each parallel shift, the charges should be shifted along the output register horizontally onto the output amplifier. Hence there should be n serial shifts after each parallel shift, where n is the number of columns.

Soon after the completion of the exposure, one should transfer the charges, one row at a time and pixels in a row, till the complete array is read. The charge transfer mechanism for three-phase CCD is illustrated in Figure (8.6). During the exposure time the phase two electrode is kept at the positive potential whereas the other two are at lower potential. The three-phase electrodes are clocked during the charge transfer period as described in this figure.

At time, t_1, only phase two is positive and hence all the electrons are

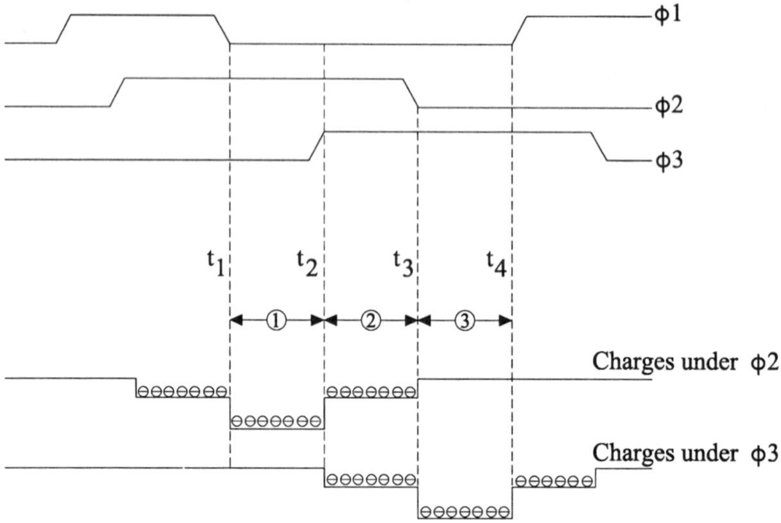

Fig. 8.6 Sequence of charge transfer from Φ_2 to Φ_3: (1) charges under Φ_2 well, (2) charges shared between Φ_2 and Φ_3 wells, and (3) charges under Φ_3.

under phase two. At time, t_2, both the phases two and three are at the same positive potential and the electrons are distributed equally under these phases. At time, t_3, phase two potential is going lower whereas phase three is positive and hence electrons start leaving phase two and cluster under phase three. At time, t_4, when the phase three is alone positive, the electron are fully under phase three. The electrons that were collected under phase two are now under phase three by this sequence. The repetition of this clock sequence results in the transfer of electrons across the columns onto the output register. The instant in which an electron reaches the output register characterizes the position of the element on the array and its intensity is amplified and digitized. This is done for all the arrays simultaneously so that one obtains a matrix of numbers which represents the distribution of intensities over the entire field.

8.2.1 *Readout procedure*

The light intensity is transformed into electrical signal by a detector. It is reiterated that the incoming photons have an inherent noise, known as photon noise, due to which this signal contains an intrinsic noise. It contains a contribution of detector noise that includes dark current, read-out, and amplifier noise. In an optimized readout procedure, such noise is limited by electronic noise in the amplifiers. The minimum signal that can be detected is the square of the read noise.

The pixels of the output register are of bigger size to hold more electrons so that as many number of rows can be added (binned[13]) to the output register. After shifting the charges serially along the output register the charge is moved across the gate of the output field effect transistor. If q is the charge and C is the node capacitance across the output gate, the voltage is developed $V = q/C$. For one electron charge and 0.1 pf capacitance the voltage is 1.6 microvolt.

Before the charge from a pixel is transferred to the node capacitor, it is recharged to a fixed potential by pulsing the reset transistor connected to it. The uncertainty involved in this process is $1/2(k_BT/C)$, in which k_B is the Boltzmann's constant, T the temperature, and C the capacitance, in the value of the voltage across the capacitor. This introduces noise in the measurement of the charge transferred. In addition the output transistor has intrinsic noise which increases as $1/f$ at low frequencies.

Both the noises can be minimized using a signal processing technique called double correlated sampling[14]. Such a technique removes an unwanted electrical signal, associated with resetting of the tiny on-chip CCD output amplifier, which would otherwise compromise the performance of the detector. It involves the making a double measurement of the output voltage before and after a charge transfer and forming a difference to eliminate electrical signals which were the same, i.e., correlated. The output of the integrator is connected to a fast A/D converter from which the signal is measured as a digital number by a computer. The gain of the signal processing chain is selected so as to cover the range of the ADC used as well as the full well capacity of the CCD pixel. The gain in the integrated amplifier, G is related to the variation in voltage between the reference level

[13]Pixel binning is a clocking scheme used to combine the charge collected by several adjacent CCD pixels. It is designed to reduce noise and improve the signal to noise ratio and frame rate of digital cameras.

[14]Sampling refers to how many pixels are used to produce details. A CCD image is made-up of tiny square shaped pixels. Each pixel has a brightness value that is assigned a shade of gray colour by the display routine. Since the pixels are square, the edges of features in the image will have a stair step appearance. The more pixels and shadows of gray that are used, the smoother the edges will be.

and the signal level (in volts), ΔV,

$$G = \frac{\Delta V C_s}{qN}, \tag{8.29}$$

where q is the electronic charge (1.6×10^{-19} C), N the number of electrons per charge packet, and C_s the capacitance of the output diode (Farads).

8.2.2 *Characteristic features*

The CCD is characterized by high system efficiency, large dynamical range, and linear response compared to other detectors. The output of the CCD camera depends on both the system spectral response and the color temperature of the illuminating source.

8.2.2.1 *Quantum efficiency*

The most appealing aspect of CCDs over other detectors is their great efficiency. Most of the CCDs that have been made are capable of registering above 50% across a broad spectral range from near-infrared (1-5 μm), visible, UV, extreme UV to soft x -ray. Peak efficiency may exceed over 80% for some wavelength of light; a back-illuminated CCD has efficiency of ~85-90% around 600 nm. In addition, the CCD is responsive to wavelengths in the region 400 nm to 1100 nm, where most of the other detectors have low QE.

In the front illuminated device, the electrodes and gates are in the path of the incident light and they absorb or reflect photons in the UV region and the spectral range becomes limited. There is absorption in the bulk substrate of 1-2 mm thickness, region below the front-sensitive area and photons absorbed in this area do not form part of the signal. To improve the UV response the CCD is given phosphor coatings which absorb UV photons and remits visible photons.

In order to enhance the quantum efficiency, CCDs are thinned out from the back to $\approx$ 10 to 15 μm, and the illumination is from the back, which means there is quantum efficiency enhancement since there is no loss either in the bulk substrate or in the electrode structure. Because of this thinning, the CCD starts showing interference effects from 700 nm up. It is to be noted that a thinned CCD also requires anti reflection coatings with Hafnium oxide, aluminum oxide, lead fluoride, zinc sulfide and silicon monoxide to reduce the reflection losses which are 60 percent in UV and 30% in visible. After thinning Silicon oxidizes and a thin layer of SiO_2 is formed at the back. The presence of impurities in this oxide layer causes a net positive charge. This results in a backside potential well along with one at the electrode. The signal electrons diffuse into it and recombine resulting in low quantum efficiency. Hence the back surface must be treated to compensate for the backside potential well. The abruptly broken bonds at the silicon lattice due to thinning are to be tied-up. First an oxide layer is grown at the back. A backside discharge mechanism such as UV flooding, corona charging, gas charging, flash gate, biased flash gate are then used to direct the signal electrons towards the electrodes.

8.2.2.2 *Charge Transfer efficiency*

The charge transfer efficiency (CTE) is the ratio of the electrons transferred and measured at the output to the electrons collected in the pixels. In the CCD architecture (surface channel operation) discussed above, the charges collected are transferred at the interface between the substrate and the SiO_2 insulating layer. The electrons get trapped at the lattice irregularities near the surface. The result is very poor charge coupling and severe image smear. To overcome this surface trapping, buried channel operation was introduced. An n-type layer is introduced between the p-type substrate and the insulating layer. This n-type layer creates a complex potential well with a potential maximum generated at slightly below the Si-SiO_2 interface where the signal electrons are collected and transferred. This is referred to as the buried channel CCD. Since this process takes place inside the bulk of the silicon, the charge transfer is very efficient as the trapping sites become much less.

8.2.2.3 *Gain*

The CCD camera gain may be determined precisely by measuring signal and variance in a flat-fielded (pixel-to-pixel sensitivity difference) frame. The variance of the flat-fielded frame should be halved to account for the increase of the noise by square root of two in the difference frame. In spite of the negligible value of the read noise compared with the variance, an input guess is applied at the read noise. Different regions in the frame are selected randomly. A number of gain values are generated as well. These values are plotted in a histogram form. The value of the gain corresponding to the peak of the histogram is known as the system gain. The values of gain that are obtained from regions with defects, traps etc., give rise to erroneous values and fall outside the main histogram peak.

8.2.2.4 *Dark current*

Performance of CCD depends on the dark current of the device. In the CCDs the main sources for the dark current results from (i) thermal generation and diffusion[15] in the neutral bulk substrate, (ii) thermal generation of electrons in the depletion region, and (iii) thermal generation of electrons in the surface states at the Si-SiO_2 interface. Among these, the contribution from the surface states is large. Most of the electrons, which are generated thermally deep in the CCD (neutral bulk), are diffused and recombined before they come under influence of the nearest potential well. The thermal electrons generated in the deep depleted regions are further diffused, and some of them may be collected by the neighboring pixels. The dark current generation due to the surface states depends on the density of the surface states at the interface and the density of the free carriers that populate the interface. This contribution of dark current can be substantially reduced by the passivation techniques[16] at the time of fabrication of the CCDs, or by operating the device in the inversion mode. When a gate is biased such that the surface potential of the phase is equal to the substrate potential, the n-channel at the Si-SiO_2 interface gets inverted, i.e., the holes from the near by channel stops are attracted and pinned at the surface. This pinning condition eliminates further hopping of the electrons from the valance band to the conduction band, and there by, reduces the dark current. If the two barrier phases in a three phase CCD are biased into the inversion sate (partial inversion operation), the rate of dark current generation decreases two-third of the non-inverted mode of operation.

Dark current builds up with time and the acquired frame would become saturated, if the device is not cooled even for a few seconds integration time; cooling the device reduces the dark noise considerably, typically < 100 counts s^{-1}. The CCD is cooled to temperatures between -60° centigrade (C) and -160° C depending on the application. For slow scan mode

---

[15]When a photo site is subjected to excessively strong illumination, the accumulated charges can become so numerous that they spill on to adjacent photo elements. A saturated pixel produces a characteristic diffusion similar to the halo surrounding bright stars on a photographic plate. In addition, the number of charges accumulated in a saturated well can be such that its contents cannot be emptied in one or more transfers. A trail starting at the saturated point then appears in the direction of the transfer of rows. This effect, called blooming, is often the signature of a CCD image.

[16]Passivation is the process of making a material passive in relation to another material prior to use the materials together. In the case of the CCDs, such a technique is used to reduce the number of the interface states by growing a thin layer of oxide to tie-up the dangling bonds at Si-SiO_2 interface.

operation of large CCD, cooling using liquid nitrogen is used. For smaller and faster acquisition systems thermoelectric cooling is used. The dark current may be measured by taking exposures with the shutter closed, subtracting of which from the observed image provides the real number of electrons due to incident light. However, the intrinsic dark current is negligibly small for the short-exposure used in interferometric experiments, but the thermal background signal may pose a problem if the detector sees large areas at room temperature.

8.2.3 *Calibration of CCD*

The linearity[17] of a CCD can be measured by illuminating it with a very stable luminous source and by tracing the detectors response as a function of integration time. The limitation, of course, comes from the heterogeneities of response, which is due to unidentical elements. A few hot pixels are abnormally receptive, while a few do not work. A few more partly variable phenomena occur, such as:

(1) thermal agitation in a CCD producing free electrons in a manner that varies from one pixel to another as well as non-zero electronic noise, and
(2) sensitivity difference from one pixel to another.

The CCD image (raw) needs to be corrected for CCD bias, thermal noise, pixel-to-pixel variation of efficiency, and sky background. The actual stellar counts, $\mathrm{D}(\vec{x})$, can be determined by,

$$\mathrm{D}(\vec{x}) = \frac{\mathrm{R}(\vec{x}) - \mathrm{B}(\vec{x})}{\mathrm{F}(\vec{x}) - \mathrm{S}(\vec{x})}, \tag{8.30}$$

in which $\vec{x} = x, y$ 2-D position vector, $\mathrm{R}(\vec{x})$ the raw CCD image, $\mathrm{B}(\vec{x})$ the bias image, $\mathrm{F}(\vec{x})$ the flat-field image, and $\mathrm{S}(\vec{x})$ the sky background.

The electronic bias needs to be subtracted to eliminate signal registered by the detector in complete darkness. The required bias image, $\mathrm{B}(\vec{x})$, is constructed by averaging a series of zero exposure images. Such exposures are averaged out in order to mitigate random noise. A pixel by pixel bias subtraction is done to obtain a bias subtracted data. The calibration for thermal noise is also carried out by similar manner, but with same exposure time and at the same temperature as for the actual astronomical

[17]Charge that is generated, collected and transferred to the sense node should be proportional to the number of photons that strike the CCD.

observations being carried out. This offset map is used in the reduction of the actual astronomical observations. In order to fulfill the condition of performing the observations at the same temperature it may be necessary to produce several offset maps for different observing conditions before and after the observing session.

The pixel-to-pixel variations are compensated by the calibration that is performed by taking the average of some exposures of a calibrated uniform light, called flat-field exposure. A pair of identical such frames close to full well capacity ($\sim$ 80%) are required to be obtained. By subtracting these frames pixel by pixel, pixel non-uniformity is removed. Usually a small window of 20$\times$20 pixels in the flat-fielded frame is selected to compute variance. The mean signal counts from the original two frames are obtained. The large scale response variations are removed through division by flat-field factor, F. The image factor is constructed from several flat-field images obtained by exposing CCD to a specially uniform continuum source. Such an uniform light may be obtained either by using sky light at twilight or by using an artificially flash light.

Flat-field images are debiased individually and combined by weighted averaging. The main advantage of using several well exposed images is to construct the template image, which reduces statistical errors introduced during the division. The normalized flat field map is used reduce the astronomical observations. The recorded image is often contaminated by sky background, $\mathrm{S}(\vec{x})$, that needs to be subtracted out. Such a background is derived from the debiased and flat-fielded image by smoothly interpolating sky data. In the case of high resolution stellar spectra, a least square low order polynomial fit to the sky data at side of the object spectrum is to be obtained. The sky background interpolated for the position spectra is subtracted from the debiased and flat-fielded image to obtain the stellar counts, $\mathrm{D}(\vec{x})$. The 1-D stellar spectrum is extracted out from the 2-D image by summing, contributions from a range of spatial pixels containing object spectrum. Such a spectrum are calibrated to wavelength scale using the coefficient obtained by fitting a low order polynomial to the comparison spectrum (known wavelength). This 1-D wavelength calibrated data is used for fitting the continuum in order to determine equivalent widths (see section 10.2.6) of stellar lines.

Another point to be noted is to avoid saturated image. Saturation occurs due to large signal generated electrons filling the storage. The discharge is deferred so that the recognition of pixel is biased resulting in displacement of image. To add to this misery, a ghost image of the saturated field may remain for several hours on the CCD. In order to eliminate such an effect, a pre-flash that makes the effect uniform over the whole array, should be done. All calibrations may be repeated after a pre-flash.

8.2.4 *Intensified CCD*

Since the light is passing through the atmospheric turbulence where the resolution of a large telescope is dictated by the atmospheric coherence length, r_0, the minimum number of photons, n_p/cm^2, is derived by equating the equations (7.93 and 7.94),

$$n_p = \frac{4\pi^2}{0.134} d^{-5/3} r_0^{-1/3}, \tag{8.31}$$

with the value of r_0 is taken into account at the sensor wavelength and d is the spot size.

From the two equations (7.96) and (8.26), one derives,

$$10^{-0.4m} = \frac{3.68 \times 10^{-2}\, d^{-5/3} r_0^{-1/3}}{\Delta\tau\eta_{tr} \int \eta_d(\lambda) d\lambda}. \tag{8.32}$$

As stated earlier, the high angular resolution imaging requires to take images with short-exposure ($<$ 20 msecs) where the S/N ratio in each frame is low therefore image intensification becomes a necessity.

A frame-transfer CCD is composed of two parallel registers and a single serial register. The parallel register next to the serial register is opaque to light and is referred to as the storage array while the other parallel register having the same format as the storage array is called image array. After the integration cycle, the charge is transferred quickly from the light sensitive pixels to the covered portion for data storage and the image is read from the storage area when the next integration starts. The frame transfer CCDs are usually operated without shutter at television frame rates. By removing the opaque plate on the storage array, this can be used as full frame imager by clocking the parallel gates of the two arrays together.

A frame-transfer intensified CCD (ICCD) detector consists of a micro-channel plate (MCP) coupled to a CCD camera. The proximity focused MCP has photo-multiplier like ultraviolet (UV) to near-IR response. The output photons are directed to the CCD by fibre optic coupling and operate at commercial video rate with an exposure of 20 ms per frame. The video frame grabber cards digitize and store the images in the memory buffer of the card. Depending on the buffer size, the number of interlaced[18] frames stored in the personal computer (PC) can vary from 2 to 32 (Saha et al., 1997a).

CCD has one or two read amplifiers and all the pixel charges are to serially pass through them. In order to increase the frame rates from the CCD, different read modes like frame-transfer and kinetic modes are normally used. Because of the architecture of CCD, even to read 10×10 pixel occupied by a single star, one has to read whole device. This increases the reading time thus limiting the number of frames that can be read with this system. In kinetic mode, region of interest say 100×100 pixel can be read, which may be digitized by A/D converter after the charge is read from CCD, the remaining area charges are dumped out from CCD after reading without being digitized. Let τ_r be the time taken to read the CCD, therefore

$$\tau_r = N_x\, N_y(\tau_{sr} + \tau_v) + N_x\, \tau_i, \tag{8.33}$$

where N_x, and N_y are the number of pixels in x and y direction of CCD respectively, τ_{sr} the time required to shift one pixel out to shift register, τ_v the time taken to digitize one pixel, τ_i the time to shift one line into shift register, and τ_s the time needed to discard a pixel charge.

For a 1 MHz CCD controller, about 80 frames per second can be read from the CCD if 100×100 region is chosen. Another drawback of such a system is the poor gain statistics resulting in the introduction of a noise factor between 2 and 3.5. Since such a systems has fixed integration time, it is subjected to limitations in detecting fast photon-event pairs. Non-detectability of a pair of photons closer than a minimum separation by the detector yields a loss in high frequency information; this, in turn, produces a hole in the center of the autocorrelation – Centreur hole, resulting in the degradation of the power spectra or bispectra (Fourier transform of triple correlation) of speckle images. The other development in the CCD sensor is the interline-transfer CCD[19] with greatly reduced cell size has been the major factor in the successful production of compact, low cost, and high quality image capturing equipment including video cameras, digital still

[18]Interlaced scan, makes two passes, and records alternate lines on each pass, so as to enable to obtain two images simultaneously.

[19]The interline transfer CCD has a parallel register that is sub divided so that the opaque storage register fits between the image register columns. The charge that is collected under the image registers is transferred to the opaque storage register at readout time. The serial register lies under the opaque registers. The readout procedure is similar to the full frame CCD.

camera etc.

8.3 Photon-counting sensors

Modern telescopes fitted with new generation photon-counting sensor have led to major advances in observational astronomy, high angular resolution imaging in particular. Such a detector is able to count how many photons it has received, but is unable to provide any information about the angle of incidence for individual photon. Classical charge-coupled device (CCD) is not considered to be a photon-counting detector, even if one photon produces one electron with $\eta_d \approx 0.9$, and with an A/D converter set that one step (ADU) corresponds to one electron, because their readout noise is too important to determine whether a photon or no photon at all has been received in an image element when the light level is low (less than 1 photon per pixel and per frame). Gen-I intensifiers did not allow photon-counting even if the phosphor was placed before a TV camera to quantify the signal since

- the pulse height distribution (PHD) of such an intensifier has a negative exponential shape with no peak, and
- the output energy that is generated by any photo-event is statistically weaker than the electronic noise of a TV camera.

In order to detect the photons individually, it is essential to increase the gain of an intensifying device to reach the quantum limit. As far as measuring the photon positions in an $(\vec{x})$, in which $\vec{x}(= x, y)$ is the 2-D position vector, in the focal plane is concerned, the image resolution of a single element photon sensor (photo-multiplier) should be increased by miniaturizing and multiplying the basis element. Actually, modern photon-counting cameras inherit from both approaches that were used alternatively, for example, from photo-multipliers to MCPs, or from low gain MCP-equipped imaging devices to photon-counting cameras.

An important problem in designing a photon-counting cameras, which needs to be addressed, is to convert as fast as possible the position in the image plane of an incoming photon into a set of coordinates (x, y) that are digital signals.

Iredale et al. (1969) addressed the problem of photon position encoding in the image intensifiers for one dimension case (x coordinate to be estimated). They presented the results of three possible optical setups (see

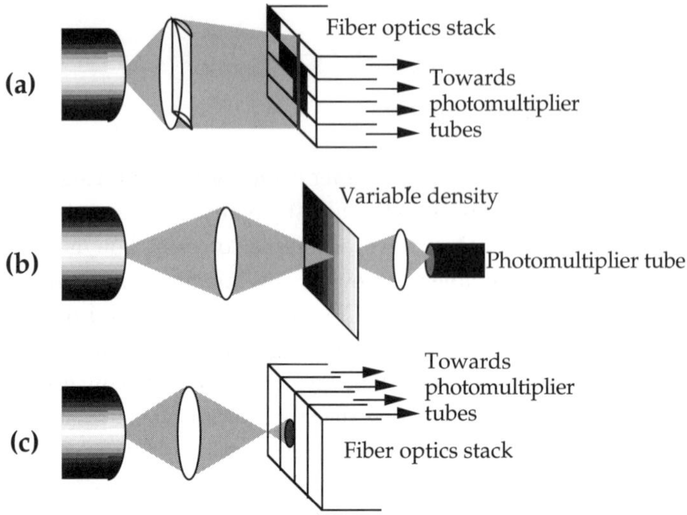

Fig. 8.7 Three systems for measuring the photon position by Iredale et al. (1969).

Figure 8.7) of which the last set-up (Figure 8.7c) was considered to be the most accurate, and proposed to extend it for 2-D imaging:

(1) The spot on the intensifier output (corresponding to a photo-event) was re-imaged as a line along y (by means of a cylindrical lens) onto a binary code mask that was located at the entrance of a stack of fiber optics, each fiber having a rectangular cross-section (Figure 8.7a). The output of each fiber fed a PMT. Therefore, the PMT outputs gave the binary value of the photo-event coordinate.
(2) The spot was re-imaged onto a neutral density filter whose attenuation was varying along x (Figure 8.7b). A PMT behind this filter gave a signal with an amplitude proportional to the photo-event x-position.
(3) The spot was re-imaged with a certain defocus on a stack of fibers connected to PMTs (Figure 8.7c) as in Figure (8.7a). By combination of the analog signals at the PMT outputs, x was found out.

It is to be noted that the problem of photon position was addressed by Anger (1952) for medical gamma imaging. Because of the large area of Na I(T1) scintillators that convert each gamma photon into a burst of visible photons, it is possible to mount a PMT array downstream of a scintillator (see Figure 8.12a). The secondary photons spread on the photocathodes of the PMTs. The combination of the analog signals given by the

PMTs provides the (x, y) coordinates of the gamma photon.

8.3.1 *CCD-based photon-counting system*

Blazit et al., (1977a) had used a photon-counting system which is coupled with a micro-channel image intensifier to a commercial television camera. This camera operates at the fixed scan line (312) with 20 ms exposure. A digital correlator discriminates the photon events and computes their positions in the digital window. It calculates the vector differences between the photon positions in the frame and integrates in memory a histogram of these difference vectors. Later, Blazit (1986) has developed another version of the photon counting camera system (CP40), that consists of a mosaic of four 288×384-pixel CCD chips (Thomson TH 7861) with a common stack of a 40 mm diameter Gen I (Varo) image intensifier and a MCP; the combination consists in a cascade of a Gen-I intensifier followed by a Gen-II that intensifies the spot at the exit of the latter (Figure 8.8).

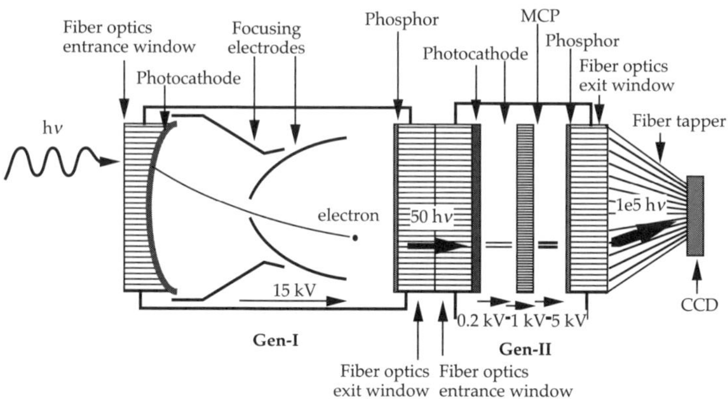

Fig. 8.8 Classical design of an intensified-CCD photon-counting camera. The represented Gen-II intensifier features proximity focusing (which reduces the image distortion).

Coordinates are extracted in real time from the CCD frames by a dedicated electronics and sent to a computer system, either for speckle imaging or for dispersed fringe imaging for long baseline interferometry. The read out speed of the CCDs was 50 frames per second (FPS). The maximum count rate for artifact-free images was about 25,000 photons/s. The read-out of this system is standard, 20 ms. The amplified image is split into four quadrants through a fibre optics reducer and four fibre optics cylinders. Each of these quadrants is read out with a CCD device at the video rate (50 Hz). This camera is associated with the CP40 processor – a hardware photon centroiding processor to compute the photo-centre of each event with an accuracy of 0.25 pixel. The major shortcomings of such system arise from the (i) calculations of the coordinates which are hardware-limited with an accuracy of 0.25 pixel, and (ii) limited dynamic range of the detector.

Recently, Thiebaut et al. (2003) have built the "CPng" (new generation) camera featuring a Gen-III (AsGa photo-cathode) image intensifier, coupled to a Gen-II image intensifier, and a 262 FPS CCD camera (Dalsa) with a 532×516-pixel resolution. The processing electronics consists of a real-time computer which extracts the photo-event positions. The software can extract these positions at sub-pixel resolution and offers a 2000×2000-pixel resolution. The maximum count rate of the CPng is around 10^6 photons per second.

8.3.2 *Digicon*

The important feature of the 'Digicon' tube, invented by the team at Beaver (1971), University of San Diego, is that it is about one of the first electronic alternatives to Lallemand tube for astrophysics and provided photon counting ability. Digicon was not a real imaging device, since it measured the photons only in the one dimension. Its main application was, therefore, spectrometry. Based on the principle of a Gen I intensifier, its originality comes due to the fact that instead of bombarding a photographic plate, accelerated electrons in the Digicon collides with an array of photo-cathodes. The signal provided by each diode, the binnarized signal resulting from the collision incremented a 16-bit registers. A Digicon, with a larger number of diodes (viz., an improved resolution) has been employed in the Hubble space telescope (HST). It is to be noted that Herrmann and Kunze (1969) introduced a photon-counting spectrometer working in the UV and featuring an array of 40 miniaturized photo-multipliers.

From the principle of Digicon, Cuby et al. (1988) investigated the 'electron-bombarded CCD' concept. It consists of a CCD array placed in a vacuum tube with photo-cathode. Electrons are accelerated by a 25 kV voltage to make them strike the CCD pixels. Each accelerated electron liberates a charge of around 7500 electrons in the CCD. With an unique diode, the characteristics of the PHD were PV $= 0.33$ and NFWHM $= 0.22$. Since these devices were not used for high resolution astronomy, they were replaced by high performance CCDs.

8.3.3 *Precision analog photon address (PAPA) camera*

The Precision Analog Photon Address (PAPA) camera, a 2-D photon-counting detector, is based on a high gain image intensifier and a set of PMTs. It allows recording of the address (position) and time of arrival of each detected photon (Papaliolios and Mertz 1982). The front-end of the camera is a high gain image intensifier which produces a bright spot (brighter than those caused by a photo-event) on its output phosphor for events detected by the photo-cathode. The back face (phosphor) of the intensifier is then re-imaged by an optical system which is made up of a large collimating lens and an array of smaller lenses. Each of the small lenses produces a separate image of the phosphor on a mask to provide position information of the detected photon. Behind each mask is a field lens which relays the pupil of the small lens onto a small photo-multiplier (PMT).

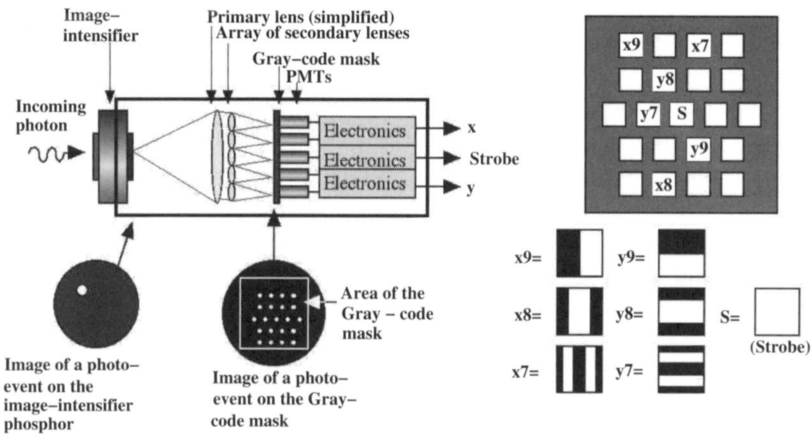

Fig. 8.9 The PAPA camera. Coding mask elements are shown on the right.

A set of 19 PMTs used out of which 9 + 9 PMTs provides a format of 512×512 pixels optical configuration. The 19th tube acts as an event strobe, registering a digital pulse if the spot in the phosphor is detected by the instrument. Nine tubes are used to obtain positional information for an event in one direction, while the other nine are used for that in the orthogonal direction. If the photon image falls on clear area, an event is registered by the photo-tubes. The masks use grey code (see Figure 8.9), which ensures that mask stripes do not have edges located in the same place in the field.

Each mask provides a Gray-code bit, either of the x or the y photo-event coordinate. The re-imaged spot may either be blocked or not by a mask, the PMT thus giving a signal that is binnarized to yield a value, either 0 or 1. This value corresponds to a Gray-code bit of x or y. One of the secondary lens + PMT system has no mask and is used to detect the presence of a photo-event by sampling the outputs of the other PMTs. For a $2^N \times 2^N$-pixel resolution, $2N + 1$ secondary lens + mask + PMT sets are required. With the PAPA detector, the time of arrival of each event is recorded, so photons may be grouped into frames in a way which maximizes the S/N ratio in the integrated power spectrum.

8.3.4 *Position sensing detectors*

A position-sensing detector (PSD) is a photoelectric device, which converts an incident light spot into continuous position data. Many industrial manufacturers and laboratories around the world use PSDs in their daily work. PSDs are able to characterize lasers and align optical systems during the manufacturing process. When used in conjunction with lasers they can be used for industrial alignment, calibration, and analysis of machinery. It provides outstanding resolution, fast response, excellent linearity for a wide range of light intensities and simple operating circuits. In order to measure the x and y positions from the PSD, four electrodes are attached to the detector and an algorithm then processes the four currents generated by photo absorption.

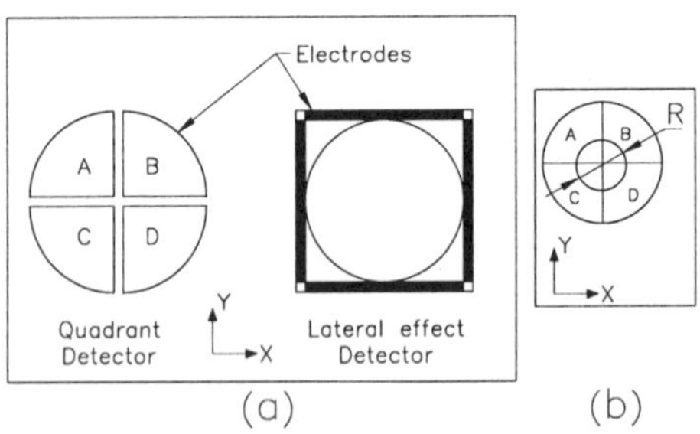

Fig. 8.10 (a) Quadrant detector, (b) Beam movement relative to the x or y direction.

Quadrant detector is a uniform disc of silicon with two gaps across its surface as shown in the Figure (8.10a). For optimum performance and resolution, the spot size should be as small as possible, while being bigger than the gap between the cells. Typically, the gap is of 10-30 μm and the active sensing area is 77 mm^2 or 100 mm^2 (depending on the exact model). When illuminated, the cells generate an output signal proportional to the magnitude of the illumination. The local areas of A and B vary with y and the local areas of C and D vary with x. The intensity on each electrode is proportional to the number of electrons received by the electrode, and therefore the local area. The intensity difference between A and B yields y, while the intensity difference between C and D yields x.

Let A, B, C, D be the four quadrants respectively, and R is the radius of the incident beam illuminating the detector. The beam position is calculated using the following formulas:

$$X = \frac{(\mathrm{B}+\mathrm{D})-(\mathrm{A}+\mathrm{C})}{P}; \qquad Y = \frac{(\mathrm{A}+\mathrm{B})-(\mathrm{C}+\mathrm{D})}{P}, \tag{8.34}$$

with P(total power) $= \mathrm{A}+\mathrm{B}+\mathrm{C}+\mathrm{D}$.

It is the electronic card which digitizes the output signal, and the host computer then processes the signal. The computer and software perform basic calculations of the position and power of the monitored beam. The output position is displayed as a fractional number or as a percentage figure, where the percentage represents the fraction of beam movement relative to the x or y direction as shown in the Figure (8.10b).

The position-sensitive photo-multipliers technology (PSPMT) uses dynodes, like classical PMTs. Such a photo-multiplier tube consists of an array of dynode chains, packed into a vacuum tube. Currents measured on electrodes at the output of the last dynodes are interpolated to find out the position of the photo-event. A photon-counting camera, based on PSPMT, installed at the exit of an image intensifier, has been built by Sinclair and Kasevich (1997). The count rate is satisfactory (500 000 photons/s), but the resolution is poor (360 μm FWHM with a 16-mm image field).

8.3.5 *Special anode cameras*

A variety of photon-counting cameras consisting of a tube with a photocathode, one or several MCPs, and a special anode to determine the photo-event (x, y) coordinates are developed. A few of them, which are used for the high resolution imaging, are elucidated.

(1) Wedge-and-strip anodes detector: The photon-counting system based on such anodes (Anger, 1966, Siegmund et al., 1983), uses conductive array structure, in which the geometrical image distortions might be eliminated. The target for the cloud of electrons coming out from the MCP is a four-electrode (A, B, C and D) anode (see Figure 8.11b) as discussed in the preceding section (8.3.4) with special shapes. It comprises multiple terminals with the x, y co-ordinate of a charge cloud determined through ratios of charge deposited onto the various terminals. The amplitudes of the signals detected on the wedge, strip and electrodes are linearly proportional to the x, y co-ordinates of the detected photon event. In this system, the spatial resolutions of the order of 40 - 70 μm FWHM and position sensitivities of 10 μm are obtained at high MCP gains. High resolution wedge-and-strip detectors can operate at event rates up to about 5×10^4 photons per second. The problems of the wedge-and-strip technique comes from the limitations on the anode capacitance, which restricts the maximum count rate 40 000 ph/s and from the defocussing (required to spread the cloud of electrons onto the anode) which is sensitive to the ambient magnetic field.

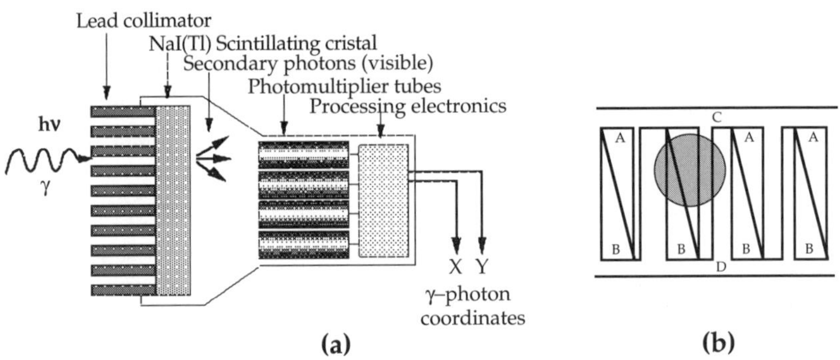

Fig. 8.11 (a) Anger gamma camera and (b) wedge-and-strip anode. The grey disc corresponds to the cloud of electrons that is spread onto a local area of the anode.

(2) Resistive anode position sensing detector: In this system, a continuous uniform resistive sheet with appropriately shaped electrodes provides the means for encoding the simultaneous location and arrival time of each detected photo-event. This is coupled to a cascaded stack of MCPs acting as the position sensitive signal amplifier. A net potential drop of about 5 kV is maintained from the cathode to the anode. Each primary photo-electron results in an avalanche of $10^7 - 10^8$ secondary

electrons onto the resistive anode. The signals resulting from the charge redistribution on the plate are amplified and fed into a high speed signal processing electronics system that produces 12 bit x, y addresses for each event. A PC based data acquisition system builds up a 1024×1024 image from this asynchronous stream of x, y values (Clampin et al. 1988). The drawback of this system is the large pixel response function; the nominal resolution of the system is about 60 μm.

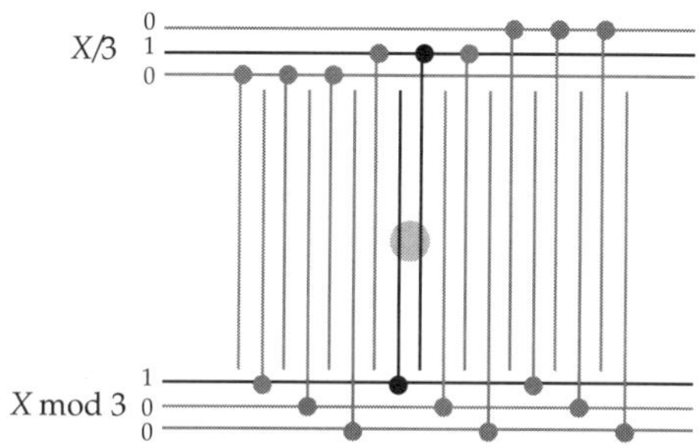

Fig. 8.12 Principle of coordinate encoding of a MAMA camera (imaginary case with $N = 3$). The grey zone represents the impact of the cloud of photons on the anode. The activated electrodes are represented in black.

(3) Multi-anode micro-channel array (MAMA): This detector allows high speed, discrete encoding of photon positions and makes use of numerous anode electrodes that identify each event's location (Timothy, 1983). The electron amplification is obtained by an MCP and the charge is collected on a crossed grid coincidence array. The idea is to slightly defocus the electron cloud at the exit of the MCP, so it falls into two wire electrodes for each coordinate (x or y). The position of the event is determined by coincidence discrimination. The resulting electron cloud hits two sets of anode arrays beneath the MCP, where one set is perpendicular in orientation to the other; the charge collected on each anode is amplified. One electrode is used to encode the coordinate divided by an integer N, and the other encodes the coordinate modulo N. Figure (8.12) displays an example with $N = 3$. Hence, the number of electrodes to encode X possible coordinate values is $N + (X \bmod N)$. To reach $X = 1024$, if $N = 32$, then 64 wire electrodes are needed (128 for a 2-D imager).

(4) Delay-line anode: This system (Sobottka and Williams, 1988) has a zigzag micro-strip transmission line etched onto a low loss, high

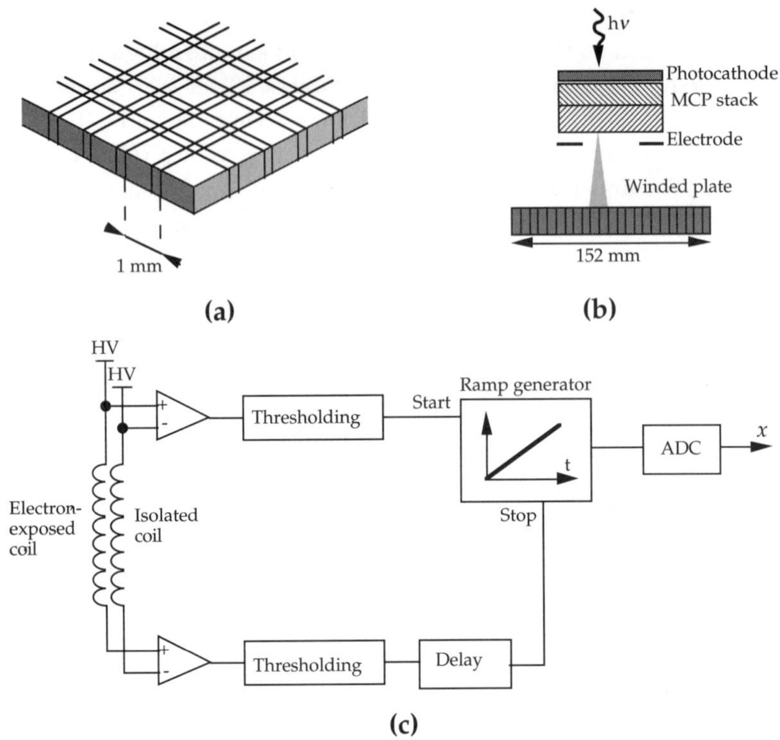

Fig. 8.13 (a) Delay-line anode, (b) camera using this anode, and (c) Readout electronics of this camera.

dielectric substrate. The position of the charge cloud event is encoded as the difference in arrival times of the charge pulse at both ends of the transmission line. Figure (8.13a) shows a a ceramic plate (152 mm × 152 mm), on which two orthogonal pairs of coils are winded. Each pair is used to encode a coordinate x or y. Within a pair, a coil is exposed to the electrons from a MCP, while the other is isolated and is used as a reference (see Figure 8.13c). The difference of current at an end of the coil pair is used to trigger a ramp generator, and at the other end to stop this generator. The voltage at the output of the generator, when it is stopped, depends on the delay between the pulses received at both ends, and therefore to the position of the electron cloud on the exposed coil. This system (Figure 8.13b) allows a high count rate (10^6 ph s^{-1}). The problem of the system is the size of the anode target, larger than any MCP and requiring a distortion-free electronic lens.

8.4 Solid state technologies

All the devices that have been described above are based on the photoelectric effect. These devices have several problems (Morel and Saha, 2005):

- The quantum efficiency is rather low (around 0.1 for multi-alkali photo-cathodes, 0.2 for AsGa photo-cathodes). It also tends to decrease with time due to the interaction of residual gas molecules with the photo-cathode (Decker, 1969).
- They present false counts due to thermoionic emission[20]. Electron emission may also be due to chemical interaction of residual gas molecules with the photo-cathode (Geiger, 1955).
- Residual gas molecules in the tube may be ionized by an electron. In this case, the positive ion may hit the photo-cathode and liberate several electrons. This phenomenon called 'ion return' causes artifacts in the image that are noticed by bright spots.
- Their constructions require a high vacuum, implying a fragility of the devices. Also very high voltage for power supply are required for operating an image intensifier, causing problems of electrical insulation.

Alternative resolutions for the detection of photons cameras with photon-counting rest on the principle of a multiplication of photo-electrons.

8.4.1 *Electron multiplying charge coupled device (EMCCD)*

Recent development of the solid state based non-intensified low light level charge coupled devices (L3CCD; Jerrom et al. 2001) using both front- and back-illuminated CCD, which can allow a signal to be detected above the noisy readout, has enabled substantial internal gain within the CCD before the signal reaches the output amplifier. After a few decades of existence of the CCD detector, this novel high-sensitivity CCD is a major breakthrough in the CCD sensor development. The electron multiplying CCD (EMCCD), is based on such a technology. It is engineered to address the challenges of ultra-low light level imaging applications. One of these applications, namely adaptive optics system requires wavefront correctors and a sensor. The sampling rate of an EMCCD scales with the turbulence in the atmosphere up to kHz and is limited by the number of photons received in a short-exposure. Optical interferometry (Labeyrie, 1975, Saha, 2002 and

[20]Thermoionic emission is a random electron emission from the photo-cathode due to the temperature.

references therein) also requires detection of very faint signals and reproduction of interferometric visibilities to high precision, therefore demands detectors and electronics with extremely low-noise.

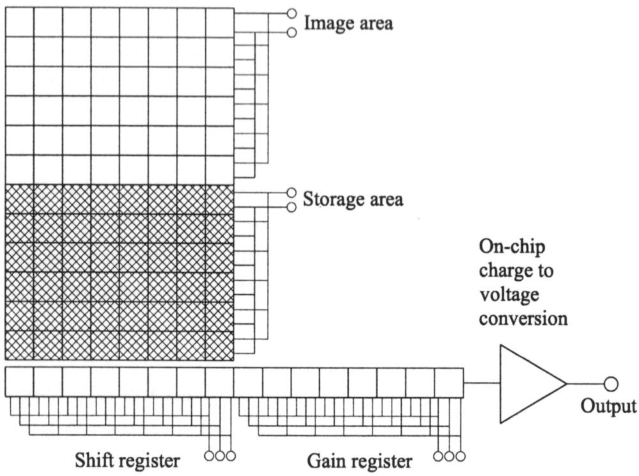

Fig. 8.14 A typical EMCCD sensor structure.

The EMCCD consists of a normal two-dimensional CCD, either in full frame or in frame-transfer format and is provided with Peltier cooling system that is comparable with liquid nitrogen cooled cryostats. The image store and readout register are of conventional design operating typically at 10 volts, but there is an extended section (see Figure 8.14) after the readout register, the multiplication register, where the multiplication or amplification takes place. After the multiplication register the charge is converted to a voltage signal with a conventional charge to voltage amplifier.

The operation of multiplication register is similar to the readout register but the clocking voltages that are much higher (typically at > 20 volts as opposed to the ~ 10 volts). At this higher voltage there is an increased probability, $\mathcal{P}$, that electrons being shifted through the multiplication register have sufficient energy to create more free electrons by impact ionization. Although the probability of secondary generation in each pixel of the multiplication register is low; typically it ranges from 0.01 to 0.016 by designing a multiplication register with many pixels the effective gain of the register can be more than $1000\times$. The gain, G, is multiplicative and for n pixels is

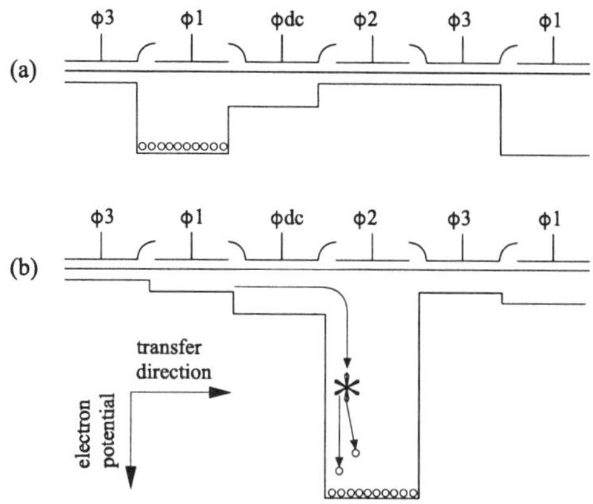

Fig. 8.15 Probability of generating secondary electrons.

expressed as,

$$G = (1 + \mathcal{P})^n, \tag{8.35}$$

The probability, $\mathcal{P}$, of generating a secondary electron is dependent on the voltage levels of the serial clock and the temperature of the CCD (Figure 8.15). The specific gain at a particular voltage and temperature may vary from sensor to sensor but the sensors all tend to follow similar trends. If one considers the effects of number of photons n_p which would generate in a pixel with quantum efficiency, η_d, a signal of N_e electrons as below,

$$N_e = \eta_d n_p. \tag{8.36}$$

As the photons follow Poisson statistics, the photon noise is given by,

$$i_{ph} = \sqrt{\eta_d n_p}. \tag{8.37}$$

The EMCCD gain amplifies the signal by G but also adds additional noise to that of the incoming photons and the excess noise known, as the noise factor, F needs to be taken into consideration. The noise factor can be expressed as,

$$F^2 = \frac{\delta_{out}^2}{G^2 \langle i_{ph} \rangle^2}. \tag{8.38}$$

The total noise, $\langle i_{tot}\rangle^2$, is calculated by adding the noise factors in quadrature as,

$$\langle i_{tot}\rangle^2 = \langle i_{ro}\rangle^2 + \langle i_d\rangle^2 + \langle i_{ph}\rangle^2, \tag{8.39}$$

in which $\langle i_{ro}\rangle^2$ is the readout noise, $\langle i_d\rangle^2$ the dark noise, and $\langle i_{ph}\rangle^2$ the noise generated by the photon signal.

Putting these terms together one generates an expression for the total detected noise, S_N, referenced to the image area,

$$S_N = \frac{\eta_d n_p}{\sqrt{F^2\left(\langle i_d\rangle^2 + \langle i_{ph}\rangle^2\right) + \langle i_{ro}\rangle^2 /G^2}}. \tag{8.40}$$

Since the dark signal is amplified by EMCCD gain, the resultant noise from the dark signal is also further increased by the noise factor. In the case of cooling the EMCCD to effectively render this negligible and substituting for the δ_s, the following expression emerges,

$$S_N = \frac{\eta_d n_p}{\sqrt{F^2\eta_d n_p + \langle i_{ro}\rangle^2 /G^2}}. \tag{8.41}$$

From this equation (8.41), it is clear that increasing gain G virtually eliminates the effect of the readout noise. This is very important at high readout rates where typically the δ_{ro} is very high. In an ideal amplifier the noise factor would be unity, however the EMCCD gain originates from a stochastic process, and theory (Hynecek and Nishiwaki, 2003) shows that for a stochastic gain process using infinite number of pixels in the gain register the noise factor should tend to $\sqrt{2}$ and this is the value observed experimentally at high gain values. It follows from this to observe an experiment with the same S/N ratio with an ideal amplifier as that with a noise factor of $\sqrt{2}$ one needs to have twice the number of photons. Alternatively this can be viewed as if the detective quantum efficiency of the sensor being half of what it actually is.

At low photon flux levels the readout noise of CCD dominates the S/N ratio and the EMCCD wins out. At higher photon flux levels the noise factor of the EMCCD reduces the S/N ratio below that of the CCD. The apparent reduction in detective quantum efficiency can be eliminated by using a true photon-counting system in which an event is recognized as a single photon. Saha and Chinnappan (2002) reported that their EMCCD camera system has the provision to change gain from 1 to 1000 by software. The noise at 1 MHz read rate is ~2 e RMS. It is a scientific grade camera with 16 bit analog to digital (A/D) conversion with 1 msec frame time; the data can be archived to a Pentium PC.

8.4.2 *Superconducting tunnel junction*

Superconducting tunnel junction (STJ) is a photon-counting sensors (Perryman et al. 1993), born from research in X-ray detectors is based on a stack (see Figure 8.16a) of different materials (Nb/Al/Al_2O_3/Al/Nb). It has the property to get a charge proportional to the energy of an incoming photon. First prototypes of STJ detectors (Peacock et al. 1996) had QE = 0.5 and a count rate of the order of 2500 photons/s. The photon-counting performances of the STJs have been improved by using niobium instead tantalum. In this case, PV$\rightarrow \infty$ and NFWHM = 0.05 (for λ = 250 nm). The spectral resolution is 8 nm at λ = 200 nm, and 80 nm at λ = 1000 nm. The main problem of the STJs is the very low temperature that they required (370 mK). Moreover, making STJ array detectors for imaging is a challenge. A 6 $\times$ 6-pixel STJ array has nevertheless been made and used in astronomy (Rando et al. 2000).

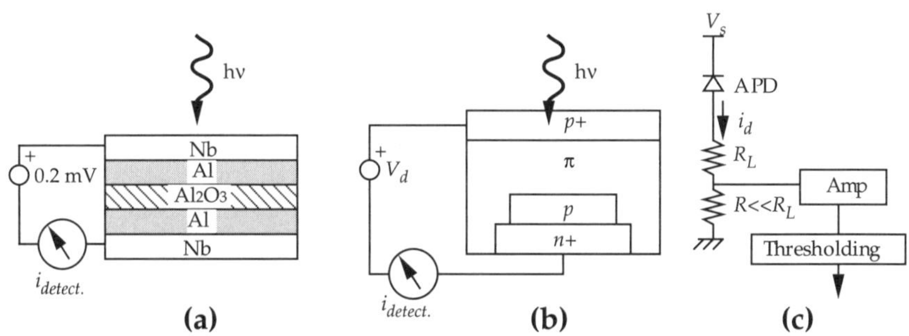

Fig. 8.16 Principles (a) of an STJ, and (b) of an APD; (c) Passive quenching of an APD.

8.4.3 *Avalanche photo-diodes*

Avalanche photo-diodes (APD; Ingerson et al. 1983) are the most common solid state photon-counting sensors. A photo-diode is a semiconductor device that allows current to flow only when exposed to light. Photo-diodes contain *p*-*n* junction, and often an intrinsic (undoped) layer between *n* and *p* layers, called *p*-*i*-*n* junction. They are packaged with either a window or optical fibre connection, in order to transfer the light to the device. Light absorbed in the depletion region or the intrinsic region generates electron-hole pairs, most of which contribute to a photocurrent. APDs are based on the ionization of a high-voltage pn^+ junction, triggered by a single photo-electron (see Figure 8.16b). By applying a high reverse bias voltage (typically 100-200 V in silicon), these diodes show an internal current gain effect ($\sim$ 100) due to impact ionization, what is known as avalanche effect. As the reverse-bias voltage increases toward the breakdown, hole-electron pairs are created by absorbed photons. An avalanche effect occurs when the hole-electron pairs acquire sufficient energy to create additional pairs when the incident photons collide with the ions, thus, a signal gain is achieved.

To be able to count photons, the APD must be used in 'Geiger mode': the voltage has to be more than around 200 V to allow a chain reaction (avalanche) of electron liberation. In such a mode, an APD is biased above its breakdown voltage for operation at very high gain (typically 10^5 to 10^6). The problem is then to stop the avalanche. In one design, called 'passive quenching' (Brown et al. 1986), a serial-mounted resistor decreases the voltage of the APD when a current from the APD crosses this resistor (Figure 8.16c). The problem of passive quenching comes from the capacitance of the system that limits the bandwidth. Another design to stop the avalanche is called 'active quenching' (Brown et al. 1987), in which the power supply of APD is controlled by a system measuring the output current.

APDs have 70% or more quantum efficiency and output a digital pulse 100 nsecs after photon detection providing very high frame rate. Number of counts from the detector is very high giving a very high S/N ratio; the dark current is low. The noise properties of an APD are affected by the semiconductor materials that the APD is made of; typical materials are used include Si, InGaAs, and Ge. Short integration time of about 200 μsecs is possible. Making integrated arrays of APDs working in photon-counting mode is a challenge because of the photon-emission phenomenon that is caused by the electron avalanche. These photons may trigger avalanche in the pixels in their vicinity.

8.5 Infrared sensors

Having longer wavelengths and lower energy than the optical light, the infrared (IR) radiation does not have sufficient energy to interact with the photographic plates that were used in optical astronomy. Early infrared observers used thermocouples, the devices which convert heat into electric current, and thermopiles, a group of thermocouples combined in a cell, to detect infrared radiation from a source. The thermocouple devices were used to detect such radiation from the Moon around the middle of the nineteenth century. Infrared radiation was detected from some bright stars namely Vegas, Arctures, as well as from the planets like Jupiter and Saturn. However, low sensitivity of these IR detectors prevented the detection of other near-IR sources. The real breakthrough in the development of sensitive IR detectors were in place by the mid 1960's. The first infrared survey of the sky was made at the Mt. Wilson Observatory employing a liquid nitrogen cooled lead-sulphide (PbS) detectors; This survey revealed as many as 20,000 IR sources from the 75% of the sky. It is to mention that such detectors can be used to study infrared in the 1 to 4 μm range. When IR radiation in this range falls on a PbS cell, it changes the resistance of the cell. This change in resistance is related to the amount of radiation that falls upon the cell. The advancement in IR detector technology made it possible to find that the centers of most galaxies emit strongly in the IR.

Development of the germanium bolometer, in the early 1960's, was a major breakthrough in IR sensor technology. This sensor was more sensitive than the earlier detectors and was capable of detecting almost entire range of IR wavelengths. A cool thin strip of germanium is kept in a container having a small opening. When IR radiation strikes the germanium through the opening, it warms the metal and changes its conductivity. The change of this conductivity is directly proportional to the amount of IR radiation entering the container. The efficiency of the germanium bolometer goes up with cooling at an extremely low temperature. Liquid helium is the best option for such a purpose, which may cool it to 4 degrees Kelvin[21]. A metal dewar is required to hold the liquid helium in which germanium bolometer can be immersed. Subsequent development in the IR sensor technology in the form of detector array (combination of several single detectors) in the 1980's pave the way to produce images containing hundreds of pixels at the same time. However, no photon-counting is possible with the current technology.

Following the early work with single element detectors, modern interferometers took the advantage of technology development in the near-IR focal plane array, such as NICMOS (Near Infrared Camera and Multi-Object Spectrograph). This detector represents vast improvement over

[21]The Kelvin temperature scale sets its zero point at absolute zero (-273.15°) on the Celsius scale.

photo-diodes and allowed various astronomical sources, young stellar objects in particular, to be observed. It has been designed for use in the HST.

NICMOS is a hybrid direct readout array integrating detector, which consists of 256×256 of 40μm square pixel size, organized in four independent 128×128 quadrants. It is fabricated in HgCdTe, with a cutoff wavelength of 2.4μm, grown on a sapphire substrate that is very rugged and provides a good thermal contraction match to silicon multiplexer (Cooper et al. 1993). Each pixel has a detector diode, which is electrically connected to its unit cell in the silicon layer. This unit cell circuits contains a Field-Effect Transistor (FET) with high gate resistance, so as to enable to read the voltage across the diode after integration without loss of any charge. The near-IR (0.8-2.5 μm) imager around a 512×512 HgCdTe array (with 18 μm pixel) is also built. The image scale of such an instrument would help in sub-arcsec imaging and reaching out to fainter limit.

Unlike a CCD, the individual pixels of the NICMOS arrays are strictly independent and can be read non-destructively. Since the array elements are independently addressed, such a sensor does not suffer from some of the artifacts that afflict CCD. However, the dark current associated with NICMOS is quite substantial compared to that produced by the new generation CCDs. The typical NICMOS3 FPAs have read noise less than 35 e- with less than 1 e-/sec detector dark current at 77 K and broadband quantum efficiency is better than 50% in the range of 0.8 to 2.5 μm.

Chapter 9

Image processing

9.1 Post-detection image reconstruction

The daunting task in astronomy is to detect fainter objects and to discern finer details. Recovery of phase information is important, whereas in other branches of physics, for example, optical interferometry, electron microscopy, optical and electron holography, wavefront sensing, and crystallography, one often wishes to recover phase. Absence of phase information in the standard autocorrelation technique prevents true image reconstruction (Saha, 1999b). Several image processing methods have been developed to reconstruct images from the blurred images caused by diffraction or atmospheric seeing, as well as to denoise images without losing any of the resolution improvement achieved by deblurring. Non-introduction of spurious artifacts during image processing is also a primary concern. Of course, the main limitation of such techniques is that it requires excessive computer processing of the image.

It is possible to improve images obtained with the adaptive optics (AO) system using such algorithms since AO compensation is not perfect because of (i) sub-aperture pupil sizes larger than the mean coherence length, (ii) time-lag between wavefront measurement and correction, (iii) limited signal-to-noise (S/N) ratio in the wavefront measurement, and (iv) differences between the reconstructed wavefront and wavefront measurement due to the geometry of the deformable mirror. Though measurements of a point source are taken to calibrate the AO point spread function (PSF), it may not be the same for the reference as for the astronomical objects of interest. Such measurements are unable to provide the exact measurement, but approximate the system's performance on the object. This is due to (i) fluctuations of the atmospheric coherence length, r_0, as well as of the atmospheric coherence time, τ_0, (ii) sensitivity of the wavefront sensor to: (a) the brightness of the object and (b) differential response of the wavefront sensor between a point like object and an extended object, and (iii) stability of the tilt-compensation between the reference and the objects of interest.

Prior to using image reconstruction algorithms, the basic operations to be performed are dead pixel removal, debiasing, flat fielding, sky or background emission subtraction, and suppression of correlated noise. Most algorithms work with corrected data. In what follows, the mathematical intricacies of the various post-detection image processing techniques are examined with emphasis set on their comparisons.

9.1.1 *Shift-and-add algorithm*

Shift-and-add (SAA) method is used for obtaining good quality images from several short-exposures with varying shifts. The method involves derivation of the differential shifts of the images. The images are shifted back to a common place. The pixel with maximum signal-to-noise (S/N) ratio in each specklegram can be co-added linearly at the same location in the resulting accumulated image (Lynds et al., 1976, Worden et al., 1976).

In a blurred image, each speckle is considered as a distorted image of the object, the brightest one being the least distorted. In this process, the position of the brightest speckle, $\vec{x}_k$, is necessary to be located in each short-exposure image (specklegram), $I_k(\vec{x})$, $[I_k(\vec{x}_k) > I_k(\vec{x})$ for all $\vec{x} \neq \vec{x}_k]$, followed by shifting the whole frame (without any rotation) so that the pixels with maximum S/N ratio can be co-added linearly at the same location in the center of the frame. The shift-and-add image, $I_{sa}(\vec{x})$, is obtained by averaging over the set of the shifted specklegrams,

$$I_{sa}(\vec{x}) = \langle I_k(\vec{x} + \vec{x}_k) \rangle \,. \tag{9.1}$$

The large variations in the brightness of the brightest pixels can be observed in a set of speckle images. The contamination level of a specklegram may not be proportional to the brightness of its brightest pixel (Bates and McDonnell, 1986). The adjusted shift-and-add image, $I_{asa}(\vec{x})$, is defined as,

$$I_{asa}(\vec{x}) = \langle w[I_k(\vec{x}_k)] I_k(\vec{x} + \vec{x}_k) \rangle \,, \tag{9.2}$$

where, $w[I_k(\vec{x}_k)]$ is the weight in relation with the brightness of the brightest pixel. The choice of the same quantity can be made as $w\{I_k(\vec{x}_k)\} = I_k(\vec{x}_k)$.

An array of impulse is constructed by putting an impulse at each of the centre of gravity with a weight proportional to the speckle intensity. This impulse array is considered to be an approximation of the instantaneous PSF. It is cross-correlated with the speckle frame. Disregarding the peaks lower than the pre-set threshold as enumerated above, the mth speckle mask, $mask_m(\vec{x})$, is defined by,

$$mask_m(\vec{x}) = \sum_{n=1}^{M} I_m(\vec{x}_{m,n})\delta(\vec{x} - \vec{x}_{m,n}). \tag{9.3}$$

The mth masked speckled image, $mI_m(\vec{x})$, is expressed as,

$$mI_m(\vec{x}) = I_m(\vec{x}) \otimes mask_m(\vec{x}), \tag{9.4}$$

in which the crossed circle, $\otimes$, denotes correlation.

The Lynds-Worden-Harvey image can be obtained by averaging $\langle mI_m(\vec{x})\rangle$. This technique contains more information in each $I_m(\vec{x})$ than $I_{asa}(\vec{x})$, and therefore, exhibits more S/N. For direct speckle imaging, the shift-and-add image, $I_{sa}(\vec{x})$, is a contaminated one containing two complications - a convolution, $S_k(\vec{x})$ and an additive residual, $C(\vec{x})$ - which means,

$$I_{sa}(\vec{x}) = O(\vec{x}) * S(\vec{x}) + C(\vec{x}), \tag{9.5}$$

where,

$$S(\vec{x}) = \sum_{k=1}^{k} \delta(\vec{x} - \vec{x}'_k)d_k, \tag{9.6}$$

in which $\vec{x}'_k$ being the constant position vectors and d_k, the positive constant.

Due to numerous possible sources of systematic error in reduction process, it is essential to calibrate $I_{sa}(\vec{x})$ with an unresolved point source and reduce it in the same way to produce $S(\vec{x})$. The estimate for the object, $O(\vec{x})$, is evaluated from the inverse Fourier transform of the following equation,

$$\widehat{O}(\vec{u}) = \frac{\widehat{I}_{sa}(\vec{u})}{\widehat{I}_0(\vec{u}) + \widehat{N}(\vec{u})}, \tag{9.7}$$

in which $\widehat{N}(\vec{u})$ stands for the noise spectrum.

This is the first approximation of the object irradiance. If this technique is applied to a point source, an image similar to Airy pattern yields. This method is found to be insensitive to the telescope aberrations. The limitation of this algorithm is not being applicable when photon noise dominates in addition to the accuracy with which speckle maxima are located. The notable drawback of the SAA method is that it can be applied only to bright objects when the photon noise is negligible, hence it is seldom used in astronomical practice.

9.1.2 *Selective image reconstruction*

Another method called 'selective image reconstruction' is used by selecting a few sharpest images from a large dataset of short-exposures, which are recorded when the atmospheric distortion is naturally at minimum (Dantowitz et al. 2000). As the atmosphere behaves like a time-varying random phase screen, with a power spectrum of irregularities characterized by spatial (atmospheric coherence length, r_0) and temporal scales, (atmospheric coherence time, τ_0), whose RMS variations are larger than those of a primary mirror of a telescope, it is improbable to obtain a perfect diffraction-limited PSF. The combined phase variations across such a telescope, $\langle\sigma\rangle$, due to both the atmosphere and the telescope mirror turn out to be less than 1 radian. The corresponding image of a celestial object may have a core that is diffraction-limited with a Strehl's ratio, $\mathcal{S}_r = e^{-<\sigma>^2}$ and angular resolution, $\theta = 1.22\lambda/D$, in which D is the aperture diameter.

Fried (1978) describes high quality short-exposure, which occur in fortuitous way, as 'lucky exposure'. If the speckle patterns change on timescales of a few milliseconds, one may expect to obtain such good images which occur relatively often during an observing run. The probability, $\mathcal{P}$ of such lucky exposures having RMS variation in wavefront phase over the aperture less than 1 rad for seeing defined by the atmospheric coherence length, r_0, is

$$\mathcal{P} \simeq 5.6e^{-0.1557(D/r_0)^2}. \tag{9.8}$$

This equation (9.8) implies that for an aperture of diameter, $7r_0$ or $8r_0$, about 0.1% of the short-exposures should be of very good quality; one needs to search several thousand specklegrams in order to get a single frame having 0.1% Strehl ratio. However, this technique requires a good site and a telescope for which the errors in the mirror figure should be small compared with the atmosphere. For a good seeing that is of the order of 0.5 arcsecs, the telescope aperture should be 2.5 m at $\lambda \sim 800$ nm.

Baldwin et al. (2001) have demonstrated the potential of selective reconstruction method, in which a number of good quality specklegrams taken at different times are re-centered and co-added. It is important to mention that in order to maximize the image quality, the images should be shifted so that the contribution from the bright speckle of each frame is brought to a common location. When these specklegrams are co-added, the contribution from the bright speckles add coherently, while the contributions from the randomly varying speckles combine incoherently. New technology detector have allowed observations to be performed at much higher rates (Baldwin et al. 2001), providing insights into the characteristics of the atmosphere.

9.1.3 *Speckle holography*

As stated in the preceding section (6.3), the speckle interferometry relies on the autocorrelation technique, which cannot provide Fourier phase information (equation 6.57). An exception is the case when a bright reference point source is available within iso-planatic patch ($\sim 7''$). It can be used as a key to reconstruct the target source in the same way as a reference coherent beam is used in holographic reconstruction (Liu and Lohmann, 1973).

The term 'holography' is defined as the process of recording and reconstructing the complete optical wavefront reflected from an object. The principle of holography is to numerically invert the process of image degradation due to atmospheric turbulence, i.e., to deconvolve the point spread function (PSF) from the observed image and reconstruct an original image of the object. Such a technique is known as speckle holography. Let the point source be represented by a Dirac impulse, $A_0\delta(\vec{x})$, at the origin and $O_1(\vec{x})$ be the nearby object to be reconstructed. The intensity distribution in the field of view is expressed as,

$$O(\vec{x}) = A_0\delta(\vec{x}) + O_1(\vec{x}). \tag{9.9}$$

A regular speckle interferometric measurement provides the squared modulus of its Fourier transform, $\widehat{O}(\vec{u})$,

$$\begin{aligned} \left|\widehat{O}(\vec{u})\right|^2 &= \left|A_0 + \widehat{O}_1(\vec{u})\right|^2 \\ &= A_0^2 + A_0\widehat{O}_1(\vec{u}) + A_0\widehat{O}_1^*(\vec{u}) + \widehat{O}_1(\vec{u})\widehat{O}_1^*(\vec{u}). \end{aligned} \tag{9.10}$$

The inverse Fourier transform provides the autocorrelation, $\mathcal{A}_c[O(\vec{x})]$, of the field of view,

$$\mathcal{A}_c\left[O(\vec{x})\right] = A_0^2\delta(\vec{x}) + A_0 O_1(\vec{x}) + A_0 O_1(-\vec{x}) + \mathcal{A}_c\left[O_1(\vec{x})\right], \tag{9.11}$$

where, $\mathcal{A}_c[O_1(\vec{x})]$ is the autocorrelation of the object.

The first and the last term in equation (9.11) are centered at the origin. If the object is far enough from the reference source, $O(\vec{x})$, and its mirror image, $O(-\vec{x})$, and therefore recovered apart from a 180° rotation ambiguity.

9.1.4 *Cross-spectrum analysis*

Another method, known as cross-spectrum analysis, is to calculate the cross-spectrum between the object and the reference source. Speckle holography is in fact an indirect way to derive such cross-spectrum. The angular distance between the two sources should be within the iso-planatic patch or the two spectral windows should be close enough. Let two speckle images, $I_1(\vec{x})$ and $I_2(\vec{x})$ be taken at same angular distance or employing different spectral windows. Their Fourier transforms are given by,

$$\left\langle \widehat{I}_1(\vec{u})\widehat{I}_2^*(\vec{u})\right\rangle = \widehat{O}_1(\vec{u})\widehat{O}_2^*(\vec{u}).\left\langle \widehat{S}_1(\vec{u})\widehat{S}_2^*(\vec{u})\right\rangle, \tag{9.12}$$

in which

$$\begin{aligned}\widehat{I}_1(\vec{u}) &= \widehat{O}_1(\vec{u}).\widehat{S}_1(\vec{u}),\\ \widehat{I}_2(\vec{u}) &= \widehat{O}_2(\vec{u}).\widehat{S}_2(\vec{u}),\end{aligned} \tag{9.13}$$

$\widehat{I}_1(\vec{u})$ and $\widehat{I}_2(\vec{u})$ are the transfer function of image irradiance distributions, $\widehat{O}_1(\vec{u})$ and $\widehat{O}_2(\vec{u})$ are the transfer function of the object distributions associated with the channel, and $\widehat{S}_1(\vec{u})$ and $\widehat{S}_2(\vec{u})$ the related instantaneous transfer functions.

Like speckle interferometry, the transfer function for the cross-spectrum, $< \widehat{S}_1(\vec{u})\widehat{S}_2^*(\vec{u}) >$, is calculated on reference unresolved source for which, $\widehat{O}_1(\vec{u}) = \widehat{O}_2(\vec{u}) = 1$. If the angular distance between the two sources, i.e., target and reference is within iso-planatic patch or two spectral windows are within the allowed spectral bandwidth, i.e., $\Delta\lambda/\lambda \ll r_0/D$, the transfer function is equivalent to the transfer function for speckle interferometry. An object is reconstructed through one channel provided the object is known at the second channel. From equation (9.12),

$$\widehat{O}_1(\vec{u}) = \frac{\left\langle \widehat{I}_1(\vec{u})\widehat{I}_2^*(\vec{u})\right\rangle}{\widehat{O}_2^*(\vec{u})\left\langle \widehat{S}_1(\vec{u})\widehat{S}_2^*(\vec{u})\right\rangle}. \tag{9.14}$$

In the case of well isolated reference speckle pattern, the cross-spectrum can be computed directly and an image can be reconstructed using the equation (9.14), or for a real speckle holography transfer function, $< |\widehat{\mathcal{S}}(\vec{u})|^2 >$,

$$\widehat{O}_1(\vec{u}) = \frac{\left\langle \widehat{I}_1(\vec{u}).\widehat{I}_2^*(\vec{u})\right\rangle}{\widehat{O}_2^*(\vec{u})\left\langle \left|\widehat{S}_2(\vec{u})\right|^2\right\rangle} = \widehat{O}_2(\vec{u})\frac{\left\langle \widehat{I}_1(\vec{u}).\widehat{I}_2^*(\vec{u})\right\rangle}{\left\langle \left|\widehat{I}_2(\vec{u})\right|^2\right\rangle}. \tag{9.15}$$

The equation (9.15) shows that the speckle holography transfer function, $< |\widehat{\mathcal{S}}(\vec{u})|^2 >$, is real. This method is insensitive to aberrations; the noise contributions from two different detectors are uncorrelated, thereby, their contributions cancel out. The expected value of the phase of the cross-spectrum coincides with the phase-difference between the object and the reference. However, when this is not the case, one must use image processing methods by using the different light levels of the speckle clouds (reference and target) to extract the object. Aime et al., (1985) have used cross-spectral analysis to correlate the solar photospheric brightness and velocity fields.

9.1.5 *Differential speckle interferometry*

Differential speckle interferometry (DSI) is a method to record speckle patterns of a stellar object in different wave modes simultaneously and to compute the average cross-spectrum or cross-correlation of pairs of speckle images (Beckers, 1982). The difference between these patterns can be measured independently from the seeing variations. By analogy with long-exposure observations of the colour difference of the coordinates of the star, the differential displacement of speckles can be measured with much higher accuracy than their size.

This technique provides a new astrophysical parameter: the vector representing the variation of the object photocenter as a function of the wavelength. In the case of objects much smaller than the Airy pattern, DSI measures the displacement between two speckle patterns even when the shift is much smaller than the speckle size. Beckers (1982) suggested that a pair of two-dimensional (2-D) speckle images be formed using narrow-band (1 Å) interference filters and a Wollaston prism in combination with a light chopper. However, recording images simultaneously in different spectral windows requires a 2-D detector with a very large number of channels.

Petrov and Cuevas (1991) proposed a scheme, where the speckle image of a star is projected onto a narrow entrance slit of the spectrograph, which transforms each speckle into a separate spectrum. Analysis of spatial information is restricted to only one coordinate along the spectrograph slit. The second coordinate of the detector is used as a spectral axis. This technique can be used to resolve close binary system below the diffraction limit of the telescope, as well as to determine (i) the relative orbit orientations in multiple system, (ii) the masses of double-lined spectroscopic system (see section 10.3), and (iii) the masses of long period binary systems.

Differential interferometry estimates the ratio, $\widehat{O}_1(\vec{u})/\widehat{O}_2(\vec{u})$. The differential image, $D_I(\vec{x})$, can be obtained by performing inverse Fourier transform of this ratio,

$$D_I(\vec{x}) = \mathcal{F}^{-1}\left[\frac{\left\langle \widehat{I}_1(\vec{u})\widehat{I}_2^*(\vec{u})\right\rangle}{\left\langle \left|\widehat{I}_2(\vec{u})\right|^2\right\rangle}\right]. \tag{9.16}$$

$D_I(\vec{x})$ is self calibrating for seeing and represents an image of the object in the emission feature having the resolution of the object imaged in the continuum (Hebden et al., 1986). Petrov et al., (1986) found an increase in S/N ratio as the band-passes of the two components of the dual specklegram increase.

9.1.6 *Knox-Thomson technique (KT)*

Knox and Thompson (1974) proposed a method for determining the phase of the object transform to the diffraction limit of the telescope, in order to recover the actual object transform. This method is a small modification of the autocorrelation technique, involving the centering in each specklegram with respect to its centroids. In order to extract the phase, they proposed to use the statistical autocorrelation of the Fourier transform of the instantaneous image intensity.

In lieu of the image energy spectrum, $< |\widehat{I}(\vec{u})|^2 >$, let the general second order moment be the cross spectrum, $< \widehat{I}(\vec{u}_1)\widehat{I}^*(\vec{u}_2) >$. The cross-spectrum takes significant values only if $|\vec{u}_1 - \vec{u}_2| < r_0/\lambda$. The typical value of $|\Delta\vec{u}|$ is

$\sim$ 0.2 - 0.5 r_0/λ. Invoking equation (5.99), a 2-D irradiance distribution, $I(\vec{x})$ and its Fourier transform, $\widehat{I}(\vec{u})$, is defined by the equation,

$$\widehat{I}(\vec{u}) = \int_{-\infty}^{\infty} I(\vec{x}) e^{-i2\pi\vec{u}\cdot\vec{x}} d\vec{x}. \tag{9.17}$$

where $\vec{x}(= x, y)$ represents 2-D position vectors, $\vec{u}(= u, v)$ the 2-D spatial frequency vectors.

Unlike the autocorrelation technique (section 6.3.1.), KT technique defines the correlation of $I(\vec{x})$ and $I(\vec{x})$ multiplied by a complex exponential factor with a spatial frequency larger than zero. In image space, the correlations of $I(\vec{x})$, are derived as,

$$I^{KT}(\vec{x}_1, \Delta\vec{u}) = \int_{-\infty}^{\infty} I^*(\vec{x}) I(\vec{x} + \vec{x}_1) e^{i2\pi\Delta\vec{u}\cdot\vec{x}} d\vec{x}, \tag{9.18}$$

where, $\vec{x}_1 = \vec{x}_{1x} + \vec{x}_{1y}$ are the 2-D spatial co-ordinate vectors.

In Fourier space, $\widehat{I}(\vec{u})$ gives the following relationship,

$$\widehat{I}^{KT}(\vec{u}_1, \Delta\vec{u}) = \widehat{I}(\vec{u}_1)\widehat{I}^*(\vec{u}_1 + \Delta\vec{u}), \tag{9.19}$$

where, $\vec{u}_1 = \vec{u}_{1x} + \vec{u}_{1y}$, and $\Delta\vec{u} = \Delta\vec{u}_x + \Delta\vec{u}_y$ are the 2-D spatial frequency vectors. $\Delta\vec{u}$ is a small, constant offset spatial frequency.

This technique consists of evaluating the three sub-planes in Fourier space corresponding to,

$$\Delta\vec{u} = \Delta\vec{u}_x; \qquad \Delta\vec{u} = \Delta\vec{u}_y; \qquad \Delta\vec{u} = \Delta\vec{u}_y + \Delta\vec{u}_x. \tag{9.20}$$

When digitized images are used, $\Delta\vec{u}$ generally corresponds to the fundamental sampling vector interval (Knox, 1976). A number of sub-planes can be used by taking different values of $\Delta\vec{u}$. Invoking equation (5.98), into equation (9.19), the following relationship is obtained,

$$\widehat{I}(\vec{u}_1)\widehat{I}^*(\vec{u}_1 + \Delta\vec{u}) = \widehat{O}(\vec{u}_1)\widehat{O}^*(\vec{u}_1 + \Delta\vec{u})\widehat{S}(\vec{u}_1)\widehat{S}^*(\vec{u}_1 + \Delta\vec{u}). \tag{9.21}$$

The modulus of the object Fourier transform can be found following the classical speckle interferometry procedure. Both the modulus and the phase provide the diffraction-limited image of the object. If the frequency separation becomes zero, $\Delta\vec{u} = 0$, equation (9.19) is identical to equation (6.53), and the phase information is lost. But for a finite $\Delta\vec{u} < r_0/\lambda$, the phase information is still present. The argument of the equation (9.19) provides the phase-difference between the two spatial frequencies separated by $\Delta\vec{u}$ and can be expressed as,

$$arg\left|\widehat{I}^{KT}(\vec{u}_1, \Delta\vec{u})\right| = \theta^{KT}(\vec{u}_1, \Delta\vec{u}) = \psi(\vec{u}_1) - \psi(\vec{u}_1 + \Delta\vec{u}). \tag{9.22}$$

The equation (9.21) is expressed as,

$$\widehat{I}^{KT}(\vec{u}_1, \Delta\vec{u}) = |\widehat{O}(\vec{u}_1)||\widehat{O}(\vec{u}_1 + \Delta\vec{u})||\widehat{S}(\vec{u}_1)||\widehat{S}(\vec{u}_1 + \Delta\vec{u})| \\ \times e^{i[\theta_O^{KT}(\vec{u}_1, \Delta\vec{u}) + \theta_S^{KT}(\vec{u}_1, \Delta\vec{u})]}. \tag{9.23}$$

In a single image realization the object phase-difference is corrupted by the random phase-differences due to the atmosphere-telescope OTF,

$$e^{i\theta_S^{KT}(\vec{u}_1, \Delta\vec{u})} = e^{i[\psi_S(\vec{u}_1) - \psi_S(\vec{u}_1 + \Delta\vec{u})]}. \tag{9.24}$$

If equation (9.23) is averaged over a large number of frames, the feature $(\Delta\vec{\psi}_S) = 0$. If $\Delta\vec{u}$ is small, $< \widehat{S}(\vec{u}_1)\widehat{S}(\vec{u}_1 + \Delta\vec{u}) >$, has a significant value for all pairs of points along the $\vec{u}$-axis up to the diffraction limit of the telescope, i.e., $|\widehat{O}(\vec{u}_1 + \Delta\vec{u})| \approx |\widehat{O}(\vec{u}_1)|$, etc. and so,

$$\left\langle \widehat{I}^{KT}(\vec{u}_1, \Delta\vec{u}) \right\rangle = |\widehat{O}(\vec{u}_1)||\widehat{O}(\vec{u}_1 + \Delta\vec{u})|e^{i[\theta_O^{KT}(\vec{u}_1, \Delta\vec{u})]} \\ \left\langle \widehat{S}(\vec{u}_1)\widehat{S}^*(\vec{u}_1 + \Delta\vec{u}) \right\rangle, \tag{9.25}$$

from which, together with equation (6.53), the object phase-spectrum, $\theta_O^{KT}(\vec{u}_1, \Delta\vec{u})$, can be determined.

The complex correlation, $\widehat{O}(\vec{u}_1)\widehat{O}(\vec{u}_1 + \Delta\vec{u})$ in the equation (9.21) is written in the form:

$$\widehat{O}(\vec{u}_1)\widehat{O}^*(\vec{u}_1 + \Delta\vec{u}) = \left|\widehat{O}(\vec{u}_1)\right| e^{i\psi_O(\vec{u}_1)} \left|\widehat{O}^*(\vec{u}_1 + \Delta\vec{u})\right| e^{-i\psi_O(\vec{u}_1 + \Delta\vec{u})}. \tag{9.26}$$

The difference in phase between points in the object phase-spectrum is encoded in the term,

$$e^{i\theta_O^{KT}(\vec{u}_1, \Delta\vec{u})} = e^{i[\psi_O(\vec{u}_1) - \psi_O(\vec{u}_1 + \Delta\vec{u})]}, \tag{9.27}$$

of the equation (9.23), which leads to,

$$\Delta\psi_O(\vec{u}_1) \equiv \psi_O(\vec{u}_1 + \Delta\vec{u}) - \psi_O(\vec{u}_1), \tag{9.28}$$

or

$$\psi_O(\vec{u}_1 + \Delta\vec{u}) = \psi_O(\vec{u}_1) + \Delta\psi_O(\vec{u}_1). \tag{9.29}$$

This is the recursive relation for calculation of the object Fourier phase. The phase difference between two nearby points of the object Fourier transform can be found from the measurements of the complex correlation between two points in the image Fourier transform. The phase of $\widehat{O}(\vec{u})$ at a particular point in the object Fourier plane is then given by the sum of the phase differences from the origin to that point.

The main requirement for the KT algorithm is that the diameter of the seeing disc, θ_s, should be less than half the field size, since it sets the scale of phase differences in the KT implementation. In addition, the number of frames required for a good S/N ratio increases as the fourth power of the seeing. Therefore, it is imperative to have good seeing. Another point to be noted is that correction for the photon noise bias is a crucial step in the implementation of this algorithm. The photon-counting detectors may be critical in permitting exact estimation and correction of these biases. The KT method is also found to be sensitive to odd order aberrations, e.g., coma but not defocusing, astigmatism etc., (Barakat and Nisenson, 1981).

9.1.7 *Triple-correlation technique*

Several algorithms have been developed to retrieve the diffraction limited phase of a degraded image, of which the triple-correlation (TC) technique, known as speckle masking method (Weigelt, 1977, Lohmann et al. 1983) is used by many observers. As stated earlier, a second-order moment (power spectrum) analysis provides only the modulus of the object Fourier transform, whereas a third-order moment (bispectrum) analysis yields the phase allowing the object to be fully reconstructed. A more recent attempt to go beyond the third order, i.e., fourth-order moment (trispectrum), provides a far more sensitive test than the bispectrum for some possible sources of non-Gaussianity, however, its implementation in optical imaging is a difficult computational task.

A triple-correlation is obtained by multiplying a shifted object, $I(\vec{x}+\vec{x}_1)$, with the original object, $I(\vec{x})$, followed by cross correlating the result with the original one (for example, in the case of a close binary star, the shift is equal to the angular separation between the stars, masking one of the two components of each double speckle), i.e.,

$$I^{TC}(\vec{x}_1, \vec{x}_2) = \int_{-\infty}^{\infty} I(\vec{x})I(\vec{x}+\vec{x}_1)I(\vec{x}+\vec{x}_2)d\vec{x}, \tag{9.30}$$

where $\vec{x}_j = \vec{x}_{jx} + \vec{x}_{jy}$ are the 2-D spatial coordinate vectors.

The nth-order correlation is given by,

$$I^{NC}(\vec{x}_1, \vec{x}_2, \cdots \vec{x}_N) = \int_{-\infty}^{\infty} I(\vec{x})I(\vec{x}+\vec{x}_1)I(\vec{x}+\vec{x}_2)\cdots I(\vec{x}+\vec{x}_N)d\vec{x}, \quad (9.31)$$

In the frequency domain, the equations (9.30 and 9.31) become,

$$\widehat{I}^{TC}(\vec{u}_1, \vec{u}_2) = \mathcal{F}\left[I^{TC}(\vec{x}_1, \vec{x}_2)\right]_{u_1, u_2} \quad (9.32)$$

$$\widehat{I}^{NC}(\vec{u}_1, \cdots \vec{u}_N) = \mathcal{F}\left[I^{NC}(\vec{x}_1, \vec{x}_2, \cdots \vec{x}_N)\right]_{u_1, \cdots u_N}, \quad (9.33)$$

where $\mathcal{F}$ is the Fourier transform.

The RHS of equation (9.33) represents an n-dimensional Fourier transformation, and the Fourier transform of triple-correlation. i.e., $\widehat{I}^{TC}(\vec{u}_1, \vec{u}_2)$, is known as the bispectrum. The above definitions may be extended to multidimensional signals. If a signal is a stochastic process and Gaussian, all odd-order correlation may turn out to be zero. However, for non-Gaussian signals, the triple-correlation is not always zero, it has many useful and practical properties of interest. In what follows, the triple-correlation method is elucidated.

Triple correlation (or speckle masking) method in astronomical application is a generalization of closure phase technique where the number of closure phases is small compared to those available from bispectrum. The advantages of such an algorithm are that it is insensitive to

- the atmospherically induced random phase errors,
- the random motion of the image centroid,
- the permanent phase errors introduced by telescope aberrations; any linear phase term in the object phase cancels out as well, and
- the images are not required to be shifted to common centroid prior to computing the bispectrum.

The other notable advantages are:

- it provides information about the object phases with better S/N ratio from a limited number of frames, and
- it serves as the means of image recovery with diluted coherent arrays.

The disadvantage of this technique is that it demands severe constraints on the computing facilities with 2-D data since the calculations are four dimensional (4-D). It requires extensive evaluation time and data storage requirements, if the correlations are performed by using digitized images on a computer. The ensemble averaged bispectrum is expressed as,

$$\widehat{I}^{TC}(\vec{u}_1,\vec{u}_2) = \left\langle \widehat{I}(\vec{u}_1)\widehat{I}^*(\vec{u}_1+\vec{u}_2)\widehat{I}(\vec{u}_2)\right\rangle, \tag{9.34}$$

where $\vec{u}_{j=1,2} = \vec{u}_{jx} + \vec{x}_{jy}$ represent the bispectrum, $\widehat{I}(\vec{u}_j)$ and $\widehat{I}^*(\vec{u}_1+\vec{u}_2)$ denote Fourier transforms of $I(\vec{x})$, i.e.,

$$\widehat{I}(\vec{u}_j) = \int_{-\infty}^{\infty} I(\vec{x})e^{-i2\pi\vec{u}_j\cdot\vec{x}}d\vec{x}, \qquad \text{and} \tag{9.35}$$

$$\widehat{I}^*(\vec{u}_1+\vec{u}_2) = \int_{-\infty}^{\infty} I(\vec{x})e^{-i2\pi(\vec{u}_1+\vec{u}_2)\cdot\vec{x}}d\vec{x}. \tag{9.36}$$

The argument of equation (9.34) is expressed as,

$$\begin{aligned} arg\left|\widehat{I}^{TC}(\vec{u}_1,\vec{u}_2)\right| &= \theta^{TC}(\vec{u}_1,\vec{u}_2) \\ &= \psi(\vec{u}_1) - \psi(\vec{u}_1+\vec{u}_2) + \psi(\vec{u}_2). \end{aligned} \tag{9.37}$$

The equation (9.37) gives the phase-difference. Invoking equation (5.98) into equation (9.34), the object bispectrum emerges as,

$$\widehat{I}^{TC}(\vec{u}_1,\vec{u}_2) = \widehat{O}(\vec{u}_1)\widehat{O}^*(\vec{u}_1+\vec{u}_2)\widehat{O}(\vec{u}_2)\left\langle \widehat{S}(\vec{u}_1)\widehat{S}^*(\vec{u}_1+\vec{u}_2)\widehat{S}(\vec{u}_2)\right\rangle. \tag{9.38}$$

The relationship in equation (9.38) implies that the image bispectrum is equal to the object bispectrum times a bispectrum transfer function, $< \widehat{S}(\vec{u}_1)\widehat{S}^*(\vec{u}_1+\vec{u}_2)\widehat{S}(\vec{u}_2) >$. The object bispectrum is given by,

$$\begin{aligned} \widehat{O}^{TC}(\vec{u}_1,\vec{u}_2) &= \widehat{O}(\vec{u}_1)\widehat{O}^*(\vec{u}_1+\vec{u}_2)\widehat{O}(\vec{u}_2) \\ &= \frac{\left\langle \widehat{I}(\vec{u}_1)\widehat{I}^*(\vec{u}_1+\vec{u}_2)\widehat{I}(\vec{u}_2)\right\rangle}{\left\langle \widehat{S}(\vec{u}_1)\widehat{S}^*(\vec{u}_1+\vec{u}_2)\widehat{S}(\vec{u}_2)\right\rangle}. \end{aligned} \tag{9.39}$$

The modulus $|\widehat{O}(\vec{u})|$ and phase $\psi(\vec{u})$ of the object Fourier transform $\widehat{O}(\vec{u})$ can be evaluated from the object bispectrum $\widehat{O}^{TC}(\vec{u}_1,\vec{u}_2)$. The object phase-spectrum, is encoded in the term,

$$e^{i\theta_O^{TC}(\vec{u}_1,\vec{u}_2)} = e^{i[\psi_O(\vec{u}_1) - \psi_O(\vec{u}_1+\vec{u}_2) + \psi_O(\vec{u}_2)]}, \tag{9.40}$$

of equation (9.34). The S/N ratio of the phase recovery contains S/N ratio for bispectrum, as well as a factor representing improvement due to redundancy of phase-information stored in the bispectrum.

The object phase-difference is corrupted by the random phase-differences due to the atmosphere-telescope OTF,

$$e^{i\theta_S^{TC}(\vec{u}_1,\vec{u}_2)} = e^{i[\psi_S(\vec{u}_1) - \psi_S(\vec{u}_1+\vec{u}_2) + \psi_S(\vec{u}_2)]}, \tag{9.41}$$

in a single image realization. If sufficient number of specklegrams are averaged, one can overcome this shortcoming. Let $\theta_O^{TC}(\vec{u}_1,\vec{u}_2)$ be the phase of the object bispectrum; then,

$$\widehat{O}(\vec{u}) = \left|\widehat{O}(\vec{u})\right| e^{i\psi(\vec{u})}, \qquad \text{and} \tag{9.42}$$

$$\widehat{O}^{TC}(\vec{u}_1,\vec{u}_2) = \left|\widehat{O}(\vec{u}_1,\vec{u}_2)\right| e^{i\theta_O^{TC}(\vec{u}_1,\vec{u}_2)}. \tag{9.43}$$

The equations (9.42) and (9.43) may be inserted into equation (9.39), yielding the relations,

$$\widehat{O}^{TC}(\vec{u}_1,\vec{u}_2) = \left|\widehat{O}(\vec{u}_1)\right| \left|\widehat{O}(\vec{u}_2)\right| \left|\widehat{O}(\vec{u}_1+\vec{u}_2)\right| \times e^{i[\psi_O(\vec{u}_1) - \psi_O(\vec{u}_1+\vec{u}_2) + \psi_O(\vec{u}_2)]} \rightarrow \tag{9.44}$$

$$\theta_O^{TC}(\vec{u}_1,\vec{u}_2) = \psi_O(\vec{u}_1) - \psi_O(\vec{u}_1+\vec{u}_2) + \psi_O(\vec{u}_2). \tag{9.45}$$

The equation (9.45) is a recursive relation for evaluating the phase of the object Fourier transform at coordinate $\vec{u} = \vec{u}_1 + \vec{u}_2$. The reconstruction of the object phase-spectrum from the phase of the bispectrum is recursive in nature. The object phase-spectrum at $(\vec{u}_1 + \vec{u}_2)$ is expressed as,

$$\psi_O(\vec{u}_1+\vec{u}_2) = \psi_O(\vec{u}) = \psi_O(\vec{u}_1) + \psi_O(\vec{u}_2) - \theta_O^{TC}(\vec{u}_1,\vec{u}_2). \tag{9.46}$$

If the object spectrum at $\vec{u}_1$ and $\vec{u}_2$ is known, the object phase-spectrum at $(\vec{u}_1 + \vec{u}_2)$ can be computed. The bispectrum phases are mod 2π, therefore, the recursive reconstruction in equation (9.45) may lead to π phase mismatches between the computed phase-spectrum values along different paths to the same point in frequency space. Northcott et al., (1988) opined that phases from different paths to the same cannot be averaged to reduce noise under this condition. A variation of the nature of computing argument of the term, $e^{i\psi_O(\vec{u}_1+\vec{u}_2)}$, is needed to obtain the object phase-spectrum and the equation (9.46) is translated to,

$$e^{i\psi_O(\vec{u}_1+\vec{u}_2)} = e^{i[\psi_O(\vec{u}_1) + \psi_O(\vec{u}_2) - \theta_O^{TC}(\vec{u}_1,\vec{u}_2)]}. \tag{9.47}$$

The values obtained using the unit amplitude phasor recursive reconstructor (equation 9.47) are insensitive to the π phase ambiguities (Ayers et al., 1988, Northcott et al., 1988)

9.1.7.1 *Deciphering phase from bispectrum*

The bispectrum of a 2-D intensity distribution is a 4-D function. Due to this extension into 4-D space it is possible that the phase information can survive. It is found experimentally (Lohmann et al. 1983) that the transfer function $< \widehat{S}(\vec{u}_1)\widehat{S}^*(\vec{u}_1 + \vec{u}_2)\widehat{S}(\vec{u}_2) >$, is real and greater than zero up to the telescope cut off frequency. Due to the reality of this transfer function, the phase of the complex bispectra of the object is identical to the phase of the average bispectrum of the object specklegrams,

$$\text{phase}\left|\left\langle \widehat{I}(\vec{u}_1, \vec{u}_2)\right\rangle\right| = \text{phase}\left|\widehat{O}(\vec{u}_1, \vec{u}_2)\right|. \tag{9.48}$$

The phase information of the object Fourier transform may be obtained directly from the average bispectrum, $< \widehat{I}(\vec{u}_1, \vec{u}_2) >$, without compensation of the transfer function. The measurement of the transfer function $< \widehat{S}(\vec{u}_1)\widehat{S}^*(\vec{u}_1 + \vec{u}_2)\widehat{S}(\vec{u}_2) >$, may be calculated by evaluating specklegrams of an astronomical point source. Thus the object bispectrum is obtained from the equation (9.39), and the object triple-correlation $O(\vec{x}_1, \vec{x}_2)$ is obtained by inverse Fourier transforming of $\widehat{O}(\vec{u}_1, \vec{u}_2)$.

Another way is to evaluate the uncorrelated specklegram of the object (Gaussian speckle masking method), where no astronomical point source is essential to be measured. The advantage of this method over previous one is that it can be performed in the correlation domain, as well as in Fourier domain. The method described by Lohmann et al. (1983) runs as follows. For a Gaussian model of the atmosphere, one obtains

$$\begin{aligned} I(\vec{x}_1, \vec{x}_2 = \vec{s}) - I_{nmm}(\vec{x}_1, \vec{x}_2 = \vec{s}) - I_{nmn}(\vec{x}, \vec{x}_1 = \vec{s}) \\ -I_{nnm}(\vec{x}_1, \vec{x}_2 = \vec{s}) + 2I_{nmk}(\vec{x}_1, \vec{x}_2 = \vec{s}) \\ = O(\vec{x}_1, \vec{x}_2 = \vec{s}) \longrightarrow O(\vec{x}), \end{aligned} \tag{9.49}$$

with

$$I_{nmk}(\vec{x}_1, \vec{x}_2) = \int_{-\infty}^{\infty} I(\vec{x})I_m(\vec{x} + \vec{x}_1)I_k(\vec{x} + \vec{x}_2),$$

in which $I(\vec{x})$ denote the intensity distribution of specklegrams and different I_{nmk} indices denote statistically independent specklegrams.

If identical specklegrams are correlated the notations turn out to be, $I_n(\vec{x}_1, \vec{x}_2) = I_{nnn}(\vec{x}_1, \vec{x}_2)$. Here the uncorrelated specklegrams are triple-correlated. The equation (9.49) can be used to calculate the triple correlation $O(\vec{x}_1, \vec{x}_2 = \vec{s})$ for any masking vector $\vec{s}$. If $\vec{s}$ is selected suitably, a true image of the object can be obtained. In the case of complicated objects it is useful to choose a set of many different masking vectors in order to improve the signal to noise ratio. The information about suitable masking vectors is obtained from the object autocorrelation.

The image can be reconstructed by implementing recursive image reconstruction method in frequency domain, which is derived by assuming that a sampled version $\widehat{O}_{p,q}$ of the object bispectra is available,

$$\widehat{O}_{p,q} = \widehat{O}_p.\widehat{O}_q.\widehat{O}_{-p-q} \quad p, q = -N.....+N, \tag{9.50}$$

with $\vec{u}_1 = p.\Delta\vec{u}$, $\vec{u}_2 = q.\Delta\vec{u}$, and $P = -N......+N$, where $\Delta\vec{u}$ is a suitable sampling distance in the Fourier domain.

In this method, the modulus and phase of the complex object bispectrum $\widehat{O}_p$ are calculated separately. Therefore $\widehat{O}_p$ is split into:

$$\widehat{O}_p = |\widehat{O}_p| e^{i\psi_p}. \tag{9.51}$$

In the recursive image reconstruction method the modulus of the object Fourier transform is obtained as in speckle interferometry. Speckle interferometry data are produced by setting $p = 0$ (or equivalently by $q = 0$ or $p = -q$) in $\widehat{O}_{p,q}$ i.e.,

$$\widehat{O}_{0.q} = \widehat{O}_0.\widehat{O}_q.\widehat{O}_{-q} = const. \left|\widehat{O}_q\right|^2, \tag{9.52}$$

where the fact that the spectrum of a real object is Hermitian i.e., $\widehat{O}_q = \widehat{O}^*_{-q}$ is used.

By substituting the equation (9.51) into equation (9.50), the phase of the complex spectrum of the object is reconstructed,

$$\left|\widehat{O}_{p,q}\right| e^{i\psi_{p+q}} = \left|\widehat{O}_p\right| e^{i\psi_p} \quad \left|\widehat{\mathrm{O}}_{\mathrm{q}}\right| e^{i\psi_q} \quad \left|\widehat{\mathrm{O}}_{-\mathrm{p}-\mathrm{q}}\right| e^{i\psi_{-p-q}}. \tag{9.53}$$

For $p + q = r$ and separate phase and modulus parts, the equation for phase would be,

$$e^{i\psi_r} = e^{i(\psi_{r-q} + \psi_q - \beta_{r-q,q})}, \tag{9.54}$$

where $\beta_{r-q,q}$ is the phase of

$$\widehat{O}_{r-q,q} = \left|\widehat{O}_{r-q,q}\right| e^{i\beta_{r-q,q}}. \tag{9.55}$$

From this equation (9.55) the phase factors $e^{i\psi_r}$ can be calculated recursively. It is sufficient to calculate $\widehat{O}_r$ for positive r since $\widehat{O}_r$ is Hermitian, and therefore $\psi_r = \psi_{-r}$. By setting $q = 1$ in the equation (9.54) yields,

$$e^{i\psi_r} = e^{i(\psi_1 + \psi_{r-1} - \beta_{r-1,1})}, \tag{9.56}$$

where $\psi_0 = \psi_1 = 0$, $r = 2.....N$. ψ_0 is equal to zero because absolute position in the reconstructed image is of no interest.

In order to explain this let the algorithm be initiated with $r = 0$. Both ψ_0 and $\beta_{0,1}$ are zero due to the reality of $O(\vec{x})$ and $O(\vec{x}_1, \vec{x}_2)$. The recursive procedure is as follows,

$$\psi_2 = 2\psi_1 - \beta_{1,1} \tag{9.57}$$

$$\begin{aligned}\psi_3 &= \psi_2 - \psi_1 - \beta_{2,1} \\ &= 3\psi_1 - \beta_{1,1} - \beta_{2,1}\end{aligned} \tag{9.58}$$

$$\cdots\cdots\cdots\cdots\cdots\cdots\cdots\cdots$$

$$\psi_r = r\psi_1 - \beta_{1,1} - \beta_{2,1} - .....\beta_{r-1,1}. \tag{9.59}$$

Apparently, the Fourier phase ψ_r and $\widehat{O}_r$ is derived from the phases of the bispectrum, except for the linear term $r\psi_1$. This unknown linear phase term corresponds to the unknown position of the object $O(\vec{x} - \vec{x}_0)$. For the reason explained above it is found that ψ_1 remains indeterminable.

The recursion given in the equation (9.56) uses the phase information contained in a single line ($q = 1$) of the object bispectrum, $\widehat{O}_{p,q}$. Additional phase informations is obtained by setting $q = 2...N$ in equation (9.54). Thus, each phase ψ_r has $(r-1)/2$ independent representations if r is odd and $r/2$ representations if r is even. These different representations of the ψ_r can be averaged. In order to find the recursion formula, the equation (9.54) is used which is insensitive to noise because of summation,

$$e^{i\psi_r} = const. \sum_{0<q\leq r/2} e^{i(\psi_q + \psi_{r-q} - \beta_{r-q,q})}, \tag{9.60}$$

where $\psi_0 = \psi_1 = 0$ and $r = 2...N$ and the index q in the summation has been selected such that all information contained in one octant of the object bispectrum is used to reconstruct the phases ψ_r.

The remaining octants of the bispectrum do not supply additional information because of the inherent symmetry of bispectra. Figure (9.1) illustrates the complex bispectrum $\widehat{O}_{p,q}$ of an object containing information about the modulus and the phase of the object spectrum $|\widehat{O}_r|e^{(i\psi_r)}$. By combining the phase factors with the modulus of the object Fourier transform (equation 9.52), the recursive image is reconstructed. The inverse

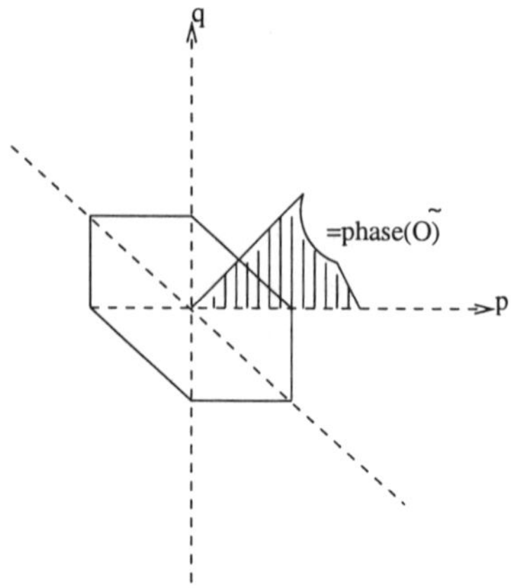

Fig. 9.1 Complex bispectrum $\widehat{O}_{p,q}$ of an object. The modulus information $|\widehat{O}_r|$ may be reconstructed from one of the axes $p = 0$, $q = 0$ or $p = q$. The phase information, $e^{(i\psi_r)}$, is contained in the area in between these distinguished axes. Owing to the eightfold symmetry of the bispectrum of a real function, one octant of the bispectrum contains non-redundant information, as indicated by the area filled by lines (Lohmann et al. 1983).

Fourier transform of the complex object spectrum yields the true image of the object.

Saha et al. (1999b) have developed a code based on the unit amplitude phasor recursive re-constructor. The algorithm written in Interactive Data Language (IDL) takes about an hour for processing 10 frames of size 128×128 using the SPARC ULTRA workstation. The memory needed for the calculation exceeds 160 MB if the array size is more than the said number. Since the bispectrum is a 4-D function, it is difficult to represent it in a 3-D co-ordinate system. Therefore, the calculated values are stored in 1-D array and used later to calculate the phase by keeping track of the component frequencies (Saha et al. 1999b). Assuming $\psi(0,0) = 0, \psi(0,\pm 1) = 0$ and $\psi(\pm 1, 0) = 0$, the phases are calculated by the unitary amplitude method. Though the memory required is independent of the dimensionality of array, the time required to access an element in a 1-D array is much smaller than that in a 4-D array. In order to reduce the high frequency noise, Wiener filter parameter (equation 6.64) has been implemented in reconstruction stage. The bispectrum method has been tested with a computer simulated image by using a code developed by Saha et al. (1999b). An example of the algorithm for 4×4 array is demonstrated in Appendix C. Figure (9.2) depicts (a) simulated binary system, (b) 2-D representation of

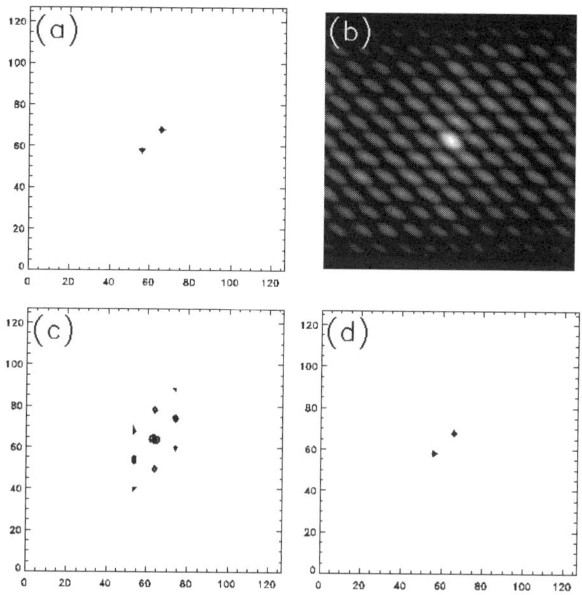

Fig. 9.2 Fourier phase of a simulated binary system: (a) 2-D maps of a simulated binary system, (b) 2-D representation of its 4-D bispectrum, (c) its triple-correlation and (d) its reconstructed image.

a 4-D bispectrum, (c) triple-correlation and (d) reconstructed image of the same binary system.

9.1.7.2 *Relationship between KT and TC*

The relationships between the Knox-Thomson (KT) and triple-correlation (TC) methods become apparent (Ayers et al. 1988) if the equation (9.21) is recast by substituting $\vec{u}_2 = \Delta\vec{u}$. A single plane of the bispectrum is expressed as,

$$\widehat{I}^{TC}(\vec{u}_1, \Delta\vec{u}) = \widehat{I}(\vec{u}_1)\widehat{I}^*(\vec{u}_1 + \Delta\vec{u})\widehat{I}(\Delta\vec{u}). \tag{9.61}$$

By inverse Fourier transforming, the corresponding image-space expression is derived, which gives rise to double correlation,

$$\begin{aligned} I^{TC}(\vec{x}_1, \Delta\vec{u}) &= \widehat{I}(\Delta\vec{u}) \int_{-\infty}^{\infty} I^*(\vec{x})I(\vec{x} + \vec{x}_1)e^{i2\pi\Delta\vec{u}\cdot\vec{x}}d\vec{x} \\ &= \widehat{I}(\Delta\vec{u})I^{KT}(\vec{x}_1, \Delta\vec{u}). \end{aligned} \tag{9.62}$$

The double correlation may be understood by considering the autocorrelation process (see section 6.3.1), which corresponds to the double correlation represented by equation (9.18) with the replacement of $\Delta\vec{u}$ by zero.

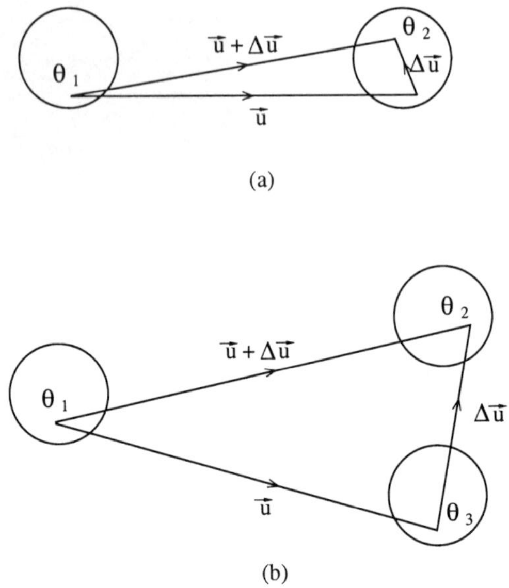

Fig. 9.3 Pupil sub-apertures of diameter r_0, (a) approximate phase-closure achieved in KT method, (b) complete phase-closure is achieved in TC method.

The resulting representations in the image and Fourier domains are respectively written as,

$$I^{\mathcal{A}_c}(\vec{x}_1) = I^{KT}(\vec{x}_1, 0) = \int_{-\infty}^{\infty} I^*(\vec{x})I(\vec{x} + \vec{x}_1)d\vec{x}, \tag{9.63}$$

$$\widehat{I}^{\mathcal{A}_c}(\vec{u}_1) = \widehat{I}^{KT}(\vec{u}_1, 0) = \widehat{I}(\vec{u}_1)\widehat{I}^*(\vec{u}_1). \tag{9.64}$$

In the KT method (equation 9.18), the correlation of $I(\vec{x})$ and $I(\vec{x})$ is multiplied by a complex exponential factor with a spatial frequency greater than zero. Such a method permits the retention of Fourier phase difference information (equation 9.19). Similarly, the triple-correlation method of the equation

$$\begin{aligned} I^{TC}(\vec{x}_1, \Delta\vec{u}) &= \widehat{I}(\Delta\vec{u}) \int_{-\infty}^{\infty} I^*(\vec{x})I(\vec{x} + \vec{x}_1)e^{i2\pi\Delta\vec{u}\cdot\vec{x}}d\vec{x} \\ &= \widehat{I}(\Delta\vec{u})I^{KT}(\vec{x}_1, \Delta u), \end{aligned} \tag{9.65}$$

allows the retention of phase difference information between two spatial frequencies separated by the vector, $\Delta\vec{u}$, if the phase at the frequency, $\vec{u}_2 = \Delta\vec{u}$, is known.

Comparing the two equations (9.19) and (9.65), it is observed that for a fixed spatial frequency difference, $\Delta\vec{u}$, the bispectrum plane corresponds to a weighted version, a signal depending factor, of the KT product in Fourier space. On comparing equations (9.22) and (9.37), the shift invariant property of the bispectrum can be expressed as,

$$\theta^{TC}(\vec{u}_1, \Delta\vec{u}) = \theta^{KT}(\vec{u}_1, \Delta\vec{u}) + \psi(\Delta\vec{u}). \tag{9.66}$$

As stated earlier in the preceding sub-section (9.1.6) that in the KT method, the transform $\widehat{I}(\vec{u})$ interferes with itself after translation by a small shift vector $\Delta\vec{u}$. The approximate phase-closure in this case is achieved by two vectors, $\vec{u}$ and $\vec{u} + \Delta\vec{u}$, assuming that the pupil phase is constant over $\Delta\vec{u}$. The major Fourier component of the fringe pattern is averaged with a component at a frequency displaced by $\Delta\vec{u}$. If this vector does not force the vector difference $-\vec{u} - \Delta\vec{u}$ to be outside the spatial frequency bandwidth of the fringe pattern, it preserves the Fourier phase-difference information in the averaged signal. The atmospheric phase effectively forms a closed loop.

$$\begin{aligned} arg\left|\widehat{I}(\vec{u})\widehat{I}(-\vec{u}-\Delta\vec{u})\right| &= \psi(\vec{u}) + \theta_1 - \theta_2 + \psi(-\vec{u}-\Delta\vec{u}) - \theta_1 + \theta_2 \\ &= \psi(\vec{u}) - \psi(\vec{u}+\Delta\vec{u}), \qquad \text{and,} \end{aligned} \tag{9.67}$$

$$\left\langle \widehat{I}(\vec{u})\widehat{I}(-\vec{u}-\Delta\vec{u}) \right\rangle \neq 0. \tag{9.68}$$

Let this system of two apertures be extended to three and the value of Δu made greater than r_0/λ, so that,

$$\begin{aligned} arg\left|\widehat{I}(\vec{u})\widehat{I}(-\vec{u}-\Delta\vec{u})\right| &= \psi(\vec{u}) + \theta_1 - \theta_2 - \psi(-\vec{u}-\Delta\vec{u}) - \theta_1 + \theta_3 \\ &= \psi(\vec{u}) - \psi(\vec{u}+\Delta\vec{u}) - \theta_2 + \theta_3, \qquad \text{and,} \end{aligned} \tag{9.69}$$

$$\left\langle \widehat{I}(\vec{u})\widehat{I}(-\vec{u}-\Delta\vec{u}) \right\rangle = 0. \tag{9.70}$$

The atmospheric phase contribution is not closed in this case. KT is limited to frequency differences $\Delta\vec{u} < r_0/\lambda$. Figure (9.3) depicts the diagrammatic representation of pupil sub-apertures of diameter r_0; (a) approximate phase-closure is achieved in KT method, and (b) complete phase-closure in TC method. In the bispectrum method, a third vector, $\Delta\vec{u} > r_0/\lambda$, is added to form phase-closure. When $\lambda\Delta\vec{u} > r_0$, the third vector is essential; the KT method fails with this arrangement. If the bispectrum average is performed, the phase is closed and Fourier phase-difference information is preserved.

$$\begin{aligned} arg\left|\widehat{I}(\vec{u})\widehat{I}(\Delta\vec{u})\widehat{I}(-\vec{u}-\Delta\vec{u})\right| &= \psi(\vec{u}) + \theta_1 - \theta_2 + \psi(\Delta\vec{u}) + \theta_2 - \theta_3 \\ &\quad + \psi(-\vec{u}-\Delta\vec{u}) - \theta_1 + \theta_3 \\ &= \psi(\vec{u}) - \psi(\vec{u}+\Delta\vec{u}) + \psi(\Delta\vec{u}). \end{aligned} \tag{9.71}$$

Thus, the bispectrum method can give phase information for phase-differences $\Delta u > r_0/\lambda$. TC method is similar to the phase-closure technique. Closure phases are insensitive to the atmospherically induced random phase errors, as well as to the permanent phase errors introduced by telescope aberrations. Any linear phase term in the object phase cancels out in the closure phase. In the case of a TC method, the information resides within isolated patches in Fourier space, while the KT method is not applicable under these conditions.

A comparative study was made by Weitzel et al., (1992) with the real data and concluded that KT and TC (bispectrum) give the same results for a binary system with a separation greater than the diffraction limit of the telescope. But they found some improvement for a binary system at a separation about the telescope diffraction limit in applying bispectrum method.

9.2 Iterative deconvolution techniques

Deconvolution method is an important topic, which applies to imaging in general, covering methods spanning from simple linear deconvolution algorithms to complex non-linear algorithms. It is an iterative non-linear process, in which 'a priori information' plays an essential role. Unlike a direct method, which attempts to solve a linear system of equations $Ax = b$ by finding the inverse of matrix A, an iterative method attempts to solve an equation or a system of equations by finding successive approximations to the solution starting from an initial guess. Such a technique is useful for problems involving a large number of variables. It can be simplified to the minimization/maximization of a criterion by using an iterative numerical method (Gerchberg and Saxton, 1972) that bounces back and forth between the image-domain and Fourier-domain constraints until two images are found that produce the input image when convolved together (Ayers and Dainty, 1988). The descriptions of various algorithms are briefly explained as follows.

9.2.1 *Fienup algorithm*

Fienup (1978) algorithm reconstructs an object using only the modulus of its Fourier transform (FT). Such an algorithm applies image plane constraints including positivity and finite support to adjust the phases while keeping the original amplitudes. At the kth iteration[1], $\mathcal{G}_k(\vec{x})$, an estimate of the object FT, is compared with the measured one and made to conform with the modulus at all Fourier frequencies. The inverse transform of the result yields an image $\mathcal{G}'_k(\vec{x})$. This iteration is completed by forming a new estimate of the object that conforms to certain object-domain constraints, e.g., positivity and finite extent, such that,

$$\widehat{\mathcal{G}}_k(\vec{u}) = |\widehat{\mathcal{G}}_k(\vec{u})|e^{i\phi_k(\vec{u})} = \mathcal{F}[\mathcal{G}_k(\vec{x})], \tag{9.72}$$

$$\widehat{\mathcal{G}}'_k(\vec{u}) = |\widehat{\mathcal{I}}(\vec{u})|e^{i\phi_k(\vec{u})}, \tag{9.73}$$

$$\mathcal{G}'_k(\vec{x}) = \mathcal{F}^{-1}[\widehat{\mathcal{G}}'_k(\vec{u})], \tag{9.74}$$

$$\begin{aligned} \mathcal{G}_{k+1}(\vec{x}) &= \mathcal{G}'_k(\vec{x}), \qquad \vec{x} \notin \gamma, \\ &= 0, \qquad\qquad \vec{x} \in \gamma, \end{aligned} \tag{9.75}$$

where the region γ is the set of all points at which $\mathcal{G}'_k(\vec{x})$ violates the object-domain constraints, and $\mathcal{G}_k(\vec{x})$, $\widehat{\mathcal{G}}'_k(\vec{u})$, and ϕ_k are estimates of $\mathcal{I}(\vec{x})$, $\widehat{\mathcal{I}}(\vec{u})$, and the phase ψ of $\widehat{\mathcal{I}}(\vec{u})$, respectively.

The above procedure may be accelerated if an estimate, $\mathcal{G}_{k+1}(\vec{x})$, is formed as,

$$\begin{aligned} \mathcal{G}_{k+1}(\vec{x}) &= \mathcal{G}_k(\vec{x}), \qquad\qquad \vec{x} \notin \gamma, \\ &= \mathcal{G}_k(\vec{x}) - \beta\mathcal{G}'_k(\vec{x}), \qquad \vec{x} \in \gamma, \end{aligned} \tag{9.76}$$

where β is a constant feedback parameter.

Fienup algorithm may be applied to clean up the images after the Fourier reconstruction such as Knox-Thomson method or triple-correlation method. This algorithm was originally suggested for obtaining phases if the amplitudes are known. For simple objects, the autocorrelation images provide the accurate amplitude estimate. Nisenson (1988) reported that starting with KT phase estimates and high S/N deconvolved amplitudes, Fienup algorithm converges in a few cycles, reducing background with little

[1] Iteration in computing is a repetition of process within an algorithm, which can be approximated using recursive techniques in functional programming language.

effect on the photometric accuracy. However, the success of the procedure is dependent on the S/N ratio in the amplitude spectrum.

9.2.2 *Blind iterative deconvolution (BID) technique*

Blind iterative deconvolution (BID) technique combines constrained iterative techniques (Gerchberg and Saxton, 1972, Fienup, 1978) with blind deconvolution (Lane and Bates, 1987). Essentially, it consists of using very limited information about the image, like positivity and image size, to iteratively arrive at a deconvolved image of the object, starting from a blind guess of either the object or both the convolving function. The iterative loop is repeated enforcing image-domain and Fourier-domain constraints until two images are found that produce the input image when convolved together (Bates and McDonnell, 1986, Ayers and Dainty, 1988). The image-domain constraints of non-negativity is generally used in iterative algorithms associated with optical processing to find effective supports of the object and or PSF from a specklegram.

The algorithm has the degraded image, $I(\vec{x})$, as the operand. An initial estimate of the PSF, $S(\vec{x})$, has to be provided. The degraded image is deconvolved from the guess PSF by Wiener filtering (equation 6.64), which is an operation of multiplying a suitable Wiener filter (constructed from the Fourier transform, $\widehat{S}(\vec{u})$, of the PSF) with the Fourier transform, $\widehat{I}(\vec{u})$, of the degraded image. This filtered deconvolution takes the form,

$$\widehat{O}(\vec{u}) = \widehat{I}(\vec{u})\frac{\widehat{W}(\vec{u})}{\widehat{S}(\vec{u})}, \tag{9.77}$$

in which $\widehat{W}(\vec{u})$, is the Wiener filter (see equation 6.64).

The noise term, $\widehat{N}(\vec{u})$ can be replaced with a constant estimated as the rms fluctuation of the high frequency region in the spectrum where the object power is negligible. The Wiener filtering spectrum, $\widehat{O}(\vec{u})$, takes the form:

$$\widehat{O}(\vec{u}) = \widehat{I}(\vec{u})\frac{\widehat{S}^*(\vec{u})}{\widehat{S}(\vec{u})\widehat{S}^*(\vec{u}) + \widehat{N}(\vec{u})\widehat{N}^*(\vec{u})}. \tag{9.78}$$

The result, $\widehat{O}(\vec{u})$, is transformed back to image space, the negatives in the image and the positives outside a prescribed domain (called object support) are set to zero. The average of negative intensities within the support are subtracted from all pixels. The process is repeated until the

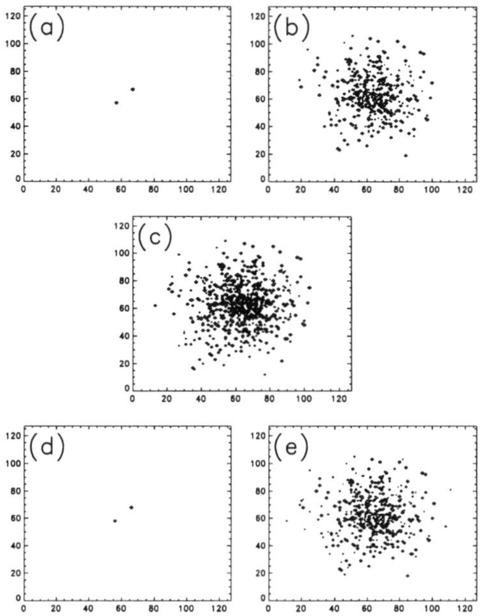

Fig. 9.4 2-D maps of the simulated image of (a) binary star, (b) simulated atmospheric PSF, (c) the convolved functions of these two (PSF), (d) the retrieved image of the binary star, (e) reconstructed atmospheric PSF.

negative intensities decrease below the noise. A new estimate of the PSF is next obtained by Wiener filtering the original image, $I(\vec{x})$, with a filter constructed from the constrained object, $O(\vec{x})$; this completes one iteration. This entire process is repeated until the derived values of $O(\vec{x})$ and $S(\vec{x})$ converge to reliable solutions.

Before applying this scheme of BID to the real data, Saha (1999b) had tested the algorithm with computer simulated convolved functions of binary stars and the PSF caused by the atmosphere and the telescope. The reconstruction of the Fourier phase of these are depicted in Figure (9.4). The starting guess for the PSF used for this calculation was a Gaussian with random noise. The output image, as well as the output PSF are obtained after 225 iterations.

This technique is indeed an ideal one to process the degraded images of extended objects, viz., (i) Planets (Saha et al., 1997), and (ii) Sun (Nisenson 1992). In order to reconstruct an extended object, a series of short-exposure specklegrams of the collision of comet Shoemaker-Levy (SL) 9 with Jupiter during the period of July 17-24, 1994 using the Nasmyth focus of 1.2 meter

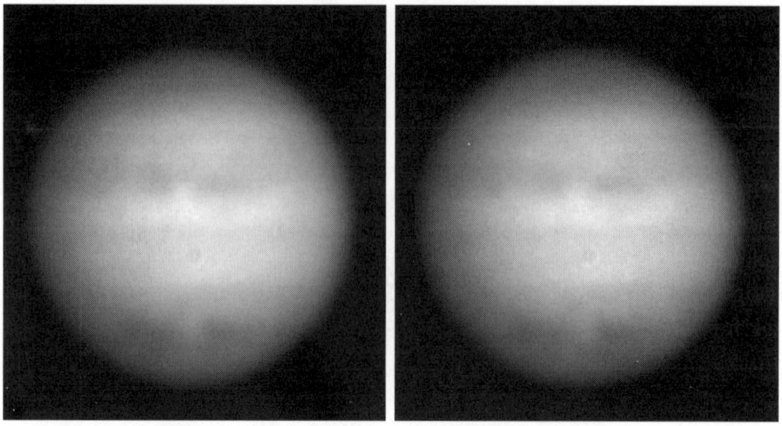

(a) (b)

Fig. 9.5 (a) Specklegram of Jupiter obtained on 24th July, 1994, and (b) its reconstructed image with BID.

telescope of Japal-Rangapur Observatory, Hyderabad, have been obtained by Saha et al. (1997). The BID technique was employed to remove the atmospherically induced point spread function (PSF) from these images to obtain diffraction limited informations of the impact sites on Jupiter. Figure (9.5a) depicts a specklegram of the Jupiter obtained on 24th. July 1994, through the green filter centered at 5500 Å, with FWHM of 300 Å. Figure (9.5b) shows the deconvolved image of the same.

The uniqueness and convergence properties of the deconvolution algorithm are uncertain for the evaluation of the reconstructed images if one used the BID method directly. The comparative analysis of the recovery reveals that both the morphology and the relative intensities are present in the retrieved diffraction-limited image and PSF. The results are vulnerable to the choice of various parameters like the support radius, the level of high frequency suppression during the Wiener filtering, etc. The availability of prior knowledge on the object through autocorrelation of the degraded image is very useful for specifying the object support radius (Saha and Venkatakrishnan, 1997, Saha, 1999b).

A major advantage of the BID compared with the rest of the methods described above is the ability to retrieve the diffraction-limited image of an object from a single specklegram without the reference star data. The notable disadvantage of the BID technique is that it requires high S/N values in the data. Nevertheless, it appears to be a powerful tool for image restoration. The restoration for speckle data at the 12th magnitude level assuming 15% photo-cathode sensitivity of the detector had been demonstrated (Nisenson et al., 1990). The technique can be improved by using multiple frames simultaneously as convergence constraints which may help in extending the magnitude level to fainter ($m_v > 12$).

9.2.3 *Richardson-Lucy algorithm*

Richardson-Lucy (RL) algorithm (Richardson, 1972, Lucy, 1974) converges to the maximum-likelihood (see Appendix B) solution for Poisson statistics in the data which is appropriate for optical data with noise from counting statistics. It forces the restored image to be non-negative and conserves flux both globally and locally at each iteration. The RL iteration can be derived from the imaging equation and the Poisson statistics. The discrete convolution of the object and the PSF is given by,

$$I_i = \sum_j S_{ij} O_j, \tag{9.79}$$

with $\sum_j S_{ij} = 1$ for all j, I is the blurred image, O is the unblurred object, and S_{ij} is the PSF, the fraction of light coming from true location j that gets scattered into observed pixel i.

From the Bayes' theorem (see Appendix B) that uses the probabilistic approach, one may write,

$$\mathcal{P}(I_i|O_j) = S_{ij}, \tag{9.80}$$

and the object distribution is expressed iteratively as,

$$O_j = O_j \sum_i \left(\frac{S_{ij} I_i}{\sum_k S_{jk} O_k} \right), \tag{9.81}$$

so that the Richardson-Lucy kernel approaches unity as the iterations progress. However, despite its advantages, the RL method suffers from:

- noise amplification, which is a generic problem with many types of algorithms that use maximum-likelihood approaches including error metric minimizing schemes, and
- difficult to stop iteration in regions where a smooth model fits the data adequately, while continuing to iterate in region where there are sharp features.

For small iterations, this algorithm produces spatial frequency components which are not sufficiently filtered by optical transfer function (OTF), i.e., the low spatial frequencies and for the spatial frequencies that are strongly filtered by the OTF require many iterations to reconstruct, i.e., the high spatial frequencies. In the presence of noise, the implication is that following many iterations the differences are small and are likely to be due to noise amplification.

9.2.4 *Maximum entropy method (MEM)*

Maximum entropy method (MEM; Ables, 1974) is a deconvolution algorithm which functions by minimizing a smoothness function (entropy) in an image. Such an algorithm is commonly employed in astronomical synthesis imaging and works better on extended sources. It is employed in a variety of other fields like medical imaging, and crystallography as well. This procedure governs the estimation of probability distributions when limited information is available. In addition, it treats all the polarization component images simultaneously and guarantees essential conditions on the image. It makes use of the highest spatial frequency information by finding the smoothest image consistent with the interferometric data. While enforcing positivity and conserving the total flux in the frame, smoothness is estimated by the 'entropy' S.

Though the concept of entropy comes from physics, it has a more fundamental meaning in information theory, such as Shannon's theorem. The entropy of a discrete message space is a measure of the amount of uncertainty in a probability distribution. An important property of entropy is that it is maximized when all the messages in the message space are equiprobable. The entropy function, S

$$S = -\sum_{i=1}^{N} \mathcal{P}_j \log \mathcal{P}_j, \tag{9.82}$$

is associated with a probability distribution $\mathcal{P}_1, \mathcal{P}_2, \cdots, \mathcal{P}_N$, each $\mathcal{P}_j \geq 0$ and $\sum_{j=1}^{N} \mathcal{P}_j = 1$.

The equation (9.82) relates the information-theoretic entropy. The principle of maximum entropy (Jaynes, 1982) governs the estimation of probability distribution from the available limited information. The best estimate is the function that fits the constraints and has maximum entropy.

In the MEM application, the resolution depends on the S/N ratio, which should be specified. The definition of entropy normalized to the flux in an image,

$$S_{im} = -\sum_i h_j \log \frac{h_j}{m_j}, \tag{9.83}$$

where $\vec{h} = [h_j]$ represents the image to be restored, and $\vec{m} = [m_j]$ is known as prior image.

It can be shown that $S \leq 0$; the equality holds if $\vec{h} = \vec{m}$. The value of S is a measure of the similarity between $\vec{h}$ and $\vec{m}$ if the entropy S is maximized without any data constraints. With data constraints, the maximum of S may become less than its absolute maximum value zero, meaning that $\vec{h}$ has been modified and deviated from its prior model $\vec{m}$.

MEM solves the multi-dimensional constraints minimization problem. It uses only those measured data and derives a brightness distribution which is the most random, i.e., has the maximum entropy S of any brightness distribution consistent with the measured data. Maintaining an adequate fit to the data, it reconstructs the final image that fits the data within the noise level.

9.2.5 *Pixon*

A non-linear iterative image reconstruction algorithm, called Pixon method has been developed by Puetter and Yahil (1999) that provides statistically unbiased photometry and robust rejection of spurious sources. Such an algorithm improves image and video quality in order to see the smallest of details, identify important information and extract clear images. It can be applied in various fields such as, astronomy, defence, medical imaging, microscopy, and security. The major advantages of the Pixon image processing are:

- multiframe analysis to improve the S/N ratio and resolution,
- multispectral analysis to optimize image processing by taking advantage of information obtained at different wavelengths,
- conservation of statistical flux with non-introduction of bias, and
- user control of the trade-off between noise suppression and resolution improvement.

Unlike other Bayesian methods (see Appendix B), Pixon technique does not assign explicit prior probabilities to image models. It minimizes complexity by smoothing the image model locally. The model is then described using a few irregular-sized pixons containing similar amounts of information, rather than many regular pixels containing variable signal-to-noise data. Eke (2001) opined that it has the ability to detect sources in low SNR data and to deconvolve a telescope beam in order to recover the internal structure of a source.

9.2.6 *Miscellaneous iterative algorithms*

(1) Magain, Courbin, Sohy (MCS; Magain et al. 1998) algorithm is based on the principle that sampled data cannot be fully deconvolved without violating the sampling theorem (Shannon, 1949) that determines the maximal sampling interval allowed so that an entire function can be reconstructed from sampled data. The sampled image should be deconvolved by a narrower function instead of the total PSF so that the resolution of the deconvolved image is compatible with the adopted sampling. The positivity constraint unlike the traditional deconvolution methods, is not mandatory; accurate astrometric and photometric informations of the astronomical objects can be obtained.

(2) The myopic iterative step preserving algorithm, (MISTRAL; Conan et al. 1998), is based on the Bayesian theorem. It incorporates a positivity constraint and some a priori knowledge of the object (an estimate of its local mean and a model for the power spectral density etc.). It also allows one to account for the noise in the image, the imprecise knowledge of the PSF. MISTRAL has produced a few spectacular images after processing AO images.

9.3 Phase retrieval

Measurement of physical properties based on the retrieval of phase information encoded in an interference pattern is required in the fields of engineering including geodesic and defence applications like target recognition, remote sensing techniques that have a range of ecological and geophysical applications. The interferometric and profilometric[2] techniques require such methods as well for determining mechanical properties such as, strain or deformation of materials. The problem of retrieving underlying phase distribution from a given interferogram is common to the interferometric

[2]A profilometer is a non-contact optical profiling instrument designed to measure the absolute figure to nanometer accuracy of long strip flat, spherical and aspherical surfaces.

methods. The phase retrieval techniques are also essential in the field of medicine, particularly for the magnetic resonance imaging, which has great medical importance for mapping internal structures of the human body.

Phase retrieval is concerned with extracting the phase of a wavefront using intensity measurements and additional constraints. This is traditionally done using interferometry. This technique has much in common with deconvolution technique. It is useful in the measurement of optical surfaces systems and has been applied to estimate the aberrations of the Hubble space telescope (HST; Fienup et al. 1993). Among others, knowledge of precise aberrations allows for (i) designing correction optics to fix the HST, (ii) optimizing alignment of secondary mirror of optical telescope assembly (OTA), (iii) monitoring telescope shrinkage and focus, and (iv) computing analytic PSFs for image deconvolution.

It is to reiterate that the point spread function (PSF) for an optical system is determined by the amplitude and phase of the spherical wavefront at the focus, which is given by,

$$\begin{aligned} S(\vec{x}) &= \left|\int_{-\infty}^{\infty} \widehat{A}(\vec{u})e^{i2\pi\vec{u}\cdot\vec{x} + \widehat{\psi}_{ab}(\vec{u})}d\vec{u}\right|^2 \\ &= \left|\int_{-\infty}^{\infty} \widehat{S}(\vec{u})e^{i2\pi\vec{u}\cdot\vec{x}}d\vec{u}\right|^2, \end{aligned} \tag{9.84}$$

where

$$\widehat{S}(\vec{u}) = \widehat{A}(\vec{u})e^{i2\pi\widehat{\psi}_{ab}(\vec{u})}, \tag{9.85}$$

the complex pupil function.

As discussed in section (3.6.3), the wavefronts in pupil plane and focal plane are related by a Fourier transform. The equation (9.84) assumes that the wavefront is not too curved over the pupil; if the curvature is larger, one uses a Fresnel transform rather than a Fourier transform. Phase retrieval problem consists of determining the phase error, $\psi_{ab}(\vec{x})$, from noise intensity measurements, $\widehat{I}(\vec{u}) = |\widehat{S}(\vec{u})|^2$, and prior information on the statistics of the phase and the noise.

The standard algorithm used in phase retrieval are the iterative transform algorithms introduced by Gerchberg and Saxton (1972) and developed by Fienup (1982). Such an algorithm uses the measured intensity, $\widehat{I}(\vec{u}) = |\widehat{S}(\vec{u})|^2$, and knowledge of the size of the aperture to constrain the solution. Enforcing magnitude constraints in both domains is the error reduction algorithm. The algorithm minimizes the mean squared error, E:

$$E = \sum_{u} \widehat{W}(\vec{u})[|\widehat{I}_j(\vec{u})| - |\widehat{S}_j(\vec{u})|]^2, \tag{9.86}$$

in which $|\widehat{I}_j(\vec{u})|$ is the squared root of measured intensity in the jth plane, $|\widehat{S}_j(\vec{u})|$ the model of the aberrated wavefront, $\widehat{\psi}_{ab}(\vec{u})$, using a low order Zernike polynomials in $\vec{u}$, and $|\widehat{W}(\vec{u})|$ the weighting function (from the real data), which eliminates the contributions of bad pixels.

The equation relating the phase and the observed PSF is non-linear in the phase-retrieval problem making this technique considerably more difficult than image restoration. The main problem arises from the maximum likelihood approaches in which the phase tends to get stuck at local maxima of the likelihood rather than finding the best solution globally.

9.3.1 *Phase-unwrapping*

Phase unwrapping is a mathematical problem-solving technique increasingly used in optical interferometry, adaptive optics, and medical imaging. This method is the reconstruction of a function on a grid given its values modulo 2π of the function on the grid. As stated earlier in chapter 3 that all intensity maxima and minima represent points in which the local phase is separated from other minima or maxima as $2\pi\lambda$. The unwrapped phase is obtained in two dimensions by summing up the phase gradient along paths. A closed contour integral of the phase gradient surrounding a single residue does not vanish, therefore, the wrapped phase gradient becomes path dependent, thus an ambiguity arises. One way to classify unwrapping algorithms is to remove path ambiguity. However, in the presence of phase residues, reconstructing a complete continuous phase-surface is improbable, albeit one tries to achieve a phase surface as continuous as possible. In atmospheric optical propagation phase residues, as stated earlier in chapter 5, arise by interference of irregular scattered portions of the optical wave due to refractive index inhomogeneities. In adaptive optics system, the residual discontinuities in the unwrapped phase introduce an unavoidable correction error in deformable mirror reconstruction systems; minimum discontinuity unwrapped phases do not yield minimum correction error (Fried and Vaughn, 1992).

A 2-D Fourier transform produces an array of complex numbers with a central peak and two side lobes. A window is required to be set to isolate one of the side lobes. Hence, the reduced data set is Fourier transformed back yielding a map of the amplitude and phase of the fringes (Roddier and Roddier, 1988). In order to enhance the computational efficiency, fast Fourier transform[3] (FFT) is used.

It is to be noted that high-frequency components of small phase amplitude may be measured as long as the slope does not exceed the spatial frequency, f. However, the limitation on the highest allowable slope arises from Nyquist limit[4]. The width of the detector elements may also contribute to a reduction of fringe contrast in the case of high density fringe.

Recall the equation (6.74), where it is observed that the phase information is contained either in $\widehat{C}(f - f_0, y)$ or equivalently in $\widehat{C}^*(f - f_0, y)$. By knowing the spatial carrier frequency of f_0 *apriori*, one may choose either of the spectra say, $\widehat{C}(f - f_0, y)$ and translate it by $\pm f_0$ on the frequency axis toward the origin to get $\widehat{C}(f, y)$. The unwanted low frequency modulation and the high frequency noise has been filtered out at this stage. Setting zero, the other frequencies and shifting the result by an amount f_0 to eliminate the carrier frequency. Again using inverse Fourier transform of $\widehat{C}(f, y)$ with respect to f one obtains $C(\vec{r})$. Thus, the phase distribution $\psi(\vec{r})$ is given by,

$$\psi(\vec{r}) + 2\pi\vec{f_0}\cdot\vec{r} = \tan^{-1}\left[\frac{\Im\left[C(\vec{r})\right]}{\Re\left[C(\vec{r})\right]}\right] \tag{9.87}$$

$$= \tan^{-1}\left[\Im\left\{C(\vec{r})\right\}, \Re\left\{C(\vec{r})\right\}\right] \tag{9.88}$$

$$= \tan^{-1}\left[C(\vec{r})\right], \tag{9.89}$$

where $\Im$ denotes imaginary part.

The phase obtained ranges from -π to $+\pi$ and the ambiguity is corrected appropriately by adopting suitable phase unwrapping technique. Since the arc tangent is a periodic function, $\psi(\vec{r})$ is determined to within a multiple of π. The common aspect of interferogram analysis leads to the necessity of phase unwrapping to remove the ambiguity caused by this loss of information and to recreate a continuous wave front. The phase unwrapping process is described by the following equation,

$$\psi(\vec{r}) = \psi'(\vec{r}) + m(\vec{r}), \tag{9.90}$$

[3]Fast Fourier transform (FFT) algorithm optimizes the computation of the discrete Fourier transform (DFT; Appendix B). Mathematically, it is identical to the DFT. It reduces the number of computations needed for N points from $2N^2$ to $2N\log N$.

[4]Nyquist limit, also known as Nyquist frequency, is the highest frequency of a signal that can be coded at a given sampling rate in order to reconstruct the signal. A sampled waveform needs at least two sampled points per cycle.

in which $m(\vec{r}) = 2\pi n(\vec{r})$ and $n(\vec{r})$ is a solution for the integer number field.

The discontinuities in $\psi'(\vec{r})$ are limited to a finite number of points, baring which $\psi(\vec{r})$ and $\psi'(\vec{r})$ are related by a constant offset, and have same derivatives; discontinuities are present wherever the magnitude of the derivative exceeds a given threshold. If such a magnitude exceeds 0.5 (waves), a discontinuity is considered to be present. For 1-D, the discrete derivative of $\psi'(x)$ is given by,

$$\frac{\psi'(x)}{dx} = \psi'(x+1) - \psi'(x), \tag{9.91}$$

with $x = 0, 1, 2, \cdots$.

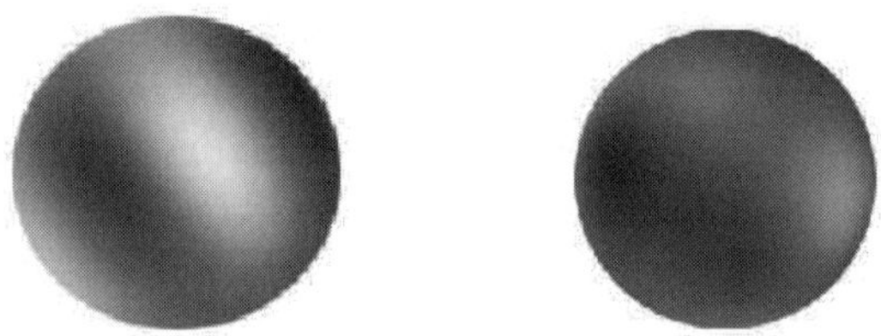

Fig. 9.6 Phase map derived from Shack-Hartmann sensor (left panel) and from polarization shear interferometer (right panel; Courtesy: J. P. Lancelot).

In order to implement two dimensional unwrapping, one-dimensional unwrapping may be extended to rows and columns of data. Figure (9.6) depicts the retrieved phase of shear fringes, which is obtained using the algorithm developed by Lancelot (2006) similar to the analysis described by Bone et al (1986). The algorithm detects discontinuities of magnitude 2π in every chosen path in the image of the wrapped phase. The algorithm then adds or subtracts C values of 2π to each point of discontinuity.

9.3.2 *Phase-diversity*

As stated earlier, an image of stellar source is the combination of two unknown quantities such as the object and and the aberrations caused by the instruments and atmospheric turbulence. It is improbable to separate the unknown quantities with single measurement. However, with two measurements, they can be quantified. The phase-diversity method is a post-collection technique that uses a number of intensity distributions encoded by known aberrations for restoring high spatial resolution detail while imaging in the presence of atmospheric turbulence. The phase is retrieved from the analysis of two simultaneous images of a stellar source, one in-focus and the other one out-of-focus (defocused); the defocused image should generally be with some known aberration (Gonsalves, 1982, Paxman et al. 1992). Given this information, an exact solution for both the unknown aberrations, as well as the unknown object may be obtained provided the data are free of noise, which is unlikely. Therefore, this method contains a noise model and tries to give an optimum estimate of the object with respect to this noise.

Let $I_1(\vec{x})$ and $I_2(\vec{x})$ be the measured intensity at the focal plane and the diversity image at the known defocus point as displayed in Figure (9.7) respectively, given by,

$$I_1(\vec{x}) = O(\vec{x}) \star S_1(\vec{x}); \qquad I_2(\vec{x}) = O(\vec{x}) \star S_2(\vec{x}), \tag{9.92}$$

in which $O(\vec{x})$ is the object irradiance distribution, $S_1(\vec{x})$ and $S_2(\vec{x})$ the respective point spread functions at the focal point and at the defocus point.

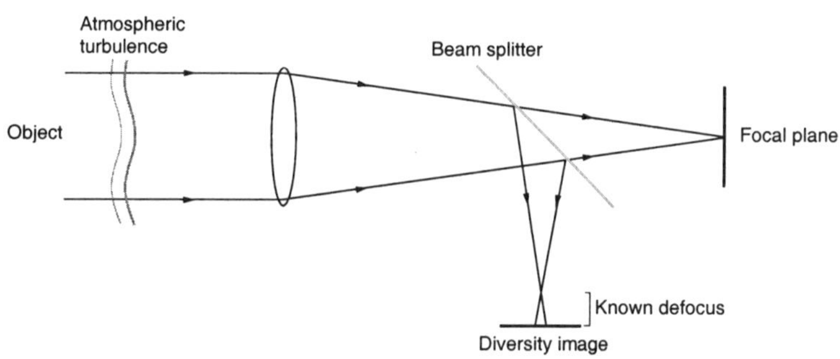

Fig. 9.7 Phase diversity method.

In this operation, the PSFs, $S_1(\vec{x})$ and $S_2(\vec{x})$ are related by a known diversity, hence only two unknown quantities such as the object intensity, $O(\vec{x})$ and the PSF $S_1(\vec{x})$ are to be determined. The impulse response at the focal plane is,

$$\widehat{S}_1(\vec{u}) = \left|\widehat{S}_1(\vec{u})\right| e^{i\psi(\vec{u})}, \tag{9.93}$$

while impulse response of the diversed image is,

$$\widehat{S}_2(\vec{u}) = \left|\widehat{S}_2(\vec{u})\right| e^{i\psi(\vec{u}) + \widehat{\psi}_{ab}(\vec{u})}, \tag{9.94}$$

where $\widehat{\psi}_{ab}(\vec{u})$ is the introduced phase aberration.

Phase-diversity restores both the target object of interest, as well as the complex wavefront phases , which represents the unknowns for the PSFs, at the pupil. The phases can be represented as either zonal (pixel by pixel) or modal (for example Zernike mode) with fewer unknowns. Modelling scintillation may improve the phase-diversity performance for both the object and phase recovery. The object spectrum may be written in terms of wavefront phases,

$$\widehat{O}(\vec{u}) = \frac{\left[\widehat{I}_1(\vec{u})\widehat{S}_1^*(\vec{u}) + \widehat{I}_2(\vec{u})\widehat{S}_2^*(\vec{u})\right]}{\left[\left|\widehat{I}_1(\vec{u})\right|^2 + \left|\widehat{I}_2(\vec{u})\right|^2\right]}. \tag{9.95}$$

Phase-diversity speckle method is an extension of this technique, whereby a time sequence of short-exposure image pairs is detected at different positions in focus and out of focus near the focal-plane. Incident energy is split into two channels by a simple beam splitter: one is collected at a conventional focal-plane, the other is defocused (by a known amount) and a second detector array permits the instantaneous collection of the latter. Reconstructions from phase-diverse speckle data involving multiple realisation of turbulence are found to be superior to that derived from traditional phase-diversity data, which involve a single realisation of turbulence. It is pertinent to note that for any single realisation of atmospheric wavefront, the optical transfer function (OTF) would have zeros in the Fourier plane at some spatial frequencies. Information about the object at these frequencies is lost, yielding a wrong identification of the companion star.

Phase-diversity method is less susceptible to the systematic errors caused by the optical hardware, and is found to be more appealing in astronomy (Seldin and Paxman, 1994). Another advantage of such a system is that point spread function measurement is not required. A few attempts employing this technique have been made at the Swedish Vacuum Solar Telescope, located at La Palma. Much of the solar photosphere is a low contrast object, with 10-15% RMS contrast granulation pattern. In such a situation, Poisson distributed noise can be approximated with additive white (Gaussian noise). This makes simplification in the phase-diversity formulation.

Chapter 10

Astronomy fundamentals

10.1 Black body radiation

Radiation emitted by a body as a result of its temperature is called thermal radiation. All bodies emit thermal radiation to their surroundings and absorb such radiations from them where the rates depend upon the temperature difference. The nature of the spectrum of thermal radiation is continuous for matter in a condensed state (i.e. solid and liquid) and is strongly dependent on temperature. Generally speaking, the detailed nature of the radiation spectrum depends somewhat on the composition of the body. However, experiments show that there is one class of hot bodies, called black bodies, that emits thermal spectra of universal character.

Though several laws governing the emission of radiation from a black body that absorbs all the thermal radiation incident upon them (in the ideal case) and re-emits the radiation completely, were available, yet, in spite of various attempts, no connection between them could be discovered until Max Planck (1858-1947) in 1901 succeeded in writing down a general expression for the energy spectrum of radiation emitted by a black body. He had suggested that radiative transfers would occur through packets of energy named quanta. The assumption is that the electromagnetic radiation (energy) is emitted or absorbed in a discontinuous manner rather than continuous manner. He introduced in his treatment of the cavity radiation, which laid the basis for the development of modern quantum physics. He had formulated the principle function of radiative transfer, which now bears his name, the Planck's function. The advancement of the theory of light is accelerated soon after the initiation of quantum theory by him. He showed that earlier laws were either special cases of the general expression or could be derived from it. In what follows, the properties of thermal radiation and other related phenomena are elucidated in brief.

10.1.1 *Cavity radiation*

At the beginning of the last century, Rayleigh and Jeans made a calculation of the energy density of cavity (or black body) radiation using classical arguments. Their result was not successful to justify the experimental results pointing out serious conflict between classical physics and experimental observations. Rayleigh and Jeans considered a cavity with metallic walls heated uniformly to temperature T. The walls emit electromagnetic radiation in the thermal range of frequencies. This happens due to the accelerated motions of the electrons in the metallic walls arising from the thermal agitation. In this treatment, the oscillations of the electrons were not considered, instead, attention was focused on the behavior of the standing electromagnetic waves (see section 2.2.2) inside the cavity.

The approach to find out a suitable distribution function capable of successfully describing the above properties starts from an alternative but equivalent consideration of a blackbody as an object containing a cavity which is connected to the outside by a small hole. Radiation incident upon the hole from the outside enters the cavity and gets absorbed at the walls of the cavity while getting reflected back and forth by the walls. If the area of the hole is very small compared to the area of the inner surface of the cavity, the amount of incident radiation reflected back through the hole is negligible and the hole appears to have the properties of the surface of a blackbody.

If the walls of the cavity are at a uniform temperature T, the thermal radiation emitted by the walls fill the cavity and a small amount of the radiation is sampled out by the hole. Since the hole must have the properties of the surface of a blackbody, the cavity radiation must also have a blackbody spectrum characteristic of the temperature T on the walls. The spectrum of the cavity radiation is specified in terms of the energy density, $E_\nu(T)$, which is defined as the energy contained in a unit volume of the cavity at temperature T in the frequency interval ν to $\nu+d\nu$. This quantity is related to the spectral radiancy[1] of the spectrum emitted by the hole on the cavity by the following expression:

$$E_\nu(T) \propto B_\nu(T), \tag{10.1}$$

[1]Spectral radiancy is the rate at which energy is radiated per unit surface area for a small wavelength interval. This is measured in energy flux density units of watts per unit wavelength.

in which T is the temperature in Kelvin (K), $B_\nu(T)$ the spectral distribution of blackbody radiation, known as spectral radiancy, and is defined in such a way that $B_\nu(T)d\nu$ equals the energy emitted per unit time in radiation of frequency in the interval of ν to $\nu + d\nu$ from a unit area of surface at absolute temperature, T.

Thus, the distribution function for describing the blackbody spectrum can be obtained by constructing a theory of cavity radiation. The properties of blackbody radiation can be taken into account in order to derive the properties of light emitted by a thermal source. The electromagnetic theory shows that these waves are standing in nature with nodes at the metallic surfaces. By using geometrical arguments, the number of standing waves in the frequency interval ν to $\nu + d\nu$ is counted in order to see how this number depends on ν. Let us consider a cavity with physical dimension, L, which supports standing waves of electromagnetic field. A variety of nodes with the wave vector, $\vec{\kappa} = \pi\nu\vec{r}/L$, in which $\vec{\kappa}(=\kappa_x,\ \kappa_y,\ \kappa_z)$ and $\vec{r}(=x, y, z)$ the position vector, may be supported by this cavity. The dependence was found out to be of the form,

$$N(\nu)d\nu = \frac{8\pi V}{c^3}\nu^2 d\nu, \tag{10.2}$$

with $N(\nu)$ as the number of nodes within the frequency range ν to $\nu + d\nu$ and V the volume of the cavity. .

The average total energy, $\bar{\mathcal{E}}$, of these waves in thermal equilibrium at a temperature T can be found out using classical kinetic theory and equipartition law[2], i.e., assigning an average energy,

$$\bar{\mathcal{E}} = k_B T, \tag{10.3}$$

in which $K_B(= 1.380662 \times 10^{-23}\mathrm{JK}^{-1})$ is the Boltzmann's constant, to each mode of vibration.

Finally, the number of standing waves in the frequency interval times the average energy of the modes divided by the volume of the cavity gives the average energy per unit volume in the frequency interval ν to $\nu + d\nu$, i.e., the energy density $E_\nu(T)$. The final form, as derived by Rayleigh and Jeans turns out to be,

$$E_\nu(T)d\nu = \frac{8\pi\nu^2 k_B T}{c^3}d\nu. \tag{10.4}$$

[2] In thermodynamics, the equipartition theorem states that the mean internal energy associated with each degree of freedom of a mono atomic ideal gas is the same.

Equation (10.4) is called the Rayleigh-Jeans formula for blackbody radiation. Comparing this distribution function with the experimental results, the discrepancy becomes evident. In the limit of low frequencies the classical spectrum approaches the experimental results but as the frequency increases the theoretical energy density approaches infinity. However, the experimental energy density remains finite always and goes to zero at very high frequencies. This deviation of the theoretical prediction from the real behaviour is known as the ultraviolet catastrophe.

10.1.2 *Planck's law*

In his attempt to resolve the discrepancy between theory and experiment Planck (1901) considered the possibility of a violation of the law of equipartition of energy. The behavior of the average energy, $\bar{\mathcal{E}}$, with the variation of ν suggested that $\bar{\mathcal{E}}$ should depend on frequency, ν, rather than being independent of ν as suggested by the equipartition law. The average total energy approaches k_BT as the frequency approaches zero, i.e., $\bar{\mathcal{E}} \rightarrow k_BT$ when $\nu \rightarrow 0$, while it approaches zero as the frequency approaches infinity, i.e., $\bar{\mathcal{E}} \rightarrow 0$ when $\nu \rightarrow \infty$).

The equipartition law arises from a comprehensive result of classical kinetic theory, known as Boltzmann probability distribution. The special form of Boltzmann distribution relevant in this case is given by,

$$\mathcal{P}(\mathcal{E}) = \frac{e^{-\mathcal{E}/k_BT}}{k_BT}, \tag{10.5}$$

in which $\mathcal{P}(\mathcal{E})$ is the Boltzmann probability distribution and k_B represents the Boltzmann constant.

The system consists of a large number of entities of the same kind in thermal equilibrium at temperature T. In a case, where the system is comprised of a large number of simple harmonic oscillating standing waves in thermal equilibrium in a blackbody cavity, according to the classical statistics the average energy is evaluated to be:

$$\bar{\mathcal{E}} = \frac{\displaystyle\int_0^\infty \mathcal{E}\mathcal{P}(\mathcal{E})d\mathcal{E}}{\displaystyle\int_0^\infty \mathcal{P}(\mathcal{E})d\mathcal{E}}, \tag{10.6}$$

in which $\mathcal{P}(\mathcal{E})d\mathcal{E}$ is the probability of finding a given entity of a system with energy in the interval between $\mathcal{E}$ and $\mathcal{E} + d\mathcal{E}$, when the density of states is independent of $\mathcal{E}$.

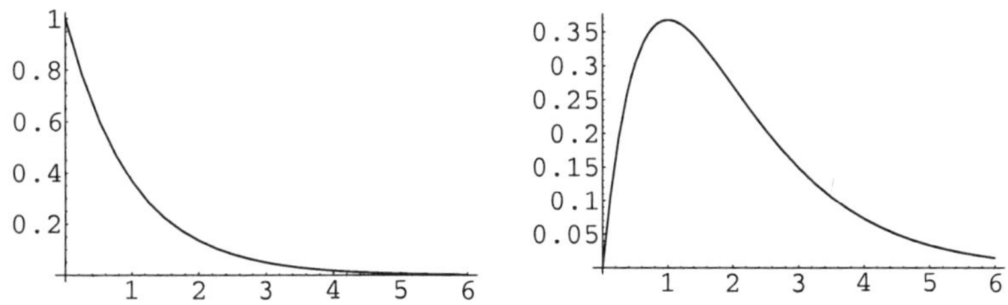

Fig. 10.1 Boltzmann probability distribution: the average value of the energy, E for the distribution is $\bar{\mathcal{E}} = k_B T$ (left panel), a plot of $\mathcal{E}\mathcal{P}(\mathcal{E})$ (right panel).

The integral, when evaluated using the Boltzmann distribution gives the law of equipartition (see Figure 10.1):

$$\bar{\mathcal{E}} = k_B T. \tag{10.7}$$

Planck's greatest contribution was to consider the energy $\mathcal{E}$ as a discrete variable instead of a continuous variable, which definitely is from the point of view of classical physics. In the calculation leading from $\mathcal{P}(\mathcal{E})$ to $\bar{\mathcal{E}}$, the integration would be replaced by a sum. He assumed that the energy, $\mathcal{E}$ could take on only certain discrete values and the values are uniformly distributed i.e., $\mathcal{E} = 0, \Delta\mathcal{E}, 2\Delta\mathcal{E}, 3\Delta\mathcal{E}\cdots$, in which $\Delta\mathcal{E}$ is the uniform interval between successive allowed values of the energy. For, $\Delta\mathcal{E} \ll k_B T$, the discreteness in energy is insignificant and hence the average energy $\bar{\mathcal{E}}$ should be given by the classical result $\bar{\mathcal{E}} \simeq k_B T$. Again this result is true in the case of blackbody spectrum for small frequency, ν. For large values of ν, however, $\bar{\mathcal{E}}$ is not given by the equipartition law and the discreteness must be significant in this domain, i.e., $\Delta\mathcal{E}$ is an increasing function of ν. Planck assumed these quantities to be proportional,

$$\begin{aligned} \Delta\mathcal{E} &\propto \nu, \qquad \text{or,} \\ \Delta\mathcal{E} &= h\nu, \end{aligned} \tag{10.8}$$

where $h(= 6.626196 \times 10^{-34}\text{joules (J)})$ is the proportionality constant called Planck's constant.

Like the speed of light, c, that sets an upper limit on any physical velocity, the Planck's constant, h, turned out to be a universal constant which sets a lower limit on action[3]. This constant had opened up new

[3]Action is defined as linear momentum multiplied by distance.

vistas of physics. It is pertinent to note that according to the Maxwell-Boltzmann distribution[4], the modes are thermally excited,

$$\mathcal{P}(\mathcal{E}_n) = \frac{e^{-(\mathcal{E}_n/k_BT)}}{\sum e^{-(\mathcal{E}_n/k_BT)}}, \tag{10.9}$$

in which $\mathcal{E}_n = (n+1/2)h\nu$ is the condition for the energy of a mode, hence a probability distribution for n photons in a single mode is given by,

$$\mathcal{P}_n(\nu) = \frac{1 - e^{-(h\nu/k_BT)}}{\sum e^{-(h\nu/k_BT)}}. \tag{10.10}$$

Thus, the average number $n(\nu, T)$ of photons per unit mode or frequency interval in thermal equilibrium can be expressed as,

$$n(\nu) = \sum_{n=0}^{\infty} n\mathcal{P}_n(\nu) = \frac{1}{e^{(h\nu/k_BT)} - 1}. \tag{10.11}$$

Equation (10.11) is the black body formula explaining the presence of the black body spectrum of a solid at temperature, T. Planck (1901) obtained the expression for the energy, $\bar{\mathcal{E}}$ as,

$$\bar{\mathcal{E}}(\nu) = \frac{h\nu}{e^{(h\nu/k_BT)} - 1}. \tag{10.12}$$

Since $e^{h\nu/k_BT} \to 1 + h\nu/k_BT$ for $h\nu/k_BT \to 0$, we find that in this limit, $\bar{\mathcal{E}}(\nu) \to k_BT$, while in the limit, $h\nu/k_BT \to \infty$, $e^{h\nu/k_BT} \to \infty$, and $\bar{\mathcal{E}} \to 0$.

Finally, the spectral energy density in the blackbody spectrum is expressed as,

$$E_\nu(T)d\nu = \frac{4\pi}{c}B_\nu(T)d\nu = \frac{8\pi\nu^2}{c^3}\frac{h\nu}{e^{(h\nu/k_BT)} - 1}d\nu, \tag{10.13}$$

which has units of energy per unit volume per unit frequency ($\mathrm{J\,m^{-3}\,Hz^{-1}}$).

Figure (10.2) displays the intensity distributions of blackbodies at various temperatures as a function of wavelength. It can be seen that the

[4]Maxwell-Boltzmann distribution gives the average number of indistinguishable particles of a system in equilibrium at temperature, T, found in a state of energy, $\mathcal{E}$, which is given by,

$$n(\mathcal{E}) = \frac{1}{Ae^{(\mathcal{E} - \mu)/(k_BT)} - 1}.$$

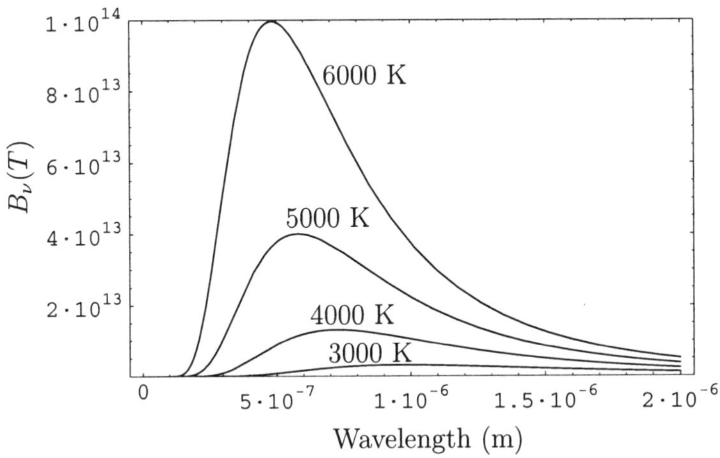

Fig. 10.2 Intensity distributions of black body radiation at various temperatures, viz., 3000°, 4000°, 5000° and 6000°. The abscissa contains the wavelengths and the ordinate depicts the intensity.

wavelength at which the curve is a maximum decreases with the increase of temperature.

10.1.3 *Application of blackbody radiation concepts to stellar emission*

Let the equation (10.13) be rewritten in terms of the spectral radiancy, $B_\nu(T)$, as a function of the frequency, ν and the temperature, T, as,

$$B_\nu(T) = \frac{2h\nu^3}{c^2}\frac{1}{e^{(h\nu/k_BT)} - 1}, \tag{10.14}$$

which has units of per unit time per unit surface area per unit solid angle per unit frequency ($\mathrm{J\,s^{-1}\,m^{-2}\,Sr^{-1}\,Hz^{-1}}$).

Stars have an overall spectrum, sources of opacity both lines as well as continuum process, which is governed largely by the equation (10.14). The spectral radiancy distribution function for a blackbody of a given area provides the following characteristics (Eisberg and Resnick, 1974):

(1) The power radiated in an interval of $d\nu$ maximizes at a particular value of ν for a definite temperature and the power radiated decreases rapidly as the fixed interval goes farther from this particular ν.

(2) The total power radiated in all frequencies increases with increasing temperature. Observations confirm that this increase is not linear. The law describing the non-linear dependence of the total power on the temperature is called the Stefan's law. Let the Planck's law be expressed as a function of wavelength, $B_\nu d\nu = -B_\lambda d\lambda$. The negative sign appears since the wavelength decreases with increasing frequency, and $\nu = c/\lambda$, in which c is the velocity of light. Hence,

$$B_\lambda = -B_\nu \frac{d\nu}{d\lambda} = B_\nu \frac{c}{\lambda^2}. \tag{10.15}$$

where $d\nu/d\lambda = -c/\lambda^2$.
Hence the equation (10.14) can be recast as,

$$B_\lambda(T) = \frac{2hc^2}{\lambda^2} \frac{1}{e^{(h\nu/k_B T)} - 1}. \tag{10.16}$$

The total brightness distribution is obtained as,

$$B(T) = \int_0^\infty B_\nu d\nu = \frac{2h}{c^2} \int_0^\infty \frac{\nu^3 d\nu}{e^{(h\nu/k_B T)} - 1}. \tag{10.17}$$

Let the integration variable be changed to $x = h\nu/k_B T$, so that the equation (10.17) is recast as,

$$B(T) = \frac{2h}{c^2} \frac{k_B^4 T^4}{h^4} \int_0^\infty \frac{x^3 dx}{e^x - 1} = AT^4, \tag{10.18}$$

with $A = \dfrac{2k_B^4}{c^2 h^3} \dfrac{\pi^4}{15}$ and $d\nu = (k_B T/h)\, dx$. The definite integral here is a real number.
The relation between the luminosity and temperature of a star can be obtained from the Stefan-Boltzmann law. The flux density F for isotropic radiation of intensity, B is:

$$F = \pi B = \sigma T^4, \tag{10.19}$$

where $\sigma (= 5.67 \times 10^{-8} \mathrm{W\, m^{-2}\, K^{-4}})$ is the Stefan-Boltzmann constant.

(3) The frequency at which the radiated power is most intense increases with increasing temperature. From the curve (see Figure 10.2), it is observed that the wavelength of the maximum intensity decreases with increasing total intensity. The mathematical law describing this radiation of ν_{max} with T is called Wein's displacement law,

$$\nu_{max} \propto T, \qquad \text{or,}$$

$$\lambda_{max} T = b = \text{const}, \tag{10.20}$$

where $b(= 0.0028978 \text{ K m})$ is the Wien's displacement constant.

The Wien Law gives the wavelength of the peak of the radiation distribution and explains the shift of the peak to shorter wavelengths as the temperature increases. When the wavelength, $\lambda \approx \lambda_{max}$ or $hc/\lambda k_B T \gg 1$, one obtains, $e^{hc/\lambda k_B T} \gg 1$, so that Wien's approximation is,

$$B_\lambda(T) \approx \frac{2hc^2}{\lambda^5} e^{-hc/\lambda k_B T}, \tag{10.21}$$

while in the case of the wavelength being greater than the maximum wavelength, for example, $\lambda \gg \lambda_{max}$, or $hc/\lambda k_B T \ll 1$, one obtains, $e^{hc/\lambda k_B T} \approx 1 = hc/\lambda k_B T$. Such a condition provides the Rayleigh-Jeans approximation,

$$B_\lambda(T) = \frac{2ck_B T}{\lambda^4}. \tag{10.22}$$

Equation (10.22) is useful in radio astronomy; it gives the brightness temperature of a radio source.

The Stefan-Boltzmann Law gives the total energy being emitted at all wavelengths by the blackbody, which is the area under the Planck's law curve (see Figure 10.2). It explains the growth in the height of the curve as the temperature increases, which is very abrupt, since it varies as the fourth power of the temperature.

10.1.4 *Radiation mechanism*

Radiation is a process of emission or absorption of energy of particles. Niels Bohr (1886-1961) developed a planetary model, referred to as Bohr model, for the hydrogen atom in 1915. It is an approximation to quantum mechanics, which states that negatively charged electrons confined to atomic shells encircle a small positively charged nucleus; electrostatic forces provide attraction. An electron jump between orbits must be accompanied by an emitted or absorbed amount of electromagnetic energy, $h\nu$, in which h is the Planck's constant and ν the frequency. The energy of the particles in the Bohr atom is restricted to certain discrete values. The energy is quantized, which means that only certain orbits with certain radii are allowed. The electromagnetic field interacts with its surroundings in discrete energy, $E = h\nu$, or multiples thereof and that energy differences in atomic jumps are emitted in these discrete amounts of energy (quantum atomic levels) as electromagnetic radiation:

$$h\nu = E_{n_2} - E_{n_1}. \tag{10.23}$$

Following the Coulomb's law, $F = e^2/(4\pi\epsilon_0 r_n^2)$, in which F is the force pulling the electron towards the proton, ϵ the vacuum permittivity, e the charge of the electron, and r_n the distance between the electron and the proton, and Newton's second law, $F = ma$, where a is the acceleration of a particle moving in a circular orbit of radius r_n and m the mass of the electron, the total energy of an electron in the orbit n is derived as,

$$\begin{aligned} E_n &= \frac{1}{2}mv_n^2 - \frac{e^2}{4\pi\epsilon_0 r_n} \\ &= -\frac{me^4}{32\pi^2\epsilon_0^2\hbar^2}\frac{1}{n^2} = -C\frac{1}{n^2}, \end{aligned} \tag{10.24}$$

where v is the speed of the electron, C a constant, and $\hbar = h/2\pi$.

The quantized energy levels for a hydrogen atom are labeled by an integer n, called a quantum number. The lowest energy state is generally termed the ground state ($n = 1$), corresponds to the lowest value of the angular momentum. In this orbit the energy of electron of hydrogen atom is - 13.58 eV (electron volts). The other orbits correspond to higher energies. The successive states possessing more energy than the ground state are known as the first excited state, the second excited state, and so on. Beyond an energy called the ionization potential the single electron of the hydrogen atom is no longer bound to the atom. The transition quantum energy from $E_{n_2} \rightarrow E_{n_1}$ is:

$$h\nu = C\left(\frac{1}{n_1^2} - \frac{1}{n_2^2}\right), \tag{10.25}$$

In terms of wavelength, λ, this may be written as,

$$\frac{1}{\lambda} = \frac{C}{hc}\left(\frac{1}{n_1^2} - \frac{1}{n_2^2}\right) = R_H\left(\frac{1}{n_1^2} - \frac{1}{n_2^2}\right), \tag{10.26}$$

in which $R_H(= 1.097 \times 10^7\ \mathrm{m}^{-1})$ is the Rydberg constant.

10.1.4.1 *Atomic transition*

The removal of energy from a beam of photons as it passes through matter is governed by four atomic processes, such as:

(1) Bound-bound transition: It is a kind of atomic process where an electron moves from one bound state to another due to the absorption of a photon. In the absorption process, a photon is captured by an atom raising an electron to a higher level and is completely lost because the electron loses energy by collisional processes or returns to a different level, either by emission of a photon of a different frequency or by the capture of another photon. If the energy of the two orbits is E_1 and E_2, a photon of frequency ν_{bb} produces a transition if,

$$h\nu_{bb} = E_2 - E_1, \tag{10.27}$$

Bound-bound processes are responsible for the spectral lines visible in stellar spectra, which are formed in the atmospheres of stars. In stellar interiors, however, these processes are not of great importance as most of the atoms are ionized; a small fraction contain electrons in bound orbits.

(2) Bound-free transition: Most of the photons in stellar atmosphere are so energetic that they are likely to cause bound-free absorption. As most of the heavy elements are not fully ionized, but are in a high state of ionization, they can absorb radiation and get further ionized. In this process, an electron moves from a bound state to an ionized state (photo-ionization). A photon of frequency ν_{bf} would convert a bound electron of energy E_1 into a free electron of energy E_3 if,

$$h\nu_{bf} = E_3 - E_1. \tag{10.28}$$

Bound-free processes lead to continuous absorption in stellar atmosphere, however, in stellar interiors, the importance of these processes is reduced due to the rarity of bound electrons. The emission of a photon by a free-bound transition, corresponds to a recombination process, in which an atom captures a free electron. It depends also both on temperature and on density or pressure, the electron pressure, P_e, in particular.

(3) Free-free transition: In the interiors of stars, both H and He are completely ionized. So the absorption can occur by changes in the kinetic energies of free electrons in the presence of protons and helium nuclei. This process is known as free-free transition, where the acceleration of an unbound (or free) electron by a proton or atomic nucleus results in the emission of electromagnetic radiation. A continuum of transitions may be obtained if the electron goes from one free-state to another

free-state, i.e.,

$$h\nu_{ff} = E_4 - E_3. \tag{10.29}$$

Since there is no restriction on the energy of photon which can induce a free-free transition, it contributes to the continuous absorption process which operates in both stellar atmospheres and stellar interiors. In both bound-free and free-free absorption, low energy photons are likely to be absorbed than high energy photons.

In addition to the above absorption processes, it is likely that the direction of travel of a photon is altered by interaction with an electron or ion. This process, known as scattering, occurs if the energy of the photon satisfies

$$h\nu \ll mc^2, \tag{10.30}$$

where m is the mass of the particle doing the scattering, the particle is scarcely moved by the collision.

10.1.4.2 *Hydrogen spectra*

A series of emission or absorption lines in the visible part of the hydrogen spectrum, known as Balmer series, are named after their discoverer, Johann J. Balmer, who examined the four wavelengths of hydrogen lines in 1885. These lines are produced by transitions ($E_n \to E_2$) between the second (or first excited) state and higher energy states of the hydrogen atom. For example, the transition from E_3 to E_2 is called Hα (656.3 nm) and is the first line of Balmer series. The other lines are Hβ (486.1 nm), Hγ (434.1 nm), and Hδ (410.2 nm). The series of lines in the hydrogen spectrum associated with transitions to or from the first energy level or ground state ($E_n \to E_1$) gives the Lyman series, which are in the ultraviolet region (UV; Lα at 121.6 nm; Lβ at 102.6 nm) with a series limit at 91.2 nm.

The other series are the Paschen series ($E_n \to E_3$) in the short wave infrared, Brackett series ($E_n \to E_4$), and Pfund series ($E_n \to E_5$) in the long wave infrared.

The spin-flip transition of neutral hydrogen in its two closely spaced energy levels in the ground state leads to the emission of a photon with a wavelength of 21.1 cm, which corresponds to a frequency of 1420.4 MHz. This line falls within the radio spectrum and is used in radio astronomy. Most of the interstellar medium is in relatively cold gas clouds (temperature is of the order of 10-100° K), where hydrogen exists in either atomic or molecular form. It is improbable to have any measurable emission at optical wavelengths. The astronomical study of the interstellar medium at 21 cm is used to observe atomic hydrogen gas directly, from one end of the Galaxy to the other. A measurement of the Doppler shift of this line radiation also yields the component velocity of the atomic hydrogen along the line of sight.

10.2 Astronomical measurements

A star is a massive luminous ball of plasma in space that produces tremendous amount of light and other forms of energy. The evolution of each star is controlled by their total mass, composition, and age. Measurements of its physical parameters like luminosity, effective temperature, angular size, distance etc., are of paramount importance to fundamental astrophysics. Measuring the position of stars in the sky at various times permits the determination of their projected motion as well. In what follows, a few such parameters are elucidated in brief.

10.2.1 *Flux density and luminosity*

The brightness of a celestial object can be derived in terms of observed flux density[5], $F_\nu(= \text{W m}^{-2})$, the energy received per unit time per unit telescope area per unit frequency. The flux, F is defined as the total amount of radiation crossing a unit area of surface in all directions and in unit frequency interval. Figure (10.3) depicts the relation between the intensity, I_ν and the energy entering the solid angle element, $d\Omega$. The amount of energy with frequency in the range, $\nu, \nu + d\nu$, passing through a surface element, dA, in time dt is,

$$dE_\nu = I_\nu \cos\theta \, dA \, d\nu \, d\Omega \, dt, \tag{10.31}$$

where I_ν is the specific intensity of radiation – the amount of energy passing a unit area of surface, placed perpendicular to the direction of propagation, per unit time, per unit solid angle, per unit frequency interval – over the entire solid angle element, $d\Omega$, which is equal to a surface element on a unit sphere; in spherical coordinates it is given by, $d\Omega = \sin\theta \; d\theta \; d\phi$, in which ϕ is the azimuthal angle, and θ the polar angle between the solid angle $d\Omega$

[5]In radio astronomy, flux densities are often expressed in Janskys; one Jansky (1 Jy) equals to 10^{-26} W m^{-2} Hz^{-1}.

and the normal to the surface.

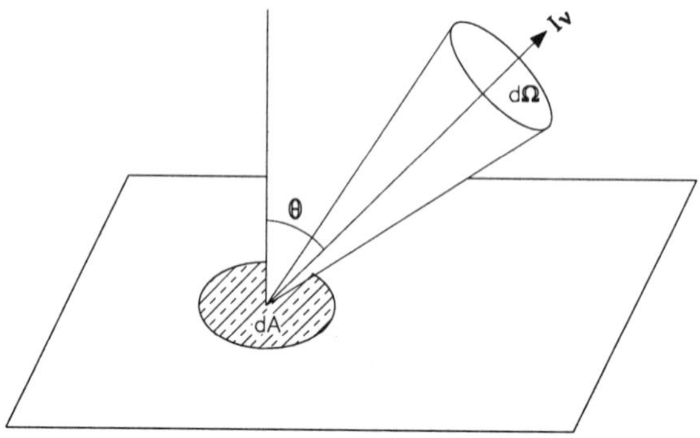

Fig. 10.3 Radiation from a small aperture.

The specific intensity, I_ν, has units of $\mathrm{W\,m^{-2}\,s^{-1}\,Hz^{-1}}$ per steradian[6], which is a function of position, direction, and time. The quantity including all possible frequencies is called the total intensity, I, and is obtained by integrating I_ν, over all frequencies,

$$I = \int_0^\infty I_\nu d\nu. \tag{10.32}$$

It is to reiterate that while observing a radiation source, one measures the energy collected by the detector during a period of time, which equals the flux density integrated over the radiation collecting area of the instrument and the time interval. The flux density, F_ν at a frequency ν is expressed in terms of the intensity as,

$$F_\nu = \frac{1}{dA\,d\nu\,dt}\int_{\mathrm{S}} dE_\nu. \tag{10.33}$$

Considering the amount of energy emitted by this small area in the direction of θ from the normal, the flux is the integration of $I_\nu \cos\theta$,

$$F_\nu = \int_{\text{all angles}} I_\nu \cos\theta \; d\Omega. \tag{10.34}$$

[6]Steradian is a measure of angular spans and is used to define two dimensional angular spans in three dimensional space. It is the solid angle subtended at the center of a sphere of radius r by a portion of the surface of the sphere having an area r^2. The total solid angle of a sphere is 4π steradian ($\mathrm{S_r}$).

The projection of the surface element dA as observed from the direction θ is $dA\cos\theta$, which explains the factor $\cos\theta$. If the intensity is independent of direction, the energy dE_ν is directly proportional to the surface element perpendicular to the direction of the radiation. The total flux density is given analogously,

$$F = \int_{\mathrm{S}} I \cos\theta d\Omega, \tag{10.35}$$

and for the isotropic radiation where the total intensity, I, is independent of the direction, it is written as,

$$F = I \int_{\mathrm{S}} \cos\theta d\Omega, \tag{10.36}$$

On replacing $d\Omega$ from equation (10.36),

$$F = I \int_{\theta=0}^{\pi} \int_{\phi=0}^{2\pi} \cos\theta \sin\theta \, d\theta \, d\phi = 0. \tag{10.37}$$

Equation (10.37) reveals that there is no unit flux of radiation, which means that there are equal amount of radiation entering and leaving the surface. In the case of isotropic radiation, the amount of radiation leaving the surface,

$$F = I \int_{\theta=0}^{\pi/2} \int_{\phi=0}^{2\pi} \cos\theta \sin\theta \, d\theta \, d\phi = \pi I. \tag{10.38}$$

The amount of energy (total flux) a star radiates per unit time in the form of electromagnetic radiation is called luminosity. It is expressed in watts, or in terms of solar luminosity, $L_\odot (= 3.845 \times 10^{26}$ W). Unlike the observed apparent brightness that is related to distance with an inverse square relationship, luminosity is an intrinsic constant independent of distance. The flux emitted by a star into a solid angle, Ω is related to its luminosity, $L_\star$,

$$L_\star = \Omega r^2 F, \tag{10.39}$$

where $F(= L_\star/4\pi r^2)$ is the flux density observed at a distance, r at the detector (energy/sec/m^2); the energy is diluted over the surface area of the sphere of radius, r.

If a star radiates isotropically, its radiation at a distance r is distributed evenly on a spherical surface whose area is $4\pi r^2$. If the flux density of the radiation passing through the surface is F, the luminosity is written as,

$$L_\star = 4\pi r^2 F. \tag{10.40}$$

Stellar luminosity should be made over the whole electromagnetic spectrum, but the Earth's atmosphere attenuates the infrared (IR) and blocks off the radiation short-ward of 3000 Å. The most accurate measurements of total stellar irradiance may be made from the spacecraft. The stellar luminosity can be used to determine the distance of the star; this method is known spectroscopic parallax.

10.2.2 *Magnitude scale*

A basic observable quantity for a star or other celestial quantities is its magnitude, which is defined as the ratio of brightness. The stellar magnitude system was introduced by the Greek astronomer Hipparchus as early as 2nd Century BC, who had divided the visible stars with six classes, according to their apparent brightness. He produced a catalogue of over 1000 stars, ranking them by 'magnitudes' one through six, from the brightest (first magnitude) to the dimmest (still visible to the unaided eye, which are of 6th magnitude).

This type of classification was placed on a precise basis by Pogson[7] (1857). His suggestion was to make this a standard. It was observed that a star of first class is roughly 100 times brighter than a sixth magnitude star, i.e., 5 magnitude lower. The magnitude system had been based on human eye, which has a non-linear (logarithmic) response. Pogson redefined the magnitude scale, a difference of five magnitudes was exactly a factor of 100 in light flux. The light flux for a one magnitude difference was found to be $\sqrt[5]{100} = 2.512$. Henceforth, the Pogson ratio became the standard method of assigning magnitudes. This is an inverse scale with the brighter stars having smaller numerical values for their magnitudes. It is to be noted that the co-efficient is not 2.512, albeit 2.5. The magnitude of the brightest star, α CMa is negative, -1.5, and that of the Sun is -26.8.

---

[7]Pogson N. R. (1829-1891) served as the Director of the Madras Observatory, India, during 1861-1891, to which the Indian Institute of Astrophysics traces its origin. At his early career as an astronomer he had computed the orbits of asteroids (minor planets) at which time the magnitude scale has been refined. He travelled to Madras, India in 1860 from Radcliffe Observatory in Oxford, England, to become the government astronomer. At the Madras Observatory he produced several positional catalogues as well as brightness variations of variable stars. He also discovered five asteroids and six variable stars. During his career he discovered a total of eight asteroids namely, (i) Isis (42), (ii) Ariadne (43), (iii) Hestia (46), (iv) Asia (67), (v) Sappho (80), (vi) Sylvia (87), (vii) Camilla (107), and (viii) Vera (245), as well as 21 variable stars.

10.2.2.1 *Apparent magnitude*

In general, the observed magnitudes of two stars at any given wavelength, denoted by m_1 and m_2 and their corresponding flux densities are F_1 and F_2, are related by the following expression,

$$\frac{F_1}{F_2} = \left(10^{0.4}\right)^{m_2 - m_1}, \tag{10.41}$$

in which the subscripts 1 and 2 refer to two different spectral regions.

The equation (10.41) may be written as,

$$\log\left(\frac{F_1}{F_2}\right) = 0.4\,(m_2 - m_1)\,. \tag{10.42}$$

Rearranging this equation (10.42), the magnitude relation is recast as,

$$m_1 - m_2 = -2.5\log\frac{F_1}{F_2}. \tag{10.43}$$

The equation (10.43) relates the magnitudes and brightnesses of two object. The derived magnitude is known as apparent magnitude. This is the ratio of radiation received at the Earth, but does not directly provide the intrinsic flux/emitted ratio. However, this quantity is a convolution of the true brightness and the effect of distance on the observed brightness, as well as any other absorption in the light path. Measuring such a quantity depends on the instrument since the sensitivity of the detector is different at different wavelengths.

10.2.2.2 *Absolute magnitude*

The absolute magnitude is a quantity measuring the intrinsic brightness of a star. It is defined as the apparent magnitude that a star has when placed at a distance of 10 parsecs (pc) from the Earth. Since a star, at a distance, r, emits the flux into a solid angle, Ω, spread over an area Ωr^2, the flux density is inversely proportional to the square of the distance. The ratio of the flux density at a distance, r, $F(r)$ to the same at a distance of 10 pc, $F(10)$ is given by,

$$\frac{F(r)}{F(10)} = \left[\frac{10\text{ pc}}{r}\right]^2. \tag{10.44}$$

The magnitude difference at these two distances turns out to be,

$$\begin{aligned} m - M &= -2.5 \log \frac{F(r)}{F(10)} \\ &= -2.5 \log \left[\frac{10 \text{ pc}}{r}\right]^2 = 5 \log \frac{r}{10 \text{ pc}}. \end{aligned} \tag{10.45}$$

It is imperative to note that a common convention is to use a lower-case m to denote an apparent magnitude and an upper-case M to denote an absolute magnitude. If the distance at a given wavelength is expressed in parsecs, the equation (10.45) may be expressed as,

$$m - M = 5 \log r - 5. \tag{10.46}$$

10.2.2.3 *Bolometric corrections*

The bolometric magnitude, m_{bol}, is defined as the brightness of an object integrated over all frequencies, so that it takes into account the total amount of energy radiated. The intensity of light emitted by any celestial object varies strongly with wavelength. If a star is a strong infrared or ultraviolet emitter, its bolometric magnitude differs vastly from its visual magnitude[8], m_v. The bolometric correction is the difference between m_v and m_{bol}.

The apparent magnitude observed for a given star by a detector depends on the range of wavelengths to which the detector is sensitive. Hence it is necessary to measure magnitude across all wavelengths; different wavelengths require different sensors. Moreover, it is difficult to measure m_{bol} directly since a part of the radiations is absorbed by the atmosphere. The bolometric magnitude is given by,

$$m_{bol} = m_v - BC, \tag{10.47}$$

in which m_v is the star's apparent visual magnitude and BC the bolometric correction, that depends on the energy distribution of the star involving effective temperature, gravity, and chemical composition, and is independent of distance of the star.

For the radiation of stars of spectral class F5, the bolometric correction is zero. Although the visual and bolometric magnitude may be equal, the flux density corresponding to the bolometric magnitude should be higher.

[8]Human eye is most sensitive to the green-yellow region of the visible spectrum of the light; the sensitivity decreases towards either side of the spectrum, i.e., towards blue and red regions of the electromagnetic spectrum. The magnitude corresponding to the sensitivity of the eye is called visual magnitude.

The brightness of an object, whether apparent or absolute, depends on the wavelength under consideration. As discussed in section (10.1.2), the Planck's radiation laws indicated the relationship between the temperature of a blackbody and the location of the peak in the radiation distribution as a function of wavelength. The absolute bolometric magnitude, M_{bol}, of a star is a measure of its total energy emitted per second, or luminosity. The bolometric correction is the visual magnitude of an object minus its bolometric magnitude. The absolute bolometric magnitudes can be expressed in terms of the luminosity, $L = 4\pi r^2 F$, with F as the total flux at a distance $r = 10$ pc. Let $F_\odot$ be the flux for the Sun, hence

$$\begin{aligned} M_{bol} - M_{bol_\odot} &= -2.5 \log \frac{F}{F_\odot} \\ &= -2.5 \log \frac{L/4\pi r^2}{L_\odot/4\pi r_\odot^2} = -2.5 \log \frac{L}{L_\odot} \end{aligned} \tag{10.48}$$

The equation (10.48) can be used for obtaining the effective temperature, T_e, provided the distance is known and alternately, if the effective temperature is known, the distance can be measured. The absolute bolometric magnitude, $M_{bol} = 0$ corresponds to a luminosity $L_0 = 3.0 \times 10^{28}$ W.

10.2.3 *Distance scale*

In astronomy, three units are commonly used to measure stellar distances. These are:

(1) Astronomical unit (AU): It is defined by the average distance between the Earth and the Sun, which is equal to 149,597,900 km.
(2) Light year: A light year (ly), is the distance travelled in one year by light propagating in vacuum; $c = 300,000$ km s^{-1}.
(3) Parsec (pc): A parsec (parallax second) is the distance to a star that has a parallax of one arcsecond employing a baseline of 1 astronomical unit (AU). By definition,

$$1 \text{ pc} = \frac{1 \text{ AU}}{\tan 1''} = \frac{1.495979 \times 10^{11} \text{ m}}{\tan\left[\dfrac{2\pi}{60 \times 60 \times 360}\right]}, \tag{10.49}$$

which turns out to be 1 pc $= 3.08568 \times 10^{16}$ m $= 3.26156378$ ly.

By observing trigonometric parallax one may estimate the distance to a star; the larger the distance, the smaller the parallax angle. The parallax is the apparent motion of an object as seen by an observer from a particular location. It is the angle at the distance of the object subtended by the diameter of the Earth's orbit around the Sun. This method, called trigonometric parallax, is used to measure the distances to nearby stars, $r = 1/\text{parallax angle}$, in which r is the distance measured in pc. The first measurements of a stellar parallax were made for the 61 Cygni by Friedrich Bessel in 1838, whose large proper motion made it a candidate for the determination of its distance by this method. The Hipparcos satellite[9] was able to measure parallaxes as small as 0.001 arcsecs, which is equal to a distance of $r(= 1/0.001 \text{ pc} = 1000 \text{ pc})$.

The proper motion is the apparent change of position of a star on the celestial sphere, measured in arcsec yr^{-1}. The average proper motion is about ~ 0.1arcsec yr^{-1} for the visible stars; the largest is about 10.25 arcsec yr^{-1} for the Barnard's star (a red main-sequence dwarf) located about 5.97 ly away in the northernmost part (α 17^h58^m; δ $+04°, 41^m$) of Constellation Ophiuchus; its apparent visual magnitude is 9.56. The proper motion of a star results from its transverse velocity, $v_t(= 4.74\ \mu r)$, in which μ is the proper motion in arcsec yr^{-1} and r the distance in parsecs, or the velocity across the sky. In order to measure the true space velocity, $v(= \sqrt{v_r^2 + v_t^2})$, in which v_r is the radial velocity, one needs to measure the radial velocity, the proper motion, and distance.

At larger distance, the periods of Cepheids and other pulsating stars are used for estimating the distance. The Cepheids are one of the most reliable types of 'standard candles' because their period of variability has been shown to be related to their absolute luminosity by a period-luminosity relationship; their periods are very regular and range from 1 to 100 days. Measuring their apparent luminosity permits to derive their distances. Using data from the Hipparcos satellite, Feast and Catchpole (1997) determined the distances to many Galactic Cepheids via trigonometric parallax. The period-luminosity of these Cepheid stars can be used to measure the distances of stars and nearby galaxies.

The moving star clusters may also be used to determine stellar distances. Such clusters are groups of stars that are gravitationally bound. When stars are born from large clouds of molecular gas and dust, they form in groups or clusters. After the remnant gas is heated and blown away, the stars are held together by mutual gravitational attraction. There are two types of star cluster such as open clusters (or galactic clusters) and globular clusters.

---

[9]The Hipparcos satellite, launched by the European Space Agency (ESA), had a small (29 cm diameter) optical telescope dedicated to determine accurate stellar positions.

The open clusters are confined to the galactic plane and young clusters are found in spiral arms. These clusters are collections of a few hundred hot and young population I[10] stars within a region up to about 30 light years (lyr). The globular clusters posses a large number of very old stars (population II) in a region of about 10 to 30 lyr. They populate the halo or bulge of the Galaxy, for example the Palomar clusters (Abell, 1955) and other galaxies with a significant concentration toward the galactic center. Since all the stars in a cluster are formed at the same time, they all have the same age.

If a star cluster is moving through space, it may be assumed that all of its stars have the same space velocity and direction. The proper motions of individual stars within a receding cluster would be directed towards the point on the sky that the entire cluster would occupy as it shrinks to a point. This point is known as convergent point. However, if the paths of the stars appear to be converging or diverging, this is a perspective effect; the greater the degree of convergence or divergence, nearer the cluster. From the parallax angle, $\pi''(= 4.74\mu''/v_r \tan\theta)$, in which μ'' is the proper motion, and v_r the radial velocity, the distance, $r(= 1/\pi'')$, can be determined. The moving cluster method has been applied to a number of clusters including (i) the Hyades, $r \sim 45$ pc (van Altena 1974, Perryman et al. 1995), (ii) the Usra Major, $r \sim$24 pc (Eggen 1958, 1960), (iii) the Pleiades, $r \sim$115 pc (van Leeuwen and Hansen Ruiz, 1997), and (iv) the Scorpio-Centaurus group $r \sim$170 pc (Bertiau, 1958).

Another method to estimate distance to further objects is to use the supernovae (SNe Ia; see section 11.2.4.2) as standard candles, which occurs as a consequence of stellar explosion. They are among the most spectacular

[10]Stars are generally classed into two main groups such as population I and population II. The stars belonging to the former group are metal-rich containing a relatively large amount of heavy elements. These stars are young (about a few hundred million years old) and found in the galactic disc. The most metal rich stars, the youngest ones, are concentrated in the discs of the spiral galaxies. The orbits of these stars are found to be almost circular; the interstellar matter moves also in almost circular orbits in the galactic plane. The stars belonging to the latter group are very old with ages ranging from 2-14 billion years. These are metal-poor stars and are found in the spherical component, the halo and the bulge, of the Milky Way (referred as the Galaxy); the stellar density is highest near the center of the Galaxy and decreases outwards. The orbits of these stars may be very eccentric and show no preference for the galactic plane. The extreme population II (most metal-poor) stars are found in the halo and in the globular clusters, while intermediate population II stars are located in the bulge. The halo dwarfs are formed at an early stage in the evolution of the galaxy, and their chemical composition may reflect that of the early universe. The helium abundance of the early universe is a critical parameter in cosmological theories.

events since they reach the same brightness as an entire galaxy, which makes them excellent distance indicators. The peak light output from a type-Ia supernova is approximately equivalent to an absolute blue sensitive magnitude of $M_B = -19.33 \pm 0.25$. Thus, if one observes a supernova of this type in a distant galaxy and measures the peak light output, from which apparent magnitude m_b, one can use the inverse square law to infer its distance and therefore the distance of its parent galaxy. The type-Ia supernovae can measure distances out to around 1000 Mpc, which is a significant fraction of the radius of the known Universe.

10.2.4 *Extinction*

The term extinction is used in astronomy to describe the radiation losses due to absorption and scattering of starlight by matter between the emitting object and the observer. It arises both from the interstellar medium and the Earth's atmosphere and is related to a coefficient, which depends on the medium and wavelength of transmission.

10.2.4.1 *Interstellar extinction*

The main components of the interstellar medium are tenuous gas and dust particles. The radiation disappears due to absorption and is converted into heat energy in the interstellar dust grains (usually re-radiated at a different wavelength) and scatter the radiation from stars and other background objects. The extinction is greatest along the galactic plane, where most of the gas and dust are located. The equation (10.46) does not hold good since the interstellar extinction reduces the luminosity L in the solid angle Ω ($dL \leq 0$) producing a dimming of the light from stars. Extinction obeys the following law:

$$dL = -\alpha_\nu L dr, \tag{10.50}$$

where $\alpha_\nu (= k_\nu \rho)$ is a factor called opacity and has units in $\text{cm}^2\,\text{gm}^{-1}$, ν the frequency, ρ the density of matter, dr the distance travelled, and k_ν the extinction coefficient that is defined as the fraction of light lost to absorption and scattering per unit distance in a particular medium and is measured in cm^{-1}.

The opacity is a property of matter that prevents radiation from travelling through it. It depends on the temperature, density, and chemical composition of the gas. The opacity is zero for a perfect transmitting medium and approaches infinity when substance becomes opaque. It depends on the frequency of the radiation as well. Introducing the optical depth, τ, which is a measure of how much radiation is absorbed in a medium, through

$$d\tau = \alpha_\nu \, dz. \tag{10.51}$$

The optical depth is a dimensionless quantity, and is measured along the optical path, $dz(= dr \cos\theta)$, in which θ is the angle between z-axis and the direction of the propagation of the radiation. On replacing the RHS of this equation (10.51), one may write,

$$dL = -L d\tau. \tag{10.52}$$

Integrating this equation (10.52) from the source where $L = L_0$, in which L_0 is the luminosity in the absence of interstellar extinction, and $\tau = 0$ to the observer,

$$\int_{L_0}^{L} \frac{dL}{L} = -\int_0^\tau d\tau; \qquad \text{or,} \tag{10.53}$$

$$L = L_0 e^{-\tau}. \tag{10.54}$$

Expressing the fluxes as, $L = \Omega r^2 F(r)$, and $L_0 = \Omega R_\star^2 F_0$, in which $R_\star$ is the radius of the star, one obtains,

$$F(r) = \frac{R_\star^2}{r^2} F_0 e^{-\tau}. \tag{10.55}$$

For absolute magnitude the flux density is considered to be at a distance of 10 pc, therefore, the difference in magnitude becomes,

$$\begin{aligned} m - M &= 5 \log \frac{r}{10 \text{ pc}} - 2.5 \log e^{-\tau} \\ &= 5 \log \frac{r}{10 \text{ pc}} + (2.5 \log e)\tau \\ &= 5 \log \frac{r}{10 \text{ pc}} + A, \end{aligned} \tag{10.56}$$

where $A \geq 0$ is the interstellar absorption.

If the opacity is constant along the line of sight, the optical depth, $\tau = \alpha \, r$, therefore equation (10.56) turns out to be,

$$m - M = 5 \log \frac{r}{10 \text{ pc}} + \beta \, r, \tag{10.57}$$

where $\beta(= 2.5\alpha \log e)$ is the constant that provides the extinction in magnitudes per unit distance.

10.2.4.2 *Color excess*

The color of a star is defined by a ratio of brightness in two different wavelengths. As determined by the measurements at two different spectral regions give information about the star's temperature. Let the bolometric luminosity be,

$$L_{bol} = \int_0^\infty L_\lambda d\lambda. \tag{10.58}$$

In reality, one measures the flux in a certain bandpass, hence for a filter with bandpass, f_λ, one writes,

$$L_f = \int_0^\infty L_\lambda f_\lambda d\lambda. \tag{10.59}$$

Astronomers have defined several filter systems comprising of a number of standardized filters, which are being employed to measure a star's brightness in selected portion of the spectrum. The peaks for transmission for these filters are in the ultraviolet (U), blue (B), visible (V), red (R), and infrared (I) respectively. Wide-band photomultiplier tubes are available, which enables observations from U to I to be done with the same detector (Bessell 1976). The UBV system, also referred to as the Johnson system, is a photometric system for classifying stars according to their magnitude (Johnson and Morgan, 1953). It is to be noted that observation of a source extends over a range of wavelengths centered about a nominal value. The effective wavelength is a representation of this value that allows the magnitude to be considered as approximately monochromatic at this wavelength. The categories of these filters are given in Table 10.1.

Table 10.1
Categories of filters

Filters	Effective wavelength	Filter width
U	λ_e 365 nm	FWHM, 68 nm
B	λ_e 440 nm	FWHM, 98 nm
V	λ_e 550 nm	FWHM, 89 nm
R	λ_e 700 nm	FWHM, 220 nm
I	λ_e 900 nm	FWHM, 240 nm

In order to measure the color index, one observes the magnitude of an object successively through two different filters. For example, the $(B-V)$ provides the color index, which is the difference between the magnitudes in the blue and visual regions of the spectrum and the $(U-B)$ color index is the analogous difference between the ultraviolet and blue regions of the spectrum and so on. If UBV system is used, it provides only the V magnitude and the color indices $(U-B)$ and $(B-V)$. The zero point of the color indices, $(B-V)$ and $(U-B)$, are defined by six stars of spectral type $A0\ V$. These stars are α Lyrae, γ UMa, 109 Vir, α CrB, γ Oph, and HR 3314. The Sun has a $(B-V)$ index of 0.656 ± 0.005, while the average index of these stars is defined to be zero, i.e., $(B-V) = (U-B) = 0$ (Johnson and Morgan, 1953).

Color indices of the celestial objects are in general affected by interstellar extinction, the effects of which on the observed magnitudes and colors of stars should be taken into account for recovering their intrinsic properties (Karttunen et al. 2000). The extinction is wavelength dependent, which increases at shorter (blue) wavelengths than longer wavelengths (red), in a process called reddening. From the equation (10.56), the visual magnitude of a star is written as,

$$V = M_V + 5\log\frac{r}{10\ \text{pc}} + A_V, \tag{10.60}$$

while for the blue magnitude it is,

$$B = M_B + 5\log\frac{r}{10\ \text{pc}} + A_B, \tag{10.61}$$

in which M_V and M_B are the absolute visual and blue magnitudes respectively and A_V and A_B the respective extinction in the V and B passbands.

The observed color index is given by,

$$\begin{aligned} B - V &= M_B - M_V + A_B - A_V \\ &= (B-V)_0 + E_{B-V}, \end{aligned} \tag{10.62}$$

with $(B-V)_0 (= M_B - M_V)$ as the intrinsic color of the star and the color excess, E_{B-V} is,

$$E_{B-V} = (B-V) - (B-V)_0. \tag{10.63}$$

10.2.4.3 *Atmospheric extinction*

The intensity of starlight gets reduced by traverse through the terrestrial atmosphere due to scattering and absorption by air molecules and aerosols. The amount of light loss depends on the height of the star above the horizon, the wavelength of observation, and the prevailing atmospheric conditions. If the zenith distance (90° - altitude), γ, of the star is not too large, one may approximate the atmosphere by plane parallel stratified layers. The intensity of light, I, received by the telescope is given by,

$$\begin{aligned} I &= I_0 e^{-\int_0^\infty k\rho \sec\gamma dh} \\ &= I_0 e^{-\sec\gamma \int_0^\infty k\rho dh} = I_0 e^{-\alpha \sec\gamma}, \end{aligned} \tag{10.64}$$

where $\alpha = \int_0^\infty k\rho dh$, I_0 is the original intensity before atmospheric extinction, ρ the density, $dh(= \sec\gamma\, dh)$ the geometrical distance, and k the extinction co-efficient.

With the increase of zenith distance ($\gamma \geq 60°$), the refraction effects, curvature of the atmosphere, and variations of air density with height may become important. The path length through the atmosphere, called air-mass, χ, is given by,

$$\chi = \sec\gamma, \tag{10.65}$$

which is a dimensionless quantity. It is one at the zenith and it grows as the altitude above the horizon decreases. Hardie (1962) gave a refined relationship:

$$\chi = \sec\gamma - 0.0018167(\sec\gamma - 1) - 0.002875(\sec\gamma - 1)^2 - 0.0008083(\sec\gamma - 1)^3, \tag{10.66}$$

and according to Young and Irvine (1967), it is

$$\chi = \sec\gamma \left[1 - 0.0012(\sec^2\gamma - 1)\right]. \tag{10.67}$$

The observed magnitude m_λ depends on the observer's position and the zenith distance of the object. The magnitude increases linearly with the distance, and the observed magnitude, m_λ, at the Earth's surface is given by,

$$m_\lambda = m_{\lambda_0} + k_\lambda \chi, \tag{10.68}$$

in which m_{λ_0} is the magnitude of the observed object outside the atmosphere and k_λ the atmospheric extinction co-efficient.

Atmospheric extinction co-efficient is generally determined by observing the same source, through an appropriate filter, at several times during a night at varying zenith distance. When the observed magnitudes of the object are plotted as a function of the air-mass, they lie on a straight line, the slope of which gives the atmospheric extinction co-efficient, k_λ.

10.2.4.4 *Instrumental magnitudes*

Atmospheric extinction needs to be corrected when calibrating instrumental magnitudes. Let the equation (10.43) be rewritten as,

$$m_1 = m_2 - 2.5 \log F_1 + 2.5 \log F_2. \tag{10.69}$$

Suppose the magnitude of the reference star (star 2) be zero and star 1 is unknown,

$$\begin{aligned} m_1 &= q - 2.5 \log F_1 \qquad \text{or,} \\ m_\lambda &= q_\lambda - 2.5 \log F_\lambda. \end{aligned} \tag{10.70}$$

where

$$F_\lambda = \int_0^\infty \mathcal{T}_\lambda A_\lambda f_\lambda \eta_D(\lambda) F_\lambda^* d\lambda, \tag{10.71}$$

refers to the observed flux, which is related to the actual flux, F_λ^*, outside the atmosphere and q_λ the constant (Henden and Kaitchuck, 1982); the subscript 1 is dropped in favor of the wavelength, λ. Here $\mathcal{T}_\lambda$ is the efficiency of the telescope, A_λ and f_λ the transmissions of the atmosphere and filter respectively, and $\eta_D(\lambda)$ the quantum efficiency of the detector.

In practice, the detector produces an electrical output that is directly proportional to the observed stellar flux, then

$$F_\lambda = K d_\lambda, \tag{10.72}$$

in which d_λ is the practical measurement (i.e., current or counts s^{-1}) and K the constant of proportionality.

Inserting equation (10.72) into equation (10.70), one writes,

$$\begin{aligned} m_\lambda &= q_\lambda - 2.5 \log(K d_\lambda) \\ &= q_\lambda - 2.5 \log K - 2.5 \log d_\lambda = q'_\lambda - 2.5 \log d_\lambda, \end{aligned} \tag{10.73}$$

with

$$q'_\lambda = q_\lambda - 2.5 \log K, \tag{10.74}$$

as the instrumental zero point constant.

The equation (10.73) relates the actual measurement, d_λ, to the constant q'_λ, as well as to the instrumental magnitude. In order to determine the color index of a star, one may write,

$$\begin{aligned} m_{\lambda_1} - m_{\lambda_2} &= q'_{\lambda_1} - q'_{\lambda_2} - 2.5 \log d_{\lambda_1} + 2.5 \log d_{\lambda_2} \\ &= q_{\lambda_{12}} - 2.5 \log \left(\frac{d_{\lambda_1}}{d_{\lambda_2}} \right), \end{aligned} \tag{10.75}$$

in which the zero point constants are collected into a single term, $q_{\lambda_{12}}$; the quantity, $m_{\lambda_1} - m_{\lambda_2}$, is in the instrumental system.

10.2.4.5 *Color and magnitude transformation*

The measured magnitude described in equation (10.68) may required to be corrected and therefore modified in the following manner,

$$m_{\lambda_0} = m_\lambda - (k'_\lambda + k''_\lambda c)\, \chi, \tag{10.76}$$

where k'_λ and k''_λ refer to the principal and second order extinction coefficients respectively, and

$$c = c_0 + (k'_c + k''_c c)\, \chi, \tag{10.77}$$

c the measured color index, c_0 the color index as observed from above the atmosphere, k'_c the main extinction coefficient and k''_c the color correction coefficient.

These extinction coefficients, k'_λ, k''_λ, k'_c, and k''_c are determined observationally. Following correcting of atmospheric extinction, the observed magnitude is transformed to a standardized magnitude, M_λ, by,

$$M_\lambda = m_{\lambda_o} + \beta_\lambda C + \gamma_\lambda, \tag{10.78}$$

where β_λ and γ_λ are the respective color coefficient and zero-point constant of the instrument, and the standardized color index is,

$$C = \delta\, c_0 + \gamma_c, \tag{10.79}$$

with δ as the color coefficient, γ_c the zero-point constant. These coefficients and constants may be determined for each filter system by the observation of standard stars[11].

10.2.4.6 *UBV transformation equations*

Through the observation of a set of known standard stars, an observer may take instrumental measurements of program stars and transform them to the standard UBV system (Henden and Kaitchuck, 1982). It is customary to change the symbols used in the transformation equation to indicate the use of the UBV system. The equation (10.73) is replaced by,

$$u = -2.5 \log d_u; \qquad b = -2.5 \log d_b; \qquad v = -2.5 \log d_v, \tag{10.80}$$

where u, b, v and d_u, d_b, d_v represent the instrumental magnitude and measurements through U, B, and V filters respectively; the lower case u, b, v refer to the instrumental system, while the upper case U, B, V refer to the standard system.

The constants q' have been dropped because they can be absorbed by the zero-point constant in the transformation equations. Similarly, the equation (10.75) is replaced by,

$$(u - b) = -2.5 \log \frac{d_u}{d_b}; \qquad (b - v) = -2.5 \log \frac{d_b}{d_v}. \tag{10.81}$$

The magnitude and colors corrected for atmospheric extinction given by the respective equations (10.76) and (10.77) become,

$$\begin{aligned} v_0 &= v - k'_v \chi \\ (u - b)_0 &= (u - b) - k'_{ub} \chi \\ (b - v)_0 &= (b - v)(1 - k''_{bv} \chi) - k'_{bv} \chi. \end{aligned} \tag{10.82}$$

In the UBV system, k''_{ub} is defined to be zero and k''_v is very small, so one may neglect. The equations (10.78) and (10.79) may be transformed into,

$$\begin{aligned} V &= v_0 + \epsilon(B - V) + \zeta_v, \\ (B - V) &= \mu(b - v)_0 + \zeta_{bv}, \\ (U - B) &= \psi(u - b)_0 + \zeta_{ub}, \end{aligned} \tag{10.83}$$

[11]A star of known flux density, magnitude, and colors is called standard star. The primary magnitude standard for the UBV systems are a set of 10 bright stars of magnitude 2 to 5. The magnitudes of these stars define the UBV color system. A standard star should not be a variable star.

in which ϵ, μ, ψ are the transformation coefficients and $\zeta_v, \zeta_{bv}, \zeta_{uv}$ the zero-point constants. These six values are required to be found by observations of the standard stars.

Another system, called Strömgren photometric system (Strömgren, 1956), is a four color, *uvby*, medium band photometric system. It is devised to measure the temperature, gravity, and reddening of early-type stars, although it was extended later to other kinds of stars (Bessell, 2005). The u (ultraviolet) filter (λ_0 350 nm, FWHM 40 nm) measures both blanketing and the Balmer discontinuity. The violet (v) filter (λ_0 410 nm, FWHM 20 nm) is centered in a region of strong blanketing. The b (blue) filter (λ_0 470 nm, FWHM 10 nm) is centered about 30 nm to the red of the B filter of the UBV system to reduce the effects of line blanketing. The y (yellow) filter (central wavelength λ_0 550 nm with FWHM 20 nm) matches the visual magnitude and corresponds well with V magnitudes. The red limit is set by the filter and not by the detector as in the case of the UBV system. In addition to determining the effective temperature, the Strömgren system also provides measure of the strength of metal lines. This system is independent of any one detector and requires no second order terms in extinction or transformation equations (Henden and Kaitchuck, 1982). Several indices are derived to measure (i) the depression owing to metal lines around 410 nm,

$$m_1 = (v - b) - (b - y), \tag{10.84}$$

(ii) the Balmer discontinuity,

$$c_1 = (u - v) - (v - b) = u - 2v + b, \tag{10.85}$$

and (iii) the strength of the Hβ line, called Hβ photometry, which is a luminosity indicator in stars of spectral type O to A and a temperature indicator in types A to G,

$$\beta = \beta_w - \beta_n, \tag{10.86}$$

where β_n and β_w represent the narrow and wideband (for measuring the adjacent continuum) filters centered on the Hβ line.

The ratio of the measurements through these filters indicates the strength of Hβ with respect to the continuum. On subtraction of the term $2v$ in equation (10.85) essentially cancels the blanketing, leaving behind the effects of the Balmer discontinuity.

10.2.5 *Stellar temperature*

Stars have an overall spectrum of emission whose pattern is governed largely by the black body equation (10.14); each temperature corresponds to such a spectrum. In Figure (10.2), it is plotted as function of wavelength. Each curve is for a given temperature, and the peak of the curve represents the maximum emission. If the star is assumed to radiate like a blackbody, from the Stefan-Boltzmann law (equation 10.19), one may write the relation between the luminosity and temperature of a star as,

$$L = 4\pi R_\star^2 F = 4\pi\sigma R_\star^2 T^4, \tag{10.87}$$

where $R_\star$ is the radius of the star.

Equation (10.87) reveals that the luminosity, radius, and temperature of a star are interdependent quantities. They are related to the absolute bolometric magnitude as well. The equation (10.48) is rewritten in terms of radii and temperatures:

$$\begin{aligned} M_{bol} - M_{bol_\odot} &= -2.5\log\frac{R_\star^2 T_\star^4}{R_\odot^2 T_\odot^4} \\ &= -5\log\frac{R_\star}{R_\odot} - 10\log\frac{T_\star}{T_\odot}. \end{aligned} \tag{10.88}$$

where $R_\odot(= 6.96 \times 10^8$ m) is the radius of the Sun.

Depending on the type of observational data used, astronomers define various kinds of temperature.

10.2.5.1 *Effective temperature*

The effective temperature, T_e, of a star indicates the amount of radiant energy it radiates per unit of surface area. It can be measured even if the star is not resolved. If the distance to a star is known, its effective temperature, T_e, is defined by the luminosity and stellar radius, i.e., $L = 4\pi R_\star^2\sigma T_e^4$. It relates to the total luminosity. The most direct and model-independent way of measuring effective temperatures is, thus, the combination of the total energy output with angular diameters. The total amount of energy passing through a sphere of radius, 1.495×10^8 km, i.e., the mean radius of the Earth's orbit, should be equal to the total energy output of the Sun, which is $6.25 \times 10^{10}\,\mathrm{ergs\,cm^{-2}\,s^{-1}}$. The flux density at a distance, r, is,

$$F' = \frac{L}{4\pi r^2} = \frac{R_\star^2}{r^2}F = \left(\frac{\theta_\star}{2}\right)^2 \sigma T_e^4, \tag{10.89}$$

where $\theta_\star = 2R_\star/r$ is the angular diameter of the star, which can be measured by means of interferometric technique.

In this equation (10.89), the dependence on distance drops out, hence the effective temperature may be derived without reference to the stellar distance. So, the effective temperature, T_e is recast as,

$$T_e = \left(\frac{4F}{\theta_\star^2 \sigma}\right)^{1/4}. \tag{10.90}$$

It is to be noted that the effective temperature of the Sun is 5760° K. These parameters are useful for the input to numerical models of stellar characteristics.

10.2.5.2 *Brightness temperature*

Brightness temperature is a measure of the intensity of radiation thermally emitted by an object, which is an equivalent black body, providing the same power output per Å as the source at a wavelength, λ. It is a parameter that expresses the rate of energy radiated at certain wavelength. A quantity that can be measured for a number of stars is the radiation cm^{-2} s^{-1} Å^{-1} for selected points in the continuum radiation. In the isotropic case, the observed flux density, F'_λ is given by,

$$F'_\lambda = \frac{R_\star^2}{r^2} F_\lambda = \left(\frac{\theta_\star}{2}\right)^2 \pi B_\lambda(T_b), \tag{10.91}$$

where $F_\lambda = \pi B_\lambda(T_b)$, T_b is the brightness temperature.

10.2.5.3 *Color temperature*

Color temperature characterizes the spectral properties of a light source. In the case of a star, it is determined by comparison of the spectral distribution of the star's radiation at two different wavelengths such as λ_1 and λ_2. It can be empirically calibrated to the effective temperature or another temperature. Assuming the intensity distribution follows Planck's law, the ratio of these flux densities are the same as the ratio obtained from this law. Hence,

$$\begin{aligned} \frac{F'_{\lambda_1}}{F'_{\lambda_2}} &= \frac{B_{\lambda_1}(T)}{B_{\lambda_2}(T)} \\ &= \left(\frac{\lambda_2}{\lambda_1}\right)^5 \frac{e^{(hc/\lambda_2 k_B T)} - 1}{e^{(hc/\lambda_1 k_B T)} - 1}. \end{aligned} \tag{10.92}$$

where c is the velocity of light, $k_B(= 1.380662 \times 10^{-23} J/K)$ is the Boltzmann constant, and $h(= 6.626196 \times 10^{34}\text{Joules J})$ the Planck's constant.

The observed flux densities corresponds to certain magnitudes, m_{λ_1} and m_{λ_2}, i.e.,

$$m_{\lambda_1} - m_{\lambda_2} = -2.5 \log \left(\frac{F'_{\lambda_1}}{F'_{\lambda_2}} \right) + C, \tag{10.93}$$

with C as the constant which is a consequence of the different zero points of the magnitude scales.

From the Wien's approximation in the visible spectrum of light, the magnitude difference is given by,

$$\begin{aligned} m_{\lambda_1} - m_{\lambda_2} &= -2.5 \log \left(\frac{B_{\lambda_1}}{B_{\lambda_2}} \right) + C \\ &= -2.5 \log \left(\frac{\lambda_2}{\lambda_1} \right)^5 + 2.5 \frac{hc}{k_B T} \left(\frac{1}{\lambda_1} - \frac{1}{\lambda_2} \right) \log e + C. \end{aligned} \tag{10.94}$$

This can be expressed as,

$$m_{\lambda_1} - m_{\lambda_2} = a + b/T_c, \tag{10.95}$$

where T_c is the color temperature.

The equation (10.95) is derived for monochromatic radiation, however, this can be applied for broad band magnitudes. Color indices of $B - V$, $U - B$ have been used to estimate color temperature. In such a situation, the two wavelengths are essentially the effective wavelengths of U, B, and V bands.

10.2.5.4 *Kinetic temperature*

Kinetic temperature is a measure of the average random motion of electrons in a system. Let v_x be the component velocity in the line of sight. The number of atoms, $N(v_x)dv_x$, with velocities in the range, v_x and $v_x + dv_x$, is given by,

$$N(v_x)dv_x = \frac{N}{v_\sigma \sqrt{\pi}} e^{-(v_x/v_\sigma)^2} dv_x, \tag{10.96}$$

where N is number of atoms and $v_\sigma(= \sqrt{2k_BT/\mu m_H})$ the root mean square velocity, T the absolute temperature, m_H the hydrogen atomic mass, and k_B the Boltzmann constant.

The kinetic motions of the atoms produces Doppler shifts $\Delta\lambda = v_x\lambda/c$, in which c is the speed of light, and broadens the line, known as Doppler broadening. It is the broadening of the line caused by the thermal, turbulent or mass motions of atoms along the line of sight. It sets the limit on resolution for the observed spectra in the visible and ultraviolet range. The broadening function is given by,

$$I(\Delta\lambda) = \frac{I_0}{\Delta_D\sqrt{\pi}} e^{-(\Delta\lambda/\Delta\lambda_D)^2}, \tag{10.97}$$

with $\Delta\lambda_D = v_\sigma\lambda/c$. It provides the half-width of the line $\Delta\lambda_{1/2}$, where the intensity drops to half of its value at the center, as

$$\Delta\lambda_{1/2} = 2\Delta\lambda_D\sqrt{\ln 2}. \tag{10.98}$$

By measuring the line profile, and particularly the half-width of a line, one may determine $\Delta\lambda_D$. This gives the v_σ and then the temperature, known as kinetic temperature, T_k.

10.2.5.5 *Excitation temperature*

The excitation temperature, T_{exe} is defined as a temperature, which if substituted into the Boltzmann distribution, provides the observed population numbers of an atomic or molecular species. The relative level population can be characterized by an excitation temperature. On applying Boltzmann's equation, the excitation temperature, T_{exe} may be derived from a comparison of the number of atoms in various excited state:

$$N_n = N_1 \frac{g_n}{g_1} e^{-\chi_n/k_B T_{exc}}, \tag{10.99}$$

in which N_1 and N_n are the population number of atoms in the ground and nth excited state, respectively, g_1 and g_n the statistical weights of these levels $(2J+1)$ for most atoms and $2n^2$ for hydrogen), χ_n the lower excitation potential of the nth excited state, and T_{exc} the excitation temperature in K.

The total number of atoms in all states are:

$$\begin{aligned} N = \sum N_n &= \frac{N_1}{g_1}\left[g_1 + g_2 e^{-\chi_2/k_B T_{exc}} + g_3 e^{-\chi_3/k_B T_{exc}} + \ldots\right] \\ &= \frac{N_1}{g_1}\mathrm{B}(T), \end{aligned} \tag{10.100}$$

where

$$B(T) = g_1 + g_2 e^{-\chi_2/k_B T_{exc}} + g_3 e^{-\chi_3/k_B T_{exc}} + \cdots, \tag{10.101}$$

is called the partition function, which represents statistical weight for the atom as a whole. In terms of such function, one writes,

$$\frac{N_n}{N} = \frac{g_n}{B(T)} e^{-\chi_n/k_B T_{exc}}, \qquad \text{or,} \tag{10.102}$$

$$\log \frac{N_n}{N} = \log \frac{g_n}{B(T)} - \frac{5040}{T}\chi_n. \tag{10.103}$$

For low temperature, $B(T) \approx g_1$. If the distribution of atoms in different levels is a consequence of mutual collisions, the excitation temperature, T_{exc} turns out to be equal to the kinetic temperature, T_k.

10.2.5.6 *Ionization temperature*

The famous Saha's ionization equation was formulated by Saha M. N. (1894-1956) in 1920, which gives the relative abundance on a ionized atom in an ionization state, i to the next, $i + 1$. The Saha's equation provides the relation:

$$\frac{N_{i+1} N_e}{N_i} = 2\frac{(2\pi m_e k_B T_i)^{3/2}}{h^3} \frac{B_{i+1}}{B_i} e^{-\chi_i/k_B T_i}, \tag{10.104}$$

where N_i and N_{i+1} are the number of atoms in two successive stages of ionization, B_i and B_{i+1} their respective partition functions, which can be derived with the aid of a term table for the atom or ion of interest as a function of the temperature and electron pressure, N_e the electron number density, χ_i the ionization potential for state i, m_e the mass of an electron, and T_i the ionization temperature that can be found by comparing the number of atoms in different states of ionization.

Expressing in logarithmic form the Saha's equation shows the dependence on pressure and temperature,

$$\log P_e \frac{N_{i+1}}{N_i} = -0.48 + \log \frac{2B_{i+1}}{B_i} + 2.5 \log T - \frac{5040}{T}\chi_I, \tag{10.105}$$

in which $P_e (= N_e k_B T)$ is the electron pressure in dynes cm^{-2} and χ_I the ionization potential in electron volts.

In most stellar spectra, one often finds lines due to atoms in two different ionization states. The degree of ionization increases with increasing temperature through third term of the right hand side of equation (10.105) and increases linearly with the decrease of electron pressure; it decreases with increasing ionization potential exponentially as well.

10.2.6 *Stellar spectra*

A graphic or photographic representation of the distribution of energy with wavelength, referred to as spectrogram, is of great interest in general. A spectral line appears as a curve with a shape defined by the line profile. It is caused by absorption of radiation between discrete levels of different types of atoms or ions. A discrete absorption is referred as line at some wavelengths as opposed to continuous absorption. The outermost layer of a star contains elements that absorb the radiations from the continuous spectrum. Each of these elements has its own characteristic pattern of lines. The strength of a spectral line is the amount of absorption measured in units of equivalent width (EW). Such a line appears as a curve with a shape defined by the velocity field and various line forming mechanism. The equivalent width, W_ν, is the width of a rectangle centered on a spectral line, which has the same area as the line; it takes out the same amount of flux as does the line integrated over the entire profile. Mathematically for normalization, one writes,

$$W_\nu = \int_{\text{line}} \frac{I_c - I_\nu}{I_c} d\nu, \tag{10.106}$$

where I_c is the intensity at the continuum, which needs to be interpolated from its value outside the line and I_ν is the intensity at frequency ν within the line.

The equivalent width may be expressed in wavelength units by the relation, $|\Delta\lambda| = c|\Delta\nu|/\nu^2$, in which $\lambda = c/\nu$ and c is the velocity of light. Thus,

$$W_\lambda = \frac{c}{\nu^2} W_\nu = \frac{\lambda^2}{c} W_\nu. \tag{10.107}$$

The equivalent width depends on the number of atoms in the atmosphere along the line of sight that are in a state to absorb a photon; the more atoms there are, the stronger and broader the spectral line is. The number of absorbing atoms per gram of stellar material can be determined from the measurement of such widths. The width of a line depends on various factors:

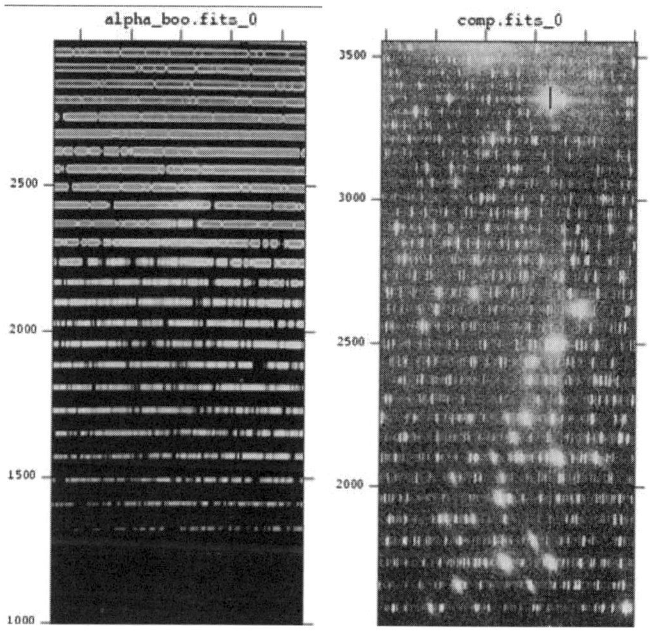

(a) (b)

Fig. 10.4 (a) High resolution two dimensional image of a K giant α Boo (Arcturus) spectrum displayed in 2k×4k CCD. Spectrum has 62 orders covering wavelength range 4000 - 10,000 Å with small gaps in between the orders. The numerous narrow dark lines are absorption lines superposed on the continuum. The image is taken with the Fiber-coupled VBT echelle spectrometer. (b) ThAr spectrum for wavelength calibration of programme stars; the white bright lines are emission lines for which accurate laboratory wavelengths are available. (Courtesy: B. E. Reddy).

- natural broadening that arises from the uncertainty in energy of the states involved in the transition,
- thermal or Doppler broadening due to motions of atoms; the distribution of velocities can be found from the Maxwell distribution,
- collisional broadening resulting from collisions between atoms, which depends on the frequency of collisions and hence on the density of the gas, and other factors like Stark broadening and Zeeman broadening etc.

The percentage of absorbing atoms or ions is compared with spectral line, which provide the abundance of the atom producing the line and can be estimated by quantum mechanics and a model of atmospheric line formation theory. Spectrum is mainly two types such as:

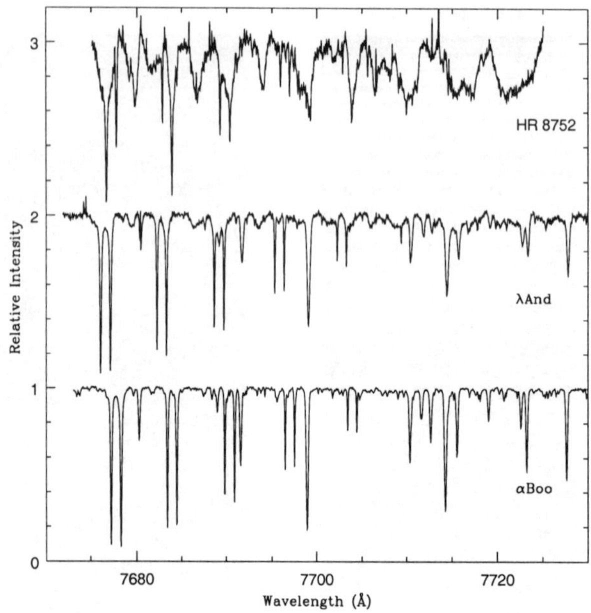

Fig. 10.5 Absorption spectra of three stars, HR 8752, λ And, and α Boo taken in the region of KI λ 7699 by means of the VBT Echelle spectrometer (Courtesy: N. K. Rao).

(1) Continuous spectrum: Continuous spectrum possesses energy at all wavelengths. Most continuous spectra are from hot, dense objects. Such a spectrum is also called a thermal spectrum, because hot, dense objects emit electromagnetic radiation at all wavelengths or colors. In the emerging spectrum of the star, bound-free and free-free processes (see section 11.2.4) give rise to continuum spectrum and scattering modify the continuum. When light with a continuous spectra is passed through a relatively cool gas, the wavelengths characteristic of the gas are absorbed, hence the spectra appear as dark lines, called Fraunhofer lines, superposed on a continuous background. Bound-bound transitions (see section 11.2.4) give rise to line spectrum. The absorption lines (see Figures 10.5a and 10.6) arise when the atoms, which are in lower state, absorb energy and get excited to higher energy levels.

(2) Discrete spectrum: Discrete spectrum possesses energy at certain wavelengths. A discrete spectrum is more complex because it depends on temperature, chemical composition of the object, the gas density, surface gravity, speed, etc. Hot gas under low pressure produces an emission line (see Figure 10.5b), as well as continuum spectrum. The lines

are brighter than the background spectrum which could provide information about flares (see section 11.1.1.2) and other atmospheric activity in the star. These lines are produced if the atoms return from the higher energy states to a lower state by emitting radiations. Line strengths of selected forbidden emission lines could provide the temperature, while intensity ratio of two suitable pair of emission lines can also yield density of plasma. The Doppler-shift in the line profile tells about the line-of-sight velocity as well. In the stellar atmosphere,

- the spectra may turn from photospheric absorption to chromospheric emission in the Sun,
- the forbidden lines from metastable atomic states as produced in regions with very low densities, for example planetary nebulae,
- the fluorescence emission lines[12] in complex spectra, like Fe II, occur in stars like symbiotic stars, that are rare objects, which have spectra indicating the presence of both cool and hot objects, and
- the stimulated emission lines appear in extended low-density clouds close to hot, eruptive stars, such as Eta Carinae.

10.2.6.1 *Hertzsprung-Russell (HR) diagram*

Stars are classified on the basis of their observed absorption line spectrum, particularly with regard to the appearance or disappearance of certain discrete features. The stars are divided into mainly O, B, A, F, G, K, M classes. These are further subdivided into subclasses denoted by numbers such as, $0, 1, \cdots, 9$; often decimals are used, for example B0.5. For normal stars, called main sequence stars, the sequence of star types is given in Table 10.2.

There are R, N, S type stars as well. The R, N type stars, also known as carbon (C) stars; S shows oxides of ZrO, LaO etc. The L type stars are the brown dwarfs. The types have been arranged into groups according to their surface temperatures from the hot surfaces to the cool. The O, B, and A-type stars are generally referred to as early, i.e., hot, spectral types, while cool stars such as G, K, and M-type are called as late type stars.

The relationship between absolute magnitude, luminosity, spectral classification, and surface temperature of stars can be depicted in a diagram, called Hertzsprung-Russell (HR) diagram, which was created in 1910 by

[12]Fluorescence lines appear when higher energy photons excite a sample, which would emit lower energy photons when the atom cascades down in the energy levels.

E. Hertzsprung and H. N. Russell. This diagram is used to study stellar evolution. The location of a star in such a diagram depends on the rate at which it is generating energy nuclear fusion in its core and the structure of the star itself. If spectral class of a star is known, the HR diagram (see Figure 10.6) may be used to provide a good estimate of its intrinsic luminosity. Traditionally HR diagram is spectral type and the other type is called color-magnitude diagram.

Table 10.2
Main-sequence stars

Type of stars	Spectral features	$\frac{M}{M_\odot}$	Temperature (°K)	$\log \frac{L}{L_\odot}$
O	He II, He I, N III Si IV, C III, O III	120	$> 25,000$	6.15
B	He I, H, C III C II, Si III, O II	17.5	11,000 - 25,000	4.72
A	H I, Ca II & H, Fe I, Fe II, Mg II, Si II	2.9	7,500 - 11,000	1.73
F	Ca II H & K, CH, Fe I, Fe II, Cr II, Ca I	1.6	6,000 - 7,500	0.81
G	CH, CN, Ca II, Fe I, Hδ, Ca I	1.05	5,000 - 6,000	0.18
K	CH, TiO, CN, MgH, Cr I, Fe I, Ti I	0.79	3,500 - 5,000	-0.38
M	TiO, CN, LaO, VO	0.51	$\leq 3,500$	-1.11
C	C_2, CN, CH, CO		$\leq$3,000	
S	ZrO, YO, LaO, CO, Ba		$\leq$3,000	

Most of the stars are concentrated in the region along a band, called main-sequence, which stretches from upper left corner to the lower right corner.

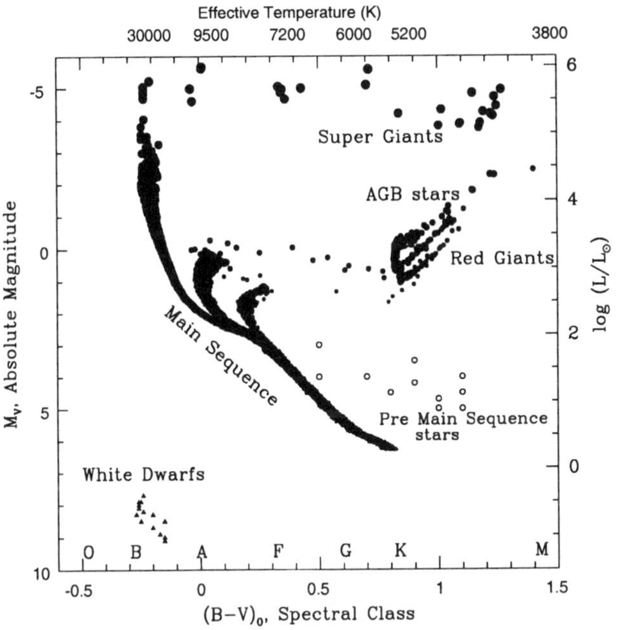

Fig. 10.6 The Hertzsprung-Russell (HR) diagram for Population I stars. Various stellar evolutionary stages are marked. This is a synthetic Colour-Magnitude diagram generated using the Padova evolutionary models for 100, 500 and 1000 Myr stellar population (Courtesy: A. Subramaniam).

Stars located on this band are known as main-sequence stars or dwarf stars; the hotter the stars, the brighter. The coolest dwarfs are the red dwarfs. Stars along the main-sequence seem to follow mass-luminosity relations. From the visual and wide eclipsing binaries given in Allen's (1976) tables, it is observed from a plot $\log(L_\star/L_\odot)$ against $\log(M_\star/M_\odot)$, in which $M_\odot (= 1.989 \times 10^{30}$ kg) is the solar mass and $M_\star$ the mass of the star of interest, for the visual and wide eclipsing binaries that for main-sequence stars, the luminosity varies as $(L_\star/L_\odot) = (M_\star/M_\odot)^{3.5}$, for high mass stars, while the relation is $(L_\star/L_\odot) = (M_\star/M_\odot)^{2.6}$ for the low mass stars with $M_\star < 0.3\ M_\odot$; Similarly, a plot of $\log(R_s tar/R_\odot)$ against $\log(M_\star/M_\odot)$ shows a mass-radius relation $(R_\star/R_\odot) = (M_\star/M_\odot)^{0.75}$. Star on the main-sequence may spend almost 90% of its lifetime. Stars of a solar mass may spend several billion years as a main-sequence star, while a massive star with

$\sim 40\ M_\odot$ may spend about a million years on the main-sequence. The Sun is 4.5 billion years old.

There are other prominent sequences such as giants, supergiants (above the giant sequence) sequences (see Figure 10.6), which lie above the main-sequence. The stars lie in those sequences, have the similar color or spectrum as the dwarfs in the main-sequence. The gap between the main-sequence and the giant sequence is referred to as Hertzsprung gap. The asymptotic branch (AGB branch) rises from the horizontal branch (where the absolute magnitude is about zero) and approaches the bright end of the red giant branch. The very small stars falling in the lower left corner are called white dwarfs.

10.2.6.2 *Spectral classification*

Classification of the stars based on their spectral features was found to be a powerful tool for understanding stars. In 1863, Angelo Secchi created crudely order of spectra and defined different spectral classes.

(1) Harvard spectral classifications: This spectral classification scheme was developed at Harvard Observatory in the early 20th century. Henry Draper begun this work in 1872. The Henry Draper (HD) catalogue was published in 1918-24, which contained spectra of 225,000 stars down to ninth magnitude. This scheme was based on the strengths of hydrogen Balmer absorption lines in stellar spectra. Now, the classification scheme relies on (i) the absence of lines, (ii) strengths or equivalent width (EW) of lines, (iii) the ratios of line strengths such as K-lines of Ca II compared to those of Balmer series. The important lines, e.g., (i) the hydrogen Balmer lines, (ii) lines of neutral and singly ionized helium, (iii) iron lines, (iv) the H and K doublet of ionized calcium at 396.8 nm and 393.3 nm, (v) the G band due to the CH molecules, (vi) several metal lines around 431 nm, (vii) the neutral calcium line at 422.7 nm, and (viii) the lines of titanium oxide (TiO) are taken into consideration. The main characteristics of the different spectral classes of stars are:

   - Type O: This type of stars are characterized by the lines from ionized atoms, such as singly ionized helium (He II) lines either in emission or absorption, and neutral helium (He I). The ionized He is maximum in early O-type star and He I and H I increases in later types. Doubly ionized nitrogen (N III) in emission, silicon (Si IV), carbon (C III), are visible, but H I lines are weak and increasing in later types. The rotational velocity is $\sim 130 - 200$ km s^{-1}. N III and He II are visible in emission at Of star.

- Type B: These stars are characterized by neutral He lines; He I (403 nm), lines in absorption are strongest at B2 and get weaker thereafter from B3 and completely disappear at B9. The singly ionized helium lines are disappearing and H lines begin to increase in strength. Other lines of elements such as the K line of Ca II, C II, C III, N II, Si III, N III, Si IV, O II, Mg II, and Si II-lines become traceable at type B3 and the neutral hydrogen lines are getting stronger. They posses large rotational velocity ~ 450 km s^{-1}. It is pertinent to note that in some O- and B-type stars, the hydrogen absorption lines have weak emission components either at the line center or in its wings. The B-type stars surrounded by an extended circumstellar envelope of hydrogen gas are referred to as Be or shell stars. Such stars are hot and fast rotating stars. The emission lines in H in their spectrum are formed in a rotationally flattened gas shell around the star. The shell and Be stars depict irregular variations, related to structural changes in the shell. In a given stellar field approximately 20% of the B stars are in fact Be stars. This percentage may go up in some young clusters where up to 60-70% of the B stars display the Be phenomenon, i.e., Balmer lines in emission and infrared excess. These stars are very bright and luminous compared to B-type stars due to the presence of their circumstellar envelope. In young clusters with many Be stars, the luminosity function may seem to contain massive stars, leading to an artificially top-heavy initial mass function[13] (IMF). Generally Be stars have high rotational velocities, which is of the order of ~ 350 km s^{-1}. The strongest emission line profiles of P Cygni have one or more absorption lines on the short wavelength side of

[13]Initial mass function is a relationship that specifies the distribution of masses created by the process of star formation. This function infers the number of stars of a given mass in a population of stars, by providing the number of stars of mass $M_\star$ per pc^3 and per unit mass. Generally, there are a few massive stars and many low mass stars. For masses $M_\star \geq 1\ M_\odot$, the number of stars formed per unit mass range $\xi(M_\star)$, is given by the power law,

$$\xi(M_\star) = \xi_0 M_\star^{-2.35},$$

in which $(M_\star)$ is the mass of a star; a star's mass determines both its lifetime and its contribution to enrich the interstellar medium with heavy elements at the time of its death.

emission line.

- Type A: These stars have strong neutral hydrogen, H I, lines, particularly AO type stars and dominate the entire spectrum, decreasing thereafter. He I lines are not seen. The metallic lines increase from A0 to A9 and Ca II H & K can also be traceable; Ca II K is half strong as (Ca II + Hϵ) lines in A5 stars. Among the other lines of Fe I, Fe II, Cr I, Cr II, Ti I, Ti II are also available. These stars rotate rapidly, but less than B-type stars. The peculiar A-type stars or (Ap[14] stars.) are strongly magnetized stars, where lines are split into several components by the Zeeman effect[15]. The lines of certain elements such as magnesium (Mg), silicon (Si), chromium (Cr), strontium (Sr), and europium (Eu) are enhanced in the Ap stars. There are lines namely, mercury (Hg), gallium (Ga) may also be seen. Another type of stars called Am stars have anomalous element abundances. The lines of rare earths and heaviest elements are strong in their spectra.
- Type F: In this category of stars, H I lines are weaker, while Ca II, H & K are strong. Many other metallic lines, for example, Fe I, Fe II, Cr II, Ti II, Ca I, and Na I become noticeable and get stronger. CH molecule (G-band) lines are visible at F3-type stars. The rotational velocity of these stars are less than 70 $\mathrm{km\,s^{-1}}$.
- Type G: The absorption lines of neutral metallic atoms and ions

[14] Additional nomenclatures are used as well to indicate peculiar features of the spectrum. Accordingly, lowercase letters are added to the end of a spectral type. These are (i) comp stands for composite spectrum, in which two spectral types are blended, indicating that the star is an unresolved binary, (ii) e – emission lines (usually hydrogen), (iii) [e] – forbidden emission lines, (iv) f – N III and He II emission, (v) He wk – weak He lines, (vi) k – spectra with interstellar absorption features, (vii) m – metallic, (viii) n – broad (nebulous) absorption lines due to fast rotation, (ix) nn – very broad lines due to very fast rotation, (x) neb – nebula's spectrum is mixed with the star's, (xi) p – peculiar spectrum, strong spectral lines due to metal, (xii) pq – peculiar spectrum, similar to the spectra of novae, (xiii) q – red and blue shift lines, (xiv) s – narrowly sharp absorption lines, (xv) ss – very narrow lines, (xvi) sh – shell star; B - F main-sequence star with emission lines from a shell of gas, (xvii) v – variable spectral features, (xviii) w – weak lines, and (xix) wl – weak lines (metal-poor star).

[15] When a single spectral line is subjected to a powerful magnetic field, it splits into more than one, a phenomena is called Zeeman effect, analogous to the Stark effect (the splitting of a spectral line into several components in the presence of an electric field); the spacing of these lines depends on the magnitude of the field. The effect is due to the distortion of the electron orbitals. The energy of a particular atomic state depends on the value of magnetic quantum number. A state of total quantum number breaks up into several substates, and their energies are slightly more or slightly less than the energy of state in the absence of magnetic field.

grow in strength in this type of stars. The H I lines get weaker, albeit Ca II H and K lines are very strong; they are strongest at G0. The metallic lines increase both in number and in intensity. The spectral type is established using Fe (λ 4143) & Hδ. The molecular bands of CH and Cyanogen (CN) are visible in giant stars. The other elements such as Ca II, Fe I, Hδ, Ca I, Fe I, H, Cr, Y II, Sr II are seen. The rotational velocity is a few km s^{-1}, which is typically the Sun's velocity.

- Type K: The H lines are weak in this kind of stars, though strong and numerous metallic lines dominate. The Ca II lines and G-band (CH molecule) are also very strong. The TiO, MgH lines appear at K5. The other lines, viz., Cr I, Fe I, Ti I, Ca I, Sr II, and Ti II are noticeable.
- Type M: The spectra are very complex in this type of stars; continuum is hardly seen. The molecular absorption bands of titanium oxide (TiO) becomes stronger. The other elements like CN, LaO, VO are also seen. A number of giant stars appear to be K or M type stars, albeit depict significant excess spectral features of carbon compounds, known as carbon stars. These stars, referred to as C-type stars, have C_2, SiC_2, C_3, CN, and CH strong molecular bands. The presence of these carbon compounds tend to absorb the blue portion of the spectrum, giving R- and N-type giants a distinctive red colour. The R-type stars posses hotter surfaces which otherwise resemble K-type stars. The late type giants, S-type stars (K5-M) show ZrO, LaO, CO, Ba, TiO molecular bands. These stars have cooler surfaces and resemble M-type stars.

It is found that the spectra of giant and supergiant G and K-type stars display K and H lines of Ca II in emission originating in the stellar chromosphere. Wilson and Bappu (1957) showed the existence of a remarkable correlation between the width of the emission in the core of the K line of Ca II and the absolute visual magnitude of late-type stars; the widths of the Ca II K emission cores increase with increasing stellar intrinsic brightness. Hence, they opined that Ca II emission line widths can be used as luminosity indicators.

(2) Yerkes spectral classifications: Unlike the Harvard classification, which is based on photospheric temperature, this scheme, also known as MKK (Morgan, Keenan, and Kellman, the authors of this classification) catalogue, measures the shape and nature of certain spectral lines to deduce

surface gravity of stars. The spectral type is determined from the spectral line strengths. This classification is based on the visual scrutiny of slit spectra with a dispersion of 11.5 nm/mm (Karttunen et al., 2000). A number of different luminosity classes are distinguished. These are: (i) Ia-0, most extreme supergiants or hypergiants, (ii) Ia, luminous supergiants, Iab, moderate supergiants, Ib, less luminous supergiants, (iii) II, bright giants, (iv) III, normal giants, (v) IV, subgiants, (vi) V, the main-sequence stars (dwarfs); the Sun may be specified as a G2V type star, (vii) VI, subdwarfs, and (vii) VII, white dwarfs.

The luminosity class is determined from spectral lines, which depend on the surface gravity. The luminosity effect in the stellar spectrum may be employed to distinguish between stars of different luminosities. The neutral hydrogen lines are deeper and narrower for high luminosity stars that are in the category of B to F spectral types. The lines from ionized elements are relatively stronger in high luminosity stars. The giants stars are redder the dwarfs of the same spectral class. There is a strong CN absorption band in the spectra of giant stars, which is almost absent in dwarfs.

10.2.6.3 *Utility of stellar spectrum*

In general by observing the stellar spectrum, one understands the physical conditions of the star. Certain lines are stronger than the rest of the lines at a given temperature, albeit less intense at a temperature either higher or lower than this. The line spectrum of a star provides the state of matter in the reversing layer. An atmosphere is considered to be in local thermodynamical equilibrium if the collisional processes dominate over radiation processes, and population of electrons and ions can be described by a thermal energy distribution.

Among others, analysis of stellar spectra may provide the temperature. From the analysis of spectral characteristics, as well as abundance analysis one infers the stellar evolutionary process. Some of the other information, which may be obtained from study of spectra are:

(1) Metallicity: The term 'metal' in astronomy is considered to be any element besides H and He. Stellar spectra depict the proportion of elements heavier than helium in the atmospheres. Metallicity is a measure of amount of heavy elements other than hydrogen and helium present in an object of interest. In general, it is given in terms of the relative amount of iron and hydrogen present, as determined by analyzing ab-

sorption lines in a stellar spectrum, relative to the solar value. The ratio of the amount of iron to the amount of hydrogen in the object, $(Fe/H)_\star$ is divided by the ratio of the amount of iron to the amount of hydrogen in the Sun, $(Fe/H)_\odot$. This value, denoted by, $[Fe/H]$, derived from the logarithmic formula:

$$[Fe/H] = \log \frac{(Fe/H)_\star}{(Fe/H)_\odot}. \tag{10.108}$$

The metallicity $[Fe/H] = -1$ denotes the abundance of heavy elements in the star is one-tenth that found in the Sun, while $[Fe/H] = +1$, denotes the heavy element abundance is ten times the metal content of the Sun.

(2) Chemical composition: The absorption spectrum of a star may be used to identify the chemical composition of the stellar atmosphere, that is type of atoms that make up the gaseous outer layer of the star. Moreover, if one element has relatively great abundance, its characteristic spectral line is strong. The chemical composition of the atmosphere can be determined from the strength of the spectral lines.

(3) Pressure, density and surface gravity: Spectral lines form all over the atmosphere. One assumes hydrostatic equilibrium and calculates pressure gradients etc. The surface gravity is the acceleration due to gravity on the surface of the celestial object, which is a function of mass and radius:

$$g_\star = \frac{GM_\star}{R_\star^2}, \tag{10.109}$$

in which $G(= 6.672 \times 10^{-11}\ \mathrm{m^{-3}\,kg^{-1}\,s^{-2}})$ is the gravitational constant, $M_\star$ the mass of the star, and $R_\star$ the radius of the star.
The surface gravity of a giant star is much lower than for a dwarf star since the radius of a giant star is much larger than a dwarf. Given the lower gravity, gas pressures and densities are much lower in giant stars than in dwarfs. These differences manifest themselves in different spectral line shapes. The density is a measure of mass per unit volume; the higher an object's density, the higher its mass per volume. Knowing the mass and the radius of an object, the mean density can be derived. The pressure and density are related to temperature through perfect gas law.

(4) Microturbulence: It arises from the small scale motions (up to 5 km/s) of the absorbing atoms over the thermal velocities. These motions in a

stellar atmosphere broaden the stars spectral lines and may contribute to their equivalent width. Microturbulence is prominent in saturated lines more distinctly. These lines are sensitive to microturbulence. The resultant broadening is given by,

$$\Delta\nu_D = \frac{\nu_0}{c}\left[\frac{2k_BT}{m} + v_m^2\right]^{1/2}, \tag{10.110}$$

where $\Delta\nu_D$ is the total broadening due to thermal and microturbulent motions, ν_0 the central frequency of the line, m the mass of the atom, and v_m the microturbulent velocity.

(5) Stellar magnetic field: Magnetic fields in the Sun and other late-type stars, are believed to play a key role on their interior, their atmospheres and their circumstellar environment, by influencing the transport processes of chemical elements and angular momentum. By studying the topology of magnetic fields, namely large- and small-scale structures, one may understand their physical origins, if they are produced within stellar plasma through hydrodynamical processes or represent a fossil remnant from a previous evolutionary stage like those of chemically peculiar stars; the potential impact of these magnetic fields on long-term stellar evolution may also be studied. With a high resolution spectro polarimeters, one can detect stellar magnetic fields through the Zeeman effect they generate in the shape and polarization state of spectral line profiles.

(6) Stellar motion: The stars are in motion and their lines are therefore Doppler shifted. The amount of the shift, depending on the velocity, its radial velocity, v_r. The radial velocity is defined as the velocity of a celestial object in the direction of the line of sight, it may be detected by looking for Doppler shifts in the star's spectral lines. The radial velocity, v_r, is given by,

$$v_r = \frac{c\Delta\lambda}{\lambda_r}, \tag{10.111}$$

in which c is the speed of light and $\Delta\lambda$ the wavelength shift, can be determined.

The spectral lines is shifted towards the blue if the star is approaching; towards the red if it is receding. An observer can measure it accurately by taking a high-resolution spectrum and comparing the measured wavelengths of known spectral lines to wavelengths from laboratory measurements. Such a method has also been used to detect exo-solar planets.

(7) Stellar rotation: The line profile also reflects rotational broadening, from which it is possible to derive how quickly the star is rotating.

10.3 Binary stars

Binary star is a system of two close stars moving around each other in space and gravitationally bound together. In most of the cases, the two members are of unequal brightness. The brighter, and generally more massive, star is called the 'primary', while the fainter is called the 'companion' or secondary.

10.3.1 *Masses of stars*

Determinations of stellar masses are based on an application of Kepler's third law of orbital motions, which explains that the ratio of the squares of the revolutionary periods for two bodies is equal to the ratio of the cubes of their semimajor axes:

$$\frac{P_1^2}{P_2^2} = \frac{a_1^3}{a_2^3}, \tag{10.112}$$

where a_1, a_2 are the semi-major axes of two orbits and P_1, P_2 the corresponding orbital periods.

Binaries are characterized by the masses of their components, M_1, M_2, orbital period, eccentricity[16], e, and the spins of the components. Unlike the case of the solar system, where one ignores the mass of the planet, $M_\oplus$, i.e., $M_\odot + M_\oplus \simeq M_\odot$, since the mass of the Sun, $M_\odot$, is much bigger, here the masses of both objects are included. Therefore, in lieu of $GM_\odot P_\oplus^2 = 4\pi^2 a_\oplus^2$, where $P_\oplus$ is the period of revolution of a planet around the Sun, and $a_\oplus$ the mean distance from the planet to the Sun, one writes as,

$$P^2 = \frac{4\pi^2 a^3}{G(M_1 + M_2)}, \tag{10.113}$$

in which P is the period, G the gravitational constant, $a(= a_1 + a_2)$ the semi-major axis of the relative orbit, measured in AU, and $M_1 + M_2$ the combined masses of the two bodies. By the definition of star's parallax Π,

[16]Eccentricity is a quantity defined for a conic section that can be given in terms of semimajor and semiminor axes. It can also be interpreted as the fraction of the distance along the semimajor axis at which the focus lies, i.e., $e = c/a$, where c is the distance from the center of the conic section to the focus and a the semimajor axis.

one gets,

$$M_1 + M_2 = \frac{a^3}{\Pi^3 P^2}. \qquad (10.114)$$

This equation (10.114) enables to determine the sum of the masses of a binary star when the parallax and its orbit are known (Smart, 1947).

10.3.2 *Types of binary systems*

Binary systems are classified into four types on the basis of the techniques that were adopted to discover them. However, selection affects limit the accuracy of binary surveys using any one particular technique. The spectroscopic searches are insensitive to wide orbits and visual searches are insensitive to distant and short period systems. Double stars with nearly equal magnitudes are nearly twice as bright as either component, resulting in skewed statistics in a magnitude limited spectroscopic survey.

Some binary systems are close to the observer and their components can be individually resolved through a telescope; their separation is larger than about $0.1''$. The stars in such a system, known as visual binaries. In other cases, the indication of binary system is the Doppler shift of the emitted light. Systems in this case, known as spectroscopic binaries. If the orbital plane is nearly along the line of sight of the observer, the two stars partially or fully occult each other regularly, and the system is called an eclipsing binary, for example Algol system[17] and β Lyrae[18]. An eclipsing binary system (see section 10.3.2.3) offers a direct method to gauge the distance to galaxies to an accuracy of 5% (Bonanos, 2006). Another type of binaries, referred to as astrometric binaries, that appear to orbit around an empty space. Any binary star can belong to several of these classes, for example, several spectroscopic binaries are also eclipsing binaries.

[17]The Algol system (β Persei) is a spectroscopic binary system with spherical or slightly ellipsoidal components. It varies regularly in magnitude from 2.3 to 3.5 over a period of a few days. This system is a multiple (trinary) star comprising of (i) Algol A (primary), a blue B8-type main sequence star, (ii) Algol B (sub-giant), K2-type star that is larger than the primary star, and (iii) Algol C, A5-class that orbits the close binary pair, Algol AB. The average separation between Algol AB system and the Algol C is about 2.69 AUs, which makes an orbit of 1.86 years. Algol A and B form a close binary system, the eclipsing binary, that are separated by 10.4 million km. This eclipsing system is semi-detached with the sub-giant filling its Roche-lobe and transferring the material at a modest rate to its more massive companion star (Pustylnik, 1995).

[18]β Lyrae is an eclipsing contact binary star system made up of a B7V type star and a main-sequence A8V star. Its components are tidally-distorted by mutual gravitation (Robinson et al., 1984); its brightness changes continuously.

10.3.2.1 *Visual binaries*

Any two closely-spaced stars may appear as a double star. The apparent alignment of these stars are not close enough to be gravitationally bound. Visual binary stars are gravitationally bound to each other but otherwise do not interact. The relative positions of these components can be plotted from long-term observations, from which their orbits can be derived. The relative position of the components changes over the years as they move in their orbit. Optical double stars have an apparent alignment of stars that are not actually close enough to be gravitationally bound. Although they appear to be located next to one another as seen from Earth, these stars may be light years apart.

10.3.2.2 *Spectroscopic binaries*

Spectroscopic binaries are close together that they appear as a single star. Some of them are spatially unresolved by the telescopes. The spectrum of such systems can show up the existence of two stars, since their spectrum lines are amalgamated. Many such binary systems have been detected from the periodic Doppler shifts of the wavelengths of lines seen in the spectrum, as the stars move through their orbits around the center of mass. There are two types of spectroscopic binaries:

(1) Double-lined spectroscopic binary system: In this system, features from both stars are visible in the spectrum; two sets of lines are visible. These lines show a periodic back and forth shift in wavelength, but are in opposite direction relative to the center of mass of the system.
(2) Single-lined spectroscopic binary system: In the spectrum of this spectroscopic system, all measurable lines move in phase with one another. A single set of lines is seen since one component is much brighter than the other.

From an analysis of the radial velocity (see Figure 10.7) of one or both the components as a function of time, one may determine the elements of the binary orbit. The orbital plane is inclined to the plane of the sky by an angle, i, which cannot be determined by spectroscopic data alone, since the observed radial velocity v_r yields the projection of the orbital velocity v along the line of sight, i.e.,

$$v_r = v \sin i. \tag{10.115}$$

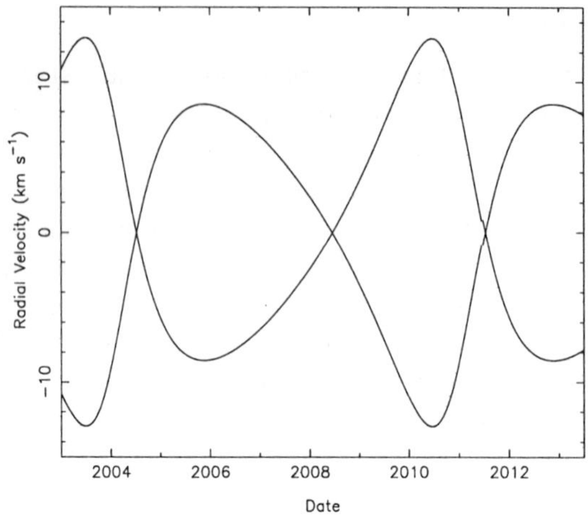

Fig. 10.7 Radial velocity curves of a binary system.

It is possible to determine the mass provided (i) the system is a double-lined spectroscopic binary and (ii) it is an eclipsing binary. Let the orbits of both the stars be circular. From the observed radial velocities, one may determine the projected radii of the two orbits,

$$a_j = \frac{v_j P}{2\pi} = \frac{v_{r_j} P}{2\pi \sin i}, \tag{10.116}$$

where $j = 1, 2$ and v_{r_j} are the amplitudes of the observed oscillations in the radial velocities of both the stars and P the orbital period.

From the definition of center of mass $M_1 a_1 = M_2 a_2$, the mass ratio is,

$$\begin{aligned} \frac{M_1 \vec{r}_1 + M_2 \vec{r}_2}{M_1 + M_2} &\rightarrow \frac{M_1}{M_2} \\ &= \frac{v_{r_2}}{v_{r_1}} = \frac{r_2}{r_1} = \frac{a_2}{a_1}, \end{aligned} \tag{10.117}$$

which is independent of the inclination angle[19], i. Here a_1 and a_2 are the radii of the orbits, r_1 and r_2 the respective distances between the center of mass and the centers of the individual objects (see Figure 10.8); $\vec{r}_1$ and $\vec{r}_2$ are oppositely directed.

[19]Inclination is the angle between the line of sight and the normal of the orbital plane. Values range from 0° to 180°; for $0° \leq i < 90°$, the motion is called direct. The companion then moves in the direction of increasing position angle, i.e., anticlockwise. For $90° < i \leq 180°$, the motion is known as retrograde.

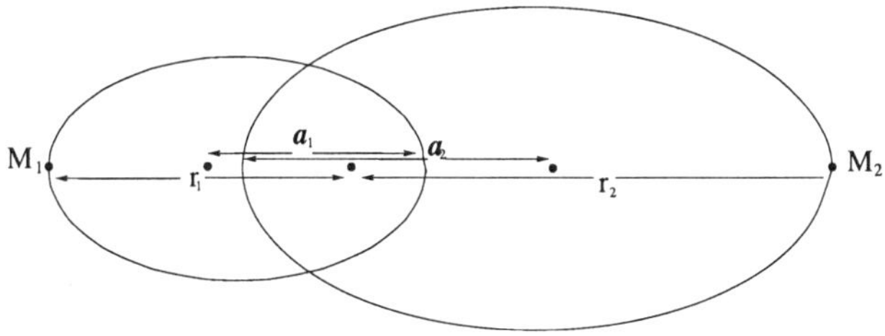

Fig. 10.8 Component of a binary system move around their common center of mass.

It may be possible to estimate the perfect mass of this system if (i) both stars are visible, (ii) their angular velocity is sufficiently high to allow a reasonable fraction of the orbit to be mapped, and (iii) the orbital plane is perpendicular to the line of sight. Defining mass function, $f(M)$, by,

$$f(M) = \frac{(a_1 \sin i)^3}{P^2}, \tag{10.118}$$

and writing $a = a_1 + a_2$, one finds,

$$a_1 = \frac{aM_2}{M_1 + M_2}. \tag{10.119}$$

The mass function is a fundamental equation for the determination of binary system parameters deriving from Kepler's second and third laws. It relates the masses of the individual components, M_1, M_2 and the inclination angle, i, through two observable quantities, the orbital period and the radial velocity which can be obtained from radial velocity curve; individual masses can be obtained if the inclination i is known. According to equation (10.116), the observed orbital velocity is written as,

$$v_{r_1} = \frac{2\pi a_1 \sin i}{P}. \tag{10.120}$$

Substituting (10.113), it is obtained

$$v_{r_1} = \frac{2\pi a}{P}\frac{M_2 \sin i}{M_1 + M_2}. \tag{10.121}$$

Therefore the mass function is expressed as,

$$f(M) = \frac{M_2^3 \sin^3 i}{(M_1 + M_2)^2} = \frac{v_{r_1}^3 P}{2\pi G}. \tag{10.122}$$

The mass function $f(M)$ provides the lower limit to the mass, i.e., at the extreme case when the mass of the companion is neglected (Casares,

2001) and the binary is seen edge on ($i = 90°$). Usually it is improbable to uncover the inclination angle. However, for large samples of a given type of star it may be appropriate to take the average inclination to determine the average mass.

$$\begin{aligned}\frac{1}{\pi/2}\int_0^{\pi/2}\sin^3 i\, di &= -\frac{2}{3\pi}\left(2+\sin^2 i\right)\cos i|_0^{\pi/2} \\ &= \frac{4}{3\pi} \approx 0.42.\end{aligned} \tag{10.123}$$

In reality, it is difficult to measure systems with $i \sim 0°$, since the radial velocity is small. This introduces a selection effect and means that the average value of $\sin^3 i$ in real samples is larger, and is of the order of $\sin^3 i =\sim 0.667 = 2/3$ (Aitken, 1964). For the single-lined spectroscopic binaries, P and v_{r_1} are observed, hence the masses of the components or the total mass cannot be determined.

10.3.2.3 *Eclipsing binaries*

Eclipsing (or photometric) binaries appear as a single star, but based on its brightness variation and spectroscopic observations, one may infer that it is two stars in close orbit around one another. If the two stars have their orbital planes lying along the observer's line of sight, they block each other from the sight during each orbital period, thus causing dips in the light curve[20]. The primary minimum occurs when the component with the higher surface luminosity is eclipsed by its fainter companion. The light curves obtained using photometry contain valuable information about the stellar size, shape, limb-darkening, mass exchange, and surface spots. The stages of eclipse may be described as:

- if the projected separation, ρ, between the two stars is greater than their combined radius, $(R_1 + R_2)$, in which R_1, R_2 are the radii of the primary and secondary components respectively, no eclipse takes place,
- if the separation, ρ is smaller than the combined radius, $(R_1 + R_2)$ or greater than $\sqrt{R_1^2 - R_2^2}$, one observes a shallow eclipse,
- while deep eclipse can be envisaged if the above condition is reversed, i.e., $\sqrt{R_1^2 - R_2^2} > \rho > (R_1 - R_2)$, and
- an annular eclipse is seen if the separation is less than the difference in diameter of these two stars, i.e. $\rho < (R_1 - R_2)$.

[20]A brightness against time plot for a variable star is called light curve.

The above conditions are valid if R_1 is greater than R_2.

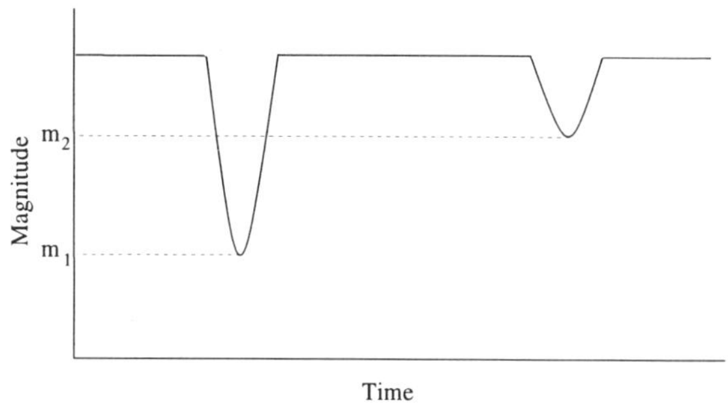

Fig. 10.9 Typical light curve of an eclipsing binary system.

The shape of the light curve (see Figure 10.9) of the eclipsing binaries depends mostly on the relative brightness of the two stars. Unless both the components are identical, the deeper curve one takes as the primary eclipse. One period of a binary system has two minima. If the effective temperatures of these components are T_{e1} and T_{e2}, and their radius is R, their luminosities are given by,

$$L_1 = 4\pi R^2 \sigma T_{e1}^4; \qquad L_2 = 4\pi R^2 \sigma T_{e2}^4. \tag{10.124}$$

The maximum brightness on light curve corresponds to the total intensity $L = L_1 + L_2$. The intensity drop is defined by the flux multiplied by the area covered due to eclipse. In terms of absolute bolometric magnitudes (see equation 10.48), the depth of the primary minimum is derived as,

$$\begin{aligned} m_1 - m &= 2.5 \log \frac{L}{L_1} = 2.5 \log \frac{4\pi R^2 \sigma (T_{e1}^4 + T_{e2}^4)}{4\pi R^2 \sigma T_{e1}^4} \\ &= 2.5 \log \left[1 + \left(\frac{T_{e2}}{T_{e1}} \right)^4 \right]. \end{aligned} \tag{10.125}$$

Similarly, the depth of the secondary minimum is,

$$m_2 - m = 2.5 \log \left[1 + \left(\frac{T_{e1}}{T_{e2}} \right)^4 \right]. \tag{10.126}$$

Since both the stars are in close orbit around one another, one of them may draw material off the surface of the other through Roche-lobe[21]. For

[21]The Roche-lobe is the region of space around a star in a binary system within which orbiting material is gravitationally bound to it. The uppermost part of the stellar

instance, W UMa variables[22], are tidally[23] distorted stars in contact binaries. A large-scale energy transfer from the larger, more massive component to the smaller, less massive one results in almost equalizing surface temperatures over the entire system. The components of such a contact binary rotate very rapidly ($v \sin i \sim 100 - 150$ km s^{-1}) as a result of spin-orbit synchronization due to strong tidal interactions between the stars.

10.3.2.4 *Astrometric binaries*

Astrometric binaries are the binaries that are too close to be resolved or the secondary is much fainter than the primary that one is unable to distinguish them visually. The presence of the faint component is deduced by observing a wobble (oscillatory motion) in the position of the bright component caused by the transverse component of a companion's motion. Such a perturbation takes place due to gravitational influence from its unseen component on the primary star. This periodical short motion has a radial counterpart measurable by spectrometry. In astrometric binaries, the orbit of the visible object about the center of mass can be observed. If the mass of this object is estimated from its luminosity, the mass of the invisible companion can also be estimated.

---

atmospheres forms a common envelope. As the friction of the envelope brakes the orbital motion, the stars may eventually merge (Voss and Tauris, 2003). At the Roche lobe surface, counteracting gravitational forces due to both stars effectively cancel each other out.

[22]W UMa variables are binaries consisting of two solar type components sharing a common outer envelope. These are the prototype of a class of contact binary variables and are classified as yellow F-type main-sequence dwarfs. Their masses range between 0.62 $M_\odot$ and 0.99 $M_\odot$, and radii varies from 0.83 $R_\odot$ to 1.14 $R_\odot$. Unlike with normal eclipsing binaries, the contact nature makes it difficult to guess precisely when an eclipse of one component by the other begins or ends. During an eclipse its apparent magnitude ranges between 7.75 and 8.48 over a period of 8 hours. These variables depict continuous light variations. Spectra of many such binaries show H and Ca II K emission lines, which are seen during eclipses (Struve, 1950). There are two subclasses of W UMa stars, namely (i) A-type and (ii) W-type systems. The former have longer periods, and are hotter having larger total mass. They posses a smaller mass-ratio and are in better contact. The primary star is hotter or almost the same temperature as the secondary, while in the case of the latter type, the the secondary appears to be hotter and the temperature difference is larger.

[23]Tidal force is a secondary effect of the gravitational force and comes into play when the latter force acting on a body varies from one side to another. This can lead to distortion of the shape of the body without any change in volume and sometimes even to breaking up of the system on which the former force acts.

10.3.3 *Binary star orbits*

The position of the companion of a binary system with respect to the primary is specified by two coordinates, namely (i) the angular separation, and (ii) the position angle. Figure (10.10) represents part of the celestial sphere in which A is the primary and B is the companion. Here AN defines the direction of the north celestial pole, which is part of the meridian through A. The angle NAB, denoted by θ, is the position angle of B with respect to A, which is measured from $0°$ to $360°$ towards east as shown. The angular distance between A and B is termed as the separation and is denoted by ρ, thus ρ and θ define the position of the companion B with respect to the primary, A.

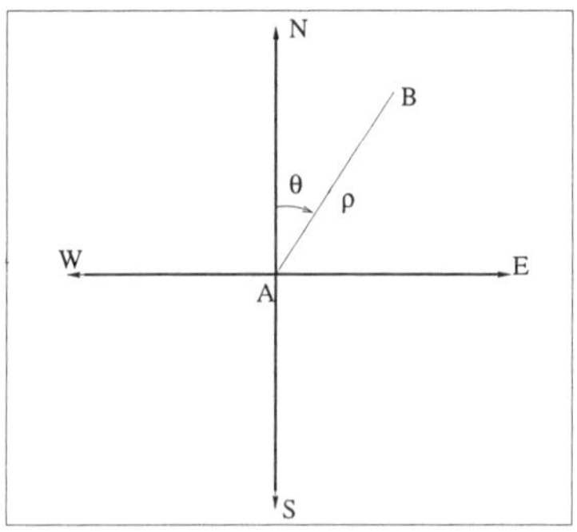

Fig. 10.10 Describing position of companion.

Due to mutual gravitational attraction, both the stars move around a common center (barycenter) of mass of the system, following Kepler's first Law, which states that the orbit of a planet is an ellipse with the Sun at one of the focii. Mathematically,

$$r = \frac{l}{1 + e\,\cos\upsilon}, \tag{10.127}$$

where υ, known as true anomaly, is the angle between the radius vector $\vec{r}$ and a constant vector (eccentricity) lying in the orbital plane, $\vec{e}$, which is considered to be the reference direction, $l[= a(1 - e^2)]$ the semi latus rectum, $\vec{r} \cdot \vec{e} = r\,e\,\cos\theta$, and a the semi-major axis of the orbit.

The center of mass of a binary system is nearer to the more massive star, but the motion of the secondary with respect to the primary would describe an elliptic orbit. This is the true orbit, the plane of which is not generally coincident with the plane of the sky[24] at the position of the primary, and its plane is the true orbital plane (Smart, 1947).

Each star follows Kepler's second law on its own, sweeping out equal areas in equal times within its own orbit[25], according to which the rate of description of area swept out in the infinitesimal interval, Δt, i.e., $(r^2\Delta\theta)/2$, divided by Δt. Mathematically,

$$r^2\frac{d\theta}{dt} = h, \tag{10.128}$$

in which $h(=$ constant), is twice the rate of description of area by the radius vector.

Since the entire area of the ellipse is πab, which is described in the interval defined by the period P, one finds,

$$\frac{2\pi a^2\sqrt{(1-e^2)}}{P} = h, \tag{10.129}$$

where the mean motion per year, $\mu = 2\pi/P$.

Finding the orbital elements of a binary system is of paramount importance in the study of binary stars since it is the only way to obtain the masses of the individual stars in that system. From this the elements of true orbit can be calculated. The absolute size of the orbit can be found if the distance of the binary is known (for example via parallax).

10.3.3.1 *Apparent orbit*

The orbit obtained by observations is the projection of the true orbit on the plane of the sky. The projection of the true orbit on the plane of the sky is referred to as apparent orbit. Both these orbits are ellipses (Smart, 1947). In general, the true orbital plane is distinct from the plane perpendicular to the line of sight. This plane is inclined against the plane of the sky with angle i. Hence instead of measuring a semi-major axis length a, one measures $a\cos i$, in which i is the inclination angle. This projection distorts the ellipse: the centre of mass is not at the observed focus and the

---

[24] Plane of the sky is the phrase, which means the tangent plane to the celestial sphere at the position of the sky.

[25] A planet in the solar system executes elliptical motion around the Sun with constantly changing angular speed as it moves about its orbit. The point of nearest approach of the planet to the Sun is called perihelion, while the point of furthest separation is known as aphelion. Hence, according to the Kepler's second law, the planet moves fastest when it is near perihelion and slowest when it is near aphelion.

observed eccentricity is different from the true one. This makes it possible to determine i if the orbit is known precisely enough.

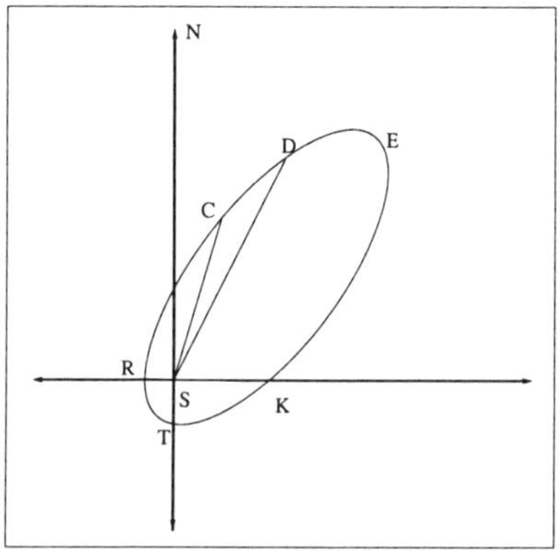

Fig. 10.11 Apparent orbit of a binary star

The apparent orbit may be determined if one determines the size of the apparent ellipse (its semi-major axis), eccentricity, position angle of the major axis, and the two coordinates of the center of the ellipse with respect to the primary star. Let the ellipse in Figure 10.11 represent the apparent orbit and S the primary; S is generally not at a focus of this ellipse. If SN denotes the direction defining position angle $\theta = 0°$ and SR that of $\theta = 90°$, the general equation of the ellipse referred to SN and SR as x and y axes respectively is given by,

$$Ax^2 + 2Hxy + By^2 + 2Gx + 2Fy + 1 = 0. \tag{10.130}$$

Equation (10.130) has five independent constants, namely, A, H, B, G, F. If the companion is at C, an observation gives ρ and θ, from which the rectangular coordinates x and y of C are derived as,

$$x = \rho \cos\theta; \qquad y = \rho \sin\theta. \tag{10.131}$$

Theoretically, five such observations spread over the orbit are sufficient to determine the five constants, A, $B, \cdots F$, of equation (10.130), however, owing to unavoidable errors in measuring ρ and θ, the ellipse cannot be determined accurately in this way. Accurate orbit cannot be found with a few observations. A large number of observations spread over many years are required to obtain a series of points such as C, D, E, $\cdots$ on the ellipse.

10.3.3.2 *Orbit determination*

Various methods are available to determine the elements of the orbit of a binary system, each with its own merits. Hartkopf et al. (1989, 1996), used a method based on 3-D grid search technique, which uses visual measurements along with interferometric data to calculate binary system orbits. If the period, P, eccentricity, e, and the time of periastron[26] passing, τ, are known roughly, the four Thiele-Innes elements (A, F, B, and G) and therefore the geometric elements, viz., semi-major axis, a_0, orbital inclination, i, the longitude of ascending node[27], Ω, the argument of periastron passage[28], ω can be determined by least square method. Once the apparent orbit is plotted with this method, P, e, and τ are obtained without much error. These values may be used to obtain more accurate orbit in Hartkopf's method. This method is relatively straightforward in its mathematical formulation. Given (P, e, τ) and a set of observations (t, x_i, y_i), the eccentric anomalies E are found via the equation,

$$M = E - e \sin E. \tag{10.132}$$

where

$$M = \frac{2\pi}{P}(t - \tau),$$

is the mean anomaly of the companion at time t. Once E is obtained, normalized rectangular coordinates X_i and Y_i are determined by a set of equations,

$$X_i = \cos E - e, \tag{10.133}$$

$$Y_i = \sqrt{(1 - e^2)} \sin E. \tag{10.134}$$

The four Thiele-Innes elements A, F, B, and G (Heintz, 1978) are found by a least squares solution of the equations,

$$x_i = AX_i + FY_i, \tag{10.135}$$

$$y_i = BX_i + GY_i. \tag{10.136}$$

[26]Periastron is the point in the orbital motion of a binary star system when the two stars are closest together, while the other extremity of the major axis is called apastron.

[27]The ascending node is the node where the object moves North from the southern hemisphere to the northern, while the descending (or south) node is where the object moves back South.

[28]Argument of periastron is the angle between the node and the periastron, measured in the plane of the true orbit and in the direction of the motion of the companion.

Once Thiele-Innes elements are obtained, the orbital elements can be deduced from it. However, Hartkopf's method requires a previous knowledge of the period of the system.

Another method known as, Kowalsky's method is used to determine the elements of a binary system. From a set of points, (x_i, y_i), five constants A, H, B, G and F (see equation 10.130) are derived. By applying least square method, i.e., minimizing sum of squares of residual with respect to each constant we obtain five equations. These equations are written using matrices as,

$$\begin{pmatrix} \sum x_i^4 & 2\sum x_i^3 y_i & \sum x_i^2 y_i^2 & 2\sum x_i^3 & 2\sum x_i^2 y_i \\ \sum x_i^3 y_i & 2\sum x_i^2 y_i^2 & \sum x_i y_i^3 & 2\sum x_i^2 y_i & 2\sum x_i y_i^2 \\ \sum x_i^2 y_i^2 & 2\sum x_i y_i^3 & \sum y_i^4 & 2\sum x_i y_i^2 & 2\sum y_i^3 \\ \sum x_i^3 & 2\sum x_i^2 y_i & \sum x_i y_i^2 & 2\sum x_i^2 & 2\sum x_i y_i \\ \sum x_i^2 y_i & 2\sum x_i y_i^2 & \sum y_i^3 & 2\sum x_i y_i & 2\sum y_i^2 \end{pmatrix} \begin{pmatrix} A \\ H \\ B \\ G \\ F \end{pmatrix} = - \begin{pmatrix} \sum x_i^2 \\ \sum x_i y_i \\ \sum y_i^2 \\ \sum x_i \\ \sum y_i \end{pmatrix}. \tag{10.137}$$

Representing first matrix by U, the second by V and the third by W, one writes,

$$UV = W, \tag{10.138}$$

and therefore, one may invert the matrix directly and solve using,

$$V = U^{-1}W. \tag{10.139}$$

The elements of matrix V provide the constants of the apparent orbit. In the reduction of the values of the unknowns, a triangular matrix formed by the diagonal elements (of the symmetric-square coefficient matrix) and those below them are used. An additional column matrix with the number of rows equal to the number of unknowns and initial elements equal to -1 is also used in the reduction procedure. Finally the values of the unknowns are directly given by the elements of the column matrix. The derived coefficients of the general second degree equation are then used to calculate the parameters of the apparent ellipse along with some parameters of the true ellipse. The true orbital parameters, the semimajor axis, eccentricity, longitude of the ascending node, longitude of the periastron passage and inclination of the orbital plane with respect to the line of sight are computed from the coefficients using Kowalsky's method. These elements, in turn, are used to compute the mean anomaly. From the linear relationship between the time of observation and mean anomaly, the time of periastron passage and the orbital period are determined using a least square technique, provided proper cycle information is made available to the input data.

In order to improve the accuracy, all the true orbital elements, derived as above, are then taken as initial guess values in an iterative, non-linear least square solution and the final values of all the orbital parameters are determined simultaneously. The program improves the parameter values by successive iterations. There are certain broad similarities between the two methods in terms of the least square fitting technique and the iterative approach to the improvement of the accuracy of result. In the iterative technique, the derived values of the constants are used as input. The position of the secondary component (ρ, θ) can be expressed as functions of the constants A, H, B... and time. The technique involves Taylor expansion of the functions about the input values. The increments of the constants are found out and are added to the initial values. These new values are again considered as inputs and the whole procedure is repeated until the values converge to the sixth place of decimal. But, the methods are essentially different in the solution techniques used and in the nature of input data. In Kowalsky's method, unlike Hartkopf's method, only the observations along with their respective epochs need to be given as inputs. No apriori estimation of orbital parameters is required. Since for most of the binary systems, the period is not available, this method can be used to get a good estimate of the period of the system.

An algorithm based on least square method is used (Saha et al. 2007) to obtain the plots (see Figure 10.12) and orbital calculations. The normal equations, in all cases are solved using cracovian matrix[29] elimination technique (Kopal, 1959). This method provides the same result as that given by the matrix inversion method, but involves a fewer number of steps. The orbit determination method presented here is the first one to use cracovian matrix elimination technique in an orbital program. The method has a system of giving different weightage to data obtained from different sources, but, in this the same weightage (unity) has been attributed to all interferometric data, speckle and non-speckle alike. Only observations with very high residues are eliminated from the data by assigning zero weightage to

[29]Cracovian matrices undergo 'column-by-column' (or row-by-row) matrix multiplication, which is non-associative in contrast with the usual 'row-by-column' matrix multiplication that is associative (Banachiewicz, 1955, Kocinsli, 2002). Cracovians were introduced into geodesic and astronomic calculations, spherical astronomy, celestial mechanics, determining orbits in particular.

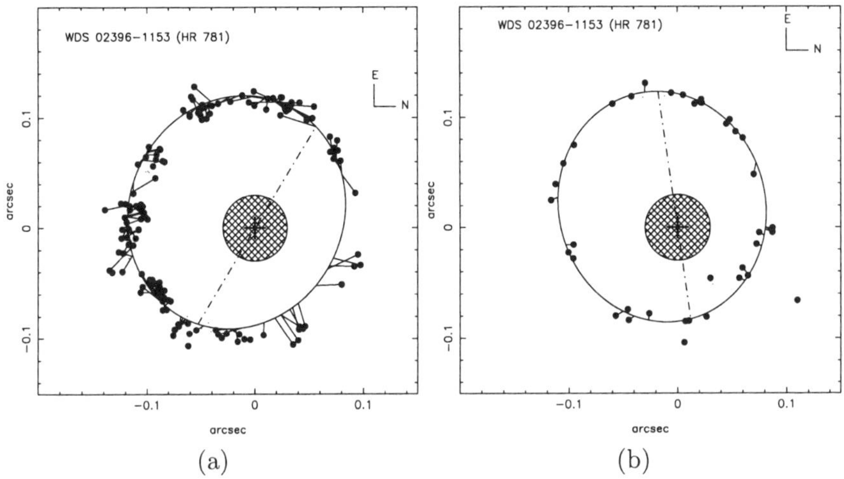

Fig. 10.12 (a) Non-speckle orbit of HR781 (ϵ Ceti in Cetus constellation) and (b) orbit of HR781 based on speckle interferometric measurements.

them. The probable errors in the orbital elements are obtained from the probable errors in the coefficients. The standard deviation of the fit is derived from,

$$\sigma = \sqrt{\frac{w_i(\text{sum of squares of residuals in x and y})}{(n - L)}}, \tag{10.140}$$

in which w_i is the weight of the observation, n the number of observations, and L the number of unknowns solved for.

The probable errors, p_e, in the unknowns are estimated using,

$$p_e = 0.6745\sigma\, w_i. \tag{10.141}$$

The elements of the last row of the triangular matrix following the final reduction provide the squares of the w_i.

10.4 Conventional instruments at telescopes

Hand-drawing from eye observations had been used in astronomy since Galileo. A limited amount of information about the celestial objects by this process was obtained till the end of the 19th century. The invention of photographic emulsion, followed by the development of photo-electric photometry had made considerable contribution in the field of observational astronomy. With the introduction of modern detectors, CCDs, during the latter part of the last century, stellar and galactic astronomy became rich in harvest. In the last few decades, several large telescopes with sophisticated equipment also came into existence. In what follows, a few such equipment, baring interferometers that are being used at the focus of a telescope to observe the various characteristics of a celestial object are illustrated.

10.4.1 *Imaging with CCD*

Till a few decades ago, astronomers used photographic technique to record images or spectra of celestial objects. Such a technique was employed in astronomy as early as 1850, when W. Bond and J. Whipple took a Daguerreotype of Vega. Silver bromide dry emulsions were used first around 1880. Though the photographic film was an inefficient detector, it had served as an imaging medium till a few decades ago other than the human eye. By exposing photographic plates for long periods, it became possible to observe much fainter objects than were accessible to visual observations. However, the magnitudes determined by the photographic plate are not, the same as those determined by the eye. This is because the sensitivity of the eye reaches a peak in the yellow-green portion of the spectrum, whereas the peak sensitivity of the basic photographic emulsion is in the blue region of the spectrum; the red sensitive emulsions are also available. Nevertheless, the panchromatic photographic plates may yield photo-visual magnitudes, which roughly agree with visual magnitudes by placing a yellow filter in front of the film. The greatest advantage of photography over visual observations was that it offered a permanent record with a vast multiplexing ability. It could record images of hundreds of thousands of objects on a single plate. However, a few percent of the photons reaching the film contribute to the recorded image. Its dynamic range is very low. It cannot record brightness differing by more than a factor of a few hundreds. Owing to low quantum efficiency of such emulsion, it requires a lot of intensity to expose a photographic plate. The 'dark background' effect becomes prominent; if a very faint object is observed, and irrespective of the exposure time, the object is drowned in a background from the plate brighter than the object.

Astronomers faced another problem concerning the measurement of the flux at each point of the plate whether it represented a field image or a spectrum. In order to address this problem, the microphotometer was developed by Pickering (1910). The photographic plates were scanned by such an instrument which locally measured, using a photo-electric cell, the intensity from the illuminated plate. With a developed version of such an instrument became a valuable tool for many investigations besides astronomy. Many branches of applied sciences such as digital cartography, electron microscopy, medicine, radiography, remote sensing are also benefited.

Fig. 10.13 *BVR* color image of the whirlpool galaxy M 51 taken at the 2 m Himalayan Chandra telescope (HCT), Hanley, India. A type II Plateau supernova SN 2005 cs is also seen (Courtesy: G. C. Anupama).

The imaging of celestial objects can be done at the prime focus or at the Cassegrain focus of a telescope. A typical imaging unit consists of a filter assembly accommodating several filters at a time and operated manually or with remote control facility. The filters may be U, B, V, R, I filters, and narrow band filters, namely 656.3 nm. Deployment of modern CCDs provided an order of magnitude increase in sensitivity. A CCD can be directly mounted on telescopes replacing both photographic plate and microphotometer system. Introduction of CCDs (see section 8.2) as light detectors have revolutionized astronomical imaging. Since the quantum efficiency of such sensors is much higher than the photographic emulsion, they have enabled astronomers to study very faint objects. Figure (10.13) displays an image of the whirlpool galaxy M 51.

10.4.2 *Photometer*

Photometry is the measurement of flux or intensity of a celestial object at several wavelengths; its spectral distributions are also measured, the term is known as spectrophotometry. If the distance of the measured object is known, photometry may provide information about the total energy emitted by the object, its size, temperature, and other physical properties. A source of radiative energy may be characterized by its spectral energy distribution, E_λ, which specifies the time rate of energy the source emits per unit wavelength interval. The total power emitted by a source is given by the integral of the spectral energy distribution,

$$P = \int_{0.36\mu m}^{0.8\mu m} E_\lambda d\lambda, \qquad \text{W}/\mu\text{m}. \tag{10.142}$$

Equation (10.142) is known as the radiant flux of the source, and is expressed in watts (W). The brightness sensation evoked by a light source with spectral energy distribution, E_λ, is specified by its luminous flux, F_ν,

$$F_\nu = K_m \int_{0.36\mu m}^{0.8\mu m} E_\lambda V(\lambda) d\lambda, \qquad \text{lumens (lm)} \tag{10.143}$$

where $K_m = 685$ lm/W[30] of is the scaling constant and $V(\lambda)$ the relative luminous efficiency.

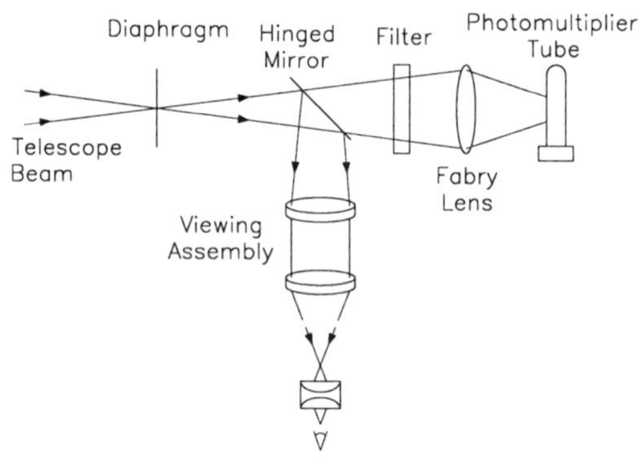

Fig. 10.14 Schematic diagram of a photometer.

A photometer measures the light intensity of a stellar object by directing its light on to a photosensitive cell such as a photo-multiplier tube. The additional requirements are (i) a field lens (Fabry lens), and (ii) a set of specialized optical filters. The photometer is usually placed at the Cassegrain

[30]An infinitesimally narrowband source of light possessing 1 W at the peak wavelength of 555 nm of the relative luminous efficiency curve yields in luminous flux of 685 lm.

focus behind the primary mirror. Figure (10.14) depicts a schematic layout of a photo-electric photometer. A small diaphragm is kept in the focal plane to stop down a star and minimize background light from the sky and other stars. Such a diaphragm must have several openings ranging from a large opening for initially centering the star to the smallest. In order to center the star in the diaphragm, an illuminated dual-cross hair post-view eyepiece and first surface flip mirror are required. A flip prism may also be employed in place of the flip mirror.

An assembly consisting of a movable mirror, a pair of lenses, and an eyepiece, whose purpose is to allow the observer to view the star in the diaphragm, is required in order to achieve proper centering. When the mirror is swung into the light path, the diverging light cone is directed toward the first lens. The focal length of this lens is equal to its distance from the diaphragm. The second lens is a small telescope objective that re-focuses the light. The eyepiece gives a magnified view of the diaphragm. Once the star is centered, the mirror is swung out of the way and light passes through the filter. The choice of the filter is dictated by the spectral region to be measured. The Fabry Lens refracts the light rays onto a photo-cathode of the PMT. This lens spreads the light on the photocathode and minimizes the photocathode surface variations. The photocathode is located, in general, exactly at the exit pupil of Fabry lens so that the image of the primary mirror on the cathode is in good focus. A detector, usually a photomultiplier tube, is housed in its own sub-compartment with a dark slide. The output current is intensified further by a preamplifier, before it can be measured and recorded by a device such as strip chart recorder or in digital form on disc. Figure (10.15) displays the light and $B - V$ color curves of AR Puppis.

A photometer is required to be calibrated, for which two basic procedures are generally employed. These are:

(1) Standard stars method: The purpose of this procedure is to calibrate a given local photometric system to a standard (or reference) system, based on detailed comparisons of published magnitude and color values of standard stars, with corresponding measurements made with local equipment. For a variable star observation, a reference star close to the actual target should be observed at regular intervals in order to derive a model for the slow changes in the atmospheric extinction, as well as for the background brightness that undergoes changes very fast.

(2) Differential photometry: In this technique, a second star of nearly the

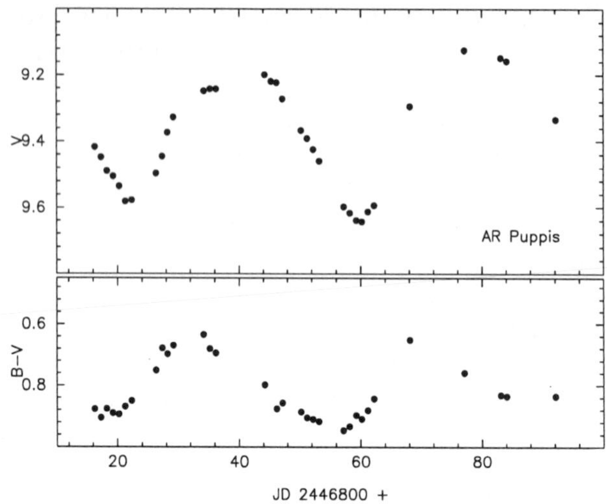

Fig. 10.15 Top panel: light curve of a star, AR Puppis; bottom panel: its $B - V$ color curve; data obtained with 34 cm telescope at VBO, Kavalur, India. (Courtesy: A. V. Raveendran).

same color and brightness as the variable star, is used as a companion star. This companion should be close enough so that an observer may switch rapidly between the two stars. The advantage of this closeness is that the extinction correction can often be ignored, since both stars are seen through identical atmospheric layers. All changes in the variable star are perceived as magnitude differences between it and the comparison star, which can be calculated by using the equation,

$$m_\star - m_c = -2.5 \log \frac{d_\star}{d_c}, \tag{10.144}$$

in which $d_\star$ and d_c represent the practical measurement (i.e., current or counts s^{-1}) of the variable and the comparison stars minus sky background respectively. The disadvantage of this project is that it is improbable to specify the actual magnitude or colors of the variable star, unless one standardize the comparison star.

10.4.3 *Spectrometer*

Spectrometer is a device that displays the radiation of a source and records it on a detector. Its purpose is to measure:

- the accurate wavelengths of emission and absorption lines in order to get line of sight component of velocities,
- the relative strengths and or equivalent widths of emission or absorption lines to have insight about composition and chemical abundances of different elements, their presence in ionization states and temperature, and
- shapes and structure of emission and or absorption line profiles, which provides information about pressure, density, rotation, and magnetic field.

It also measures the spectral energy distribution of continuum radiation, which helps to understand the physical mechanisms and to derive the temperature of the source.

A simple spectrograph can be developed with a prism placed in front of a telescope. Such spectrum can be registered on a photographic plate or on a CCD. This kind of device is known as objective prism spectrograph. In order to increase the width of the spectrum, the telescope may be moved slightly perpendicular to the spectrum. With such an instrument, a large number of spectra can be photographed for spectral classification. For precise information, the slit spectrograph, which has a narrow slit in the focal plane of the telescope is used. The light is guided through the slit to a collimator that reflects or refracts the light beams into a parallel beam, following which the light is dispersed into a spectrum by a prism or grating, and focused with a camera onto a CCD. A laboratory spectra is required to be exposed along with the stellar spectrum to determine the precise wavelengths.

A diffraction grating or grism can also be used to form the spectrum. Either a reflection grating or a transmission grating is used to develop a spectrograph. In the case of the former, no light is absorbed by the glass as in the case of the latter. In general a reflection grating is illuminated by parallel light that can be obtained by placing a slit at the focus of a collimating lens. The reflected beam from the grating is focussed by an imaging lens to form a desired spectrum. For astronomical spectrographs, the reciprocal linear dispersion, $d\lambda/dx$, in which x is the linear distance along the spectrum from some reference point, usually has the value in the range,

$$10^{-7} < \frac{d\lambda}{dx} < 5 \times 10^{-5}. \tag{10.145}$$

The resolving power of a spectrograph setup is dictated by the spectral resolution of the grating, the resolving power of its optics, and the projected width of the entrance slit. Detector may also play a role depending on the size of the detector elements and the linear dispersion of the spectrograph setup. If the slit-width is larger than certain value, it degrades the resolution. The physical size of the resolution element of the detector (typically 2 pixels) dictates the minimum physical size of the widths. The slit width, $s(d\lambda/dx)$, in wavelength terms, is known as spectral purity of the spectrograph. The physical width of the entrance slit of the spectrograph has a physical maximum width, s or less not to degrade the spectral resolution, where

$$s = \frac{\lambda f_1}{N\, d \cos\theta}, \tag{10.146}$$

in which θ is the angle of the exit beam to the plane of the grating, f_1 the focal length of the collimator, λ the wavelength of the radiation, N the total number of grooves on the grating, and d the grating constant (number of grooves per unit length).

In a well corrected system, let l be the length of the grating. The width of the beam becomes, $D = l\cos\theta$, which is the diffraction-limit of slit-width. The spectrum is formed from an infinite number of monochromatic images of the entrance slit. The width of these images S is given by,

$$S = s\frac{f_2}{f_1}, \tag{10.147}$$

where f_2 is the focal length of the imaging elements.

Figure (10.16) depicts the schematic diagram of the Echelle[31] spectrograph (Rao et al. 2004) at 2.34 m Vainu Bappu telescope (VBT), Vainu Bappu Observatory, Kavalur, India. It is a fiber-fed instrument, where the spectrograph is housed in a temperature and humidity controlled isolated Coudé laboratory and kept on a stable platform. Unlike Coudé conventional scheme, where about seven reflections cause loss of about 10% light at each reflection are required to bring light from the prime focus to the laboratory, the fiber-fed spectrograph transmits light through an optical

[31]Echelle grating uses high angle of incidences and operates in higher orders of the grating. An Echelle format arranges the spectrum in a series of orders whose spacing is very small, which are said to be collapsed. In order to distangle them, an optical dispersing element like a prism or a grating, called cross-disperser, stretches the spectrum in perpendicular direction to the dispersion.

fiber with reduced loss of light from the said telescope focus to the Coudé laboratory.

At VBT, the $f/3.25$ beam from the prime focus is fed to the input end of the optical fiber of 100 μm diameter. The fiber brings the light to the spectrograph, which is a Littrow configuration with the same optical element (of focal length 75.5 mm) serving as collimator and camera. The light from the prime focus transported by optical fiber is converted to a $f/5$ beam of size 151 mm with the help of a focal converter. This beam is fed to the collimator-camera system at an off-axis angle of 0.5° by a folding mirror. Since a prism has high throughput efficiency and gives rise to a more uniform order spacing, it is used as predisperser in the first pass and as order separator in the other pass. It has a length of 165 mm, two sides of 188 mm length, and a base length of 128 mm.

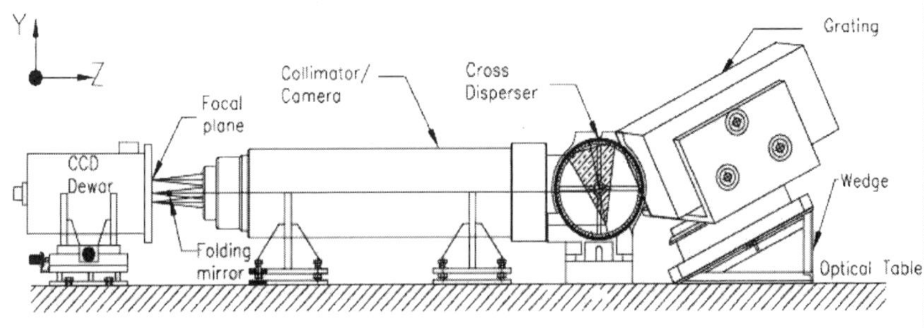

Fig. 10.16 Schematic diagram of a fiber-fed Echelle spectrograph (Rao et al. 2004; Courtesy: N. K. Rao).

An echelle grating (with 52.6 grooves/mm) with ruled surface of 408 $mm \times 208$ mm and blaze angle of 70° is used, which provides nearly uniform intensity over orders 50 to 80 covering entire optical range. Such a grating disperses the beam in the horizontal direction and the dispersed beam on its way out following dispersion passes through the cross-disperser (prism). The dispersed beam get stretched in the vertical direction. The instrument provides a resolving power of 72,000 (4 km s^{-1}) with a 60 μm slit and provides continuous coverage with gaps for $\lambda < 1$ μm. The wavelength coverage in a single order varies from roughly 35Å at $\lambda = 4050$Å (range 4030Å- 4065Å) to 70Å(8465Å- 8535Å). The doubly dispersed beam is allowed to pass through the collimator-camera lens system once again. The camera focuses the dispersed beam on to the $2K \times 4K$ CCD system having 15 μm pixel.

10.5 Occultation technique

Any planetary body of a notable size, say Moon, moving along its orbital path, passes in front of a background star or any celestial object, the light coming from the latter is occulted[32]. During such events observers can obtain accurate diameters or information on their upper atmospheres. The Moon does not possess atmosphere and stars have no appreciable angular size. When the leading or trailing edge of the Moon's shadow crosses an observer, the star disappears or reappears instantaneously. The diffracted shadow bands from the lunar dark limb race across the landscape of the Earth at high speed. The observed intensity drops to zero in a very short time. This phenomenon is referred to as lunar occultation.

Stars can be occulted by planets; Uranus rings were discovered when a planet occulted a star in the late 1970s. It is also possible for one planet to occult another planet. The mutual events[33] are occultations/ eclipses of one satellite by another. However, these mutual events are rare. The main purpose of observation of occultations by other solar system bodies such as, asteroids, comets etc. is to obtain the precise size and shape of the occulting body. The sizes and shapes of asteroids and comets are known to be very small.

Lunar occultation represents a powerful method because of the extraordinary geometric precision it provides. It remains useful for determining angular resolution at the milliarcseconds level in optical wavelengths. Lunar occultations can be performed in other wavelengths of the electromagnetic spectrum such as, radio, infrared, and X-ray as well; the EXOSAT satellite was launched to carryout X-ray lunar occultations. Radio occultations achieve resolutions of the order of a few arcsecond, while stellar angular diameters have been measured down to a few milliarcseconds during optical and IR lunar occultations. The occultation studies have various other applications such as:

- precision astrometry for determination of the lunar orbit; measurements of the Moon's position over a long time provides astrophysicists new information about its motion and orbit,
- information about star positions, about the hills and valleys on the

[32]An occultation may be defined as the total or partial obscuration of one celestial body by another. In a solar eclipse, the Sun is hidden by the Moon, but in a stellar eclipse, it is the star or a planet that is hidden.

[33]If the foreground planet is smaller in apparent size than the background planet, the event is known as mutual planetary transit.

edge of the Moon, and

- measurement of stellar diameters and discover new double stars (Nather and Evans, 1970).

The shortcomings of an occultation technique is noted from the fact that due to its singular nature, the object may not occult again until one Saros cycle later (18.6 yr), however, for several months during one Saros cycle the same star is occulted. It must be limited to the zodiacal belt of the sky (10% of the celestial sphere) as well. But it has the unique characteristic that the angular resolution is independent of the diameter of the telescope. A relatively small telescope may reach angular resolution of the order of a few milliarcseconds. The limiting magnitude does not depend on the size of the telescope aperture, since the events are recorded in the presence of sky background that gives rise to a strong photon noise. However, the spatial resolution goes down with increase in telescope size (due to sampling of 2-3 bands at a time. The use of a large telescope may effectively increase the signal-to-noise (S/N) ratio.

10.5.1 *Methodology of occultation observation*

A lunar occultation can occur depending on the relative position of the Moon, the stars, and the observer. Since the positions of the stars are known with high precision, if the position of an observer and the observation time are accurately known, it is possible to determine the position of the Moon. Moon travels across the sky at a rate of roughly $0.5''$ arcsecs per second against the stellar background as seen from the center of the Earth, yielding in a fringe passage at the rate of about 0.9 m per msec for a star occulted at the leading edge of the Moon's disc (Nather and Evans, 1970). The time scale of events may be modified by the occultation on the disc with respect to the motion vector. Another effect on such a time scale arise from the rotation of the Earth, and hence reduces the apparent lunar rate, due to varying parallax.

Observation with a lunar occultation requires the time at which stars are eclipsed by the Moon to be recorded. It works well when the precise position of the source is known, and of course, if the object structure is simple. Furthermore, only sources which lie along the orbital track of the moon may be observed. The data required for making such an observation are (i) position of the telescope, namely latitude (to $1''$ precision), longitude (to $1''$ precision), and altitude of observation point, (ii) precise time, and

(iii) lunar occultation predictions.

A photometer recording intensity from the occulted object depicts a series of alternating darker and brighter diffraction fringes, known as Fresnel diffraction patterns (see section 3.6.2), depending on the width of the pass band, provided the data is recorded at high time resolution. Such patterns, an oscillation of the stellar flux just prior to the occultation, are wavelength dependent. From an analysis of the diffraction fringe spacing and height, the angular size of the occulted star may be determined.

Although observations with increasing optical bandwidth permit for the improvement of S/N ratio, it may lead to blur the finer fringes. Of course other problems such as (i) atmospheric effects of dispersion, (ii) seeing (section 5.4.1 and 5.5.6), and (iii) atmospheric scintillation (5.5.2) may tend to smear fringes. With the criticality of time for occultation observations, dispersion due to air-mass is uncontrollable for the observer. In the case of severe scintillation, the star may drift near the edge of the diaphragm, which allows a tiny amount of image motion (see section 5.5) to have a considerable modulation effect (Nather and Evans, 1970). A very narrow-band filter provides highest order fringes and is useful for specific effects in stellar atmospheres. When a monochromatic point source is obscured by a straight edge, the expected intensity pattern is described in terms of classical Fresnel integrals,

$$\mathcal{C}(\mathrm{w}) = \int_0^{\mathrm{w}} \cos\left(\frac{\pi}{2}\tau^2\right) d\tau; \qquad \mathcal{S}(\mathrm{w}) = \int_0^{\mathrm{w}} \sin\left(\frac{\pi}{2}\tau^2\right) d\tau, \tag{10.148}$$

in which $\mathrm{w} = x(2/\lambda L)^{5/2}$ is the Fresnel number (dimensionless), λ the wavelength of light, L the distance from the Moon to Earth ($L = 384,000$ km), and x the distance in meters from the observer to the edge of the lunar shadow.

As the Moon moves across the source, the fringe pattern moves across the telescope aperture as well and a light curve is observed. The irradiance, I, is given by (Born and Wolf, 1984),

$$I(\mathrm{w}) = \frac{I_0}{2}\left[\left(\frac{1}{2} + \mathcal{C}(\mathrm{w})\right)^2 + \left(\frac{1}{2} + \mathcal{S}(\mathrm{w})\right)^2\right], \tag{10.149}$$

in which I_0 is the unobstructed irradiance.

Figure (10.17) depicts the curves generated for (a) $\mathcal{C}(\mathrm{w})$, (b) $\mathcal{S}(\mathrm{w})$, and (c) the monochromatic Fresnel diffraction patterns using the equation (10.149). An occultation light curve embeds information about the

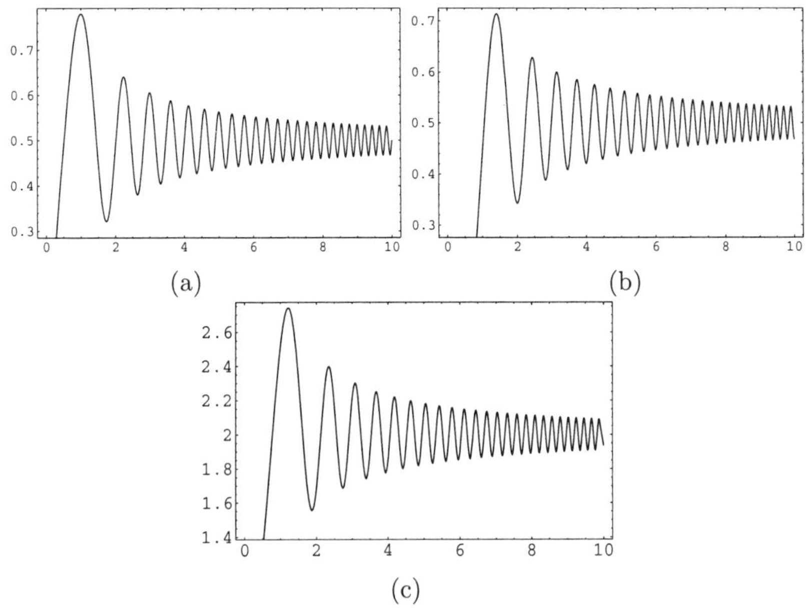

Fig. 10.17 (a) Fresnel diffraction curves generated for $\mathcal{C}(\mathrm{w})$, (b) $\mathcal{S}(\mathrm{w})$, and (c) the monochromatic Fresnel diffraction patterns using the equation.

one-dimensional (1-D) brightness profile of the source along the direction perpendicular to the lunar limb. Of course, the detection of the ideal light curve is a difficult process, since it introduces additional instrumental effects such as:

- time response of the detector over the integration time, Δt,
- combined (telescope, filters and detector) wavelength response of the system within the pass band, and
- 1-D function obtained by integrating area of the primary mirror along the direction of the lunar motion.

These situations set limits to the resolution, which may vary from one telescope to another; they can be dependent on the signal-to-noise (S/N) ratio in the light curve for the same telescope as well. In the case of the IR wavebands, the sky background has several components such as (i) the solar light reflected by the bright-limb and scattered through the atmosphere into the beam, (ii) the thermal emission from the sky, and (iii) the thermal emission from the dark-limb. These components have different intensities depending on (i) the wavelength of interest, (ii) the phase of the Moon, (iii) the zenith distance, and (iv) the atmospheric conditions. Richichi (1988) mentioned that in the K filter (2.2 μm), the background in a 15$''$ arcseconds diaphragm may be equivalent to a magnitude K $\sim 3 - 5$.

A major source of error in the occultation data analysis may be due to the uncertainty in the lunar slope at the point of occultation (Taylor, 1966), which has a direct effect on the time scale. Such an uncertainty exists for stars with large diameters, where the fringe pattern may get obscured. Irregularities at the lunar limb with sizes comparable to the first Fresnel zone at the wavelength of interest, may also cause distortion in fringe pattern. The other notable sources of noise arise from the scintillation and seeing, and more importantly from the system noise; the scintillation noise can be large and may exceed the background noise. Depending on the observing condition, the limiting magnitude may vary.

10.5.2 *Science with occultation technique*

Occultation technique has more applications than determining the Moon's motion and lunar limb profiles. It has applications in determining the asteroid profiles, astrometric and galactic parameters etc. Such a technique is sensitive enough to detect the presence of an atmosphere on Ganymede, a Jovian satellite, from its occultation of SAO 186800 (Carlson et al., 1973) as well. The occultations of stars by Saturn's largest moon Titan have yielded critical information on some of the properties of Titan's atmosphere. The lunar occultations have been used in different fields like observations of galactic center, active galactic nuclei, and other peculiar sources, however, the most widely used application in optical wavelengths has been the study of stellar systems (McAlister, 1985).

Angular diameters of many stars with an accuracy sufficient to pose tight constraints on the theory of stellar atmospheres were determined employing the lunar occultation techniques. These measurements are of particular importance to fundamental astrophysics, since effective temperature, T_e, are poorly determined particularly for the coolest spectral types.

MacMahon (1909) suggested that the time of disappearance could be interpreted in terms of the diameter of the star. An estimate by him found that a star with 0.001$''$ should disappear in 2 msecs. Eddington (1909) derived that diffraction effects at the lunar limb from a point source would limit the time of disappearance to eight times of this value. It is to be noted that the fringe spacing is a function of the wavelength of interest (see chapter 3). The critical size of an irregularity would be about 12.8 m, which is the size of the first Fresnel zone at lunar distance.

From the diffraction fringes that were observed during an occultation of α Leo, Arnulf (1936) deduced a diameter of $0.0018''$ for the star. This result is in excellent agreement with the value of $0.0013''$ obtained for its diameter by the intensity interferometer[34] at Narrabri (Hanbury Brown, 1974). Tej et al., (1999) have reported the measurements of the angular diameter of a Mira variable, R Leonis, in near IR bands at the 1.2 m telescope, Mt. Abu, India.

The notable advantage of occultation of binary stars is that of determining relative intensities and measure the separations comparable to those measured by long baseline interferometers. Of course, the separation between the components should not be less than ~10 mas and the difference in intensities of them smaller than ~4-5 m_v (Richichi, 1988). However, the data obtained by means of occultation technique are difficult to interpret since the measured separation is a projection of the true angular separation. The situation becomes acute, if non-standard data are used. Moreover, the problem related to the lunar limb irregularities can be applied to binary star measures. Observations with multiple wavelengths may distinguish lunar limb effects from observations of real binaries. Observations carried out at different locations, where the character of the local lunar limb at the time of occultation may vary, can serve to distinguish between the real and artifacts.

During a total occultation by the Moon, for two stars occulted by the leading edge of the approaching lunar limb, the velocity turns out to be equal to the lunar rate at that time, in a direction nearly perpendicular to the lunar limb. The diffraction effects cause gradual disappearances and reappearances, which is the evidence of a close binary star; binary stars produce step events. Spectroscopic binaries do not produce step events, but a merged broader fringes.

A single occultation measurement provides the vector separation of the pair, but the second star can be anywhere along the line parallel to the first

[34]Intensity interferometer computes the fluctuations of the intensities I_1, I_2 at two different points of the wavefronts. The fluctuations of the electrical signals from the two detectors are compared by a multiplier. The current output of each photo-electric detector is proportional to the instantaneous intensity I of the incident light, which is the squared modulus of the amplitude A. The fluctuation of the current output is proportional to $\Delta I = |A|^2 - < |A|^2 >$. The covariance of the intensity fluctuations,

$$\langle \Delta I_1 \Delta I_2 \rangle = |A_1 A_2^*|^2 ,$$

is the squared modulus of the covariance of the complex amplitude.

one. Hence, the minimum separation between the two stars is obtained in the direction of fringe passage. A measurement at a different location and hence position angle on the Moon provides (hence a different fringe passage) a second vector separation. On combining these two measurements serves to fix the location of the second star with respect to the first (Nather and Evans, 1970). It is to reiterate that speckle interferometry measures angular separation, ρ, and position angle, θ, for the time of observation, while the separations measured by lunar occultations are the component of the true separation in the vector direction that is perpendicular to the local lunar limb slope. This vector separation, ρ_v, is analogous to the angular separation, ρ, while the position angle (θ) is replaced by the vector direction, θ_v. The true position of the companion lies along a line perpendicular to the vector direction, θ_v at the point of vector separation, ρ_v, however it is uncertain to pinpoint its direction or the distance along this line. Nather and Evans (1970) suggested that two different vector measures obtained at the same time may define a point in the plane of the sky at the intersecting point of these two perpendiculars.

Occultation observations have resulted in the discovery of many double stars. The occultation binary star survey, is being carried out by Center for high angular resolution astronomy (CHARA) group (Mason, 1995). Although the speckle interferometric technique is being used routinely to measure close binary systems having separations in the range previously detectable during occultations, the latter observations are, of course, required to discover new close pairs since the former can be carried out for a limited number of stars. The contribution of speckle survey of occultation binaries till date, at the smallest separation region, is of the order of $< 0.025''$. The direct speckle interferometric measurement of more than 2 dozens new occultation binaries have been reported (Mason, 1995, 1996).

Another subject in the field of stellar physics is the study of the circumstellar envelopes[35], with physical dimensions exceeding a few tens of milliarcseconds, which are often present around very young and very cool stars. Any such envelopes can be revealed by a smoothed monotonically varying background superimposed on the fringe pattern of the central star. However, Richichi (1988) opined that compact components smaller than $\sim 0.5''$ can be studied since the larger ones posses light curves with timescales comparable to the frequencies of the atmospheric turbulence.

[35]The outer layers of gas in a star are called its envelope.

Chapter 11

Astronomical applications

11.1 High resolution imaging of extended objects

Interferometric observations may reveal the fundamental processes on the Sun that take place on sub-arcsecond scales concerning convection and magnetic fields (which can be measured by Hanle effect[1]). A moderate telescope of 2 meter class can facilitate simultaneous measurements of the solar atmospheric parameters and of the magnitude and direction of magnetic fields with high accuracy. Such a facility equipped with modern adaptive optics system may provide insight on important topics, viz., (i) magnetohydrodynamic (MHD) waves[2] and oscillations in the solar plasma at different

[1]The Hanle effect describes resonant line scattering polarization in a magnetic medium. It has been used in solar physics to study the magnetic structures present in prominences, to determine their field strength and distribution. With an accurate spectro-polarimeter, the Hanle diagnostic may be used for the weak magnetic fields in the solar photosphere and chromosphere. The Hanle effect can be applied as a diagnostic of circumstellar magnetic fields for early-type stars, in which it is sensitive to field strengths in the range of 1-300 gauss (G).

[2]Magnetohydrodynamic (MHD) waves occur due to the presence of magnetic field in the plasma. The magnetic fields impart magnetic pressure and tension forces which act as restoring forces for the wave propagation. Generally, MHD waves do not propagate alone. They interact with sound waves (see section 2.1, footnote 2) to produce magneto-acoustic waves. The characteristic speed of a magnetic disturbance is the Alfvén speed, v_A, defined by,

$$v_A = \sqrt{\frac{2\,P_M}{\rho}} = \sqrt{\frac{B_0^2}{\mu_0\,\rho}},$$

in which $P_M(= B^2/2\mu_0)$ is the magnetic pressure, B_0 the magnetic field strength, and μ the magnetic permeability.

These magnetic waves propagate at the Alfvén speed that depends on the magnetic field strength and the density. The relative speed of acoustic and MHD waves depends on the plasma ratio, β = gas pressure/magnetic pressure. In low density plasma such as in solar corona Alfvén speed generally exceeds the sound speed.

heights in solar atmosphere, and (ii) active region evolution. Solar small scale structures evolve fast and change position by about $0.1''$ within a fraction of a minute or so (Von der Lühe, 1989). The limitations come from

- the rapid evolution of solar granulation that prevents the collection of long sequences of specklegrams for reconstruction and
- the lack of efficient detectors to record a large number of frames within the stipulated time before the structure changes.

Though the highest resolution ground based image cannot compete with the resolution obtained by fly-by missions for solar system objects, the primary strength of ground based observations is their ability to do synoptic monitoring of solar system objects. Objects such as Pluto, Mercury, larger satellites of Jupiter can also be tackled. Speckle imaging has been successful in resolving the Pluto-Charon system (Bonneau and Foy, 1980), as well as in determining shapes of asteroids (Drummond et al. 1988). Reconstructions of high resolution features on the extended objects, viz., (i) Sun, (ii) Jupiter were also made with interferometric techniques. The evolution of cometary comas may also be studied over a very large range of heliocentric distances and in far greater spatial detail than is now possible.

11.1.1 *The Sun*

The Sun is the nearest star at a distance of 149,597,900 km from the Earth. It is a main sequence star, whose absolute magnitude, M_v, is 4.79 and effective temperature 5785° K. The surface brightness, $B_\odot$ of the Sun can be derived from the luminosity that is related to the total radiation received at the mean distance of Earth, $L_\odot/4\pi r_\odot^2$, in which $r_\odot$ is the distance between the Sun and the Earth,

$$L = 4\pi R_\odot^2 F_\odot = 4\pi r_\odot^2 F, \tag{11.1}$$

where $L_\odot(= 3.845 \times 10^{26}$ W) is the bolometric luminosity of the Sun, $F_\odot$ is the flux density on the surface of the Sun, $R_\odot(= 6.96 \times 10^8$ m) is its radius, and the flux density, F is,

$$F = F_\odot \left(\frac{R_\odot}{r_\odot}\right)^2. \tag{11.2}$$

The Sun subtends an angle at a distance $r_\odot \gg R_\odot$ as $\Omega = A/r_\odot^2 = \pi(R_\odot/r_\odot)^2 (= 6.81 \times 10^{-5}$ *sterad*), in which A is the cross-section of the

Sun and $R_\odot/r_\odot = \theta/2$, $\theta(= 32$ arcminutes). The surface brightness, $B_\odot$ of the Sun is,

$$B_\odot = \frac{F}{\Omega} = \frac{F_\odot}{\pi}. \tag{11.3}$$

On applying equation (10.40), the surface brightness turns out to be equal to the intensity, i.e., $B_\odot = I_\odot$. The surface brightness $B_\odot = S_\odot/\Omega = 2.04 \times 10^7\ \mathrm{W\,m^{-2}sterad^{-1}}$, with $S_\odot(= 1390\ \mathrm{W\,m^{-2}})$ as the flux density of the Sun on Earth, known as the solar constant. The surface gravity of the Sun given by,

$$g_\odot = \frac{GM_\odot}{R_\odot^2}, \tag{11.4}$$

is equal to 274 $\mathrm{m}\,s^{-2}$, in which $M_\odot(= 1.989 \times 10^{30}\ kg)$ is the mass of the Sun.

The mean density of the Sun is derived as,

$$\bar{\rho}_\odot = \frac{M_\odot}{V} = \frac{3M_\odot}{4\pi R_\odot^3}, \tag{11.5}$$

which turns out to be 1409 $\mathrm{kg\,m^{-3}}$.

11.1.1.1 *Solar structure*

The interior of the Sun has different regions like core, radiative, and convective zones. The atmosphere of the Sun has different regions such as photosphere, chromosphere, and corona.

(1) Core: The sun is composed primarily of hydrogen, with some helium and heavier trace elements. It produces the heat and light by thermonuclear reactions taking place inside the core. This core is highly densed having one-half of the solar mass within a region of one-fourth solar radius. This region is responsible for generating Sun's emitted energy, where protons (^{1}H) are being converted into the atoms of helium (He) nuclei by thermonuclear reactions, called nuclear fusion[3]. These nuclei are built up mainly by the proton-proton cycle but partly by the carbon (CNO) cycle that uses carbon (C), nitrogen (N) and oxygen

[3]Fusion is the process by which two atoms fuse together to form a heavier atom. It is accompanied by the release or absorption of energy depending on the masses of the nuclei involved. The energy released in most nuclear reactions is much larger than that for chemical reactions, because the binding energy that holds nuclei together is far greater than the energy that holds electrons to nuclei.

(O) as catalysts for the production of He. CNO are much larger atoms than H or He and can form into various isotopes[4]. At the end of these cycles (lasting about 10^7 yr), four hydrogen atoms (1 hydrogen atom is 1.0080 atomic mass units, AMU) are converted into one helium atom (4.0026 AMU), according to the equation,

$$4^1\mathrm{H} \rightarrow ^4\mathrm{He} + 2e^+ + 2\nu + \gamma, \qquad (11.6)$$

where the energy is released in the form of high frequency γ-rays (26.2 MeV) and two neutrinos[5] (0.5 MeV) denoted by ν, e^+ denotes a positron[6] and the superscripts denote the number of protons plus neutrons in the nucleus; the positron reacts with an electron and pair disappears with the emission of two gamma rays. A mass of 0.0294 AMU is lost and converted into energy, i.e., a fraction of about 7×10^{-3} of hydrogen gets converted into energy.

The density of the core is about $\sim$ 150 gm cm^{-3} where the temperature is about $\sim$ 15 million degrees centigrade. Both the density and temperature decrease with the increasing distance from the center. The thermonuclear burning takes place upto 0.3 $R_\odot$ solar radius. At this point the density drops to about $\sim$ 20 gm cm^{-3}, while the temperature comes down to half the central value. The core and envelope extend to radii of 0.3 $R_\odot$ and 0.9 $R_\odot$ respectively.

(2) Radiative zone: The thermonuclear reactions stop at the base of this region and the energy transfer from the core to the surface is primarily by photon diffusion[7]. The energy made in the core is in the form of photons with high-energy gamma rays, when it flows outwards. This energy is changed into less energetic photons as it moves through the radiative zone, and eventually escape from the surface of the Sun; it may need a million years to get out from the very dense and opaque

[4]An isotope is a different form of a chemical element having same number of protons in the nucleus but having different number of neutrons.

[5]The neutrino is an elementary particle of zero charge. The neutrinos are tiny, and have a very small but non-zero mass. They escape unimpeded from the core through the rest of the solar interior; they have extremely small interaction probability with other matter. It has half-integer spin ($\hbar/2$), and is therefore a fermion.

[6]The positron is the antiparticle of an electron that is like an electron in all respects except that it has a positive charge. It has an electric charge of +1, a spin of 1/2, and the same mass as an electron. If a low-energy positron collides with a low-energy electron, annihilation occurs, resulting in the production of two gamma ray photons.

[7]Photon diffusion refers to a situation where photons travel through a material with a high optical depth and very short mean free path. Photons are absorbed and re-emitted many times as they diffuse toward the surface from the stellar core.

internal layers. Outward from the core, the temperature, pressure, and density decrease rapidly. The temperature falls to two million degrees centigrade, from which the convective zone begins. The density drops from ~ 20 gm cm^{-3} to ~ 0.2 gm cm^{-3} from bottom to the top of the radiation zone. At about 0.86 $R_{\odot}$, the gas properties have changed to such an extent radiation gets absorbed more readily making the gas convectively unstable, and hence turbulent convection occurs (Gibson, 1973).

(3) Convective zone: Temperature comes down to a low value at the top of the radiation zone where convection ensues; the energy transfer is being carried out through convection. This outermost layer of the solar interior, known as convection zone, extends from a depth of about 200,000 km right up to the visible surface. The atoms in this layer have electrons. A significant number of free electrons in the gas are moving slowly, which can be captured into bound energy states by hydrogen and other nuclei to form atoms. With the increase of number of atoms, there is an increase in the opacity of the gas and resulting increase in the temperature gradient. Atoms with electrons are able to absorb and emit radiation. Because of this heating and the large temperature gradient, the motion upward is accelerated and turbulent convection results. The energy is transferred through the process of convection; the fluids coming from the radiative zone expands and cools as it leaves. As it falls down to the top of the radiative zone, it heats up and starts to rise. This process repeats, creating convection currents and the visual effect of boiling on the Sun's surface. This is called granulation (see Figure 11.1). Each element of gas carries its own parcel of energy directly to the surface. The powerful turbulence generates mechanical energy which, as sound waves, propagates through the photosphere and into the Sun's outer layers (Gibson, 1973). The combination of the events taking place in the convection zone and the differential rotation of the Sun creates the solar dynamo causing the changes that occur in the magnetic field of the Sun.

(4) Photosphere: The layer above the convective zone where light is emitted is called photosphere. It is the visible surface of the Sun, which is a very thin layer of about 100 km thick. In this layer, a photon emitted outward has a little probability of being reabsorbed or scattered. The photon is likely to escape into space through the solar transparent atmosphere. The photosphere defines sharp visible edge of the Sun. In this region, the temperature is relatively low, and because the Sun is in

hydrostatic equilibrium, the scale height of the atmosphere is small and the density drops off rapidly. The density of the negative hydrogen ion ($H + e \rightleftharpoons H^-$) that is responsible for most of the radiation emission and absorption at visible wavelengths, decreases with height sharply as well. The decrease in total density as well as density in negative hydrogen is related in chainlike fashion to the decrease of the electron density (Gibson, 1973), which is a sensitive function of the decreasing temperature. At the center of the disc of the Sun, it appears to be bright and hotter when one looks straight in, while the falling of intensity in a solar image from the center to the edge or limb of the image is a phenomena that refers to limb darkening. Limb darkening occurs as a consequence of the following effects:

- The density of the star diminishes as the distance from the center increases. The photons come, on average, from optical depth, $\tau \sim 1$. Since the Sun is spherical, the photons from the center of the disc include some from a brighter zone than those at the limb of the disc.
- The temperature of the star decreases with increasing distance outwardly. Since surface brightness scales as T^4 for thermal radiation, the limbs of stars appear fainter than the central portions of their discs.

It is to be noted that the temperature for the Sun does not uniformly drop as the radius increases, and for certain spectral lines, the optical depth is unity in a region of increasing temperature. In such a situation, one observes the phenomenon of limb brightening.

The Sun's photosphere is composed of short-lived convection cells, termed as solar granules (each approximately 700 - 1500 km with mean distance cell centers is about 1800 km), where hot gas rises up from the center and cooler gases fall in the narrow dark lanes between them (Priest, 1982). The center of a granule appears brighter than its boundary in a high resolution image. Its mean lifetime is about 8 min, but individual granules may live for about 20 min, resulting in a continually shifting boiling pattern. These granules are in continual motion. Another type of granules, known as supergranules, are larger versions of granules with 20,000 to 54,000 km in diameter having lifespans of up to 24 hours. These cells are irregular in shape. The flow of material like supergranules, as well as large scale flows and a pattern of waves and oscillations can be observed with the Doppler effect.

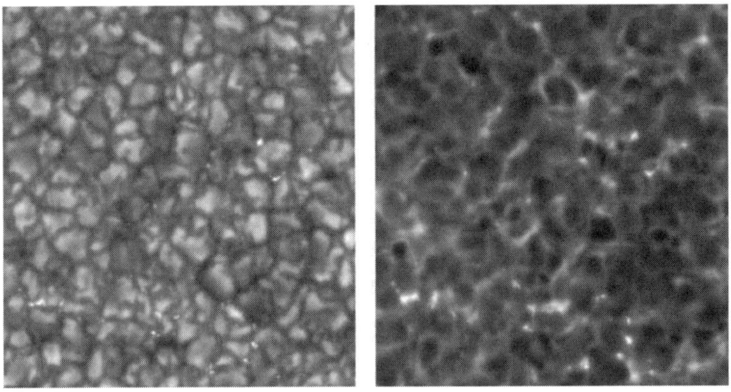

Fig. 11.1 Left panel: Solar granulation seen in the G-band (430 nm) and right panel: its extension in the Ca II H line (397 nm) of once-ionized calcium formed in the 1000 km thin chromosphere higher up in the solar atmosphere (Courtesy: HINODE Mission first results, November 2006).

(5) Chromosphere: Solar chromosphere is a thin spherical shell region that extends upwards from the top of the photosphere to heights from ten to fifteen thousand kilometers. It is an irregular layer, in which the temperature rises from 5000° K to 20,000° K at the upper boundary. At this temperature, hydrogen emits light which gives off a reddish color. It is fainter than the photosphere and can be seen during solar eclipses for a short time immediately after the second contact and immediately before the third contact, when it can be discernible as a pink flash with accompanying emission line spectrum (see Figure 11.2a). The chromosphere is highly non-uniform. Chromospheric emission in the Ca II K line reveals the network of supergranulation boundaries clearly as an irregular bright pattern. In the Hα wings the network is discernible as a dark pattern, while in the Hα core it shows up bright. At the limb one observes chromosphere as a mass of plasma jets, known as spicules. They move upwards at about 20 km s^{-1} from the high chromospheric part of the supergranule boundaries. They have typical lifetimes of 5 to 10 min, diameter 500-1200 km and are usually associated with regions of high magnetic flux. The mass flux of these spicules is about 100 times that of the solar wind.

(6) Corona: Above the chromosphere, a pearly white halo called corona extends tens of thousands of kilometers into space. It can be photographed at the time of total solar eclipse, when both the photosphere and chromosphere are covered by the disc of the Moon (see

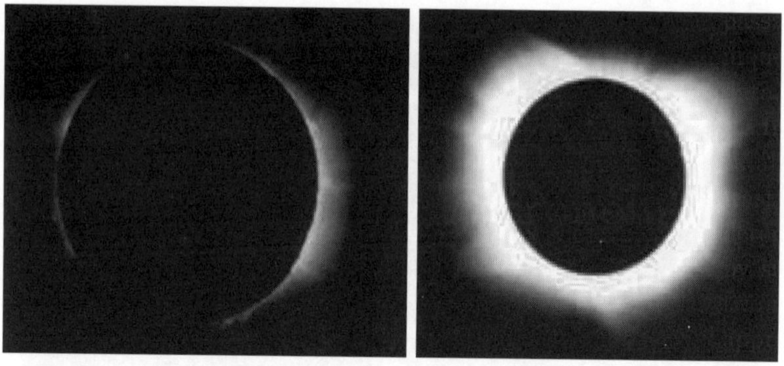

(a) (b)

Fig. 11.2 (a) Chromosphere and (b) Corona of the Sun as seen during the March 29, 2006 total solar eclipse observed at Sidi Barani, Egypt. (Courtesy: Serge Koutchmy).

Figure 11.2b); it is also observable with a coronagraph[8]. It is a luminous atmosphere of the Sun produced by the scattering of sunlight due to free electrons. The temperature of the solar corona is hotter than 1,000,000° K. The nature of the processes that heat the corona, maintain at such high temperatures, and accelerate the solar wind is a mystery in solar physics. The solar wind is a stream of charged particles consisting of mainly high energy protons (~1 keV) and electrons together with nuclei of heavier elements in smaller numbers, that are blown off continuously from the surface of the Sun at an average velocity of about 400 km s^{-1}. These particles are accelerated by the high temperatures of solar corona to velocities large enough to escape from the Sun's gravitational field. The Sun loses mass through the solar wind, what is termed as mass loss, which is about 10^{-14} $M_\odot$ per year, in which $M_\odot$ is the solar mass. The expanding solar wind pulls the solar magnetic field outward, forming the interplanetary magnetic field (IMF). The region of space where such magnetic field is dominant is known as the heliosphere. It is to be noted that the solar wind moves out almost radially from the Sun, but the rotation of the Sun gives the magnetic field a spiral form. The solar wind is responsible for distorting the symmetry of planetary magnetosphere and deflecting the tails

[8]A coronagraph is a telescopic attachment, which is designed to physically block the incident sunlight with a small occulting disc located at an intermediate focal plane of the solar images. It produces an artificial eclipse and thus reveal the faint solar corona, stars, planets and comets.

of comets[9] away from the Sun. Several models have been proposed to explain the heating of solar corona but they are yet unable to explain all its physical and dynamical properties although recent 3-D modelling (Peter et al. 2006) start to bring in some clues. It has been recognized that magnetic fields play an important role in heating up the plasma at the base of the corona, in the transition zone, most probably by flux braiding through photospheric footpoint motion and oscillations, albeit the identification of the process or processes still needs further studies. The brightness distribution of the corona provides the distribution of the electron density in the corona. It is found to vary from, $N_e = 3 \times 10^8 \, cm^{-3}$ at $d = 1.012R_\odot$, which is a height of about 8400 km above the photosphere, to $N_e = 3 \times 10^5 \, cm^{-3}$ at $d = 2R_\odot$. The corona is 10^{-12} as dense as the photosphere and is separated from the photosphere by the chromosphere and the thin layer known as transition zone where the temperature rises sharply as the density drops. The outer edges of the solar corona are being lost as solar wind. The corona is more or less confined to the equatorial regions, with coronal holes covering the polar regions during the minimum years of the solar cycle, while at solar maximum it is evenly distributed in the equatorial and polar regions. However, it is most prominent in areas with sunspot activity. Coronal holes are extended regions that are lower in density (by a factor of three) than the the rest of the corona, and are cooler, having temperature of about 1.4 - 1.8 $\times$ 10^6 K at 2 R$_\odot$, than their surroundings. They appear dark in pictures taken during a total solar eclipse or with a solar coronagraph. Coronal holes are found predominantly near the Sun's poles and may appear at any time of the solar cycle, but are common during the declining phase of the cycle. They posses an open magnetic field structure that permits charged particles to escape from the Sun and results in coronal holes being the primary source of

[9]Comets are small irregularly shaped bodies in the solar system orbiting the Sun, which exhibit coma and or tail as they approach the Sun. The tails of luminous material extend upto thousands of kilometers from the head, away from the Sun. The nucleus of these comets are composed of rock, dust, and ice. These comets have elliptical orbits bringing them close to the Sun and swinging them deeply into space, often beyond the orbit of Pluto. Most comets are believed to originate in a cloud (the Oort cloud, a spherical cloud) at large distances about 50,000 to 100,000 AU from the Sun. They are classified according to their orbital periods such as (i) short period comets that have orbits of less than 200 years, (ii) long period comets that have longer orbits, and (iii) main-belt comets orbit within the asteroid belt. Single-apparition comets have parabolic or hyperbolic orbits which may cause them to leave the solar system after a single pass by the Sun.

the solar wind and the exclusive source of its high-speed component. It is believed that when photospheric, chromospheric, and coronal magnetic field gets highly sheared, the magnetic field energy is released by reconnection with a high explosion. This results in ejection filaments causing 'Coronal mass ejection' and generates accelerated non-thermal high energy particles. When these high energy particles hit the chromosphere/ photosphere, they produce enhanced Hα emission and very hot plasma is seen in the form of double-ribbon flare and microwave and hard X-ray sources.

11.1.1.2 *Transient phenomena*

The photosphere is not featureless (Priest, 1982). A number of features namely sunspots, faculae, and granules can be envisaged in the photosphere. There are regions of dark patches, called sunspot groups (see Figure 11.3). Sunspot groups generally consist of a few big and tiny spots. In the bigger spots, one notices the central dark region called umbra and surrounded by a lighter dark area, known as penumbra. Sunspots are relatively cooler (about 4300° K compared to the rest of the photosphere, where the temperature is 5700° K; the temperature at the centers of sunspots may drop to about 3700° K.

The systematic study of sunspot dates back to the time of Galileo. From the studies of sunspot the following results were obtained. Sunspots generally occur in pairs. The sunspot magnetic flux appears first in the upwelling at the center of supergranulation cell and seen in Hα as an arch-filament system. The footpoints migrate to the cell boundary in about next four or five hours and flux tends to be concentrated most at a junction of three cells where a pore eventually appears over about 45 minutes.

Before the formation of a sunspot the enhancement in the magnetic field is seen in the Ca II K spectroheliogram as faculae that continue to exist even after the disappearance of the spot. These faculae are bright spots that form in the canyons between solar granules. They are produced by concentrations of magnetic field lines, and are mostly seen in the vicinity of sunspots. These pores are darker than the surrounding photosphere and have no penumbra. They have diameters of 700-4000 km, about 50% of photospheric brightness and field strength in excess of 1500 G. Often they last only hours or days, but sometimes one develops into a small sunspot. During its growth phase, say between 3 and 10 days, more and more magnetic flux is added to it. This is evident by the approach of

(a) (b)

Fig. 11.3 A sunspot region observed in November 2003. (a) The left picture shows the lowest visible layer of the solar atmosphere, the photosphere, in the light of CH molecules, and (b) the right picture shows the 1000 km higher layer, the chromosphere, in the Ca II H line (397 nm) of once-ionized calcium. In the photosphere, magnetic fields suppress the convective energy transport from the solar interior, which makes sunspots dark. Solar magnetic fields become the dominant force in the chromosphere, and they become brighter due to magnetic heating processes. Faculae, associated with enhanced chromospheric heating, are the manifestation of the magnetic activity. They are bright cloud like features around sunspots, which are regions of higher temperature and density within the chromosphere. (Courtesy: Dutch Open Telescope).

the spot to moving magnetic features (or magnetic knots) with speed of $0.25\ \mathrm{km\,s^{-1}}$. They have the same polarity as the spot and many appear as pores visible in white light.

Later, it develops into a rudimentary spot (where the penumbra is not clearly visible) and it develops into a leading spot (which means the bigger spot in western side of the group). Subsequently the following spots (towards the eastern side) are formed and evolve into a sunspot group. The line joining the leading spot and the following spots makes an angle of approximately $10°$ to the equator; the leading spot is at the lower latitude. After the complete growth, the sunspot group gradually diminishes. The lifetime of a typical sunspot group is about 1-2 months. From the day-today measurements of its position, it was observed that the Sun rotates on its axis once in about 27 days. The Sun does not rotate like a rigid body, it has the differential rotation; its equator rotates faster than its poles. The period of rotation near the equator is about 25 days while, near the poles, it is about 35 days. Sun's rotation axis is tilted by about $\pm 23.5°$ from the Earth's axis of rotation.

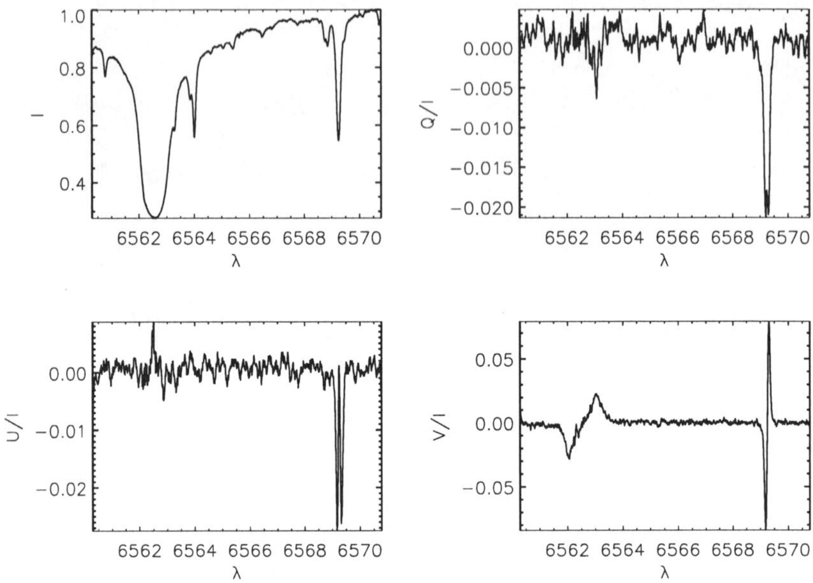

Fig. 11.4 Stokes I, Q, U, V profiles in the Fe I at 6569 Å and Hα at 6562.8 Å lines observed on 28th April, 2006, using a spectropolarimeter at Kodaikanal Solar telescope. The profiles are from a row cut in a typical sunspot umbral region. The Q and U profiles show the linear polarization and the orientation, while V profile indicates the circular polarization. Stokes I profile gives the total intensity (Courtesy: K. Nagaraju).

The relative sunspot (RSS) number is defined as $K(10\ g + f)$, in which g is the number of sunspot group present, K the observer's constant, and f the total number of sunspots. The RSS number has got a cycle with a period of approximately 11 years; solar activity follows the same cycle when the sunspots are at peak. In the beginning of the cycle the sunspot forms around $\pm$ 30° latitudes. Later, they gradually move towards the equator and during the maximum phase, these spots are observed around $\pm$ 10 – 20° latitude, while during the minimum phase, they are noticed close to the equator. The analysis of the spectra of sunspots at Kodaikanal Observatory led to the discovery of the phenomenon of radial motion in sunspots, in 1909, what is now termed as the Evershed effect[10] (Evershed,

[10] John Evershed (1864-1956) served as the Director of Kodaikanal Observatory, during

1909).

The magnetic field in a sunspot may be 1000 times stronger than in the surrounding area where it is about 1 G (gauss). In general, this can be measured using a spectropolarimetric[11] observations (see Figure 11.4). The presence of such a strong magnetic field keeps a sunspot cool. The packed magnetic field lines provide a barrier, which prevents hot gas from being convected into the sunspots. All the leading spots in each hemisphere posses the same polarity (magnetic), while the following spots posses opposite polarity. In the other hemisphere the polarity reverses. In the next cycle, the polarity changes. The magnetic field is strongest in umbral region of the sunspots, while it is weaker and more horizontal in penumbral region. From the interpretation of the spectra using the Zeeman effect, it is found that the magnetic field strength of the sunspot is around 3000 G for a bigger group. Cowling (1946) observed that the area of the spot decreases faster than the strength of the magnetic field.

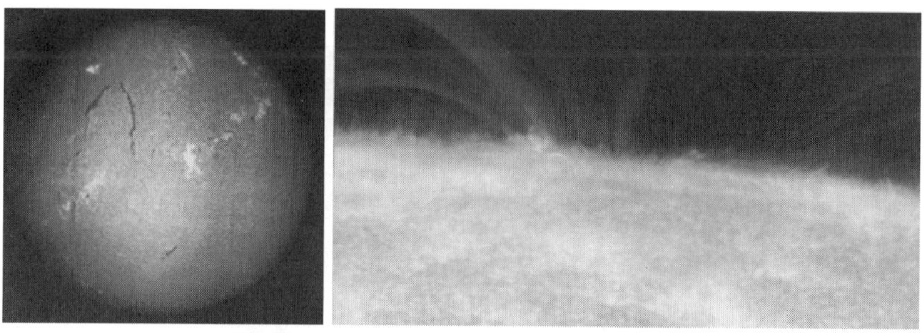

(a) (b)

Fig. 11.5 (a) Hα solar flare picture taken with spectroheliograph at the Kodaikanal Observatory, India; (b) A recent eruption observed with the HINODE satellite above a sunspot seen in the Ca II H spectral line (Courtesy: HINODE Mission first results, November 2006).

Magnetic field lines near a large group of sunspots can suddenly snap triggering a solar flare (see Figure 11.5), which represents a sudden short-lived (on a time scale of about 1 hr) eruption of hot ionized gas in a localized region of the Sun. Such a flare releases large energy (10^{29}-10^{32} erg) in the

1911-1923.

[11]A two beam spectropolarimeter employs a polarizing beam displacer, half-wave plate, and a quarter-wave plate followed by a compensator in conjunction with a high resolution spectrograph. Recording two orthogonal states of polarization simultaneously helps to reduce the seeing induced effects considerably.

form of radiation and fast particles. When particle radiation passes into the Earth's upper atmosphere, complicated geomagnetic and ionospheric storms occur due to which radio communication gets disturbed. In the neighborhood of the sunspot, the flares are observed as enhancement in monochromatic radiation of Hα or Ca II K lines, but the area of the enhancement is small compared to the total area of the Sun. Bursts of solar radiation at widely different wavelengths sometimes occur during the observation of a flare in Hα and their individual characteristics differ greatly. The duration of the emission may range from less than a minute to several hours. The growth phase is usually a matter of minutes and then the flare fades slowly (tens of minutes).

The flares occur along the magnetic neutral line. Generally, it is initiated by the appearance of parasite polarity (opposite polarity) poles very close to the sunspot. Thereby, development of high magnetic field gradient ensues. The onset of flare is not clearly understood. The flares are classified according to their area and intensity.

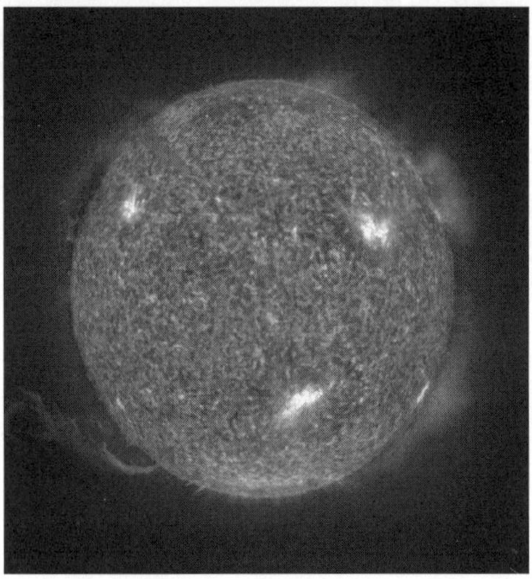

Fig. 11.6 Image of the solar prominence taken with solar and heliospheric observatory (SOHO), an ESA/NASA mission of international cooperation (Courtesy: SOHO ESA/NASA mission), and the dark strips appeared in Figure (11.5a) are the filaments; these filaments are the projection of the prominences on the disc.

Sun's magnetic field varies on a 22 year cycle due to variations in the magnetic polarity. Solar prominences (a giant glowing loop extending well beyond the corona) are a type of phenomena (see Figure 11.6); they contain plasma that is denser and cooler than the tenuous corona in which they are suspended. They are supported by magnetic force. When the prominences are seen against the disc, they are called filaments (see Figure 11.5a). Filaments (or prominences) appear along the magnetic neutral line. Generally, the quiescent prominences stay for 2-3 solar rotations. The average length is 60,000 to 600,000 Km and height ranges between 15,000 to 100,000 km, and thickness is in the range of 4,000 to 15,000 km. During the progress of sunspot cycle prominences in active region gradually move towards the equator, while prominences at higher latitude gradually move towards the pole. Around the sunspot maximum, the prominences are seen close to the pole.

11.1.1.3 *Solar interferometric observations*

Interferometric techniques can bring out high resolution information of the fundamental processes on the Sun that take place on sub-arcsecond scales concerning convection and magnetic fields (Von der Lühe and Zirker, 1988). The existence of solar features with sizes of the order of 100 km or smaller was found by means of speckle imaging (Harvey, 1972). Subsequently, solar granulation has been studied extensively with the said technique by many others (Aime et al., 1975, Von der Lühe and Dunn, 1987). From the observations of photospheric granulation from disc centre to limb at $\lambda = 550 \pm 5\ nm$, by means of speckle interferometric technique, at the Vacuum Tower Telescope (VTT), at the Observatorio del Teide (Tenerife), Wilken et al., (1997) found the decrease of the relative RMS-contrast of the centre-to-limb of the granular intensity. Time series of high spatial resolution observations with the same telescope reveal the highly dynamical evolution of sunspot fine structures, namely, umbral dots, penumbral filaments, facular points (Denker, 1998). The small-scale brightenings near sunspots, were also observed in the wings of strong chromospheric absorption lines. These structures, which are concomitant with strong magnetic fields show brightness variations close to the diffraction-limit of the VTT ($\sim$0.16$''$ at 550 nm). With the phase-diverse speckle method, Seldin et al. (1996) found that the photosphere at scales below 0.3$''$ is highly dynamic.

In order to study the intensity enhancement in the inner line wings of Hα (656.28 nm), Denker et al (1995) used a speckle interferometer to obtain images of the solar chromosphere. The set-up consists of a field stop at the prime focus of the afore-mentioned telescope reducing stray-light, two achromats sampling 0.08$''$/pixel of the detector, an interference filter with

FWHM of 3 nm and the Universal Birefringes filter with FWHM of 0.1 nm or 0.05 nm (transparency amounts to 11% at Hα). A beam splitter was inserted in the light path to feed 90% light to the slow-scan CCD camera for speckle imaging and remaining light to the video CCD camera for guiding.

Magnetic fields in the solar photosphere are usually detected by measuring the polarization in the wing of a Zeeman spectral line. Keller and Von der Lühe (1992a, 1992b) applied differential speckle interferometric technique to solar polarimetry to image the quiet granulation, as well as to make polarimetric measurements of a solar active region. The set-up used at the Swedish Vacuum Solar Telescope, La Palma, consists of

(1) an achromatic quarter-wave plate that transforms circularly polarized into linearly polarized light,
(2) two calcite plates rotated 90° relative to each other to ensure that two beams have identical path lengths, and
(3) a quarter-wave plate balancing the intensity in the two beams.

The two beams are split up by a non-polarizing beam-splitter cube; one passes through a 8.2 nm FWHM interference filter centered at 520 nm and the other passes through a Zeiss tunable filter centered in the blue wing of Fe I 525.02 nm with FWHM of about 0.015 nm. The former is used to determine the instantaneous PSF; CCD video cameras were used to detect these channels.

Keller and Johannesson (1995) have developed another method to obtain diffraction-limited spectrograms of Sun consisting of speckle polarimetry technique and a rapid spectrograph (with a reflecting slit) scanning system. Two cameras record the spectrograms and 2-d slit-jaw images simultaneously. The slit of the spectrograph scans the solar surface during the observing run.

In order to reconstruct solar images, various image processing algorithms, viz., (i) Knox-Thomson technique, (ii) speckle masking method, (iii) the technique of BID, have been applied. The spectral ratio technique (Von der Lühe, 1984), which is based on a comparison between long and short-exposure images, has been employed (Wilken et al. 1997) to derive atmospheric coherence length. Models of the speckle transfer function (Korff, 1973) and of average short-exposure MTF have also been applied to compare the observed spectral ratios with theoretical values. In this respect, the technique of BID, (see section 6.5), where a direct measurement of calibrating speckle transfer function is not required (Nisenson, 1992), has clear advantage over other techniques in retrieving the solar image.

11.1.1.4 *Solar speckle observation during eclipse*

High resolution solar images obtained during partial solar eclipse may help in estimating the seeing effect. The limb of the Moon eclipsing the Sun provides a sharp edge as a reference object. The intensity profile falls off sharply at the limb. The departure of this fall off gives an indirect estimate of the atmospheric PSF. Callados and Vàzquez, (1987) reported the measurement of PSF during the observation of solar eclipse of 30 May, 1984 using the 40 cm Newton Vacuum telescope at Observatorio dei Teide. Saha et al., (1997b) developed an experiment for the speckle reconstruction of solar features during the partial eclipse of Sun as viewed from Bangalore on 24 October, 1995. The set-up is described below.

A Carl-Zeiss 15 cm Cassegrain-Schmidt reflector was used as telescope fitted with an aluminised glass plate in front that transmits 20% and reflects back the rest. A pair of polaroids were placed in the converging beam ahead of the focal plane. A 3 nm passband filter centered at 600 nm was inserted between these polaroids. One of the polaroids was mounted on a rotatable holder (see Figure 11.7), so as to adjust the amount of light falling onto the camera. A pin-hole of 1 mm diameter was set at the focal plane for isolating a small field-of-view. A microscope objective ($\times 5$) re-imaged this pin-hole onto the EEV CCD camera operated in the TV mode. The images were planned to be acquired with exposure time of 20 ms using a Data Translation^{TM} frame-grabber card DT-2861. Unfortunately, unfavorable weather conditions at Bangalore prevented in recording any data.

The image reconstruction involves the treatment of both amplitude errors, as well as phase errors. The 20 ms exposure time is small enough to preserve phase errors. Any of the schemes for phase reconstruction that satisfactorily reproduces the lunar limb would be valid for solar features close to the limb (within iso-planatic patch). Also, the limb reconstruction would be valid only for phase distortions along one direction (in a direction normal to the lunar limb). In spite of these shortcomings, the limb data would have provided additional constraints for techniques, like BID.

Another novel experiment was conducted by Koutchmy et al., 1994, using the modern 3.6 m Canada French Hawaii telescope (CFHT) at Mauna-Kea during total solar eclipse of 11 July, 1991 to probe the solar corona. In this venture, several cameras combining fast photographic 70 mm cameras and the video-CCD cameras were employed during the totality to acquire sub-arc-second spatial resolution white light images. In order to detect coronal radiation, two video-CCD cameras were used, viz.,

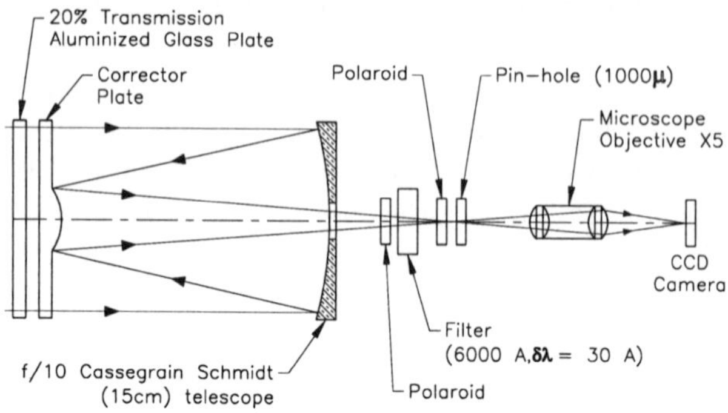

Fig. 11.7 Schematic layout of the set-up for the speckle reconstruction of solar features during the partial eclipse of Sun as viewed from Bangalore on October 24th, 1995 (Saha et al., 1997b).

(1) one fitted with a broad-band filter with $\times 2.5$ magnification ratio to record video frames during the totality so as to enable them to perform speckle interferometry image processing over the faintest coronal structures, and
(2) a second one with a 581×756 pixels CCD (with 11 μm pixel) fitted with a narrow-band filter ($\lambda = 637$ nm, FWHM $= 7$ nm).

Fine-scale irregularities along coronal loops of very large aspect ratio were observed in a time series, confirming the presence of plasmoid-like activity in the inner corona. Coronal loop is a feature in the solar corona consisting of an arch, extending upward from the photosphere for tens of thousands of kilometers. The whole of solar corona is believed to be made of coronal loops. The plasma density is thought to be the largest in the coronal loops considering all the coronal features. The density is about 10^{10} near the 1.1 solar radii, the streamers have average density of 10^9 whereas equatorial and coronal hole regions have density of the order of 5×10^8 and 10^7, respectively and density decreases with height above the limb. Bright coronal loops, in the form of coronal condensations and bright spots, are common around the time of solar maximum. Larger faint ones, lasting days or weeks, are more typical of the quiet corona. The other type of coronal loops are highly dynamic, short lived and associated with flares. The two ends of a loop, known as footprints, lie in regions of the photosphere of opposite magnetic polarity to each other. The study of formation of coronal loops, its physical and dynamical properties may provide a clue to the mechanism responsible to heat the plasma in the solar corona. These loops are believed to be thermally insulated from the surroundings because of magnetic pressure being larger than gas pressure, formed by impulsive heating and then cooling by radiative process. The observed increase in line-width of the forbidden emission lines with height above the limb in the solar corona has been attributed to the increase in non-thermal velocity caused by the coronal waves. The heating of solar corona has been explained in terms of these waves.

11.1.2 *Jupiter*

Jupiter is the largest planet in the solar system having a mass of 1.9 $\times 10^{27}$ kg with a diameter of 142,984 km across the equator. It possesses as many as 60 satellites, four of which, namely Callisto, Europa, Ganymede, and Io, were observed by Galileo in 1610 with his refracting telescope, who recorded their back and forth motions around the planet Jupiter. There exits a ring system, but invisible from the Earth.

Jupiter is a gas planet composed of hydrogen and helium, with traces of methane, water, ammonia and other compounds; the gaseous material gets denser with depth. Its interior is hot ($\sim 20,000°$ K), but not enough to ignite nuclear reactions. Hence, the energy is generated by the Kelvin-Helmholtz mechanism. Jupiter may have a core of rocky material. At deep inside, the hydrogen atoms are broken up due to the high pressure and the electrons are freed leaving behind the atoms with bare protons, which makes a state where the hydrogen turns out to metallic. The materials above the core are in the form of liquid metallic hydrogen consisting of ionized protons and electrons. This layer may have some helium and traces of various ice. The outermost layer is composed of molecular hydrogen and helium which is liquid in the interior and gaseous further out. The atmosphere contains trace amounts of methane, water vapor, ammonia, and silicon-based compounds. Traces of carbon, ethane, hydrogen sulphide, and other molecules are also present in small amounts. The outermost layer of the atmosphere contains crystals of frozen ammonia (Gautier et al. 1981, Kunde et al. 2004).

Jupiter has high velocity winds that are confined in colorful latitudinal bands. These winds blow in opposite directions in adjacent bands. Slight changes in chemical composition, as well as in temperature between these bands are responsible for the appearance of colored bands that are referred to zones; the dark ones are called belts. The Great Red Spot is a complex storm moving in a counter-clockwise direction; other smaller albeit similar spots are available as well. The material appears to rotate in four to six days at the outer boundary, while motions are small but random in direction adjacent to the center. Jupiter has a strong magnetic field, whose magnetosphere[12] extends more than 650 million km, within which Jupiter's Moons are orbiting.

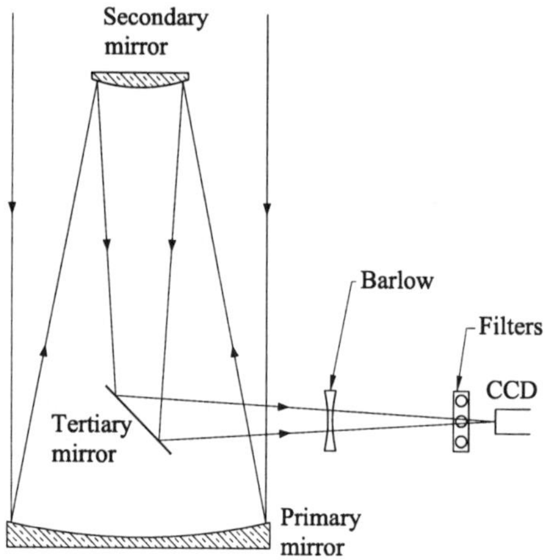

Fig. 11.8 Experimental set-up of the interferometer to record the images of Jupiter (Saha et al., 1997c).

During the period of July 16-22, 1994, the fragments of the comet Shoemaker-Levy 9 (SL 9; 1993e), after breaking up under the influence of Jupiter's tidal forces, collided with Jupiter with spectacular results. The fragments closest to Jupiter fell with a greater acceleration, due to the greater gravitational force. Observations of the crash phenomena, starting from the observations in the visible part of the electro magnetic spectrum to the radio frequencies, have been carried out extensively worldwide. With a goal of achieving features with a resolution of 0.3-0.5 arcsec., in the optical band, Saha et al. (1997c) had developed an interferometer (see Figure

[12]Jupiter's magnetosphere is not spherical; it extends a few million kilometers in the direction toward the Sun.

11.8) to record specklegrams of Jupiter at the 1.2 meter telescope, Japal-Rangapur Observatory (JRO), Osmania University, Hyderabad. The image scale at the Nasmyth focus (f/13.7) of this telescope was enlarged by a Barlow lens arrangement, with a sampling of 0.11″/pixel of the CCD. A set of 3 filters were used to image Jupiter, viz., (i) centered at 550 nm, with FWHM of 30 nm, (ii) centered at 611 nm, with FWHM of 9.9 nm, and (iii) RG 9 with a lower wavelength cut-off at 800 nm. A water-cooled bare 1024×1024 CCD was used as detector to record specklegrams of entire planetary disc of Jupiter with exposure times of 100 - 200 ms.

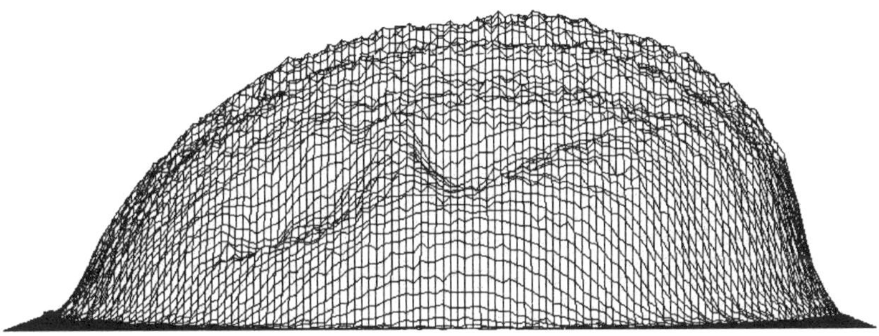

Fig. 11.9 Deconvolved image of Jupiter following the SL 9 impacts observed on 22nd July, 1994 (Saha et al., 1997c).

Reconstructions of Jupiter with sub-arcsecond resolution have been carried out by Saha et al. (1997c). They have identified the complex spots due to impacts by the fragments using the BID technique (see Figure 11.9). The main results of the construction is the enhancement in the contrast of spots. The complex spots in the East are due to impacts by fragments Q, R, S, D, G and the spots close to the centre are due to K and L impacts.

11.1.3 *Asteroids*

Asteroids are large rocky and metallic objects that orbit the Sun. They are found inside the Earth's orbit to beyond Saturn's orbit. Most of these objects are contained within the main asteroid belt that is located between the orbits of Mars and Jupiter roughly 2-4 AU away from the Sun; however some are found inside the Earth's orbit, as well as beyond Saturn's orbit. The total mass of these asteroids is estimated to be about 3.0-3.6$\times 10^{21}$ kg (Krasinsky et al. 2002). Some of these bodies have orbits crossing Earth's path and some have hit the Earth; Barringer Meteor Crater, Arizona, is an example.

Asteroids are too small to be considered as planets. The largest asteroid, which is named as Ceres, has a diameter of about 900 km, and the smallest ones are down to the size of pebbles. Twenty six asteroids have a diameter of about 200 km or more. They are classified into groups based on the characteristics of their orbits and on the details of their spectra and albedo[13] (Chapman et al. 1975). They are mainly

(1) C-type - carbonaceous, which are extremely dark (albedo 0.03); more than 75% of known asteroids are in this category,
(2) S-type - siliceous, that are relatively bright (albedo 0.10-0.22); about 17% of the known asteroids fall in this group, and
(3) M-type - metallic, which are bright with albedo 0.10-0.18.

The shape of the asteroids is irregular and is of fundamental interest. Apart from this, high resolution optical imaging data provide essential clues to infer the collisional history of asteroids which is intimately related with their origin. With modern imaging techniques, one should be able to obtain images at the inner edge of the main belt at a spatial resolution of 20 km. The expected high dynamic range can be used to either discover or place limits on possible satellites.

Detailed studies of the asteroids are essential to our understanding of the origin and evolution of our planetary system. Knowledge of an asteroid's rotation period, pole direction, size, detailed shape, bulk density, and the presence or absence of satellites may help to infer the asteroid's collisional history. Single high resolution images of the asteroids would furnish direct information about sizes, shapes and the extent of albedo variation, while multiple such images, taken over a range of rotational phases and perhaps during different apparitions, may (i) reveal the asteroid's full three dimensional shape, (ii) yield global albedo maps, furnish powerful constraints on the spin vector, and (iii) disclose the presence of satellites. Single aperture speckle interferometry has provided extensive measurements of the largest asteroids, including rotational image sequences for Vesta (Drummond et al. 1988); it has been studied recently with HST.

[13]Albedo, known as reflection coefficient, is a ratio of reflected/scattered power to incident power of the electromagnetic radiation (Rees, 1990). It is a unitless quantity of surface's (especially of a celestial body) reflectivity, which is given by, $A \equiv 1 - \epsilon$, where ϵ is the emissivity. The albedo of planets, satellites and asteroids are used to infer much about their properties. Photometry is used to study albedos, their dependence on wavelength, phase angle, and variation in time.

11.2 Stellar objects

The field of research that has benefited the most from such a technique using large-/moderate-sized single telescopes, and shall continue to benefit, is the origin and evolution of stellar systems. This evolution begins with star formation, including multiplicity, and ends with the mass loss process which recycles heavier elements into the interstellar medium. Large-scale star formation provides coupling between small-scale and large-scale processes. Stellar chemical evolution or nucleo-synthesis[14] further influences evolutionary process. High resolution observations are needed for the detection of proto-planetary discs and possibly planets (either astrometrically, through their influence on the disc, or even directly). Studies of the morphology of stellar atmospheres, the circumstellar environment of nova or supernova, young planetary nebula (YPN), long period variables (LPV), rapid variability of active galactic nuclei (AGN) etc. are also essential. The spatial distribution of circumstellar matter surrounding objects which eject mass, particularly young compact planetary nebulae (YPN) or newly formed stars in addition to T Tauri stars, late type giants or supergiants may also be explored. The technique is also being applied to studies of starburst AGN, and quasars.

11.2.1 *Measurement of stellar diameter*

Measurement of apparent angular diameters is one of the most important applications of interferometric technique. The high resolution interferometric techniques have made a dent in resolving many giant stars. The most direct way of measuring effective temperatures is the combination of bolometric fluxes with angular diameters although bolometric corrections do depend on models to an extent. These measurements are of particular importance of effective temperatures since they are relatively poorly determined especially for the coolest spectral types. With a known spectral type and luminosity class for a star, one computes its absolute magnitude; from observations its apparent magnitude, the star's probable distance is derived, and hence its probable diameter. On using the relation,

$$m - M = 5 \log r - 5 + A, \tag{11.7}$$

where m_v is the apparent visual magnitude of the star, M_V is the absolute

[14]Nucleosynthesis is a process of creating new atomic nuclei from the pre-existing nucleons.

visual magnitude of the star, r the distance (in pareses) of the star, and A the interstellar absorption (neglecting this absorption provides a conservative estimate of a star's expected diameter), the apparent distance of the star in parsecs can be derived.

Combined with photometry, the emergent flux or surface brightness of the stars can be obtained from these measurements (see equation 10.40). It is to reiterate that the effective temperature of a star is defined by its luminosity and stellar radius, provided the distance to the same is known. The measurement of angular diameter is the direct means of deriving star's effective temperature, the temperature of its visible surface, which indicates the amount of heat that it radiates per unit of surface area. Eclipsing binary system is a direct method to measure the radius of a star, both the primary and the secondary, from the time for the light curve to reach and rise from minimum. However, the eclipsing binaries are very rare since the orbits of the stars should be edge-on to our solar system.

Single aperture interferometry, speckle interferometry in particular, provides a possibility to measure diameters of the nearest, and therefore the brightest giants and supergiants like α Ori, α Tau, and Mira-variables with a formal accuracy of about 10%. Of course, one can expect better accuracy from the optical long baseline interferometric[15] observations. A recent survey of Mira variables showed detectable asymmetry (Ragland et al. 2006).

Observations carried out by means of speckle interferometry at large telescopes show a wavelength dependence of the photospheric diameter of α Ori and late type Mira variables due to the wavelength dependence of the opacity (Bonneau and Labeyrie, 1973, Labeyrie et al., 1977, Balega et al., 1982, Saha 1999b and references therein). Many young, cool stars, especially M-type stars, exhibit evidence of very strong chromospheric activity in the form of emission-line cores to their absorption features. Several supergiants have extended gaseous atmosphere, which can be imaged in their chromospheric lines. By acquiring specklegrams in the continuum and in the chromospheric emission lines (H and K lines of Ca II) simultaneously, differential image can be constructed. With the bispectrum image reconstructions of α Ori, Klückers et al., (1997) have found the evidence of asymmetry on its surface. Prior to this, Karovska and Nisenson (1992) have also found the evidence for the presence of a large bright feature on the surface. The rotation shear interferometer (Roddier and Roddier, 1988)

[15]An interferometer combines light from two or more telescopes to obtain measurements with higher resolution than could be obtained with either telescopes individually (Labeyrie, 1975, Saha 2002 and references therein).

had also been applied in the visible band to map the visibility of fringes produced by the star, and the reconstruction of the image revealed the presence of light scattered by a highly asymmetric dust envelope. Aperture synthesis using non-redundant masking technique at various telescopes depicted the presence of hotspots and other asymmetries on the surface of red supergiants and Mira variables as well (Bedding et al., 1997a, Tuthill et al., 1997). Wilson et al., (1997) have detected a complex bright structure in the surface intensity distribution of α Ori, which changes in the relative flux and positions of the spots over a period of eight weeks; a new circularly symmetric structure around the star with a diameter of $\geq 0.3''$ is found.

In essence, interferometry measures the diameter of the surface, and the atmospheres of cool stars are so tenuous and extended that they become opaque at a substantially larger radius in absorption bands of molecules such as TiO than in the continuum. Wavelength-dependent measurements of stellar diameters with spectral resolution across TiO and other molecular bands can thus provide a powerful new tool for the study of the extended atmospheres of cool stars. Measured diameter of the star is systematically larger at 712 nm than at 754 nm. The diameter ratio increases with decreasing effective temperature. The available models do not adequately describe the TiO opacity in the tenuous outer layers of the atmosphere or at the base of the wind; the interferometric data lend support to the existence of an extended molecular sphere, which has been postulated on the basis of infrared (IR) spectroscopy.

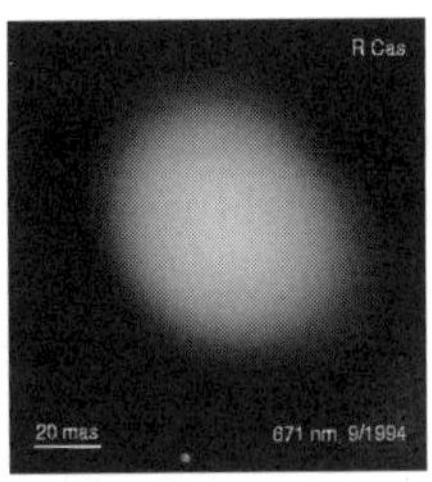

Fig. 11.10 Bispectrum reconstructed image of R Cas (Weigelt et al. 1996; Courtesy: Y. Y. Balega).

As far as Mira variables are concerned, Bedding et al. (1997a) have found that the diameter of R Doradus (57±5 mas) exceeded that of Betelgeuse; an asymmetric brightness distribution has also been detected from non-zero closure phases measurement. The diameter of a small amplitude Mira, W Hya, are reported to be 44±4 mas. Strong variations of diameter with TiO absorption depth have indeed been observed in the Mira variables o Ceti, R Leonis, and R Cas in qualitative agreement with model predictions. Time series measurements in well-defined narrow-band filters covering several pulsation cycles are required for a more detailed comparison between observations and theory. By means of speckle interferometry, Karovska et al. (1991) detected asymmetries in the extended atmosphere of o Ceti, which showed that the strength and the shape of this asymmetry changes as a function of wavelength and time. Plausible causes for the origin of the observed asymmetries, according to these authors, include instabilities in the pulsating atmosphere, nonspherical pulsation, or the interaction with the nearby companion. In addition, R Cas, a M-type red giant with a mean apparent magnitude of +9.97 (its brightness varies from +4.7 to +13.5 with a period of 430.5 days), displays an asymmetric profile in the TiO absorption bands. Figure (11.10) depicts the reconstructed images of Mira variable R Cas (Weigelt et al., 1996). It is observed that the disc of the star is non-uniform and elongated along the position angle $52^\circ \pm 7^\circ$ and $57^\circ \pm 7^\circ$ in the 700 nm (moderate TiO band absorption), as well as in the 714 nm (strong TiO band absorption) respectively. The variations in the size and position angle of the asymmetric structure occurred on a time scale of a few weeks. Comparison with the data taken a year back at almost the same phases also showed pronounced changes from cycle to cycle.

11.2.2 *Variable stars*

The variable stars undergo significant variations in apparent brightness (magnitude). They can vary in brightness due to intrinsic or external properties. In the former case, variation in brightness is because of internal physical changes, normally evidenced by pulsations or eruptions, while in the latter case, the received light fluctuates due to some process external to the star itself like eclipses due to a companion or planet or a dense cloud. In the intrinsic category, they are divided into three principal groups that run as follows.

11.2.2.1 *Pulsating variables*

Pulsating variable stars show periodic expansion and contraction of their surface layers. Pulsations may be radial or non-radial. A radial pulsating star remains spherical in shape, while a star experiencing non-radial pulsations displays periodical deformations of the spherical shape. Pulsating variables have instabilities in their envelopes, which cause them to pulsate in size and temperature over timescales of a few days to few hundred days. They show periodic changes in brightness accompanied by shifts in their spectral lines. During expansion, the visible surface of the star approaches the observer, the spectral lines are shifted towards the shorter wavelengths providing a negative radial velocity over and above the motion of the star in space, while in the contraction phase, it recedes, the spectral lines are shifted towards the longer wavelengths, giving positive radial velocity.

The diameter of a pulsating star may become double during the pulsation. The main cause of the light variation is periodic variation of the surface temperature of the star, since the luminosity of a star depends on its effective temperature, $L_\star \propto T_e^4$, where a small change in T_e can lead to a large variation in brightness. Their main characteristics are (i) regular periodicity in their light and radial velocity curves that are smooth, (ii) the star's spectral type changes with phase; it is earliest at maximum and latest at minimum, and (iii) there is a correlation between the absolute luminosity and the period, it depends on color as well.

(1) Cepheids: These stars are luminous, supergiant variables, which have instabilities in their envelopes that cause them to pulsate in size and temperature over timescales of 1-70 days. They are named after the first-such pulsating variable, δ Cephei, a prototype supergiant F star with a radius of about 70 solar radius. There are two types of Cepheids, namely (i) the original type I or classical Cepheids and (ii) the slightly dimmer Type II. Both these types are located in a region, known as the instability strip, of the HR diagram. The type I Cepheids are population I supergiants of spectral class F-K. The pulsations of the Cepheids are very regular, for example, δ Cephei, has a period of 5.37 days; its amplitude of radial velocity variation is 20 $\mathrm{km\,s^{-1}}$. The shape of the light curve is regular, depicting a fairly rapid brightening, followed by a sluggish fall off (Sandage and Gustav, 1968). These variable stars have masses between five and twenty times the solar mass. The more massive ones have extended envelopes. The type II Cepheids, that are stars of old population II, also obey period-luminosity relation. They are named after the first star identified in this group, W Virginis. This type of Cepheids are intrinsically less luminous by 1.5 - 2 magnitudes than the Type I Cepheids and have typical periods of 12-30 days. Their spectra are characterized by the lower metallicities. Type II light curves

show a characteristic bump on the decline side and they have an amplitude range of 0.3 - 1.2 magnitudes.

(2) Mira variables: The first pulsating variable discovered was the long-period variable Mira (o Ceti). Another example is R Leonis. These stars are M-type cool red giants or supergiants and occupy the high luminosity portion of the asymptotic giant branch (AGB) in the HR diagram, along with semi-regular variables. They are characterized by long periods of 80-1,000 days, usually with emission lines in their spectrum and vary from about third magnitude way down to tenth. Diameter changes, opacity changes, and possibly other processes like convection contribute to the brightness variation in these stars. The effective temperature of Mira variables is about 3000° K. They have a feeble gravitational hold on the material in their outer layers. Their great size and instability promote a wind[16], as a result of which, they loose gas in a steady flow at the rate of about 10^{-7} to $> 10^{-5}\ M_\odot\,\mathrm{yr}^{-1}$ (Knapp and Morris, 1985). The inner nuclear burning portions condensing into a burnt-out white dwarf, the rest of the materials accumulates around it as an extensive circumstellar shell.

(3) RR Lyrae: These old population II yellow or white giant pulsating variable are mostly found in globular clusters or elsewhere in the galactic halo[17]. They are characterized by their brightness change (0.3 - 2 m_v) with a regular period of a few days ($\sim 0.2 - \sim 2$ days). They follow their own period-luminosity relationship, with a mean absolute visual magnitude of +0.6; some of these stars have similar light curves to those of Cepheids. Thus, these stars can serve as distance indicator for relatively near objects (Benedict et al. 2002). RR Lyrae stars are older, and less massive than Cepheids, but are much hotter than the Sun; their typical luminosities are $\sim 45 L_\odot$. They are in a stage of their life where they have expended the hydrogen in their core, and are, at present, burning helium into carbon through nuclear fusion processes.

---

[16]A stellar wind is a continuous flow of gas that is ejected from the atmosphere of a star into space (Lamers and Cassinelli, 1999). These particles are accelerated differently, depending upon the nature of the star. In a cool star, like the Sun, the wind arises from pressure-expansion in a hot corona; during the main sequence phase, most stars blow a very modest wind. In hotter stars, the high radiative flux, drives the wind primarily by means of line scattering, which may be thought of as a transfer of momentum from the photons striking the atoms of gas. Stellar winds represent a mechanism by which the material is returned to the interstellar medium to be recycled as a new generation stars.

[17]Outside the plane of the Galaxy, an almost spherically symmetric halo extends out to 50 kpc and beyond. The halo contains very little interstellar material.

(4) Other pulsating variables: The semi-regular and irregular variables fall in this category. These stars are supergiants, often very massive young stars with unsteady pulsations in their extended outer layers. With periodicity in their pulsation, these variables are referred to semi-regular for example α Orionis, or else, they are known as irregular. The dwarf Cepheids, that are located below the RR Lyrae stars in the Cepheid instability strip in the HR diagram, are varying more rapidly than the classical Cepheids. The variations of hot massive β Cephei stars are rapid and of small amplitudes. The period of RV Tauri star that lie between the Cepheids and Mira variables in the HR diagram depends on the luminosity. The interesting features in their light curves are the alternating deep and shallow minima.

11.2.2.2 *Eruptive variables*

Eruptive variables experience sudden extreme increases in brightness due to violent outbursts caused by processes within the star. The brightness variations may also be accompanied by the flares or mass ejection events. There is a wide variety of eruptive or cataclysmic variables:

(1) Herbig Ae/Be stars: These objects, which are believed to be intermediate-mass stars, having 2-8 $M_\odot$, still in their phase of pre-main sequence contraction. They are of spectral type A and B, with strong emission lines (especially Hα and the calcium H and K lines) located in the star forming regions. These young stars are associated with bright nebulosity; some of them are located in the Orion nebula[18]. The spectral type of these stars are earlier than F0, and Balmer emission lines can be envisaged in the stellar spectrum. The excess IR radiation in comparison with normal stars are due to circumstellar dust. Some of these stars show photometric variations with an amplitude larger than 0.05 magnitudes (van den Ancker et al. 1997).

(2) R Coronae Borealis (RCrB) stars: These objects are luminous super-

[18]The Orion nebula, also known as Messier 42 (M42) or NGC 1976 (RA; $05^h35^m, 17.3^s; \delta - 05^\circ 23' 28''$), is a diffuse bright nebula situated south of Orion's Belt. It contains a gigantic dark cloud of matter, associations of stars, ionized volumes of gas and reflection nebulae. Within the Orion molecular cloud, new stars are forming. Observations with HST have yielded the major discovery of protoplanetary disks within this nebula (McCaughrean and O'dell, 1996) and brown dwarfs. Recently a pair of eclipsing binary dwarfs, 2MASS J053521840546085, have been discovered, whose masses are found to be 0.054 $M_\odot$ and 0.034 $M_\odot$ respectively, with an orbital period of 9.8 days; the more massive of the two turned out to be less luminous (Stassun et al. 2006).

giants, rich in carbon and poor in hydrogen, of spectral type F or G, which become faint suddenly by 2 to 8 magnitudes at unpredictable times. Berman (1935) showed in one of the earliest observations that RCrB itself was extremely hydrogen-deficient. The effective temperatures of these variables fall mostly around 7000° K, although a small number are as cool as 5500° K (Kilkenny and Whittet, 1984, Asplund et al. 1996); the surface gravities are also low. The ejection of material with velocities of 200 km s^{-1} is indicated by the violet displaced spectral lines. The reduction in brightness is thought to be caused by carbon-rich material puffed off the star from certain zones of the star. As the cloud moves out, it cools and eventually condenses into carbon dust particles, which absorb much of the incoming light from the star's photosphere. The normal brightness is expected when the dust is blown away by radiation pressure. At maximum brightness, these stars are found to undergo small-scale, Cepheid-like variations with fluctuations of few tenths of a magnitude. The radial velocity variations have been found with amplitudes proportional to the light variations indicating that RCrB stara are radially pulsating stars (Alexander et al. 1972).

(3) Flare stars: There are flare outbursts on the surface of the stars displaying sudden and unpredictable changes in light, some last over a timescale of a few seconds and others over an interval of minutes at irregular intervals, similar to those on the Sun but much more energetic. Within seconds the star may brighten by up to about 4-5 magnitudes and then fades away within an interval of a few minutes. An increase in radio and X-ray emission accompanies the optical outburst. These stars are termed as flare stars. For example, the UV Ceti stars that are red or orange emission-line dwarf of spectral class M are flare stars. They are young and are found in young star clusters and associations.

11.2.2.3 *Cataclysmic variables*

Cataclysmic or explosive variables undergo a dramatic change in their properties. These variables are binary systems which have a white dwarf (often referred as primary) and a cooler companion star. They have typically small orbital periods in the range 1-10 hrs. A white dwarf accretes[19] matter through Roche-lobe overflow from its companion star, which could be a main sequence star or a red giant. Since the white dwarf is very dense, the gravitational potential energy is enormous, and some of it is converted

[19] Accretion is a process which accumulates matter.

into X-rays during the accretion process. The cataclysmic variables are classified into subclasses according to properties of the outbursts:

(1) Novae: A nova occurs in a close binary system when a surface thermonuclear explosion on a white dwarf takes place and throws out its envelope of about $\sim 10^{-4}\ M_\odot$ (Prialnik, 2001). It is characterized by a rapid and unpredictable rise in brightness of several magnitudes within a few days. When dormant, a novae looks like an O-type subdwarf of absolute magnitude, $M_V = 5$. The eruptive event is followed by a steady decline back to the pre-nova magnitude over a few months, which suggests that the event causing the nova does not destroy the original star. Novae are classified into several subtypes, such as (i) ordinary novae, (ii) recurrent novae, having periodic outbursts of moderate amplitude; their spectral types at minimum are A, F with corresponding absolute magnitudes ranging from -0.5 to 1.5, (iii) dwarf novae, and (iv) nova-like variables.

(2) Supernovae (SN) (see section 11.2.4.2).

(3) Symbiotic stars: These stars are believed to be binaries, whose spectra indicate the presence of both cool and hot objects; one component may be an irregular long period cool M-type and the other is a hot compact star. Both the stars are associated with a nebular environment. Interaction of both components may lead to accretion phenomena and highly variable light curves. For example, CH Cyg (M7 giant) is a triple symbiotic system with a very complicated photometric and spectroscopic behavior (Iijima, 1998). Interferometric K-band observation of the said star yielded a uniform disk diameter of 10.4±0.6 mas (Dyck et al. 1998). The iron abundance for CH Cyg was found to be solar, $[Fe/H] = 0.0 \pm 0.19$ (Schmidt et al. 2006). Rather than material being accreted by gravitational attraction, the material is ejected from the surface of the red giant due to stellar wind. The resultant outbursts as material falls onto the white dwarf are irregular and smaller than in other eruptive variables brightening by up to three magnitudes.

In the extrinsic category, there are two types of stars:

(1) Eclipsing binaries (see section 10.3.2.3).

(2) Rapidly rotating stars: If a star rotates rapidly, it is distorted into an oblate shape, which makes the ratio of equatorial/ polar diameter greater than unity. The equatorial region is cooler than the polar region. This phenomenon occurs because the star is flattened by the

greater centrifugal force at its equator, which has the effect of making the surface temperature significantly cooler than at its polar regions. Rotating stars produce extreme starspots.

11.2.3 *Young stellar objects*

Stars form in giant clouds of gas and dust, called molecular clouds, which are composed primarily of H and He. The other notable molecules are H_2, H_2O, OH, CO, H_2CO, dust of silicates, iron etc. These clouds are approximately in a state of virial[20] equilibrium, which occurs when gas pressure equals gravity. The pressure builds up following the perfect gas law;

$$P = \frac{\bar{R}}{\bar{\mu}}\rho T, \tag{11.8}$$

in which $\bar{R}(= k_B/m_P)$ is the gas constant, T the absolute temperature, $K_B(= 1.380662 \times 10^{-23}\text{JK}^{-1})$ the Boltzmann's constant, ρ the gas density at radius r, P the pressure at the lower surface of the volume element, $\bar{\mu}(= m/m_P)$ the mean molecular weight (the average mass per particle in units of m_P), and m the mean particle mass and hence the equation becomes,

$$P = \frac{k_B}{m}\rho T. \tag{11.9}$$

Here the pressure, density, and temperature are functions of the distance r. Combining with the definition of optical depth, one obtains,

$$\frac{dP}{d\tau} = \frac{g}{k}, \tag{11.10}$$

where k is the absorption coefficient, which is a function of temperature, pressure, and chemical composition.

The change in radiation P with depth is given by,

$$\frac{dP}{dr} = \frac{k\rho\sigma}{c}T^4. \tag{11.11}$$

The outward pressure exactly balances the inward gravitational pull, a condition that is known as hydrostatic equilibrium. Assuming a cylindrical

[20]The virial theorem states that, for a stable, self-gravitating, spherical distribution of equal mass objects such as stars, galaxies, etc., the time-averaged value of the kinetic energy of the objects is equal to minus 1/2 times the time-averaged value of the gravitational potential energy.

volume element at a distance r from the center of a star, the equation of hydrostatic equilibrium is written as,

$$\frac{dP}{dr} = -\rho g = -\frac{GM_r\rho}{r^2}, \tag{11.12}$$

in which $g(= M_rG/r^2)$ is the gravitational acceleration, G is the gravitational constant, and M_r the mass contained within radius r; the mass continuity equation is expressed as,

$$\frac{dM_r}{dr} = 4\pi r^2\rho. \tag{11.13}$$

Combining equations (11.12 and 11.13), and plugging equation (10.40) one finds the expression for radiative equilibrium,

$$\frac{dT}{dr} = \frac{dT}{dP}\frac{dP}{dr} = -\frac{3kL_r\rho}{16\pi acr^2T^3}, \tag{11.14}$$

in which $a = B(0) = B(T_0)$, B is the brightness distribution, and for the convective equilibrium, according to Schwarzchild criterion, a fluid layer in hydrostatic equilibrium in a gravitational field would become unstable if the rate of change of pressure with density exceeds the corresponding adiabatic derivative,

$$\frac{d\log T}{d\log P} = \frac{\gamma - 1}{\gamma}\frac{T}{P}, \tag{11.15}$$

$\gamma(= C_P/C_v)$, C_P and C_v being the specific heats at constant pressure and constant volume respectively.

Combining with equation (11.12), one writes,

$$\frac{dT}{dr} = -\frac{\gamma - 1}{\gamma}\frac{T}{P}\frac{GM_r\rho}{r^2}. \tag{11.16}$$

The equations (11.12 - 11.16) are the fundamental equations of stellar structure.

The gravity pulls gas and dust towards a common center (the core) where the temperature increases as gas atom collisions increase. The gas pressure increases as atomic collisions and density increase. The gravitational binding energy of such a cloud is balanced by the kinetic energy of its constituent molecules. The molecular clouds are observed to have turbulent velocities imposed on all scales within the cloud. These turbulent velocities compress the gas in shocks generating filaments and clumpy structures within the cloud over a wide range of sizes and densities, and this process is known as turbulent fragmentation. When a fragment of such a cloud reaches a critical mass, called Jeans mass,

$$M_J = C\left(\frac{k_B T_k}{\bar{\mu} G}\right)^{3/2} \frac{1}{\sqrt{\rho}}, \tag{11.17}$$

in which C is the constant and T_k the kinetic temperature, it becomes gravitationally unstable and may again fragment to form a single or multiple star system.

As the star accretes more gas and dust, it tries to maintain equilibrium. On reaching equilibrium, accretion stops; failing which the star collapses. As the stellar formation progresses, the rotating cloud collapses to form a central source with a Keplerian (accretion) disc. Considering dynamics of contraction, the equation of motion for a shell of material at a distance r from the center is given by,

$$\frac{d^2 r}{dt^2} = -\frac{GM_r}{r^2} - 4\pi r^2 \frac{\partial P}{\partial M_r}. \tag{11.18}$$

The rate at which energy is produced depends on the distance to the center. The increase in the luminosity as one passes through the shell from inside, moving outwards is equal to the energy produced within the shell. The energy conservation equation is given by,

$$\frac{dL_r}{dr} = 4\pi r^2 \rho\, \varepsilon, \tag{11.19}$$

where L_r is the amount of energy passing through the surface r per unit time and ε the coefficient of energy production, which is defined as the amount of energy released in the star per unit time and mass.

In order to eliminate M_r, the equations (11.12 and 11.13) are combined, so that,

$$\frac{d}{dr}\left(\frac{r^2}{\rho}\frac{dP}{dr}\right) = -4\pi G \rho r^2. \tag{11.20}$$

In this equation (11.20), there are two unknown parameters, P and ρ, which are probably related to each other. Adding the equation of state for a polytrope that refers to a stellar model in which the pressure and density inside the star are related in the form,

$$P = K\rho^{(1+1/n)}, \tag{11.21}$$

where K is the constant and $n(=1/(\gamma-1))$ is called the polytropic index.

The equation (11.20) turns out to be,

$$\frac{P_c(n+1)}{4\pi G\rho_c^2}\frac{1}{r^2}\frac{d}{dr}\left(r^2\frac{d\theta}{dr}\right)+\theta^n=0, \tag{11.22}$$

where $\rho = \rho_c\theta^n$ and $P = P_c\theta^{n+1}$ are the central density and pressure respectively and θ is a common parameter, called the Lane-Emden function; $\theta = 1$ in the center of the star and 0 at the surface.

Introducing the dimensionless variable,

$$\xi = r\frac{4\pi G\rho_c^2}{(n+1)P_c},$$

so that $r = \alpha\xi$, in which

$$\alpha = \sqrt{\frac{P_c(n+1)}{4\pi G\rho_c^2}}, \tag{11.23}$$

one obtains,

$$\frac{1}{\xi^2}\frac{d}{d\xi}\left(\xi^2\frac{d\theta}{d\xi}\right)+\theta^n=0. \tag{11.24}$$

Equation (11.24) is the Lane-Emden equation for the gravitational potential of a self-gravitating, spherically symmetric polytropic fluid. This equation can be solved with boundary condition: $\theta = 1$ and $d\theta/d\xi = 0$ at $\xi = 0$. There are exact solutions of this equation for $n = 0, 1$, and 5,

$$\theta = \begin{cases} 1-\xi^2/6 & n=0, \\ \sin\xi/\xi & n=1, \\ 1/\sqrt{1+\xi^2/3} & n=5. \end{cases} \tag{11.25}$$

The first zero of the Emden function at ξ_1 corresponding to $\theta = 0$, i.e., $P = 0$ would represent the boundary of the star. It is found that ξ_1 increases with n starting from $\xi_1 = \sqrt{6}$ for $n = 0$ and reaching infinity for $n = 5$; $n = 0$ represents a uniform density sphere and $n = 5$ corresponds to highly centrally concentrated system.

Recalling equation (11.13), the central density can be obtained, i.e.,

$$\begin{aligned} M &= 4\pi\int_0^R r^2\rho dr \\ &= \pi\alpha^3\int_0^{\xi_1}\rho_c\theta^n\xi^2 d\xi. \end{aligned} \tag{11.26}$$

Using Lane-Emden equation, one gets,

$$M = -4\pi\alpha^3\rho_c \int_0^{\xi_1} \frac{d}{dx}\left(\xi^2 \frac{d\theta}{d\xi}\right) d\xi$$
$$= -\frac{4\pi R^3 \rho_c}{3}\left[\frac{3}{\xi}\frac{d\theta}{d\xi}\right]_{\xi_1 = R/\alpha}. \tag{11.27}$$

Writing $M = 4\pi r^3 \bar{\rho}/3$, in which $\bar{\rho}$ is the mean density, one writes the expression for the central density,

$$\rho_c = \frac{3M}{4\pi R^3} / \left[-\frac{3}{\xi}\frac{d\theta}{d\xi}\right]_{\xi_1}. \tag{11.28}$$

The equation (11.23) provides,

$$\frac{P_c(n+1)}{4\pi G \rho_c^2} = \alpha^2 = \frac{R^2}{\xi_1^2}, \tag{11.29}$$

from which, one obtains the central pressure, P_c,

$$P_c = \frac{GM^2}{R^2} / \left[4\pi(n+1)\left(\frac{d\theta}{d\xi}\right)_{x_1}^2\right]. \tag{11.30}$$

The rate at which energy is produced depends on the distance to the core. If the pressure is small one gets contraction by free fall of material towards the core, which increases the density. The release of gravitational energy heats up the material, which in turn, produce thermal pressure and stop the in-fall. A part of the gravitational energy lost in this collapse is radiated in the infrared, with the rest increasing the temperature of the core of the object. The luminosity, L, during the contraction phase is equal to the rate of change of kinetic energy, K, i.e.,

$$L = -\frac{dE}{dt} = -\frac{1}{2}\frac{d\Omega}{dt} = \frac{dK}{dt}, \tag{11.31}$$

where $E(= \Omega/2 = -K)$ is the total energy, and

$$\Omega = -\frac{3}{5-n}\frac{GM^2}{R}, \tag{11.32}$$

is the potential energy of the polytrope.

When the density and temperature are high enough, deuterium (a proton-neutron system) fusion ignition takes place, and the outward pressure of the resultant radiation slows down the collapse. Material comprising the cloud continues to fall onto the protostar. Once a protostar has become a hydrogen-burning star, a strong stellar wind forms, usually along the axis of rotation. Thus, many young stars have a bipolar flow, which may be due to the angular momentum of the infalling material. This early phase in the life of a star is known as the T Tauri phase. During this phase, the star has (i) vigorous activity (flares, eruptions), (ii) strong stellar winds, and (iii) variable and irregular light curves. Stars in this phase are usually surrounded by massive, opaque, circumstellar discs, which gradually accrete onto to the stellar surface. A fraction of the material accreted onto the star is ejected perpendicular to the plane of the disc in a highly collimated stellar jet. When the protostar becomes opaque, there is a quasi-hydrostatic equilibrium between the gravitational and pressure forces. The temperature gradient during this quasi-hydrostatic pre-main sequence contraction of the protostar depends on how the energy is transported; the transportation can take place by conduction, convection or radiation. Conduction is inefficient in the interior of normal stars, albeit can become important in the case of compact stars such as white dwarfs and neutron stars[21]. The time scale of contraction τ_g, known as gravitational time scale, can be obtained by integrating the following equation (Abhyankar, 1992),

$$\frac{dt}{dR} = -\frac{3}{2(5-n)}\frac{GM^2}{LR^2}. \tag{11.33}$$

The term dt/dR becomes negative during contraction. If the star is in radiative equilibrium, for $n \sim 3$, one gets,

$$\frac{dt}{dR} = -\frac{3}{4}\frac{GM^2}{LR^2}. \tag{11.34}$$

Here the contraction is slow, which keeps L nearly constant. Thus the time scale for the star which is in radiative equilibrium,

$$\tau_{gr} = \frac{3}{4}\frac{GM^2}{LR}. \tag{11.35}$$

[21]A neutron star is formed from the collapsed remnant of a massive star after a Type II, Type Ib, or Type Ic supernova. If the mass of a normal star is squeezed into a small enough volume (radius $\sim$ 10 km), the protons and electrons would be forced to combine to form neutrons. The interior of a neutron star is believed to consist of mostly neutrons with a small number of superconducting protons and electrons. At the center it is expected to have a very high density core mostly made up of quarks. Typically, neutron stars have masses range between more than 1.4 $M_\odot$ and less than 3 $M_\odot$. These stars have densities of about 10^{15} gm cm^{-3}. Further, they have other properties (Ryden, 2003) such as, (i) rapidly rotate with spin period in the range of $\sim$ 5 s to 1 ms, and emit periodic pulses, and hence they are believed to be pulsars, (ii) strongly magnetized (10^8 - 10^{12} G), and (iii) very hot ($\sim 10^6$ K).

If the radiative transfer of energy becomes inefficient, the absolute value of the radiative temperature gradient turns out to be very large. When the opacity is high, convection takes over. In the convective motion, analogous to the turbulent motion in a pot of water as it boils, the gas transporting heat rises upwards into cooler layers, where it loses its energy and sinks. The operation of convection depends on its viscosity and any forces, such as gravity, which tend to resist the convective motion. The upward and downward motion of gas elements mix the stellar material, and the composition of the convective parts of a star becomes homogeneous. The expression for the convective equilibrium for $n = 1.5$, is given by,

$$\frac{dt}{dR} = -\frac{3}{7}\frac{GM^2}{LR^2}. \tag{11.36}$$

Here the contraction is fast. The convective transport of energy removes the excess energy effectively, so that the luminosity of the star drops rapidly while the effective temperature, T_e remains almost constant. Invoking equation (10.90), L is replaced and followed by integrating one obtains the time scale for the star that is in convective equilibrium,

$$\begin{aligned} \tau_{gc} &= -\frac{3}{7}\frac{GM^2}{4\pi\sigma T_e^4}\int\frac{dR}{R^4} \\ &= \frac{GM^2}{28\pi\sigma T_e^4 R^3} = \frac{GM^2}{7L_* R}, \end{aligned} \tag{11.37}$$

with $L_* = 4\pi R^3 \sigma T_e^4$ as the luminosity of the present star.

The protostar continues to contract slowly and the gravitational energy released by contraction keeps the star radiating light. The contraction continues until the temperature in the core reaches a critical level[22]. At this point, a nuclear fusion is triggered, following which hydrogen begins to fuse in the core into helium by fusion reactions. Since a massive star posses

[22]Failing to reach the critical temperature in the core of a protostar, it ends up a brown dwarf. Stars with masses less than 0.08 $M_\odot$ cannot sustain significant nuclear fusion reactions in their core. When brown dwarfs are very young, they generate some energy from the fusion of deuterium, into helium nuclei. These stars do not posses radiation pressure to counter their own gravity; they contract slowly. According to the International Astronomical Union (IAU) statement (2003), objects possessing 13 Jupiter masses (for objects of solar metallicity) are called brown dwarfs, while objects under that mass (and orbiting stars or stellar remnants mass) are planets. Their surface temperatures are expected to be around 1000° K and the luminosities are of the order of 10^{-5} $L_\odot$. Many of these objects orbit solar type stars. When these stars become redgiants, they can engulf their companions (brown dwarfs). However, a brown dwarf, WD 0137-349 B, seems to have survived the primary's red giant phase (Maxted et al. 2006).

stronger gravity, which makes its core more compressed, the temperature rises in its core at which fusion can begin. This type of stars fuse their hydrogen into helium more quickly than lower mass stars and have hotter cores that release more energy, which in turn, makes them more luminous.

Following the onset of the nuclear fusion, the collapse is halted by the pressure of the heated gas and radiation which counteract gravity. The hydrostatic equilibrium is reached between opposing forces. The nuclear fusion in the form of photons exert a pressure, known as radiation pressure, on the stellar material as they travel outward from the center of the star. If the energy produced by nuclear fusion is less than the radiated energy, the star contracts causing an enhancement of central temperature and along with it an increase in energy production. When the energy produced is more than the radiated energy, the star heats up, which causes a build up of pressure, and in turn makes the star expand and lower its central temperature. Hence the star continues to radiate energy at more or less a constant rate for a long time, representing the main sequence phase of the star's life. After a star is born, the rest of the enveloping material is cleared away.

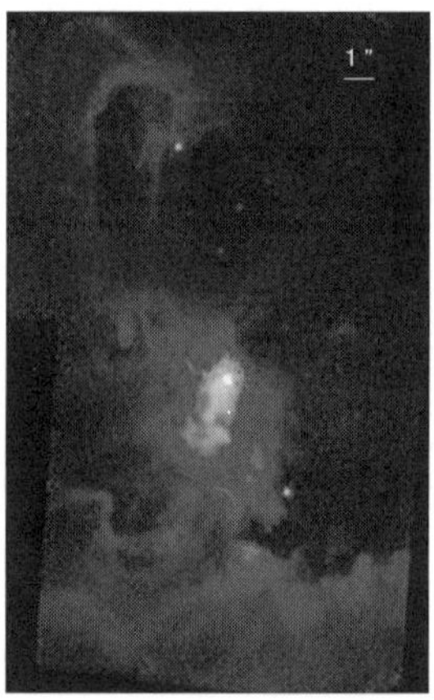

Fig. 11.11 Bispectrum reconstructed image of $13'' \times 21''$ area centered on S140 IRS 1 in K'- band; the image has a diffraction limited resolution of 76 mas (Courtesy: Y. Y. Balega).

It is to be noted that a single nebula can give birth to many stars. The evolution of young stars is from a cluster of protostars deeply embedded in a molecular clouds core, to a cluster of T Tauri stars[23] whose hot surface

[23]T Tauri stars are young (<10 million years), low-mass ($< 3\ M_\odot$) and are identified

and stellar winds heat the surrounding gas to form H II (ionized atomic hydrogen) region (young nebula), in which star formation is taking place. The H II regions are composed primarily of hydrogen and have the temperatures around 10,000° K. Generally, less than 10% of the available gases of this region is converted into stars, with the remainder of the gases dispersed by radiation pressure, supernova explosions, and strong stellar winds from the most massive stars, leaving behind open clusters. Later, the cluster breaks out, the gas is blown away, and the stars evolve.

Bispectrum speckle interferometry is employed to explore the immediate environment of deeply embedded young stellar objects. Some reconstructed images (see Figure 11.11) show a dynamic range of more than 8 magnitudes thus revealing many previously unknown complex structures around young stars. Observation of YSO, S140 IRS1, by Weigelt et al. (2002) showed the evidence for multiple outflows. In addition to the bright, elongated, and very clumpy feature pointing from the central source to the south-east, they have found several arc-like structures north-east of IRS1, extended diffuse emission south of IRS1, and four new point sources. The diffuse and fragmentary structures close to IRS1 appear to trace circumstellar material swept up by energetic outflows. Their image provides direct confirmation that two distinct bipolar outflow systems continue to be driven from IRS1 on scales between $3''$ and $100''$. A system of three arc-like structures to the north-east is also found to be consistent with cavities excavated by a precessing jet or wind-driven outflow.

11.2.4 *Circumstellar shell*

As stated in the preceding section (11.2.2), stars along the main sequence are burning hydrogen into helium in their cores (see Figure 11.12a). The

as pre-main sequence stars in the early gravitation contraction phase. They are found in nebulae or very young clusters. These stars contain low-temperature spectra with strong emission lines displaying Balmer lines and Ca H and K-lines, and broad absorption lines; their photospheric absorption spectra are similar to those of stars with spectral type F, G, K, and M (Appenzeller and Mundt, 1989 and references therein). T Tauri stars show high abundances of Li which gets destroyed by nuclear reactions in stellar interiors. The origin of the infrared excess in these stars is attributed to the presence of a circumstellar disk (Beckwith and Sargent 1993) left over from stellar formation. These stars exhibit irregular fluctuations of light (Appenzeller and Mundt, 1989), although some of them show a quasi-periodic behaviour stellar surface (Bertout 1989). Their irregular brightness changes may be due to instabilities in the disc, violent activity in their surfaces, as well as dust envelopes during the process of contraction. According to the strength of the Hα emission line, T Tauri stars are classified as (i) weak Line T Tauri Stars or (ii) classical T Tauri Stars.

proton-proton (pp) chain is the main source of energy for main sequence stars. It consists of following steps,

$$\begin{aligned} {}^{1}\mathrm{H} \;+\; {}^{1}\mathrm{H} &\rightarrow {}^{2}\mathrm{D} + e^{+} + \nu \\ {}^{2}\mathrm{D} \;+\; {}^{1}\mathrm{H} &\rightarrow {}^{3}\mathrm{He} + \gamma, \qquad \text{following two such reactions} \\ {}^{3}\mathrm{He} \;+\; {}^{3}\mathrm{He} &\rightarrow {}^{4}\mathrm{He} + \; 2\,{}^{1}\mathrm{H}. \end{aligned} \tag{11.38}$$

Two protons collide to form a deuteron, a positron and a neutrino. In the second step, the deuteron combines with a proton to form ^{3}He and a gamma ray is emitted. In third step, two ^{3}He particles combine to form one helium nucleus and two protons. Since the last step of the equation (11.38), uses ^{3}He particles, the first and second step occur twice as frequently as step three. The result of this proton-proton cycle is,

$$4\,{}^{1}\mathrm{H} \; + 2e^{-} \rightarrow {}^{4}\mathrm{He} + \; 2\nu. \tag{11.39}$$

For a star of above 1.5 $m_\odot$, the carbon (CNO) cycle becomes dominant since its reaction rates increases more rapidly with temperature. The reaction cycle is given by,

$$\begin{aligned} {}^{12}\mathrm{C} \;+\; {}^{1}\mathrm{H} &\rightarrow {}^{13}\mathrm{N} + \gamma, \\ {}^{13}\mathrm{N} &\rightarrow {}^{13}\mathrm{C} + e^{+} + \nu, \\ {}^{13}\,\mathrm{C} + \; {}^{1}\mathrm{H} &\rightarrow {}^{14}\mathrm{N} + \gamma, \\ {}^{14}\mathrm{N} \;+\; {}^{1}\mathrm{H} &\rightarrow {}^{15}\mathrm{O} + \gamma, \\ {}^{15}\mathrm{O} &\rightarrow {}^{15}\mathrm{N} + e^{+} + \nu, \\ {}^{15}\mathrm{N} \;+\; {}^{1}\mathrm{H} &\rightarrow {}^{12}\mathrm{C} + \; {}^{4}\mathrm{He} + \gamma. \end{aligned} \tag{11.40}$$

From the above reactions, one may notice that various positrons, neutrinos, and gamma-rays are produced as the decay products from the unstable isotopes. These are the source of energy in addition to the proton-proton chain gamma-rays.

The abundance of helium in the stellar interior increase as a consequence of the reactions described in equation (11.40). At the time of leaving the main sequence (see Figure 11.12b), the star possesses, (i) an isothermal He-core, (ii) an H-burning shell surrounding such a core, and (iii) an envelope. As this He-core has grown to its maximum size, it begins to contract. The contraction of the core liberates gravitational energy causing the central temperature, as well as the surrounding H-burning shell temperature, which in turn, increases in the rate of nuclear reactions in the shell. As the nuclear reactions continue in H-burning shell, the hydrogen burnt becomes part of

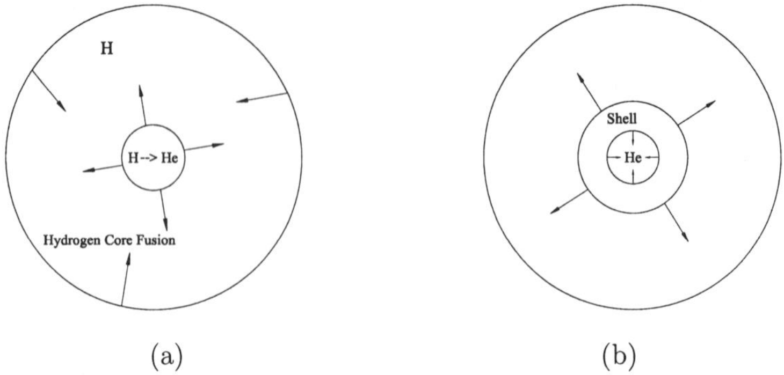

Fig. 11.12 Structure of a star at various stages of evolution; (a) main sequence stage, and (b) post main sequence stage.

the He-core and the shell moves outward in respect of mass; the temperature in the shell is maintained at H-burning level by the contraction of core. For a very low mass star, the envelope is convective. As a result, the energy produced in the shell pushes the stellar envelope outward, against the pull of gravity with the result of enhanced luminosity.

The opacity impedes the radiative flow of the generated energy towards the surface. The tendency of the gases to inhibit the flow of radiant energy is dependent on the mechanisms, namely (i) Bound-bound transition, (ii) Bound-free transition, (iii) (iii) Free-free transition, and (iv) scattering (see section 10.1.4.2). In the stellar core, the scattering by free electrons and bound-free transitions, mostly of H and He nuclei, are the major contributers to the opacity. As the convection zone is approached, bound-free transitions of heavier elements become dominant and increases the opacity.

In the case of the intermediate mass stars (of mass $\sim$0.8 - 8 $M_\odot$), the luminosity becomes constant at the initial phase, but with the drop of the surface temperature, their envelopes become convective and the luminosity rises. Thus all stars move into the red giant region in HR diagram. The mass loss increases when the star swells up to the size and low gravity of a red giant. The mass of these stars is not high enough to start further nuclear reactions and their cores end up as carbon-oxygen white dwarfs surrounded by an inner shell of helium fusion and an outer shell of hydrogen fusion; the helium fusion continue in the core until the core fuel supply is exhausted. This double-shell burning phase is referred to as the asymptotic giant branch (AGB) stage.

As the central temperature of contracting core rises to a value of the order of about 10^8, the helium can be transformed into carbon in the triple alpha reactions[24],

$$3\,{}^{4}\mathrm{He} \;\rightarrow\; {}^{12}\mathrm{C} + \gamma. \tag{11.41}$$

The temperature is high enough for helium nuclei to overcome the repulsive electrical barrier and fuse to form carbon (C); some of the carbon nuclei produced react with helium nuclei to form oxygen (O):

$${}^{12}\mathrm{C} \;+\; {}^{4}\mathrm{He} \rightarrow\; {}^{16}\mathrm{O} + \gamma. \tag{11.42}$$

The core arrives at a state of electron degeneracy obeying Pauli exclusion principle[25]. The core of a red giant star is dense enough to fill up available lower energy states. The availability of high energy states make the core to resist further compression; there is a pressure due to the electron degeneracy.

For the stars of mass less than 3 $M_\odot$, the electrons in the core become degenerate as the star contracts, while in the case of massive stars having mass more than 3 $M_\odot$, the core remains non-degenerate throughout, as a result the star settles down to a steady burning of helium[26] at the center and hydrogen burning in the shell. Of course, for the very low mass stars of mass less than 0.5 $M_\odot$, the core is not hot enough at the center to burn helium. During helium burning, the rising central temperature enhances the rate of energy production, causing large increase in luminosity with accompanying mass ejection, what is known as helium flash. Soon the temperature becomes high enough to remove degeneracy and the stars settles down in a helium burning phase.

After the helium is exhausted in the core, carbon burning sets in at the temperature $\sim (5-8)\times 10^8$. For the low mass stars of $M_\star < M_\odot$, no further reactions are feasible; the C-O core begins to contract again. Central temperatures are not high enough to burn carbon by reaction,

$$\begin{aligned} {}^{12}\mathrm{C} \;+\; {}^{12}\mathrm{C} \rightarrow\; & {}^{20}\mathrm{Ne} \;+\; {}^{4}\mathrm{He} \\ \rightarrow\; & {}^{16}\mathrm{O} \;+\; 2\,{}^{4}\mathrm{He}. \end{aligned} \tag{11.43}$$

Hence, the star turns out to be C-O white dwarf, which is the final stage of the evolution of stars with a main sequence mass $< 7-8M_\odot$ after they

[24]The triple alpha process releases about 20% as much energy as hydrogen burning.

[25]Pauli exclusion principle predicts that no two electrons in an atom can have identical quantum numbers. This principle has been used to explain the arrangement and the properties of different elements in the Periodic table.

[26]The lifetime on the helium burning main sequence is about 2 billion years.

lost their envelopes by stellar wind. The path of these through the region occupied by RR Lyrae variables in the HR diagram. The high mass stars ($M_\star > 8\ M_\odot$) have non-degenerate cores that may ignite C-O at higher temperatures in the reactions:

$$\begin{aligned} {}^{16}\mathrm{O} + {}^{16}\mathrm{O} &\rightarrow {}^{28}\mathrm{Si} + {}^{4}\mathrm{He} \\ &\rightarrow {}^{24}\mathrm{Mg} + 2\ {}^{4}\mathrm{He}. \end{aligned} \tag{11.44}$$

11.2.4.1 *Planetary nebulae*

Planetary nebulae (PNe) represent late stages of a majority of stars in a galaxy. They occur in a variety of morphologies and internal structures that contain signatures of mechanisms responsible for mass loss of their progenitors. In the case of the stars with masses between 3 $M_\odot$ and 8 $M_\odot$, carbon burning takes place in the core, albeit the helium and hydrogen burning shells continue to burn in two outer shells. The latter causes a sporadic outflow of material from the envelope, which gives rise to phenomenon displaying gaseous nebulae surrounding the hot stellar core, known as the central star or the nucleus of a PN, in their center. The resulting planetary nebula is the interaction of the ejected shell of gas and the ultraviolet light from the hot stellar remnant, which heats the gas. The nucleus photoionizes and excites the ejected gaseous shell, giving rise to a nebula with a rich emission line spectra that contains both allowed and forbidden transitions. The nebulae drift from the C-O-Mg core, which cools down to become a white dwarf, at a speed of 10 to 30 km s^{-1}. These objects represent the relatively short-lived phase, lasting for about 10,000 years, formed as a result of the stars in their AGB phase losing their outer shells before reaching the white dwarf stage (Pottasch 1984). White dwarfs are quite common, being found in binary systems and in clusters.

It is believed that planetary nebulae may retain signatures of the progenitor mass loss. In many cases, the images are very complex reflecting the complicated multicomponent structure of the circumstellar material. Images from HST revealed structures and micro-structures such as jets (both bi-polar and multi-polar), cometary globules, rings, knots, point symmetries and FLIERS (Sahai and Trauger, 1998). Basically the progenitor of a PN undergoes mass loss in several stages, viz.,

- while it is in the giant stage at a faster rate, albeit at slower speeds ($dM/dt = 10^{-6} - 10^{-7} M_\odot\,\mathrm{yr}^{-1}$ and $v = 10\ \mathrm{km\,s}^{-1}$), and
- later on after the hot central star gets detached, at a slower rate, but at

higher speeds ($dM/dt = 10^{-8} - 10^{-9} M_\odot \, \text{yr}^{-1}$; $v = 1000 - 2000$ km s^{-1}; Kwok, 2000).

Formation of PNe may be interpreted as a result of the interaction of these winds. The two-wind interaction mechanism describes PNe as a result of the ploughing action of the hot stellar wind on the slow progenitor wind.

In principle, the PNe shells are ejected in two scenarios that involve a single star or a binary system. The single star hypothesis predicts super-wind mass loss under which for a short duration of the order of 1000 yr the star loses mass at a rate of $\sim 10^{-4}$ $M_\odot$ per year. In the case of the binary system, the heavy mass loss occurs by the ejection of a common envelope. The equatorial density enhancement may be produced by rotation in a single star or by tidal interactions with companion stars if the central stars are binary system. The extent to which the density contrast exists between the equatorial and higher latitudes, decides how the subsequent mass loss, either by superwind in a single star or CE in a close binary system, would cause different observed morphologies (Soker, 1998).

Diffraction-limited images of the stars at the latest stages of their evolution were obtained mainly in the 12 μm range using large telescopes with an angular resolution up to 50 mas. The large, older and evolved planetary nebulae (PNe) show a great variety of structure (Balick, 1987) that are (i) spherically symmetric (A39), (ii) filamentary (NGC 6543[27]), (iii) bi-polar (NGC 6302) morphology that has multiple structure components, and (iv) peculiar (A35). The structure may form in the very early phases of the formation of the nebulae itself which is very compact and unresolved, for example the proto-planetary nebula (PPN). The PPN phase of stellar evolution falls between the mass-losing AGB star, and the evolved planetary nebula (Meixner, 2000). The lifetime for this phase is $\leq 10^3$ yrs and marks the time from when the star is forced off the AGB by intensive mass loss to when central star turns out to be hot enough ($T_e \sim 3 \times 10^4$ K) to photo-ionize the neutral circumstellar shell (Kwok, 1993). The spherically symmetric, pulsating mass loss from the AGB is followed by a large outburst with complex symmetries, during which 0.1 $M_\odot$ of atomic or molecular gas is lost. PPN often show a bipolar distribution, which is the result of the light from the central star emerging from the dissipating circumstellar dust envelope.

Due to the poor spatial resolution of the conventional imaging (section 10.4.1), the first $\sim$ 10^3 years of a PN spent in a phase that remains obscured

[27]NGC stands for New General Catalogue.

for structural details. In order to understand the processes that determine the structure and dynamics of the nebular matter in the PN, one needs to resolve and map the same when they are young and compact. By making maps at many epochs, as well as by following the motion of specific structural features, it would enable one to understand the dynamical processes at work. The structures could be different in different spectral lines e.g., ionization stratification in NGC 6720 (Hawley and Miller, 1977), and hence maps can be made in various atomic and ionic emission lines too.

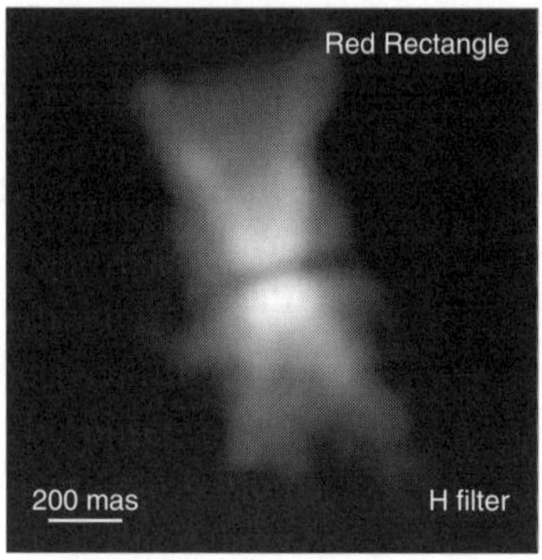

Fig. 11.13 Speckle masking reconstruction of the evolved object Red Rectangle; the resolution of this object is 75 mas for the H band (Courtesy: R. Osterbart); observations were carried out at SAO 6 m telescope (Osterbart et al., 1996).

The angular diameters of several young PNe in the Magellanic Clouds were determined with speckle interferometric technique (Wood et al. 1987). The high spatial resolution images of Red Rectangle (AFGL 915; see Figure 11.13), a reflection nebula associated with the A0 type post-AGB star, HD44179, near the Monocerotis constellation were recorded by Osterbart et al., (1997). It exhibits two lobes with the separation of $\sim 0.15''$. These authors argued that the dark lane between the lobes is due to an obscuring dust disc containing icy dust grains and hydrocarbon molecules formed in the cool outflow and the central star is a close binary system. This type of nebula displays a bipolar flow carrying a significant amount of mass away from the central stars. These stars may create a pair of jets, which might throw gas into a thick disc. Since it is viewed edge-on, the boundary edges of the cone shapes appear to form a rectangle.

Another object CW Leo, a carbon star IRC+10216, which is at present in a late phase of its evolution, a phase known as the AGB, also showed a resolved central peak surrounded by patchy circumstellar matter (Osterbart et al., 1996). This is a peculiar object with the central star being embedded in a thick dust envelope. A separation of 0.13″-0.21″ between bright clouds was noticed, implying a stochastic behaviour of the mass outflow in pulsating carbon stars. Weigelt et al., (1998) found that five individual clouds were resolved within a 0.21″ radius of the central object in their high resolution K′ band observation (see Figure 11.14). They argued that the structures were produced by circumstellar dust formation. The resolution of this object is 76 mas for the K′ band. Since CW Leo is representative of carbon-rich AGB stars, it is suggested that mass-loss in AGB stars may not be smooth and homogeneous.

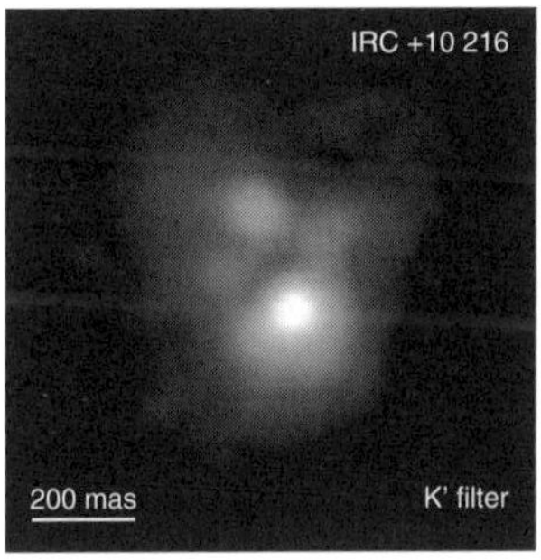

Fig. 11.14 Speckle masking reconstructed image of IRC+10216 (Courtesy: R. Osterbart).

Tuthill et al., (1999) recorded high resolution IR (1.65 μm and 2.27 μm) images of WR 104 by means of aperture masking technique at the 10 m Keck telescope, Hawaii. The reconstructed images of the same at two epochs depict a spiral pinwheel in the dust around the star with a rotation period of 220 $\pm$ 30 days. They opined that the circumstellar dust and its rotation are the consequence of a binary companion. The aspherical dust shell of the oxygen-rich AGB star AFGL 2290 (Gauger et al., 1999) has also been reported. Images of the young star, LkHα 101 in which the structure of the inner accretion disk is resolved have been reported as well (Tuthill et al. 2001). Detailed information that is needed for the modeling of the 2-D radiative transfer[28] concerning the symmetry – spherical, axial or lack of

[28]Radiative transfer is defined as the process of transmission of the electromagnetic radiation through the atmosphere. The atmospheric effect is classified into two effects such as multiplicative effects and additive effects. The former effect arises from the extinction (see section 10.2.3), while the latter comes from the emission produced by thermal radiation from the atmosphere and atmospheric scattering.

clouds, plumes etc. of the objects – can also be determined (Men'shchikov and Henning, 1997, Gauger et al. 1999).

Another interesting object is the star VY CMa, a late type M supergiant with peculiarities, mostly related to the intense circumstellar environment due to its high mass-loss rate. It displays large amplitude variability in the visible and strong dust emission and high polarization respectively in the mid- and near-IR. It is suggested that this star is considerably more luminous ($L_\star \sim 5 \times 10^5 L_\odot$) and larger ($R_\star \sim 2800 R_\odot$) than other galactic red supergiants. From surface imaging of another interesting object, the red super giant VY CMa, Wittkowski et al., (1998a) found to have non-spherical circumstellar envelope. They opined that the star was an immediate progenitor of IRC+10420, a post red supergiant during its transformation into a Wolf-Rayet (WR) star[29]. The visibility function in the 2.11 μm band image reconstruction depicted the contribution to the total flux from the dust shell to be $\sim$ 40%, and the rest from the unresolved central object (Blöcker et al., 1999); the ring like shell's intensity distribution was also noticed. From the reconstructed images of the non-redundant masking of 21-hole aperture observations of VY CMa carried out at the 10 m Keck telescope in the IR wave bands, Monnier et al., (1999) have found emission to be one-sided, inhomogeneous and asymmetric in the near IR. They were able to derive the line-of-sight optical depths of circumstellar dust shell; the results allow the bolometric luminosity of VY CMa to be estimated independent of dust shell geometry.

Among the other notable stars, the radiative transfer modeling of the supergiant NML Cyg revealed the multiple dust shell structures (Blöcker et al. 2001). Haas et al. (1997) have detected a halo of the Herbig Ae/Be star Elias I (V892 Tau), a peculiar pre-main sequence object in the Taurus dark cloud, and an unresolved core with near-IR speckle interferometry; Elias I is a close binary system with 0.05″ separation. The halo was found to be flattened and elongated in the East-West direction; the halo component

[29]The Wolf-Rayet (WR) stars are massive stars (more than 20 $M_\odot$) with a high rate of mass loss. They represent an evolutionary phase and are luminous, extremely hot stars of O-type. Massive Wolf-Rayet stars are young (population I) hydrogen-deficient stars (Hamann, 1996), as are binary systems such as υ Sgr (Jeffery, 1996). The spectra of these stars are characterized by the strong and wide emission lines (with equivalent widths upto $\sim$ 1000Å) of highly ionized elements such as, H, He, C, N, and O, which indicate an outflow velocity of a few thousand kilometers per second. Their surface composition is exotic, being dominated by helium. They are divided into two groups, namely (i) WC-type containing He, C, and O and (ii) WN-type which contain He and N. The presence or absence of hydrogen, respectively, is used to distinguish the late type WN stars (WNL) from the early type WN stars (WNE).

may scatter in bipolar lobes with a polar axis oriented East-West.

11.2.4.2 *Supernovae*

The nuclear burning proceeds towards more heavier elements for the stars more massive than 8 $M_{\odot}$, so that an iron core is developed at the center with a temperature of about $\sim 10^9$ degrees Kelvin (K). Following several steps, the burning of silicon (Si) produces nickel (Ni) and iron (Fe) by the following reactions:

$$\begin{aligned} {}^{28}\mathrm{Si} + {}^{28}\mathrm{Si} &\rightarrow {}^{56}\mathrm{Ni} + \gamma \\ {}^{56}\mathrm{Ni} &\rightarrow {}^{56}\mathrm{Fe} + 2e^{+} + 2\nu. \end{aligned} \tag{11.45}$$

With further contraction of this core, the central temperature rises to about 5×10^9 degrees, where the energy of the photons becomes large enough to destroy certain nuclei; such reactions are known as photo-dissociation. The photo-dissociation of Fe converts into He, which gives rise to an instability causing the contraction of the core to high density. The envelope is thrown out into the interstellar medium with a big explosion, a phenomenon called supernova (SN; Icko, 1986, Fender, 2002), leaving behind the star called neutron star or a black hole[30]. A typical supernova is characterized by a sudden and dramatic rise in brightness by several magnitudes (see Figure 11.15) outshining the rest of its galaxy for several days or a few weeks.

Observationally, the supernovae are classified according to the lines of different chemical elements that appear in their spectra, for example, if a spectrum contains a hydrogen line, it is classified Type II, otherwise it is Type I. There are subdivisions according to the presence or absence of lines and the shape of the light curve of the supernova. These are: Type Ia

[30]A black hole is an object with a gravitational field so powerful that light cannot escape its pull (Pasachoff, 2006). Black holes are conceived as singularities in space time. The space time metric defining the vacuum exterior of a classical black hole, and the black hole itself, is characterized by parameters such as, the mass of the black hole M_{BH}, the rotation (spin) J, and charge q. For $J = q = 0$, one obtains a Schwarzschild black hole, and for $q = 0$ and $J \neq 0$ one obtains a Kerr black hole. Black holes may be broadly classified into two categories, the stellar mass ($M_{BH} > 20\ M_{\odot}$) and super massive ($M_{BH} \geq 10^6 M_{\odot}$) black holes (Julian 1999). The birth history of the former is theoretically known with almost absolute certainty; they are the endpoint of the gravitational collapse of massive stars, while the latter may form through the monolithic collapse of early proto-spheroid gaseous mass originated at the time of galaxy formation or a number of stellar/intermediate mass ($M_{BH} \sim 10^{3-4} M_{\odot}$) black holes may merge to form it. They are expected to be present at the centers of large galaxies.

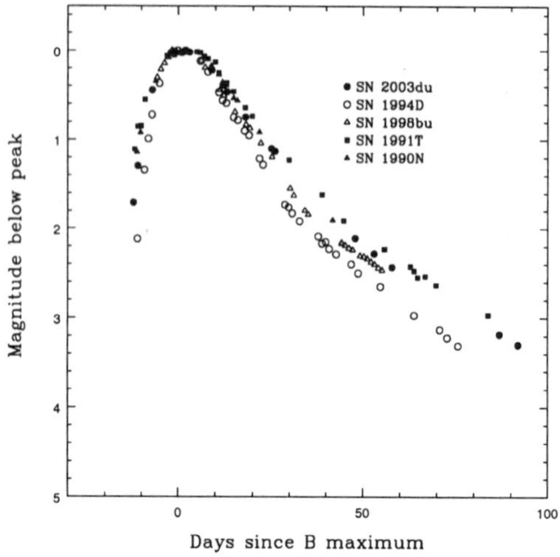

Fig. 11.15 Light curves of type Ia supernovae in the V band; the magnitudes are normalized to their respective peak (Anupama et al. 2005 and references therein; Courtesy: G. C. Anupama).

has Si II line at 615.0 nm, Type Ib contains He I line at 587.6 nm, and Type Ic possesses weak or no helium lines. The Type II supernovae are classified based on the shape of their light curves into Type II P (Plateau; see Figure 11.16) and Type II L. The former reaches a plateau in their light curve while the latter has a linear decrease in their light curve, in which it is linear in magnitude against time, or exponential in luminosity against time. The type Ia supernovae are white dwarf stars in binary systems in which mass is being transferred from an evolving companion onto the white dwarf. Two classes of models are discussed (Hoeflich, 2005). Both involve the expansion of white dwarfs to the supergiant phase.

(1) Final helium Shell Flash model: If the amount of matter transferred is enough to push the white dwarf over the Chandrasekhar mass limit[31] (Chandrasekhar, 1931) for electron-degeneracy support, the white dwarf may begin to collapse under gravity. A white dwarf may have a mass between 0.6 and 1.2 $M_\odot$ at its initial phase, and by accretion, approaches such a limit. Unlike massive stars with iron cores, such

[31] Chandrasekhar (1931) concluded that if the mass of the burnt core of a star is less than 1.4 $M_\odot$, it becomes a white dwarf. This mass limit is known as the 'Chandrasekhar mass limit'.

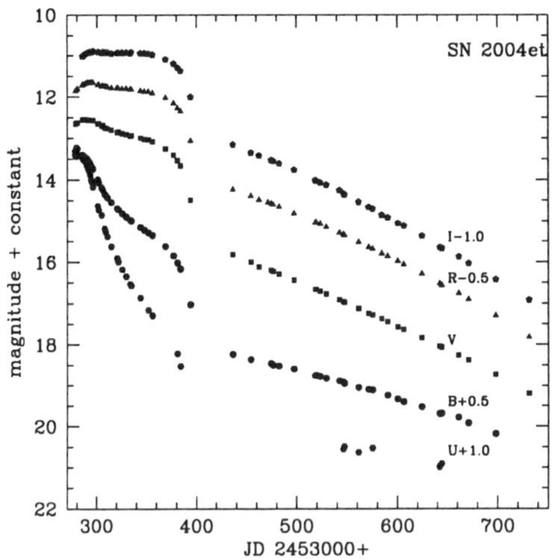

Fig. 11.16 UBVRI light curves of the type II P (plateau) supernova SN 2004et (Sahu et al. 2006). Note the almost constant magnitude phase, the plateau phase, prominently seen in the VRI bands (Courtesy: G. C. Anupama).

a dwarf has a C-O core which undergoes further nuclear reactions. Depending on the kind of the companion star, the accreted material may be either H, He or C-O rich. If H or He is accreted, nuclear burning on the surface converts it to a C-O mixture at an equal ratio in all cases. The explosion is triggered by compressional heating near the center of the white dwarf, and blow the remnant apart in a thermonuclear deflagration.

(2) Double Degenerate model: The supernova could be an explosion of a rotating configuration formed from the merging of two low-mass white dwarfs on a dynamical scale, following the loss of angular momentum due to gravitational radiation.

Supernovae are the major contributors to the chemical enrichment of the interstellar matter with heavy elements, which is the key to understand the chemical evolution of the Galaxy. The SNe Ia are an ideal laboratory for advanced radiation hydrodynamics, combustion theory and nuclear and atomic physics (Hoeflich, 2005).

Both nova and supernova (SN) have complex nature of shells viz., multiple, secondary and asymmetric; high resolution mapping may depict the events near the star and the interaction zones between gas clouds with

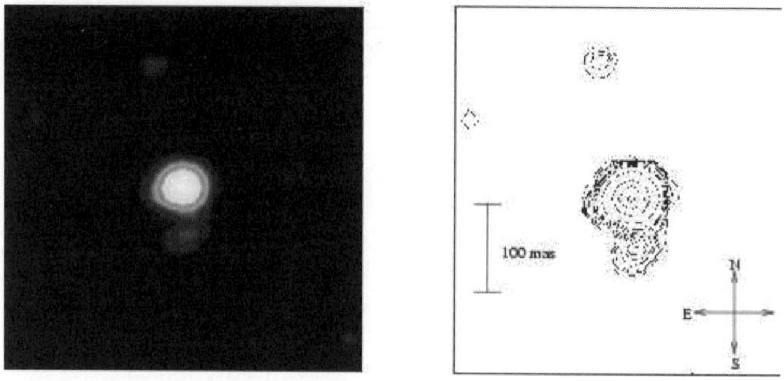

Fig. 11.17 Reconstructed image and contour plot of SN 1987A (Nisenson and Papaliolios, 1999, Courtesy: P. Nisenson).

different velocities. Soon after the explosion of the supernova SN 1987A, various groups of observers monitored routinely the expansion of the shell in different wavelengths by means of speckle imaging (Nisenson et al., 1987, Saha 1999b and references therein). It has been found that the size of this object was strongly wavelength dependent at the early epoch – pre-nebular phase indicating stratification in its envelope. A bright source at $0.06''$ away from the said SN with a magnitude difference of 2.7 at Hα had been detected. Based on the Knox Thomson algorithm, Karovska and Nisenson (1992) reported the presence of knot-like structures. They opined that the knot-like structure might be due to a light echo from material located behind the supernova. Studies by Nisenson and Papaliolios (1999) with a image reconstruction based on modified iterative transfer algorithm reveal a second spot, a fainter one (4.2 magnitude difference) on the opposite side of the SN with 160 mas separation (see Figure 11.17).

11.2.5 *Close binary systems*

Close binary stars play a fundamental role in measuring stellar masses, providing a benchmark for stellar evolution calculations; a long-term benefit of interferometric imaging is a better calibration of the main-sequence mass-luminosity relationship. High resolution imaging data in conjunction with spectroscopic data may yield component masses and a non-astrometric distance estimate. The notable shortcoming of spectroscopic surveys is that the determination of mass and distance as well as the information about binaries are missed.

Speckle interferometry (Labeyrie, 1970) has made major inroads into decreasing the gap between visual and spectroscopic binaries by achieving angular resolution down to 20 milliarcseonds (mas). Prior to the onset of such a technique, visual observers of binary stars could use the speckle structure of a binary star image in order to obtain information concerning the separation and position angles between the components. In this manner, they have utilized such a method without knowing about it. Following its application at the large and moderate telescopes, hundreds of close binary systems were resolved (Saha, 1999, 2002 and references therein). Major contributions in this respect came from the Center for High Angular Resolution Astronomy (CHARA) at Georgia State University, USA (Hartkopf et al., (1997). In a span of a little more than 20 yr, this group had observed more than 8000 objects; 75% of all published interferometric observations are of binary stars. The separation of most of the new components discovered by means of interferometric observations are found to be less than 0.25″ (McAlister et al., 1993). From an inspection of the interferometric data Mason et al., (1999b) have confirmed the binary nature of 848 objects, discovered by the Hipparcos satellites. Prieur et al. (2001) reported high angular resolution astrometric data of 43 binary stars that were also observed with same satellite.

A survey of chromospheric emission in several hundred southern stars (solar type) reveals that about 70% of them are inactive (Henry et al., 1996). In a programme of bright Galactic O-type stars for duplicity, Mason et al., (1998) could resolve 15 new components. They opined that at least one-third of the O-type stars, especially those among the members of clusters and associations, have close companions; a number of them, may even have a third companion. Among a speckle survey of several Be stars, Mason et al., (1997) were able to resolve a few binaries including a new discovery. From a survey for duplicity among white dwarf stars, McAlister et al., (1996) reported faint red companions to GD 319 and HZ 43.

Survey of visual and interferometric binary stars with orbital motions have also been reported. Leinert et al., (1997) have resolved 11 binaries by means of near IR speckle interferometry, out of 31 Herbig Ae/Be stars, of which 5 constitute sub-arc-second binaries. Reconstructing the phase of binary systems using various image processing algorithms have been made (Saha and Venkatakrishnan, 1997, Saha 1999b and references therein). Figure (11.18) demonstrates the reconstructed image of a close binary star, 41 Dra with double-lined F7V components, in the constellation of the northern hemisphere (Balega et al., 1997); the separation of the binary components was found to be about 25 mas. Based on spectral and speckle-interferometric observations of this system, model atmosphere parameters of the system components have also been derived by them. The masses of the components of 41 Dra were found to be 1.26 and 1.18 $M_{\odot}$.

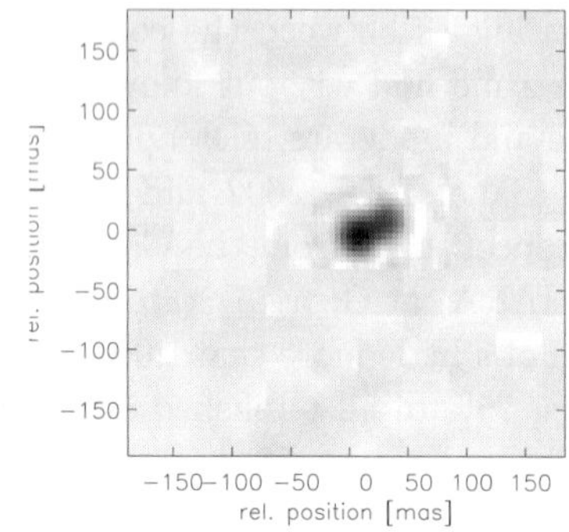

Fig. 11.18 Speckle masking reconstruction of 41 Dra (Balega et al., 1997); the separation of this system was found to be about 25 mas (Courtesy: R. Osterbart).

The most common binary orbit periods (as estimated from their separations and typical distances) lie between 10 to 30 years. Thus at the present stage, a large number of binary systems have completed one or more revolutions under speckle study and speckle data alone can be sufficient to construct the orbits. Various investigators have also calculated the orbital characteristics of many binary systems (Gies et al., 1997, Saha 1999b and references therein). Torres et al., (1997) derived individual masses for θ^1 Tau using the distance information from θ^2 Tau. They found the empirical mass-luminosity relation from the data in good agreement with the theoretical models.

Kuwamura et al., (1992) obtained a series of spectra using objective speckle spectrograph with the bandwidth spanning from 400 to 800 nm and applied shift-and-add algorithm for retrieving the diffraction-limited object prism spectra of ζ Tauri and ADS16836. They have resolved spatially two objective prism spectra corresponding to the primary and the secondary stars of ADS16836 with an angular separation of $\approx 0.5''$ using speckle spectroscopy imaging spectroscopy, Baba et al., (1994b) have observed a binary star, ϕ And (separation $0.53''$) at a moderate 1.88 m telescope; the reconstructed spectra using algorithm based on cross-correlation method revealed that the primary star (Be star) has an Hα emission line while the secondary star has an Hα absorption line.

High angular polarization measurements of the pre-main sequence binary system, Z CMa, at 2.2 μm revealed that both the components are polarized (Fischer et al. 1998); the secondary showed an unexpected large polarization degree. Robertson et al., (1999) reported from the measurements with an aperture masking technique that β Cen, a β Cephei star is a binary system with components separated by $0.015''$.

Stars from the Hyades cluster, that is 46.34 pc (151 light years) with an uncertainty of less than 0.27 pc (1 lyr) away from the Earth, may also be observed using single aperture interferometric techniques. These stars are bright with two thirds brighter than 11 m_v, the brightest ones are visible to the eye, in the constellation Taurus. Using modern spectroscopy as well as proper motion data, Stefanik and Latham (1985) identified 150 stars, all brighter than 14 m_v, which they consider to be the members of the Hyades cluster. Availability of high quality (σ_v 0.75 kms^-1) echelle data obtained by them from which they have discovered 20 binaries and identified 30 suspected binaries, is important for high resolution imaging. The principal scientific gains of the study of Hyades binaries are (i) the determination of the empirical mass-luminosity relation for the prototype population I cluster, (ii) the determination of the duplicity statistics in a well defined group of stars, and (iii) a non-astrometric distance estimate (McAlister, 1985).

Most of the late-type stars are available in the vicinity of the Sun. All known stars, within 5 pc radius from the sun are red dwarfs with $m_v > +15$. Due to the intrinsically faint nature of K- and M- dwarfs, their physical properties are not studied extensively. These dwarfs may often be close binaries which can be detected by speckle interferometric technique.

High resolution imaging of the population II stars may yield scientific results such as (i) helium abundance of the halo stars and (ii) statistics of duplicity and in general multiplicity of this ancient group of stars. Unfortunately, the helium abundance of the halo stars can not be measured spectroscopically owing to the low surface temperature of the sub-dwarfs. High resolution imaging data supplemented with existing radial velocity data or astrometric data (for the brighter star) can be used to derive the masses and hence the helium abundance.

11.2.6 *Multiple stars*

Multiple star systems are also gravitationally bound, and generally, move around each other in a stable orbit. Several multiple stars were observed by means of speckle interferometric method. The close companions of θ^1 Ori A and θ^1 Ori B (Petr et al., (1998), subsequently, an additional faint companion of the latter and a close companion of θ^1 Ori C with a separation of $\sim$ 33 mas are detected (Weigelt et al. 1999) in the IR band. These Trapezium system, θ^1 Ori ABCD, are massive O-type and early B-type stars and are located at the centre of the brightest diffuse Orion nebula, M 42. They range in brightness from magnitude 5 to magnitude 8; two fainter stars (E and F) can also be envisaged with a moderate telescope. Both the θ^1 Ori A and θ^1 Ori B stars are the eclipsing binary systems. The former is known as V 1016 Ori, some part of its light is blocked off by its companion in about every 65.43 days, while the latter has a period of 6.47 days with a magnitude range of 7.96 to 8.65. The θ^1 Ori C is a massive star having 40 $M_\odot$ with a temperature of about 40,000° K. It has the power to evaporate dusty discs around nearby new stars. Figure (11.19) displays bispectrum speckle K-band image of θ^1 Ori B, in which the faint fourth companion is seen near the center of the image (Schertl et al. 2003).

A star-like object, Luminous Blue Variable (LBV), η Carina, located in the constellation Carina (α 10 h 45.1 m, δ 59°41′), is surrounded by a large, bright nebula, known as the Eta Carinae Nebula (NGC 3372). This object was found to be a multiple object. Image reconstruction with speckle masking method of the same object showed 4 components with separations 0.11″, 0.18″ and 0.21″ (Hofmann and Weigelt, 1993). Falcke et al., (1996) recorded speckle polarimetric images of the same object with the ESO 2.2 m telescope. The polarimetric reconstructed images with 0.11″ resolution in the Hα line exhibit a compact structure elongated consistent with the presence of a circumstellar equatorial disc. Karovska et al., (1986) detected two close optical companions to the supergiant α Orionis; the separations of the closest and the furthest companions from the said star were found to be 0.06″ and 0.51″ respectively. The respective magnitude differences with respect to the primary at Hα were also found to be 3.4 and 4.6.

Ground-based conventional observations of another important luminous central object, R 136 (HD38268), of the 30 Doradus nebula in the large Magellanic cloud[32] (LMC) depict three components R136; a, b, and c, of

[32]The Galaxy (Milky Way) is a barred spiral galaxy (Alard, 2001) of the local group. The main disk of the Galaxy is about 80,000 to 100,000 ly in diameter and its mass is thought to be about 5.8×10^{11} $M_\odot$ (Battaglia et al. 2005, Karachentsev and Kashibadze, 2006) comprising 200 to 400 billion stars. It has two satellites, namely large Magellanic clouds (LMC) and small Magellanic clouds (SMC; Connors et al. 2006). The visual

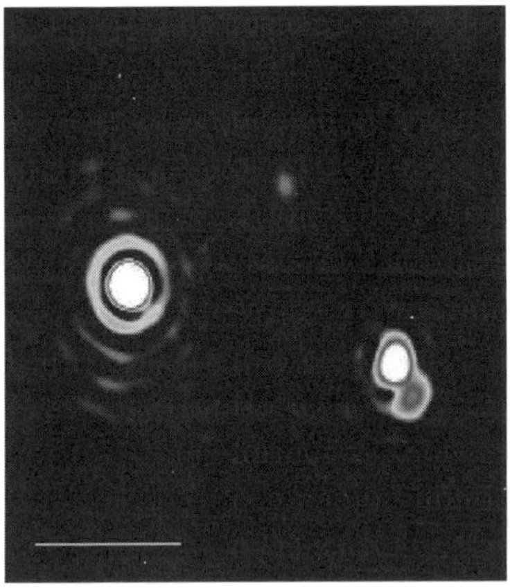

Fig. 11.19 Speckle masking reconstruction of a multiple stars, θ^1 Ori B (Schertl et al. 2003; Courtesy: Y. Y. Balega).

which R136a was thought to be the most massive star with a solar mass of $\sim 2500 M_{\odot}$ (Cassinelli et al., 1981). Later, it was found to be a dense cluster of stars with speckle interferometric observations (Weigelt and Baier, 1985).

Observations of R 64, HD32228, the dense stellar core of the OB association LH9 in the LMC, revealed 25 stellar component within a $6.4'' \times 6.4''$ field of view (Schertl et al. 1996). Specklegrams of this object were recorded through the Johnson V spectral band, as well as in the strong Wolf-Rayet emission lines between 450 and 490 nm. Several sets of speckle data through different filters, viz., (a) RG 695 nm, (b) 658 nm, (c) 545 nm, and (d) 471 nm of the central object HD97950 in the giant H II region starburst cluster NGC 3603 at the 2.2 m ESO telescope, were also recorded (Hofmann et al., 1995). The speckle masking reconstructed images depict 28 stars within the field of view of $6.3'' \times 6.3''$, down to the diffraction-limited resolution of $\sim 0.07''$ with m$_v$ in the range from 11.40 - 15.6.

11.2.7 *Extragalactic objects*

A galaxy is a gravitationally bound system of stars, neutral and ionized gas, dust, molecular clouds, and dark matter. Typical galaxies contain millions of stars, which orbit a common center of gravity. Most galaxies

brightness of the former ($\alpha = 5^h 23.6^m; \delta - 69^\circ 45^m$) is 0.1 m_v. Its apparent dimension is 650×550 arcmin and situated at a distance of about 179 kly. The visual brightness of the latter ($\alpha = 00^h 52.7^m; \delta - 72^\circ 50^m$) is 2.3 m_v. Its apparent dimension is 280×160 arcmin and situated at a distance of about 210 kly. Both these clouds are orbiting the Galaxy.

contain a large number of multiple star systems and star clusters, as well as various types of nebulae. At the center of many galaxies, there is a compact nucleus. The luminosities of the brightest galaxies may correspond to $10^{12}\ L_\odot$; a giant galaxy may have a mass of about $10^{13}\ M_\odot$ and a radius of 30 kiloparsecs (kpc). The masses of galaxies may be derived from observed velocities of stars and gas. The distribution of mass in spiral galaxies is studied by using the observed rotational velocities of the interstellar gas, which can be done either at visible wavelengths from the emission lines of ionized gas in H II regions or at radio wavelengths from the hydrogen 21 cm lines. Most galaxies are, in general, separated from one another by distances on the order of millions of light years. The space between galaxies, known as intergalactic space, is filled with a tenuous plasma with an average density less than one atom per cubic meter. There are probably more than a hundred billion galaxies in the universe. They form in various systems such as galaxy pairs, small groups, large clusters, and superclusters.

At the beginning of the last century, several galaxies of various shapes were discovered. Hubble (1936) classified these into elliptical, lenticular (or $S0$), spiral, and irregular galaxies. These galaxies are ordered in a sequence, what is referred to as, the Hubble sequence, from early to late types. They are arranged in a tuning fork sequence, the base of which represents elliptical galaxies of various types, while the spiral galaxies are arranged in two branches, the upper one represents normal spirals, and the lower one represents barred spirals. The elliptical galaxies are subdivided into $E0$, $E1, \cdots$, $E7$. The index denotes the ellipticity, ϵ of the galaxy and is related to the ellipticity by the relation,

$$n = 10\left(1 - \frac{b}{a}\right) = 10\epsilon, \tag{11.46}$$

where a and b are the semimajor and semiminor axes respectively.

An $E0$-type galaxy is almost spherical. The spiral galaxies are divided into normal and barred spirals. The density of stars in the elliptical galaxies falls off in a regular fashion as one goes outwards. The $S0$ type galaxies are placed in between the elliptical and spiral galaxies. Both elliptical and S0 galaxies are almost gas free systems (Karttunen et al. 2000). In addition to the elliptical stellar component, they posses a bright, massive disc made up of stars; in some elliptical galaxies there is also faint disc hidden behind the bulge. The distribution of surface brightness in the disc is given by,

$$I(R) = I_0 e^{R/R_0}, \tag{11.47}$$

where $I(R)$ is the surface brightness, R the radius along the major axis, I_0 the central surface brightness, and R_0 the radial scale length.

Spiral galaxies are relatively bright objects and have three basic components such as (i) the stellar disc containing the spiral arms, (ii) the halo, and (iii) the nucleus or central bulge. Some have large scale two-armed spiral pattern, while in others the spiral structure is made up of a many short filamentary arms. In addition, there is a thin disc of gas and other interstellar matter, in which stars are born, which forms the spiral structure. There are two sequences of spirals, normal Sa, Sb, and Sc and barred SBa, SBb, and SBc. The spiral arms in spiral galaxies have approximate logarithmic shape. These arms also rotate around the center, but with constant angular velocity. Most of the interstellar gas in such galaxies is in the form of molecular hydrogen; the availability of neutral hydrogen in Sa-type spirals is about 2%, while in Sc-type spirals is about 10%.

Another type of galaxies, referred as irregular galaxies (Abell, 1975), feature neither spiral nor elliptical morphology. Most of them are deformed by gravitational action. There are two major Hubble types of irregular galaxies such as $Irr\ I$ and $Irr\ II$. The former features is a continuation of the Hubble sequence towards later type beyond Sc-type galaxies. They are rich in gas and contain many young stars; they posses neutral hydrogen up to 30% or more. Both the large and small Magellanic clouds are $Irr\ I$-type dwarf galaxies. The latter types are dusty, irregular small ellipticals. Other types of dwarf galaxies are introduced, for example, dwarf spheroidal type dE. Another is the blue compact galaxies (also known as extragalactic H II regions), in which the light comes from a small region of bright, newly formed stars.

A few percent of the galaxies have unusual spectra, hence are referred as peculiar galaxies. Many of these galaxies are members of multiple systems, which have bridges, tails, and counterarms of various sizes and shapes; such peculiarities may have resulted from the interactions of two or more galaxies (Barnes and Hernquist, 1992, Weil and Hernquist, 1996). Stars in two nearby galaxies are generally accelerated due to tidal effects, which in turn leads to increase in the internal energy of this system. As the total energy is conserved, this results in loss of energy of the orbital motion of these galaxies. As a result, two galaxies, moving initially in an unbound (parabolic or hyperbolic) orbit may transform into another with a smaller eccentricity, or may form a bound orbit. Since most of the galaxies are found in pairs and multiple systems, they are bound to interact with each other frequently.

Gravitational interactions can transform the morphology of the galaxies. The galaxies with close companions experience tidal friction, which decreases their orbital radii and leads to their gradually forming a single system in equilibrium, what is known as dynamical friction. They are expected to merge in a few galactic crossing times. Giant luminous galaxies at the cores of dense clusters are supposed to have formed by the merger of smaller neighbours. Merging and disruption are two important processes in the dynamical evolution of a binary stellar system. The ratio of the times of disruption, t_d and merging, t_m for distant pairs is given by,

$$\frac{t_d}{t_m} \simeq \frac{6}{(5-n)} \frac{a}{R} \frac{M}{M_1}, \tag{11.48}$$

in which a is the orbital radius, R the radius of the galaxy, M and M_1 the masses of the stellar systems, and n the polytropic index describing the density distribution of the stellar system (Alladin and Parthasarathy, 1978).

It can be seen from equation (11.48) that if the galaxies are centrally concentrated (i.e., $n = 4$) and have similar mass, merging occurs nonrapidly than disruption. On the other hand if the masses are dissimilar, the interaction between them is likely to cause considerable disruption to the less massive companion and in this case the disruption time could be shorter than the merging time.

Every large galaxy, including the Galaxy, harbors a nuclear supermassive black hole (SMBH; Kormendy and Richardson, 1995). The extraction of gravitational energy from a SMBH accretion is assumed to power the energy generation mechanism of X-ray binaries, and of the most luminous objects such as active galactic nuclei (AGN) and the quasars (Frank et al. 2002). Accretion on to such a massive black hole transforms gravitational potential energy into radiation and outflows, emitting nearly constant energy from the optical to X-ray wavelengths; the typical AGN X-ray luminosities range from $10^{33} - 10^{39}$ W.

11.2.7.1 *Active galactic nuclei (AGN)*

Some galaxies are active, referred to as active galaxies, in which a significant portion of the total energy output from the galaxy is emitted by a source other than the stars, dust, and interstellar medium. They exhibit violent activity that is produced in the nucleus, which appear to be extremely bright at any given epoch. Their nuclei containing a large quantity of gas are called active galactic nuclei (AGN; Binney and Merrifield, 1998, Krolik, 1999).

AGN were first discovered in the 1940s as point-like sources of powerful optical emission with spectra showing very broad and strong emission lines in their nuclei, indicating large internal velocities. These lines exhibit strong Doppler broadening, which may be due to either rotational velocities of the order of several thousand km s^{-1} near a black hole or due to explosive events in the nucleus. They were also found to show significant optical variability on time-scales of months, with the emitting source being completely unresolved. AGN may also posses (i) an obscuring torus of gas and dust obscuring the broad-line region from some directions, (ii) an accretion disc[33] and corona in the immediate vicinity of the supermassive black hole (SMBH) with a mass ranging from 10^6 to 10^{10} $M_\odot$, and (iii) a relativistic jet[34] emerging out of the nucleus. The strengths, sizes, and

---

[33]Accreting matter is thrown into circular orbits around the central accretor due to angular momentum, leading to the formation of the accretion discs around a young star, a protostar, a white dwarf, a neutron star, the galactic and extra-galactic black holes (Frank et al. 2002). Accretion discs surrounding T Tauri stars are called protoplanetary discs. Typical AGN accretion discs are optically-thick and physically-thin (Shakura and Sunayev, 1973) and are thought to extend out to ~0.1 pc. The associated temperatures are in the range of $\sim 10^5 - 10^6$ K, making them sources of quasi-thermal optical and UV radiation that scatters off electrons commonly found in the ambient media of hot and ionized accretion regions. These coronal electrons are heated up by the energy feeding magnetic fields. Cool accretion disc photons thus undergo inverse-Compton scattering off the hot electrons, and emerge as high energy X-rays. Multiple scatterings within the corona increase the energy further, which result in the characteristic power-law (non-thermal) spectra extending from under 1 keV to several hundred keV.

[34]Jets are the powerful streamer of sub-atomic particles blasting away from the center of the galaxy and appear in pairs, with each one aiming in the opposite directions to each other. They seem to present in many radio galaxies and quasars, and are thought to be produced by the strong electromagnetic forces created by the matter swirling toward the SMBH. These forces pull the plasma and magnetic fields away from the black hole along its axis of rotation into a narrow jet. Inside the jet the shock waves produce high-energy electrons spiraling around the magnetic field and radiate the observed radio, optical and X-ray knots via the synchrotron process (Marshall et al. 2002). From the study of the active galaxy 3C 120, Marscher et al (2002) opined that the jets in active galaxies are powered by discs of hot gas orbiting around supermassive black holes. Similar jets, on a much smaller scale, may also develop around the accretion disc of neutron stars and stellar mass black holes. For example, the enigmatic compact star, SS 433, which is known to have a companion with an orbital period of 13.1 d, and a large disc, has two highly collimated relativistic jets moving at a velocity of $\sim$ 0.26c. Its central object could be a low mass black hole (Hillwig et al. 2004). Recent multi-wavelength campaign (Chakrabarti et al. 2005) of this object revealed that the short time-scale variations are present (28 min) on all the days in all the wavelengths, which may indicate disc instabilities causing ejection of bullet-like entities.

extents of the various ingredients vary from one AGN to another. There are some galaxies, which may have very bright nucleus similar to a large region of ionized hydrogen. These may be young galaxies where large numbers of stars are forming near the center and evolving into supernovae (starburst[35] nuclei).

Role of powerful AGN feedback through winds and ionization of the interstellar media is now seen as an integral part of the process of galaxy formation. Some of the most recent X-ray surveys are revealing unexpected populations of AGN in the distant Universe, and suggest that there may have been more than one major epoch of black hole mass accretion assembly in the history of the Universe (Gandhi, 2005). The sizes of accretion disc are thought to be of the order of light-days for typical SMBHs of mass $10^6\ M_\odot$. However, even at the distance of the nearest AGN, such sizes are too small to be resolved by the current generation of telescopes, since the resolution required is close to 1 mas (Gallimore et al. 1997); large optical interferometric arrays with very large telescopes may be able to resolve the discs for the very nearest AGN (Labeyrie, 2005, Saha 2002, and references therein).

Discrete and patchy cloud-like structures much further out from the SMBH produce the bulk of AGN optical emission line radiation, according to which AGN were first classified; they are classified according to their optical emission line properties. Two main structures such as the broad-line region (BLR), where the gas is very hot and moving fast corresponding to a velocity of $\sim 10^4$ km s^{-1}, and narrow-line region (NLR), where FWHM $<\sim 10^3$ km s^{-1}, suggesting that movement of the gas in these galaxies is slow, are discerned, based on the Doppler widths of emission line thought to be generated in each region. Essentially, those with very broad per-

[35]Starburst refers to a region of space with an intense burst of high-mass star formation. Galaxies are often observed to have a burst of star formation after a collision or close encounter between two galaxies. The Antenna galaxies (NGC 4038/NGC 4039), M82, and IC 10 are the well known starburst galaxies. The impact of such a collision generates shock waves throughout the galaxy, which push on giant clouds of interstellar gas and dust. This shock waves in turn cause the clouds to collapse and form short-lived massive stars. These stars that form from this collision use up their nuclear fuel quickly and explode in a supernova, which generate more shock waves and consequently more star formations. The formation of a starburst galaxy ends when its clouds are used up or blown away due to the explosion. Spectacular loops (size ~10 kpc) of hot gas extending into the intergalactic space, and a low surface brightness hot halo, extending out to ~18 kpc are seen in the Antennae (Fabbiano et al., 2004). This gas may be the result of super-winds driven by supernovae from the starburst. Scaled-down versions of starbursts are found in the local group of galaxies, like 30 Doradus in the LMC.

mitted recombination lines also typically possess blue broad-band optical continua and are designated type 1 AGN. Those with comparatively narrow permitted lines are called type 2 AGN and usually have red continua. Intermediate types also exist which display both narrow and broad components to the permitted lines, and all types can show significant forbidden lines, but these are always narrow.

Photo-ionization may be the principal mechanism for emission line generation, though collisional excitation can have a significant effect, especially for forbidden lines. The forbidden lines are always observed to be narrow, which suggests the existence of two distinct regions where optical emission occurs. A dense ($10^7 - 10^{10}$ cm^{-3}) BLR is inferred to exist on scales much smaller than a parsec (pc) within the deep black hole gravitational potential, leading to the observed large Doppler-broadened line-widths. The size of this region to be $0.1L_{46}^{0.5}$ pc, in which L_{46} is the photoionizing luminosity in units of 10^{46} erg s^{-1} (Peterson, 1993). The NLR extends over scales of tens to hundreds of pc, where the emitted lines are all narrow due to weaker gravity. The very existence of strong forbidden lines implies that densities in the NLR cannot exceed the critical density for collisional de-excitation, and are inferred to be low ($\sim 10^4$ cm^{-3}). The typical value of temperatures of the photoionized media for the NLR is of the order of $\sim 10^4$ K; the BLR being hotter by a factor of $\sim 1.5 - 2$ (Osterbrock, 1989).

There are two notable classes of active galaxies such as Seyfert and the radio galaxies[36]. The term, Seyferts, named after the discoverer, Carl Seyfert, traditionally refers to AGN which are fainter than $M_{\rm B} \sim -21.5 + 5\log H_0$, in which $M_{\rm B}$ is the absolute B-magnitude and H_0 the Hubble

[36]Radio galaxies emitt radio waves from their disc and halo around the central region and are extremely luminous ($10^{33} - 10^{38}$ W) at radio wavelengths between 100 MHz – 50 GHz. The radio emission is due to the synchrotron process that are associated with the relativistic plasma (Lorentz factors of $\sim 10^4$) in magnetic fields, as inferred from power-law spectrum (Bordovitsyn, 1999). Radio galaxies are generally expected to be elliptical galaxies. They display a wide range of structures in radio maps. The characteristic feature of a strong radio galaxy is a double (fairly) symmetrical, roughly ellipsoidal structures, called lobes, placed on either side of the plane of galactic disk. The observed radio double-lobe structure is determined by the interaction between twin jets and the external medium, modified by the effects of relativistic beaming (Fanaroff and Riley, 1974). Such double-lobed structure of radio galaxies was first discovered from radio interferometric observations of the powerful radio galaxy Cygnus A (Jennison and Das Gupta, 1953). Differences in the main observational characteristics of different types of AGN are due to orientation rather than intrinsic. For the optical, orientation effects give rise to differences in the obscuration and polarization properties, whereas in the radio they cause the beaming of the relativistic jet making some AGN powerful radio sources. Radio-quiet AGN are difficult to identify, especially at high redshift.

constant in units of 100 km s^{-1} Mpc^{-1}, in the optical. The Hubble constant is the ratio of the speed of recession of a galaxy to its distance from the observer, which is used to estimate the size and age of the Universe and determine the intrinsic brightness and masses of stars in nearby galaxies. It may examine the same properties in distant galaxies and galaxy clusters, and deduce the amount of dark matter present in the universe as well. Hubble (1929) found that the spectral lines of galaxies were redshifted by an amount proportional to their distances. This redshift appeared to have a larger displacement for faint galaxies. Mathematically, the Hubble's law is written as,

$$z = \frac{H_0}{c} r, \tag{11.49}$$

with $z = (\lambda_o - \lambda_r)/\lambda_r)$, as the Doppler redshift, λ_o λ_r are the observed and reference wavelengths respectively, H_0 the Hubble constant which is typically expressed in km s^{-1} Mpc^{-1}, and corresponds to the value of Hubble parameter, H that decreases over time, r the distance of the galaxy from the Earth, and c the speed of light.

For small velocities $v \ll c$, the Doppler redshift $z = v/c$, in which v is the recessional velocity due to redshift, and hence

$$v = H_0 r. \tag{11.50}$$

This equation (11.50) is known as the Hubble law[37]. For a set of observed standard candles, for example, galaxies whose absolute magnitudes are close to mean M_0, the Hubble constant corresponds to a linear relationship between the apparent magnitude, $m(= M_0 + 5\log(r/10\,pc)$ and the logarithm of the redshift, $\log z$, and hence the Hubble's law yields,

$$m = 5\log z + C, \tag{11.51}$$

where C is the constant that depends on H_0 and M_0.

Seyfert galaxies are usually spiral or irregular, while the radio galaxies are ellipticals; about 1% of the spiral galaxies are Seyfert galaxies and

[37]The Hubble law justifies the Big Bang theory, which states that the universe was created about 14 billion years ago from a cosmic explosion. Extrapolated into the past from the Hubble redshift of distance galaxies, it was observed that the universe has expanded from a state, in which all the matter and energy of space was concentrated in a tiny volume. This theory also predicts the existence of cosmic microwave background radiation (CMBR), which is expected to be due to the remnant heat from the explosion (Gamow, 1948). It received confirmation when this radiation was discovered by Penzias and Wilson (1965).

are weak radio sources. The other classes of AGN include blazars[38], BL Lacertae[39], (BL Lac) objects, and LINERs[40].

Study of the physical processes, viz., temperature, density and velocity of gas in the active regions of the AGN is an important field of observational astronomy. Optical imaging in the light of emission lines on sub-arc-second scales can reveal the structure of the narrow-line region. However, the input of high resolution imaging in the study of extragalactic objects is very modest. Though the input of speckle interferometry in the study of such objects is very modest, the scale of narrow-line regions is well resolved by the diffraction limit of a moderate-sized telescope. The time variability of AGNs ranging from minutes to decades is an important phenomena, which may also be studied with high resolution interferometric technique.

Two of the brightest Seyfert galaxies, (i) NGC 1068 (M 77), and (ii) NGC 4151, have strong emission lines. NGC 1068 (Antonucci and Miller, 1985), the brightest Seyfert galaxy (m_v 9.6), whose apparent diameter is $7' \times 6'$, was uncovered through polarized light. It is an archetype type 2 Seyfert galaxy about 50.0 million light years away in the constellation Cetus (α $02^h 42^m 40.2^s$ and δ $-00°00'48''$).

The only galaxy that was studied at different observatories, was the nucleus of NGC 1068. The core of this galaxy is very luminous, not only in optical, but in ultraviolet and X-ray as well. Observations of this object corroborated with theoretical modeling like radiative transfer calculations have made significant contributions on its structure. Ebstein et al. (1989)

---

[38]The term 'Blazars' arises from a combination of 'BL Lac' and 'quasar', and are a subset of AGN. This radio-loud extragalactic objects are optically violent variable (OVV) quasars, flat-spectrum quasars, high polarized quasars. These objects are very compact energy source associated with a supermassive black hole at the center of a host galaxy. They posses relativistic jets pointing toward the observer and exhibit extremely intense, broad and the large amplitude flux variations among all the AGN.

[39]BL Lacertae objects are the sub-class of blazars. They are the active cores of elliptical galaxies and have compact radio source with non-thermal continuous spectrum extending into the IR, visible and X-ray. These objects have spectra dominate by a featureless non-thermal continuum. BL Lacs are characterized by (i) large and rapid amplitude flux variability and (ii) strong and rapidly varying polarization. Some of these objects have jets which are emitted from the active core. BL Lac objects show no optical emission lines. Their redshifts may be determined from features in the spectral of their host galaxies. The emission line features may be swamped by the additional variable component, in which case, it may become visible when the variable component is at a low level (Vermeulen et al. 1995).

[40]LINERs (low-ionization nuclear emission-line region) are the type of objects whose optical spectra are quite distinct from those of both H II regions and classical AGNs. These objects are characterized by optical spectra generally, which reveal that ionized gas is present, but the gas is weakly ionized.

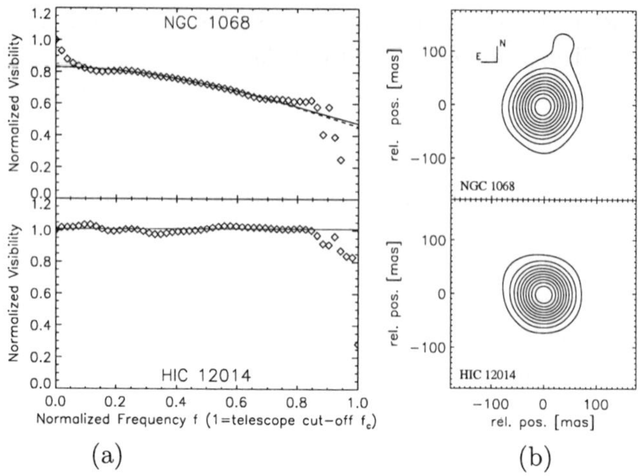

Fig. 11.20 (a) Azimuthally averaged visibilities of NGC 1068 (top) and of unresolved reference star HIC12014 (bottom). The diamonds indicate the observed visibilities, the solid lines the Gaussian fits, and the dashed line is UD fit, and (b) speckle masking reconstruction of NGC 1068 (top) and the unresolved star HIC12014 (bottom); the contours are from 6% to 100% of peak intensity. (Courtesy: M. Wittkowski).

found a bipolar structure of this object in the [O III] emission line. Near-IR observations at the Keck I telescope trace a very compact central core and extended emission with a size of the order of 10 pc on either side of an unresolved nucleus (Weinberger et al. 1999). Wittkowski et al., (1998b) have resolved this compact source with the diameter of 0.03″ in the K-band, with a FWHM size of ~2 pc for an assumed Gaussian intensity distribution. Figures (11.20a) and (11.20b) depict respectively the azimuthally averaged visibilities and the diffraction-limited speckle masking reconstruction of NGC 1068.

In the case of NGC 4151, a 10.8 m_v spiral galaxy in Canes Venatici, Shields (1999) points out the relative prominence of narrow forbidden lines in comparison with the NGC 1068. NGC 4151 is identified as the archetype type 1 Seyfert galaxy. It has also been called one of the most enigmatic of galaxies. The composite spectrum of this galaxy shows the wide variety of emission lines present, from the Lyman limit at 912 Å to the mid-infrared. The observed variability in high resolution optical observations of these objects suggests that the sizes of the emission region in a nearby active galaxy such as NGC 4151 is of the order of 0.2 mas, possibly resolvable with a baseline of several hundred meters. The ionization parameter appears to be roughly constant over a large range of luminosity, which suggests that the radius of BLR scales as $\sqrt{L}$, and when studying objects near the limit of sensitivity the distance is also roughly proportional to $\sqrt{L}$, so that the angular size of BLR is, to first order, independent of the distance of the objects (Ulrich, 1981). Ebstein et al., (1989) have resolved NGC 4151 in O III lines and from the reconstructed image they found the diameter to be $0.4''$. Zeidler et al., (1992) have resolved the $2.3''$ central region of another AGN, NGC 1346, into 4 clouds, distributed along the position angle of 20°.

11.2.7.2 *Quasars*

Quasars (QUASi-stellAR radio source, which means star-like radio sources) are the most distant luminous objects, which display a very high redshift; the highest redshift currently known for a quasar is 6.4. Some of them display rapid changes in luminosity as well. The first quasar was discovered in 1963 by radio observation; the optical emission lines (hydrogen Balmer lines) of the known radio source, 3C 273, were found to be redshifted by a large amount ($z = 0.16$; Schmidt, 1963). Quasars may be observed in radio, IR, visible, UV, X-ray, and gamma ray bands. However, the region of intense visible emission is quite small compared to the rest of the galaxy that it is embedded in, while huge regions of radio emission, produced by the quasar, can stretch out to large distances outside the galaxy. Most of the quasars are radio-quiet. Careful observations depict faint jets are emerging out from some quasars.

Many similar objects do not emit radio radiation, these objects are designated as QSO (quasi-stellar object). These objects are closely related to the active galaxies such as more luminous AGN, Seyfert galaxies or BL Lac objects. The distinction between the quasars and QSOs depends upon their radio-loudness and is largely historical. While objects with a 2-10 keV X-ray luminosity $L_{2-10} > 10^{45}$ erg s^{-1} are called quasars regardless of their optical power, however, the dividing line between Seyferts and quasars is not clearly defined; a generally accepted value is $\sim 3 \times 10^{44}$ erg s^{-1} in the 2-10 keV band.

QSOs are believed to be powered by accretion of material onto supermassive black holes in the nuclei of distant galaxies. They are found to vary in luminosity on a variety of time scales such as a few months, weeks, days, or hours, indicating that their enormous energy output originates in a very compact source. The high luminosity of quasars may be a result of friction caused by gas and dust falling into the accretion discs of supermassive black holes. Such objects exhibit properties common to active galaxies, for example, radiation is nonthermal and some are observed to have jets and lobes like those of radio galaxies.

QSOs may be gravitationally lensed by stellar objects such as, stars, galaxies, clusters of galaxies etc., located along the line of sight. Gravitational lensing occurs when the gravitational field from a massive object warps space and deflects light from a distant object behind it. The image may be magnified, distorted, or multiplied by the lens, depending upon the position of the source with respect to the lensing mass. This process is one of the predictions of Einstein's general relativity theory, which states that a large mass deforms spacetime to create gravitational fields and bend the light of path. There are three classes of gravitational lensing such as (i) strong lensing in which Einstein rings[41] (Chwolson 1924), arcs, and multiple images are formed, (ii) weak lensing, where the distortions of background objects are much smaller, and (iii) microlensing in which distortion in shape is invisible, but the amount of light received from a background object changes in time.

The aim of the high angular imagery of these QSOs is to find their structure and components. Their number and structure as a probe of the distribution of the mass in the Universe. The capability of resolving these objects in the range of $0.2''$ to $0.6''$ would allow the discovery of more lensing events. The gravitational image of the multiple QSO PG1115+08 was resolved by Foy et al. (1985); one of the bright components, discovered to be double (Hege et al. 1981), was found to be elongated that might be, according to them, due to a fifth component of the QSO.

11.2.8 *Impact of adaptive optics in astrophysics*

Adaptive optics (AO) technology has become an affordable tool at all new large astronomical telescopes. The noted advantages of such a system over the conventional techniques are the ability to recover near diffraction-limited images and to improve the point source sensitivity. Combination of AO systems with speckle imaging may enhance the results. By the end of the next decade (post 2010), observations using the AO system on a new generation very large telescope, will revolutionize the mapping of ultra-faint objects like blazars, extra-solar planets etc.; certain aspects of galactic evolution like chemical evolution in the Virgo cluster of galaxies can be studied as well.

[41] An Einstein ring is a special case of gravitational lensing, caused by the perfect alignment of two galaxies one behind the other.

Observations using AO system on large telescopes of 10 m class could surpass the resolution achievable with the present day orbital telescope. However, these need excellent seeing conditions; an exact knowledge of point spread function is necessary. Amplitude fluctuations are generally small and their effect on image degradation remains limited, and therefore, their correction is not needed, except for detection of exo-solar planets (Love and Gourlay, 1996).

Image recovery is relatively simple where the target is a point source. But the major problem of reconstructing images comes from difficulty in estimating the PSF due to the lack of a reference point source in the case of the extended objects, the Sun in particular, unlike stellar objects where this parameter can be determined from a nearby reference star. Moreover, intensive computations are generally required in post-detection image restoration techniques in solar astronomy. A few higher order solar adaptive optics systems are in use or under development (Beckers, 1999 and references therein). Images of sunspots on the solar surface were obtained with Lockheed adaptive optics system (Acton and Smithson, 1992) at the Sacramento Peak Vacuum telescope.

Adaptive optics (AO) observations have contributed to the study of the solar system, and added to the results of space borne instruments, for examples, monitoring of the volcanic activity on Io or of the cloud cover on Neptune[42] , the detection of Neptune's dark satellites and arcs, and the ongoing discovery of companions to asteroids etc.; they are now greatly contributing to the study of the Sun itself as well.

Most of the results obtained from the ground-based telescopes equipped with AO systems are in the near-IR band; while results at visible wave lengths continue to be sparse (Roddier 1999, Saha 2002 and references therein). The contributions are in the form of studying (i) planetary meteorology; images of Neptune's ring arcs are obtained (Sicardy et al. 1999) that are interpreted as gravitational effects by one or more moons, (ii) nu-

[42]Neptune, a gas planet, is the outermost and farthest planet (about 30.06 AU away from the Sun) in the solar system. A portion of its orbit lies farther from the Sun than the dwarf planet Pluto's, which is because of highly eccentric orbit of the latter. It's hazy atmosphere primarily composed of hydrogen and helium, with traces of methane (CH_4) and strong winds confined to bands of latitude and large storms or vortices. Its blue color is primarily the result of absorption of red light by CH_4 in the atmosphere. Neptune has very strong winds, measured as high as about 2,100 km h^{-1} (Suomi et al. 1991). A huge storm blows on Neptune, called 'great dark spot' which is about half the size of the Jupiter's red spot. It also has a smaller dark spot as well and a small irregular white cloud in the southern hemisphere. Neptune has 13 moons as well as rings, one of them appears to have twisted structure.

cleus of M31, (iii) young stars and multiple star systems (Bouvier et al. 1997), (iv) galactic center, (v) Seyfert galaxies, QSO host galaxies, and (vi) circumstellar environment.

Images of the objects such as, (a) the nuclear region of NGC 3690 in the interacting galaxy Arp 299, (b) the starburst/AGNs, NGC 863, NGC 7469, NGC 1365, NGC 1068, (c) the core of the globular cluster M 13. and (d) R 136 etc., are obtained from the moderate-sized telescopes. Brandl et al., (1996) have reported $0.15''$ resolution near IR imaging of the R 136 star cluster in 30 Doradus (LMC), an unusual high concentration of massive and bright O, B, Wolf-Rayet stars. Over 500 stars are detected within the field of view $12.8'' \times 12.8''$ covering a magnitude range of 11.2, of which ~ 110 are reported to be red stars.

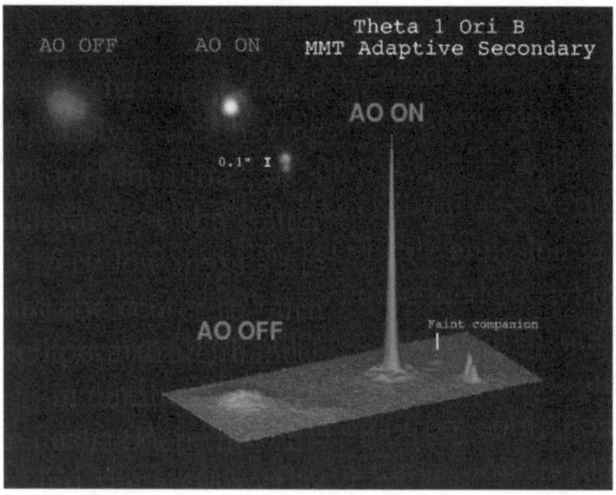

Fig. 11.21 AO image of Θ 1 Ori B; without AO, this object appears to be two stars, but with AO turned on it is revealed that the lower star is a close binary having separated by 0.1 arcseconds; the brighter one is a laser guide star, and the fainter one slightly to the right (see white arrow) is a very faint companion (Courtesy: L. Close).

AO systems can also employed for studying young stars, multiple stars, natal discs, and related inward flows, jets and related outward flows, protoplanetary discs, brown dwarfs and planets. Figure (11.21) depicts the AO image of θ^1 Ori B with a faint companion, while Figure (11.22) depicts the real time image of ADS 1585 (Close 2003) with a resolution of $0.07''$ (FWHM). These images were acquired with adaptive secondary mirror at the 6.5 meter Multi Mirror Telescope (MMT), Mt. Hopkins Observatory, Arizona, USA. A series of sequential images (Close, 2003) of real-time imaging of θ^1 Ori B, star are particularly interesting for they show the change from 0.5 arcsec (FWHM) ground-based seeing to diffraction-limited images of 0.06 arcsecs at a wavelength of $\sim 2\ \mu$m.

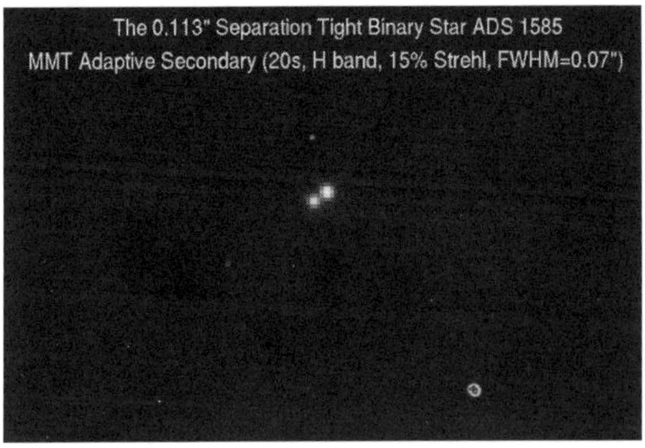

Fig. 11.22 H-band (1.65 μm) real time image of ADS 1585 (Courtesy: L. Close).

Roddier et al. (1996) have detected a binary system consisting of a K7-MO star with an M4 companion that rotates clockwise; they suggest that the system might be surrounded by a warm unresolved disc. The massive star Sanduleak-66°41 in the LMC was resolved into 12 components by Heydari and Beuzit (1994). Success in resolving companions to nearby dwarfs has been reported. The improved resolution of crowded fields like globular clusters would allow derivation of luminosity functions and spectral type, to analyze proper motions in their central area. Simon et al. (1999) have detected 292 stars in the dense Trapezium star cluster of the Orion nebula and resolved pairs to the diffraction-limit of a 2.2 m telescope. Optical and near-IR observations of the close Herbig Ae/Be binary star NX Pup, associated with the cometary globular cluster I, Schöller et al. (1996) estimated the mass and age of both the components and suggest that circumstellar matter around the former could be described by a viscous accretion disc.

Stellar populations in galaxies in near-IR region provides the peak of the spectral energy distribution for old populations. Bedding et al. (1997b) have observed the Sgr A window at the Galactic center of the Galaxy. They have produced an IR luminosity function and color-magnitude diagram for 70 stars down to $m_v \simeq$19.5 mag. These are the deepest yet measured for the galactic bulge, reaching beyond the turn-off. The marked advantage over traditional approach is the usage of near IR region, where the peak of the spectral energy distribution for old populations is found by them. Figure (11.23) depicts the ADONIS K$'$ image of the Sgr window.

Images have been obtained of the star forming region Messier 16 (Currie

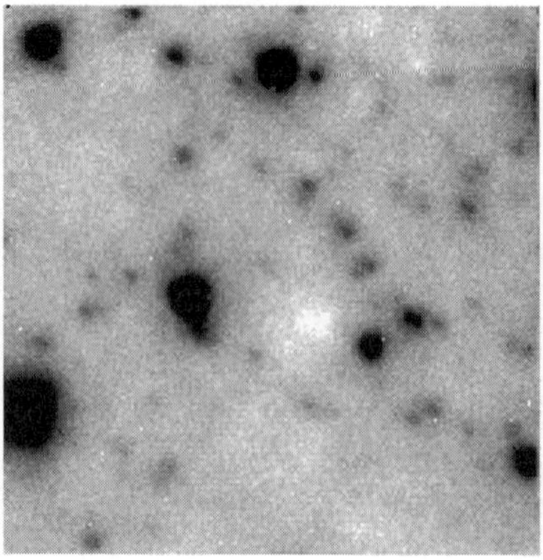

Fig. 11.23 The ADONIS K′ image of the Sgr window in the bulge of the milky way (Bedding et al., 1997b). The image is 8″ × 8″ (Courtesy: T. Bedding).

et al. 1996), the reflection nebula NGC 2023 revealing small-scale structure in the associated molecular cloud, close to the exciting star, in Orion (Rouan et al. 1997). Close et al. (1997) mapped near-IR polarimetric observations of the reflection nebula R Mon resolving a faint source, 0.69″ away from R Mon and identified it as a T Tauri star. Monnier et al. (1999) found a variety of dust condensations that include a large scattering plume, a bow shaped dust feature around the red supergiant VY CMa; a bright knot of emission 1″ away from the star is also reported. They argued in favor of the presence of chaotic and violent dust formation processes around the star. Imaging of proto-planetary nebulae (PPNe), Frosty Leo and the Red Rectangle by Roddier et al. (1995) revealed a binary star at the origin of these PPNe.

Imaging of the extragalactic objects, particularly the central area of active galaxies where cold molecular gas and star formation occur is an important program. From the images of nucleus of NGC 1068, Rouan et al. (1998), found several components that include: (i) an unresolved conspicuous core, (ii) an elongated structure, and (iii) large and small-scale spiral structures. Lai et al. (1998) have recorded images of Markarian 231, a galaxy 160 Mpc away demonstrating the limits of achievements in terms of morphological structures of distant objects.

Aretxaga et al. (1998) reported the unambiguous detection of the host galaxy of a normal radio-quiet QSO at high-redshift in K-band; detection of emission line gas within the host galaxies of high redshift QSOs has been reported as well (Hutchings et al. 2001). Observations by Ledoux et al. (1998) of broad absorption line quasar APM 08279+5255 at z=3.87 show the object consists of a double source ($\rho = 0.35'' \pm 0.02''$; intensity ratio = 1.21 ± 0.25 in H band). They proposed a gravitational lensing hypothesis which came from the uniformity of the quasar spectrum as a function of the spatial position. Search for molecular gas in high redshift normal galaxies in the foreground of the gravitationally lensed quasar Q1208+1011 has also been made (Sams et al. 1996). AO imaging of a few low and intermediate redshift quasars has been reported (Márquez et al. 2001).

11.3 Dark speckle method

Direct imaging of photon-starved sources close to a bright object such as circumstellar discs, substellar objects, extragalactic nebulosities, and extra-solar planets is a difficult task. The limitations come from the light diffracted by the telescope and instrument optics; polishing defects, spider arms, and the wavefront residual bumpiness, as well as from a host of noise, including speckle noise. These objects can be seen in ground-based images employing the light cancellation in dark speckles (Labeyrie, 1995) to remove the halo of the starlight. The aim of this method is to detect faint objects around a star when the difference in magnitude is significant. If a dark speckle is at the location of the companion in the image, the companion emits enough light to reveal itself.

Dark speckle method uses the randomly moving dark zones between speckles – 'dark speckles'. It exploits the light cancellation effect in a random coherent fields according to the Bose-Einstein statistics; highly destructive interferences that depict near black spots in the speckle pattern may occur occasionally. The dark speckle analysis involves an elaborate statistical treatment of multiple exposures each shorter than the speckle life time. In each exposure, the speckle pattern is different and dark speckles appear at different locations. A dark speckle appearing at the companion's location improves its detectability since the contaminating photon count n is decreased.

The method can be applied with a telescope equipped with an adaptive coronagraph, where residual turbulence achieves the speckle 'boiling'. The required system consists of a telescope with an AO system, a coronagraph, a Wynne corrector[43], and a fast photon-counting camera with a low dark

[43]Wynne corrector is generally installed before the focal plane of a telescope that

noise. It also requires fine sampling to exploit the darkest parts of the dark speckles, for a given threshold of detection, ϵ (Boccaletti et al. 1998a),

$$j = \frac{(\lambda/D)^2}{s^2} = 0.62\frac{R}{G}, \tag{11.52}$$

where $s(\approx 1.27\sqrt{\epsilon}\lambda/D)$ is the size of the pixel over which the light is integrated, j the number of pixel per speckle area; for a companion ten times fainter than the average speckle halo, a sampling of 6.2 pixels per speckle area is essential, R the star/companion luminosity ratio, G the gain of the adaptive optics, i.e., the ratio of intensities in the central peak and speckled halo, referred to as the adaptive optics gain by Angel (1994).

The relevance of using coronagraphy in imaging or spectroscopy of faint structure near a bright object can be noted in terms of reducing the light coming from the central star, and filtering out of the light at low spatial frequency; the remaining light at the edge of the pupil corresponds to high frequencies. A coronagraph reduces off-axis light from an on-axis source with an occulting stop in the image-plane as well as with a matched Lyot stop in the next pupil plane. While using the former stop the size of the latter pupil should be chosen carefully to find the best trade-off between the throughput and image suppression. The limitations come from the light diffracted by the telescope and instrument optics. Coronagraphy with dynamic range can be a powerful tool for direct imaging of extra-solar planets.

If a pixel of the photon-counting camera is illuminated by the star only (in the Airy rings area), because of the AO system, the number of photons in each pixel, for a given interval (frame), is statistically given by a Bose-Einstein probability distribution of the form (Goodman, 1985),

$$\mathcal{P}(n_\star) = \frac{1}{1 + \langle n_\star \rangle}\left(\frac{\langle n_\star \rangle}{1 + \langle n_\star \rangle}\right)^{n_\star}, \tag{11.53}$$

in which $\langle n_\star \rangle$ is the number of stellar photoevents per pixel per short-exposure.

The number of photons per frame in the central peak of the image of a point source obeys a classical Poisson distribution (see Appendix B),

$$\mathcal{P}(n_o) = e^{-\langle n_o \rangle}\frac{\langle n_o \rangle^{n_o}}{n_o!}. \tag{11.54}$$

---

suffers from optics degradation due to off-axis coma while aiming at wide field imaging. Essentially it is a three element (lens) system.

in which $\langle n_o \rangle$ is photo-events per pixel per short-exposure that are contributed from the companion.

For the pixels containing the image of the companion, the number of photons, resulting from both the star and the companion, is given by a different distribution (computed by mixing Bose-Einstein and Poisson distributions),

$$\mathcal{P}(n) = \frac{e^{-\langle n_o \rangle}}{1 + \langle n_\star \rangle} \sum_{i=0}^{i=n} \left[\frac{\langle n_\star \rangle^i}{1 + \langle n_\star \rangle} \frac{\langle n \rangle^{n-i}}{(n-i)!} \right], \tag{11.55}$$

where $n(= n_\star + n_o)$ is the total count of photoevents in a single pixel per short-exposure originating from the star and the planet.

One noticeable property is that the probability to get zero photons in a frame is very low for the pixels containing the image of the companion, and much higher for the pixels containing only the contribution from the star. The probability of zero photon is given by,

$$\mathcal{P}(0) = \mathcal{P}_\star(0)\mathcal{P}_o(0) = \frac{e^{-\langle n_o \rangle}}{1 + \langle n_\star \rangle}, \tag{11.56}$$

in which $\mathcal{P}_\star(n_\star)$ and $\mathcal{P}_o(n_o)$ are the probabilities to detect $n_\star$ and n_o photons per pixel per short-exposure originating from the star and the companion, respectively.

Therefore, if the 'no photon in the frame' events for each pixel is counted, and for a very large number of frames, a 'dark map' can be built that may show the pixels for which the distribution of the number of photons is not Bose-Einstein type, therefore revealing the location of a faint companion. The difference between two images in the reference frame of the coronagraph cancels the speckle pattern, while leaving positive and negative companion images at two points in the field separated by the rotation angle. Because of the incoherent image subtraction, the result is limited by the Poisson noise, which is the square root of the photon count recorded in each exposure, before the subtraction. Repeated sequences may improve the sensitivity if the pattern drifts. Following the detection of the companion, the contrast of its image can be improved by creating a permanent dark speckle in the starlight at its location, permitting to obtain low-resolution spectra of the companion. The condition for such a detection is that the number of photons received from the companion should be greater than the Poisson noise. With N exposures, a companion is detectable in a single pixel; the different photon distribution from the star and the companion defines the

S/N-ratio, and according to the central limit theorem,

$$N\mathcal{P}_\star(0)\left[1-\mathcal{P}_o(0)\right] = S/N\sqrt{N\mathcal{P}_\star(0)}. \tag{11.57}$$

Theoretical expressions of S/N-ratio for the dark speckle exposure is given by (Boccaletti et al. 1988a),

$$S/N = \frac{n'_\star}{R}\left[\frac{tT}{j+(t\,n'_\star)/G}\right]^{1/2}, \tag{11.58}$$

where T is the total observing time, t the short-exposure time, and $n'_\star$ the total number of photons s^{-1} detected from the star.

The role of the Wynne corrector is to give residual speckles the same size regardless of the wavelength. Otherwise, dark speckles at a given wavelength would be overlapped by bright speckles at other wavelengths. With the current technology, by means of the dark speckle technique, a 3.6 m telescope should allow detection of a companion with $\Delta m_k \approx$6-7 mag.

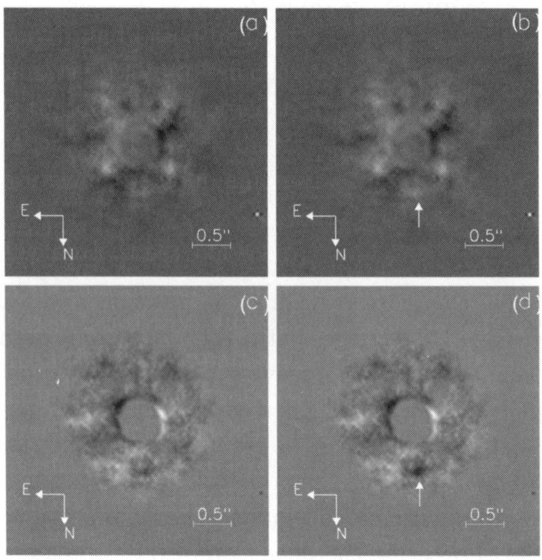

Fig. 11.24 Coronagraphic images of the star HD192876 (Courtesy: A. Boccaletti). An artificial companion is added to the data to assess the detection threshold (Δm_K=6.0 mag, ρ = 0.65"); (a) direct image : co-addition of 400×60 ms frames, (b) same as (a) with a Δm_K=6.0 mag companion (SNR 1.8), (c) dark speckle analysis, and (d) dark speckle analysis with the companion (SNR 4.8); the detection threshold is about Δm_K=7.5 mag on that image, i.e an improvement of 1.5 mag compared to the direct image (Boccaletti et al. 2001).

Boccaletti et al., (1998a) have found from the laboratory simulations the capability of detecting a stellar companion of relative intensity 10^6 at 5 Airy radii from the star using an avalanche photo-diode as detector. They also have recorded dark speckle data at the 1.52 m telescope of Haute-Provence using an AO system and detected a faint component of the spectroscopic binary star HD 144217 (Δm $= 4.8$, separation $= 0.45''$). Subsequently, Boccaletti et al., (1998b) have applied the same technique at the said telescope to observe the relatively faint companions of δ Per and η Psc and were able to estimate their position and magnitude difference. Figures (11.24) and (11.25) depict the coronagraphic images of the binary stars, HD192876 and HD222493, respectively (Boccaletti et al. 2001); the data were obtained with ADONIS in the K band (2.2 μm) on the European Southern Observatory's (ESO) 3.6 m telescope. Due to the lack of a perfect detector (no read-out noise) at near-IR band, every pixel under the defined threshold (a few times the read-out noise) is accounted as a dark speckle.

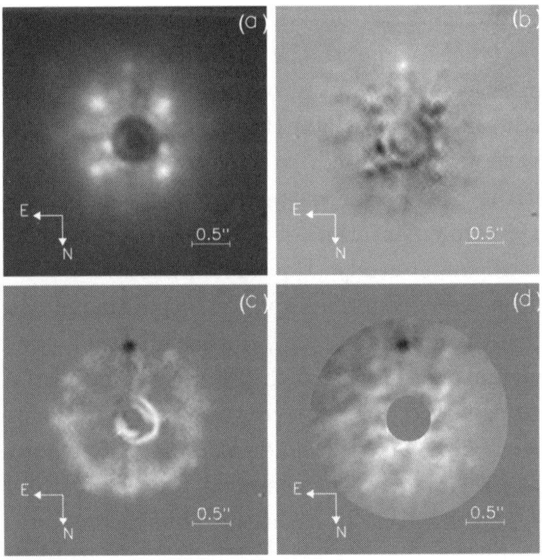

Fig. 11.25 Coronagraphic images of the binary star HD222493 (Δm_K=3.8 mag, $\rho =$ 0.89"); (a) direct image: co-addition of 600×60 ms frames, (b) subtraction of the direct image with a reference star (SNR=14.6), (c) dark speckle analysis (constant threshold) and subtraction of a reference star, and (d) dark speckle analysis (radial threshold) and subtraction of a reference star (SNR=26.7) (Boccaletti et al. 2001: Courtesy: A. Boccaletti).

Phase boiling, a relatively new technique that consists of adding a small amount of white noise to the actuators in order to get a fast temporal decorrelation of the speckles during long-exposure acquisition, may produce better results. Aime (2000) has computed the S/N ratio for two different cases: short-exposure and long-exposure. According to him, even with an electron-noise limited detector like a CCD or a near-IR camera multi-object spectrometer (NICMOS), the latter can provide better results if the halo has its residual speckles smoothed by fast residual 'seeing' acting during the long-exposure than building a dark map from short-exposures in the photon-counting mode. Artificial very fast seeing can also be generated by applying fast random noise to the actuators, at amplitude levels comparable to the residual seeing left over by the AO system.

The question is, what is easiest: dark speckle analysis or a 'hyper-turbulated' long-exposure? Labeyrie (2000) made simulations supporting Aime's (2000) results. Boccaletti (2001) has compared the dark speckle signal-to-noise ratio (SNR) with the long-exposure SNR (Angel, 1994). The speckle lifetime has to be of order 0.1 ms. Currently it is impossible to drive a DM at this frequency (10 kHz). With the 5 m Palomar telescope Boccaletti (2001) tried to smooth the speckle pattern by adding a straight-forward random noise on the actuators (the DM is equipped with 241 actuators) at maximum speed of 500 Hz. Effectively, the halo is smoothed, but its intensity is also increased, so that the companion SNR is actually decreased. Blurring the speckle pattern would probably require wavefront sensor telemetry; implementation of a hyper-turbulated long-exposure at the Palomar is still under study (Boccaletti, 2001).

High resolution stellar coronagraphy is of paramount importance in (i) detecting low mass companions, e.g., both white and brown dwarfs, dust shells around asymptotic giant branch (AGB) and post-AGB stars, (ii) observing nebulosities leading to the formation of a planetary system, ejected envelops, accretion disc, and (iii) understanding of structure (torus, disc, jets, star forming regions), and dynamical process in the environment of AGNs and QSOs. By means of coronagraphic techniques the environs of a few interesting objects have been explored. They include: (i) a very low mass companion to the astrometric binary Gliese 105 A (Golimowski et al. 1995), (ii) a warp of the circumstellar disc around the star β Pic (Mouillet et al. 1997), (iii) highly asymmetric features in AG Carina's circumstellar environment (Nota et al. 1992), (iv) bipolar nebula around the LBV R127 (Clampin et al. 1993), and (v) the remnant envelope of star formation around pre-main sequence stars (Nakajima and Golimowski, 1995).

Appendix A

Typical tables

Table I

The Maxwell's equations of electromagnetism for the time domain

Names	Equations
Faraday's law	$\nabla \times E(\vec{r},t) = -\frac{1}{c}\left[\frac{\partial B(\vec{r},t)}{\partial t}\right]$
Ampère - Maxwell law	$\nabla \times H(\vec{r},t) = \frac{1}{c}\left[4\pi J(\vec{r},t) + \frac{\partial D(\vec{r},t)}{\partial t}\right]$
Gauss' electric law	$\nabla \cdot D(\vec{r},t) = 4\pi\rho(\vec{r},t)$
Gauss' magnetic law	$\nabla \cdot B(\vec{r},t) = 0$
Equation of continuity for electric charge	$\nabla \cdot \vec{J} + \frac{\partial \rho}{\partial t} = 0$
Lorentz force expression	$\vec{F} = q\left[\vec{E} + \frac{1}{c}\vec{v} \times \vec{B}\right]$
Poynting vector	$S(\vec{r},t) = \frac{c}{4\pi}\left[E(\vec{r},t) \times H(\vec{r},t)\right]$

Table II
Normalized states of elliptically polarized wave

Polarization	Angles (γ,δ); (χ,β)	$\vec{S}$	$\vec{E}$	$\vec{D}$
Linear (H)	(0, -); (0, 0)	$\begin{bmatrix}1\\1\\0\\0\end{bmatrix}$	$\begin{bmatrix}1\\0\end{bmatrix}$	$\begin{bmatrix}1&0\\0&0\end{bmatrix}$
Vertical (V)	$\left(\frac{\pi}{2},-\right)$; $\left(0,\frac{\pi}{2}\right)$	$\begin{bmatrix}1\\-1\\0\\0\end{bmatrix}$	$\begin{bmatrix}0\\1\end{bmatrix}$	$\begin{bmatrix}0&0\\0&1\end{bmatrix}$
Linear +45°	$\left(\frac{\pi}{4},0\right)$; $\left(0,\frac{\pi}{4}\right)$	$\begin{bmatrix}1\\0\\1\\0\end{bmatrix}$	$\frac{1}{\sqrt{2}}\begin{bmatrix}1\\1\end{bmatrix}$	$\frac{1}{2}\begin{bmatrix}1&1\\1&1\end{bmatrix}$
Linear -45°	$\left(\frac{\pi}{4},\pi\right)$; $\left(0,\frac{3\pi}{4}\right)$	$\begin{bmatrix}1\\0\\-1\\0\end{bmatrix}$	$\frac{1}{\sqrt{2}}\begin{bmatrix}1\\-1\end{bmatrix}$	$\frac{1}{2}\begin{bmatrix}1&-1\\-1&1\end{bmatrix}$
RH Circular	$\left(\frac{\pi}{4},-\frac{\pi}{2}\right)$; $\left(-\frac{\pi}{4},-\right)$	$\begin{bmatrix}1\\0\\0\\1\end{bmatrix}$	$\frac{1}{\sqrt{2}}\begin{bmatrix}1\\-i\end{bmatrix}$	$\frac{1}{2}\begin{bmatrix}1&i\\-i&1\end{bmatrix}$
LH Circular	$\left(\frac{\pi}{4},\frac{\pi}{2}\right)$; $\left(\frac{\pi}{4},-\right)$	$\begin{bmatrix}1\\0\\0\\-1\end{bmatrix}$	$\frac{1}{\sqrt{2}}\begin{bmatrix}1\\i\end{bmatrix}$	$\frac{1}{2}\begin{bmatrix}1&-i\\i&1\end{bmatrix}$

Table III

Correspondence between the Zernike polynomials, Z_j for $j = 1, 2, \cdots, 8$ and the common optical aberrations. n is the radial order and m the azimuthal order. The modes are ordered such that even values of j represent the symmetric modes given by $\cos(m\theta)$ and odd j values correspond to the antisymmetric modes given by $\sin(m\theta)$.

n	$m = 0$	$m = 1$	$m = 2$
0	$Z_1 = 1$ Piston or bias		
1		$Z_2 = 2\rho\cos\theta$ Tilt x (Lateral position)	
1		$Z_3 = 2\rho\sin\theta$ Tilt y (Longitudinal position)	
2	$Z_4 = \sqrt{3}(2\rho^2 - 1)$ Defocus		$Z_5 = \sqrt{6}\rho^2\sin 2\theta$ $Z_6 = \sqrt{6}\rho^2\cos 2\theta$ Astigmatism (3rd order)
3		$Z_7 = \sqrt{8}(3\rho^3 - 2\rho)\sin\theta$ $Z_8 = \sqrt{8}(3\rho^3 - 2\rho)\cos\theta$ Coma (3rd order)	
4	Z_{11} $= \sqrt{5}(6\rho^4 - 6\rho^2 + 1)$ Spherical aberration		

Table IV

Zernike-Kolmogorov residual variance, Δ_J, after the first J Zernike modes are removed. Here D is the telescope diameter and r_0 the atmospheric coherence length. The difference given in the right column illustrates the differential improvement.

Residual variance, Δ_J	Differences
$\Delta_1 = 1.030 \left(\frac{D}{r_0}\right)^{5/3}$	
$\Delta_2 = 0.582 \left(\frac{D}{r_0}\right)^{5/3}$	$\Delta_2 - \Delta_1 = 0.449$
$\Delta_3 = 0.134 \left(\frac{D}{r_0}\right)^{5/3}$	$\Delta_3 - \Delta_2 = 0.449$
$\Delta_4 = 0.111 \left(\frac{D}{r_0}\right)^{5/3}$	$\Delta_4 - \Delta_3 = 0.0232$
$\Delta_5 = 0.088 \left(\frac{D}{r_0}\right)^{5/3}$	$\Delta_5 - \Delta_4 = 0.0232$
$\Delta_6 = 0.0648 \left(\frac{D}{r_0}\right)^{5/3}$	$\Delta_6 - \Delta_5 = 0.0232$
$\Delta_7 = 0.0587 \left(\frac{D}{r_0}\right)^{5/3}$	$\Delta_7 - \Delta_6 = 0.0062$
$\Delta_8 = 0.0525 \left(\frac{D}{r_0}\right)^{5/3}$	$\Delta_8 - \Delta_7 = 0.0062$
$\Delta_9 = 0.0463 \left(\frac{D}{r_0}\right)^{5/3}$	$\Delta_9 - \Delta_8 = 0.0062$
$\Delta_{10} = 0.0401 \left(\frac{D}{r_0}\right)^{5/3}$	$\Delta_{10} - \Delta_9 = 0.0062$
$\Delta_{11} = 0.0377 \left(\frac{D}{r_0}\right)^{5/3}$	$\Delta_{11} - \Delta_{10} = 0.0024$

Appendix B

Basic mathematics for Fourier optics

B.1 Fourier transform

The basic properties of Fourier transform (FT; J. B. J. Fourier, 1768-1830) are the indispensable instruments in optics and astronomical applications. The free wave equation is a linear homogeneous differential equation, therefore, any linear combination of its solutions is a solution as well and Fourier analysis makes use of this linearity extensively. The Fourier transform pair can be expressed in the space domain,

$$\widehat{f}(u) = \int_{-\infty}^{\infty} f(x) e^{-i2\pi ux} dx, \tag{B.1}$$

$$f(x) = \int_{-\infty}^{\infty} \widehat{f}(u) e^{i2\pi ux} du. \tag{B.2}$$

Since there is considerable symmetry within each of these pairs of equations, $\widehat{f}(u)$ and $f(x)$ are each described as the Fourier transform of each other. The equation (B.2) shows that $f(x)$ can be decomposed into an integral in u-space. The coefficients $\widehat{f}(u)$ are the weighting factors.

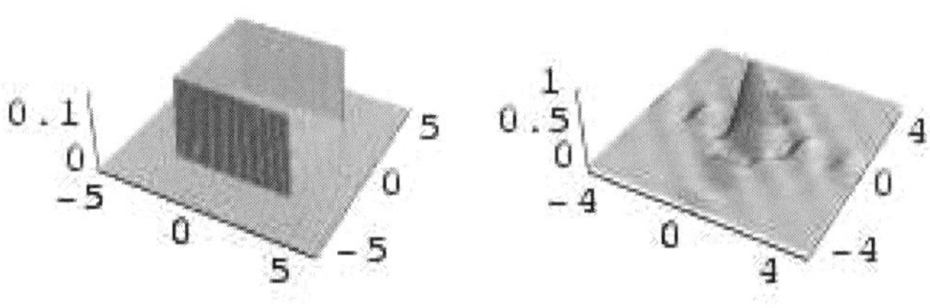

Fig. B.1 2-D Fourier transform of $\Pi(x, y)$.

The definition and properties of FT can be generalized to two and more dimensions. In the case of two dimensional FT, one writes

$$\widehat{f}(\vec{u}) = \int_{-\infty}^{\infty} f(\vec{x}) e^{-i2\pi\vec{u} \cdot \vec{x}} d\vec{x}, \tag{B.3}$$

in which $\vec{x} = (x, y)$ is the 2-D position vector and the dimensionless variable is the 2-D spatial vector $\vec{u} = (u, v) = (x/\lambda, y/\lambda)$.

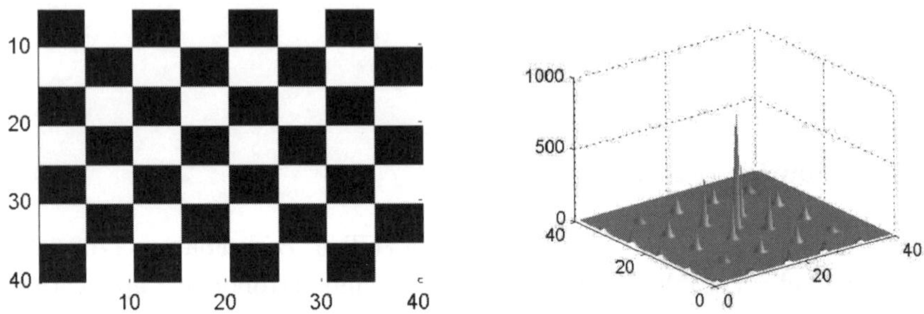

Fig. B.2 2-D Fourier transform of chess board.

It is assumed that $f(\vec{x})$ is bounded and goes to zero asymptotically as $\vec{x} \to \infty$. The inversion formula is given by,

$$f(\vec{x}) = \int_{-\infty}^{\infty} \widehat{f}(\vec{u}) e^{i2\pi\vec{u} \cdot \vec{x}} d\vec{u}, \tag{B.4}$$

B.1.1 *Basic properties and theorem*

The mathematical properties of Fourier transform of a small number of theorems that play a basic role in one form or another are given below:

(1) Fourier transform pairs:

$$e^{-\pi x^2} \rightleftharpoons e^{-\pi u^2}, \tag{B.5}$$

$$\operatorname{sinc} x \rightleftharpoons \Pi(u), \tag{B.6}$$

$$\operatorname{sinc}^2 x \rightleftharpoons \Lambda(u), \tag{B.7}$$

$$\delta(x) \rightleftharpoons 1, \tag{B.8}$$

$$\sin \pi x \rightleftharpoons \frac{i}{2}\delta\left(u + \frac{1}{2}\right) - \frac{i}{2}\delta\left(u - \frac{1}{2}\right), \tag{B.9}$$

$$\cos \pi x \rightleftharpoons \frac{1}{2}\delta\left(u + \frac{1}{2}\right) - \frac{1}{2}\delta\left(u - \frac{1}{2}\right), \tag{B.10}$$

in which

$$\operatorname{sinc} x = \frac{\sin \pi x}{\pi x}, \tag{B.11}$$

$$\Pi(x) = \begin{cases} 1 & \text{if } |x| < 1/2, \\ 0 & \text{otherwise}, \end{cases} \tag{B.12}$$

$$\Lambda(x) = \begin{cases} 1 - |x| & \text{if } |x| < 1, \\ 0 & \text{otherwise}. \end{cases} \tag{B.13}$$

(2) Parity and symmetry: The equation (B.1) is developed as,

$$\begin{aligned} \widehat{f}(u) &= \int_{-\infty}^{\infty} f(x)\cos(2\pi ux)dx - i\int_{-\infty}^{\infty} f(x)\sin(2\pi ux)dx \\ &= \mathcal{F}_c\left[f(x)\right] - i\mathcal{F}_s\left[f(x)\right], \end{aligned} \tag{B.14}$$

where $\mathcal{F}$ represents the Fourier operator and the cosine and sine functions are,

$$\begin{aligned} \mathcal{F}_c\left[f(x)\right] &= \int_{-\infty}^{\infty} f(x)\cos(2\pi ux)dx; \\ \mathcal{F}_s\left[f(x)\right] &= \int_{-\infty}^{\infty} f(x)\sin(2\pi ux)dx. \end{aligned} \tag{B.15}$$

By introducing $f_e(x)$ and $f_o(x)$ respectively for the even and the odd parts of $f(x)$,

$$f(x) = f_e(x) + f_o(x), \tag{B.16}$$

in which $f_e(x)$ is the even part of $f(x)$ and $f_o(x)$ is the odd part of $f(x)$ and thus one may write,

$$\begin{aligned} \widehat{f}(u) &= \int_{-\infty}^{\infty} f_e(x)\cos(2\pi ux)dx - i\int_{-\infty}^{\infty} f_o(x)\sin(2\pi ux)dx \\ &= 2\int_{0}^{\infty} f_e(x)\cos(2\pi ux)dx - 2i\int_{0}^{\infty} f_o(x)\sin(2\pi ux)dx \\ &= \mathcal{F}_c\left[f_e(x)\right] - i\mathcal{F}_s\left[f_o(x)\right]. \end{aligned} \tag{B.17}$$

This equation (B.17) expresses that the even part of $f(x)$ transform into the even part of $\widehat{f}(u)$ with corresponding real and imaginary parts. The odd part of $f(x)$ transform into the odd part of $\widehat{f}(u)$ with crossed real and imaginary parts. If $f(x)$ is real and having no symmetry, $\widehat{f}(u)$ is Hermitian, i.e., even real part and an odd imaginary part. It is to be noted that the term Hermitian is defined as $f(x) = f^*(-x)$.

(3) Linearity theorem: In this, the input produces a unique output. The Fourier transform of the function $f(x)$, is denoted symbolically as,

$$\mathcal{F}[f(x)] = \widehat{f}(u). \tag{B.18}$$

For two-dimensional linearity theorem, $\widehat{f}(\vec{u})$ is expressed as,

$$\mathcal{F}[f(\vec{x})] = \widehat{f}(\vec{u}). \tag{B.19}$$

The other related theorems are:

(4) Addition theorem: If $h(x) = af(x) + bg(x)$, the transform of sum of two functions is simply the sum of their individual transforms, i.e.,

$$\widehat{h}(u) = \mathcal{F}[af(x) + bg(x)] = a\widehat{f}(u) + b\widehat{g}(u). \tag{B.20}$$

where a, b are complex numbers.

(5) Similarity theorem: Unlike addition theorem, a stretching of the co-ordinates in the space domain (x, y) results in the contraction of the co-ordinates in the frequency domain (u, v) plus a change in the overall amplitude of the spectrum.

$$\mathcal{F}[f(ax)] = \frac{1}{|a|}\widehat{f}\left(\frac{u}{a}\right), \tag{B.21}$$

$$\mathcal{F}[f(ax + by)] = \frac{1}{|ab|}\widehat{f}\left(\frac{u}{a}, \frac{v}{b}\right). \tag{B.22}$$

(6) Shift theorem: A shift in the time at which the input starts is seen to cause a shift in the time at which the output starts; the shape of the input is unchanged by the shift. In shift theorem, the translation of a function in a space domain introduces a linear phase shift in the frequency domain. i.e.,

$$\mathcal{F}[f(x - a)] = \widehat{f}(u)e^{-i2\pi au}, \tag{B.23}$$

and in the case of two-dimensional space vector, one expresses,

$$\mathcal{F}[f(\vec{x} - \vec{a})] = \widehat{f}(\vec{u})e^{-i2\pi\vec{u}\cdot\vec{a}}. \tag{B.24}$$

(7) Derivative theorem for $f(x)$ can be expressed as,

$$\mathcal{F}\left[\frac{d}{dx}f(x)\right] = i2\pi u\widehat{f}(u). \tag{B.25}$$

B.1.2 *Discrete Fourier transform*

The Fourier transform of a discrete function is used for representing a sampled physical signal, in general when the number of samples N is finite. If $f(x)$ and $f(u)$ consist of sequence of N samples, the respective direct and inverse discrete Fourier transform (DFT) of a signal are defined as,

$$\widehat{f}(u) = \frac{1}{N}\sum_{x=0}^{N-1} f(x)e^{-i2\pi xu/N}, \tag{B.26}$$

$$f(x) = \sum_{u=0}^{N-1} \widehat{f}(u)e^{i2\pi xu/N}. \tag{B.27}$$

The change of notation emphasizes that the variables are discrete. The DFT assumes that the data, $f(x)$, is periodic outside the sampled range, and returns a transform, which is periodic as well,

$$\begin{aligned}\widehat{f}(u+N) &= \frac{1}{N}\sum_{x=0}^{N-1} f(x)e^{-i2\pi x(u+N)/N}, \\ &= \frac{1}{N}\sum_{x=0}^{N-1} f(x)e^{-i2\pi xu/N}e^{-i2\pi x}, \\ &= \widehat{f}(u).\end{aligned} \tag{B.28}$$

The two dimension DFT for an $N \times N$ is recast as,

$$G(u,v) = \frac{1}{N^2}\sum_{x=0}^{N-1}\sum_{y=0}^{N-1} g(x,y)e^{-i2\pi(ux+vy)/N}, \tag{B.29}$$

and the inverse operation is,

$$G(x,y) = \frac{1}{N^2}\sum_{u=0}^{N-1}\sum_{v=0}^{N-1} g(u,v)e^{i2\pi(ux+vy)/N}. \tag{B.30}$$

B.1.3 *Convolution*

Convolution simulates phenomena such as a blurring of a photograph. This blurring may be caused by poor focus, by the motion of a photographer during the exposure, or by the dirt on the lens etc. In such a blurred picture each point of object is replaced by a spread function. The spread function is disk shaped in the case of poor focus, line shaped if the photograph has moved, halo shaped if there is a dust on lens. It is known that Dirac delta function is zero everywhere except at the origin, but has an integral of unity. But generally the measurement does not produce this.

The convolution of two functions is a mathematical procedure (Goodman, 1968), or an operation that is found to arise frequently in the theory of linear systems. Let an input curve be represented by $f(x)$ in terms of a set of close delta functions which are spread. Here, the shape of the response of the system including unwanted spread, is same for all values of x (invariant for each considered delta function). Now, the value of the function $f(x)$ at x for the whole curve is defined mathematically,

$$h(x) = \int_{-\infty}^{\infty} f(x')g(x - x')dx', \tag{B.31}$$

where $h(x)$ is the output value at particular point x,

This integral is defined as convolution of $f(x)$ and $g(x)$, in which $g(x)$ is referred to as a blurring function or line spread function (LSF). The LSF is symmetric about its center and is equal to the derivative of the edge spread function (the image of an edge object). The mathematical description of convolution of two functions is of the form,

$$h(x) = f(x) \star g(x), \tag{B.32}$$

where, $g(x)$ is referred to as a blurring function and $\star$ stands for convolution.

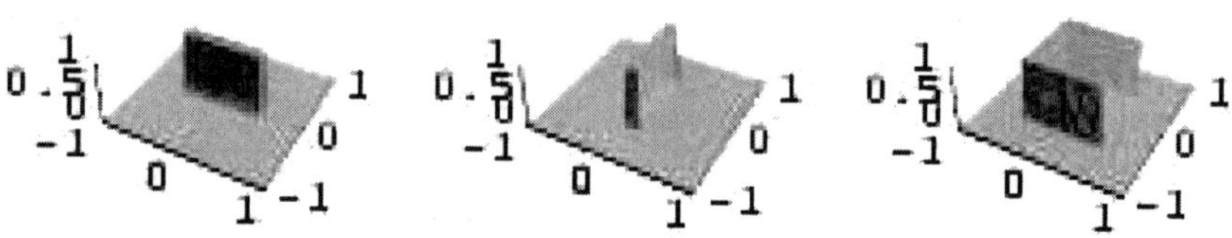

Fig. B.3 2-D convolution of two rectangular functions.

The commutative, associative and distributive over addition law for the convolution are given respectively below;

$$f(x) \star g(x) = g(x) \star f(x), \tag{B.33}$$

$$f(x) \star [g(x) \star h(x)] = [f(x) \star g(x)] \star h(x), \tag{B.34}$$

$$f(x) \star [g(x) + h(x)] = [f(x) \star g(x)] + [f(x) \star h(x)]. \tag{B.35}$$

The Fourier convolution theorem states that the convolution of two functions is the product of the Fourier transforms of the two functions, therefore, in the Fourier-plane the effect turns out to be a multiplication, point by point, of the transform of $\widehat{f}(u)$ with the transfer function $\widehat{g}(u)$.

$$\begin{aligned}\widehat{h}(u) &= \mathcal{F}\left[f(x) \star g(x)\right] \\ &= \mathcal{F}\left[f(x)\right].\mathcal{F}\left[g(x)\right] = \widehat{f}(u).\widehat{g}(u).\end{aligned} \tag{B.36}$$

In two-dimensional case, the convolution is treated as,

$$F\left[\iint_{-\infty}^{\infty} g(\xi,\eta)h(x-\xi,y-\eta)d\xi d\eta\right] = \widehat{g}(\vec{u})\widehat{h}(\vec{u}), \tag{B.37}$$

i.e.,

$$F[g(\vec{x}) \star h(\vec{x})] = F[g(\vec{x})] \cdot F[h(\vec{x})] = \widehat{g}(\vec{u}) \cdot \widehat{h}(\vec{u}). \tag{B.38}$$

B.1.4 *Autocorrelation*

Autocorrelation is a mathematical tool used in the study of functions representing observational data, particularly observations that exhibit some degree of randomness. Such a theorem extracts a signal from a background of random noise. It is the cross-correlation of a signal with itself. The original function is displaced spatially or temporally, the product of the displaced and undisplaced versions is formed, and the area under that product (corresponding to the degree of overlap) is compared by means of the integral. The autocorrelation of $f(x)$ in the plane x, is the correlation of $f(x)$ and $f(x)$ multiplied by the complex exponential factor with zero spatial frequency,

$$\mathcal{F}\left[\int_{-\infty}^{+\infty} f(x')f^*(x'-x)dx'\right] = \mathcal{F}\left[f(x) \otimes f(x)\right] = \left|\widehat{f}(u)\right|^2. \tag{B.39}$$

The process of autocorrelation involves displacement, multiplication, and integration. The 2-D autocorrelation is expressed as,

$$\mathcal{F}\left[\int_{-\infty}^{+\infty} f(\vec{x}')f^*(\vec{x}'-\vec{x})d\vec{x}'\right] = \left|\widehat{f}(\vec{u})\right|^2. \tag{B.40}$$

in which $|\widehat{f}(\vec{u})|^2$ is described as the power spectrum in terms of spatial frequency.

This is the form of the Wiener-Khintchine theorem, which allows for determination of the spectrum by way of the autocorrelation of the generating function. Such a theorem extracts a signal from a background of random noise.

The complex auto-correlation function, $\gamma_{ac}(x)$ is defined as,

$$\begin{aligned}\gamma_{ac} &= \int_{-\infty}^{\infty} f(x')f^*(x'-x)dx' \\ &= f(x) \otimes f^*(x).\end{aligned} \tag{B.41}$$

The normalized auto-correlation function is given by,

$$\gamma_{ac} = \frac{\displaystyle\int_{-\infty}^{\infty} f(x')f^*(x'-x)dx'}{\displaystyle\int_{-\infty}^{\infty} |f(x)|^2\, dx'}. \tag{B.42}$$

B.1.5 *Parseval's theorem*

The Parseval's or Power theorem is generally interpreted as a statement of conservation of energy. It says that the total energy in the real domain is equal to the total energy in the Fourier domain. In a diffraction pattern (see chapter 3), the measured quantity (the radiation power density) is proportional to $|\widehat{f}|^2$. The incident power density should be proportional to $|f|^2$. On integrating these two functions over their respective variables, i.e., u for $\widehat{f}$ and x for f, one finds,

$$\begin{aligned}\int_{-\infty}^{\infty} \left|\widehat{f}(u)\right|^2 du &= \int_{-\infty}^{\infty} \left[\int_{-\infty}^{\infty} f(x)e^{i2\pi ux}dx \int_{-\infty}^{\infty} f^*(x')e^{i2\pi ux'}dx'\right] du \\ &= \iint_{-\infty}^{\infty} f(x)f^*(x')\left[\int_{-\infty}^{\infty} e^{i2\pi u(x-x')}du\right] dxdx' \\ &= \iint_{-\infty}^{\infty} f(x)f^*(x')\delta(x'-x)dxdx' \\ &= \int_{-\infty}^{\infty} |f(x)|^2 dx.\end{aligned} \tag{B.43}$$

where $*$ stands for the conjugate.

The equation (B.43) states that the integral of the squared modulus of a function is equal to the integral of the squared modulus of its spectrum.

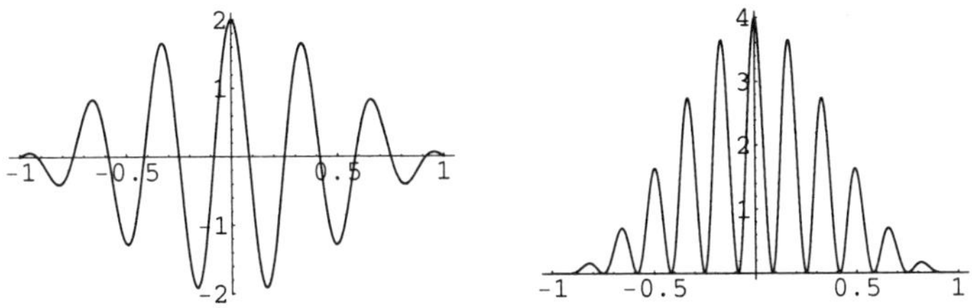

Fig. B.4 Left panel: a sinusoidal function, and right panel: its power spectrum.

This theorem, known as Rayleigh's theorem, corresponds to Parseval's theorem for Fourier series. In two-dimensional case, Parseval's theorem may be expressed as,

$$\int_{-\infty}^{\infty} \left|\widehat{f}(\vec{u})\right|^2 d\vec{u} = \int_{-\infty}^{\infty} |f(\vec{x})|^2 \, d\vec{x}. \tag{B.44}$$

B.1.6 *Some important corollaries*

A few important mathematical relations are also described:

(1) Definite integral: the definite integral of a function, $f(x)$ from $-\infty$ to ∞ is given by the central ordinate of its Fourier transform, i.e.,

$$\int_{-\infty}^{\infty} f(x)dx = \widehat{f}(0). \tag{B.45}$$

(2) First moment: The first moment of $f(x)$ about the origin is,

$$\int_{-\infty}^{\infty} xf(x)dx = \frac{i\widehat{f}'(0)}{2\pi}. \tag{B.46}$$

(3) Centroid: The centroid of $f(x)$ means the point with abscissa $\langle x \rangle$ such that the area of the function times $\langle x \rangle$ is equal to the first moment, thus,

$$\langle x \rangle = \frac{\displaystyle\int_{-\infty}^{\infty} xf(x)dx}{\displaystyle\int_{-\infty}^{\infty} f(x)dx} = \frac{i\widehat{f}'(0)}{2\pi\widehat{f}(0)}. \tag{B.47}$$

(4) Uncertainty relationship: An appropriate measurement of the width of

a function can be defined as,

$$(\Delta x)^2 = \frac{\displaystyle\int_{-\infty}^{\infty} x^2|f(x)|^2 dx}{\displaystyle\int_{-\infty}^{\infty} |f(x)|^2 dx} \tag{B.48}$$

By using Schwarz's inequality, it can be shown that the sizes of $f(x)$ and $\widehat{f}(u)$ are related by $\Delta x.\Delta u \geq 1/4\pi$.

(5) Smoothness and asymptotic behavior: A quantitative definition of the smoothness of a function is the number of its continuous successive derivatives. The asymptotic behavior of $\widehat{f}(u)$ is related to the smoothness of $f(x)$. If $f(x)$ and its first derivatives are continuous,

$$\lim_{|u|\to\infty} |u|^n \widehat{f}(u) = 0. \tag{B.49}$$

For example, the modulus of sinc$u$, the FT of $\Pi(x)$, decreases as u^{-1}, while sinc^2u, the FT of $\Lambda(x)$ decreases with u^{-2}. More generally,

$$\begin{aligned} &\widehat{f}(u) \sim u^{-m}, \\ &\widehat{g}(u) \sim u^{-n}, \qquad u \to \infty, \end{aligned} \tag{B.50}$$

it follows that

$$\widehat{f}(u)\widehat{g}(u) \sim u^{-(m+n)}, \tag{B.51}$$

Hence the convolved functions $f(x) \star f(x)$ is smoother than $f(x)$ and $f(x)$. The smoothness increases with repeated convolution.

B.1.7 *Hilbert transform*

A function may be specified either in the time domain or in the frequency domain. The Hilbert transform of a function $f(t)$ is defined to be the signal whose frequency components are all phase shifted by $-\pi/2$ radians. The real and imaginary parts of the frequency response of any physical system are related to each other by a Hilbert transform (Papoulis, 1968); this relationship is also known as Kramers-Kronig relationship. The Hilbert transform is used in complex analysis to generate complex-valued analytic functions from real functions, as well as to generate functions whose components are harmonic conjugates. It is a useful tool to describe the complex envelope of real valued carrier modulated signal in communication theory.

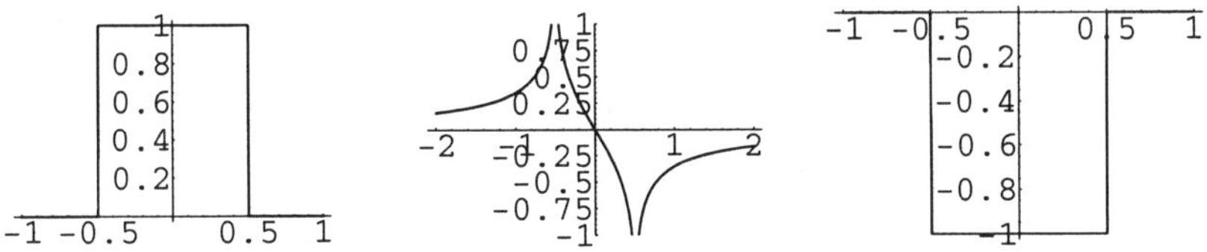

Fig. B.5 Left panel: a rectangle function, middle and right panels: its two successive Hilbert transformations.

As stated earlier, the Fourier transform specifies the function in the other domain, while the Hilbert transform arises when half the information is in the time domain and the other half is in the frequency domain. The Hilbert transform $F_{Hi}(t)$ of a signal $f(t)$ is defined as,

$$F_{Hi}(t) = \frac{1}{\pi} \int_{-\infty}^{\infty} \frac{f(\tau)d\tau}{\tau - t}. \tag{B.52}$$

The integral in equation (B.52) has the form of a convolution integral. The divergence at $t = \tau$ is permitted for by taking Cauchy principal value of the integral. The Hilbert transform $F_{Hi}(t)$ is a linear functional of $f(t)$ and is obtainable from $f(t)$ by convolution with $-1/(\pi t)$, i.e.,

$$F_{Hi} = f(t) \star \frac{-1}{\pi t}, \tag{B.53}$$

where $\star$ denotes the convolution.

The Fourier transform of $-1/(\pi t)$ is $i\, sgn\, \nu$ (see Figure 12.5), which is equal to $+i$ or $-i$ for positive and negative values of ν respectively. Therefore, the Hilbert transformation is equivalent to a kind of filtering, where the amplitudes of the spectral components are left unchanged, albeit their phases are altered by $\pi/2$, either positively or negatively according to the sign of ν. Hence,

$$\begin{aligned} f(t) &= F_{Hi} \star - \left(\frac{-1}{\pi t}\right), \\ &= \frac{-1}{\pi} \int_{-\infty}^{\infty} \frac{F_{Hi}(\tau)d\tau}{\tau - t}. \end{aligned} \tag{B.54}$$

B.2 Laplace transform

Laplace transform is an integral transform and is useful in solving linear ordinary differential equations. In conventional control theory, the system can be described by linear differential equations and the behavior is analyzed using linear control theory. Laplace transforms greatly simplifies the system analysis and are normally used because it maps linear differential equations to linear algebraic expression.

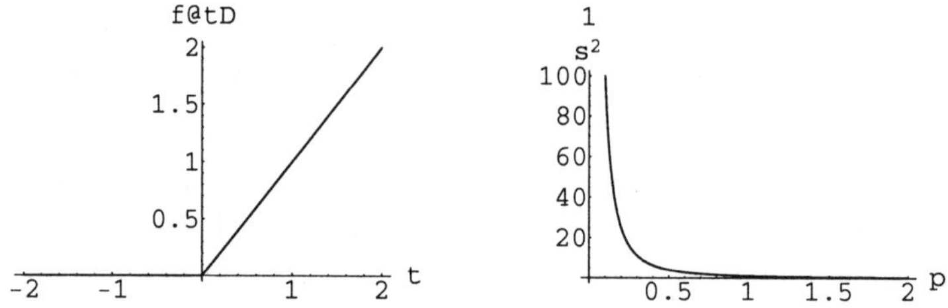

Fig. B.6 Laplace transform of Heaviside unit step, $f(t) = t\,H(s)$.

Laplace transform maps a function in the time domain, $f(t)$, defined on $0 \leq t < \infty$ to a complex function,

$$F(s) = \mathcal{L}\{f(t)\} = \int_0^\infty f(t)e^{-st}dt, \tag{B.55}$$

in which $\mathcal{L}$ stands for the Laplace transform operator and s the complex quantity, which demands a suitable contour of integration to be defined on the complex s plane.

A transform of a function exists if the integral, $\int_0^\infty |f(t)|e^{-\sigma_1 t}dt$, converges for some real, positive value of σ_1 that is a suitably chosen constant. The inverse Laplace transform is given by,

$$f(t) = \frac{1}{2\pi i}\int_{\sigma-i\infty}^{\sigma+i\infty} F(s)e^{st}ds. \tag{B.56}$$

From the definition of Laplace transform, it is noted that the integral converges if the real value of s does not go beyond certain limits in both regions (> 0 or < 0). The allowed region for the integral to converge is called the strip of convergence of the Laplace transform (Bracewell, 1965).

The transform of the first derivative of the function $f(t)$ is expressed as,

$$\begin{aligned}\mathcal{L}\{f'(t)\} &= \int_0^\infty \frac{df(t)}{dt}e^{-st}dt = \int_0^\infty e^{-st}d\{(f(t)\} \\ &= \int_0^\infty f(t)d(e^{-st}) - f(0) = s\int_0^\infty f(t)e^{-st}dt - f(0) \\ &= s\mathcal{L}\{f(t)\},\end{aligned} \tag{B.57}$$

if $f(0) = 0$. Table I describes some general properties of Laplace transform.

Table I

Laplace transform properties

Names	$f(t)$	$F(s)$
Similarity	$f(at)$	$\frac{1}{\|a\|}F\left(\frac{s}{a}\right)$
Linearity	$\alpha f(t) + \beta g(t)$	$\alpha F(s) + \beta G(s)$
Time-shift	$f(t+T)$	$e^{sT}F(s)$
Differentiation	$f'(t)$	$s\,F(s)$
Integration	$\int_0^t f(t)dt$	$\frac{1}{s}F(s)$
Reversal	$f(-t)$	$F(-s)$
Convolution	$\int_0^t f(t')g(t-t')dt'$	$F(s)G(s)$
Impulse response	$\int_0^t f(t')\delta(t-t')dt'$	$F(s)$

B.3 Probability, statistics, and random processes

Probability theory plays an important role in modern physics and wave mechanics. Most signals have a random component, for example, the slope of a wavefront or the number of photons measured in a detector element. These signals are described in terms of their probability distributions.

B.3.1 *Probability distribution*

A probability distribution (or density) assigns to every interval of the real numbers a probability, so that the probability axioms are satisfied. Every random variable gives rise to a probability distribution that contains most of the important information about the variable.

(1) Discrete probability distribution: This distribution is defined on a countable, discrete set, such as a subset of integers. The notable such distributions are the discrete uniform distribution, the Poisson distribution, the binomial distribution, and the Maxwell-Boltzmann distribution. It is a function that can take random variable, N,

$$\mathcal{P}(n) = \mathcal{P}(N = n), \tag{B.58}$$

to each of the possible discrete outcomes, x. There are two requirements for a function to be discrete probability distribution such as (i) $\mathcal{P}(n)$ is non-negative for all real n, and (ii) $\sum_n \mathcal{P}(n) = 1$. The consequence of these two properties is that $0 \le \mathcal{P}(n) \le 1$. The expected value of N, called as mean, is written as $E[N]$ or $\langle N \rangle$. The mean is given by,

$$\langle N \rangle = \frac{1}{n}\sum_{i=1}^{n} N_i, \tag{B.59}$$

therefore, the variance of N, $\langle \sigma \rangle^2$, describes the spread of the distribution around the mean; the higher the variance, the larger the spread of values. The variance is computed as the average squared deviation of each number from its mean. The variance is written as,

$$\begin{aligned} Var(N) &= \langle \sigma \rangle^2 = E[N - \langle N \rangle]^2 \\ &= \frac{1}{n-1}\sum_{i=1}^{n}(N_i - \langle N \rangle)^2 = \sum_n (n - \langle N \rangle)^2 \mathcal{P}(n), \end{aligned} \tag{B.60}$$

where $\langle N \rangle$ is the arithmetic mean, while the standard deviation is defined as the root mean square (RMS) value of the deviation from the mean, or square root of the average squared residual (or variance). It is a measure of the quality of the observations, of the values from their arithmetic mean. It is the square root of variance,

$$\langle \sigma \rangle = \sqrt{\frac{1}{n-1}\sum_{i=1}^{n}(N_i - \langle N \rangle)^2}, \tag{B.61}$$

where N_i are the values of the individual measurements, and N the total number of measurements taken.

(2) Binomial distribution: This distribution provides the discrete probability distribution, $\mathcal{P}(n|N)$,

$$\mathcal{P}(n|N) = \binom{N}{n} p^n q^{N-n} = \frac{N!}{n!\,(N-n)!} p^n (1-p)^{N-n}, \tag{B.62}$$

with $\binom{N}{n}\left(=\frac{N!}{n!\,(N-n)!}\right)$ as a binomial coefficient, p the true probability, and $q = 1 - p$ the false probability.

(3) Poisson distribution: It, a discrete probability distribution, expresses the probability of a number of events that occur in a fixed period of time provided they occur with a known average rate, and are independent of the time since the last event. The distribution of photons detected in a pixel follows Poisson distribution. Unlike binomial distribution, where it is basically the number of heads in repeated tosses of a coin, the Poisson distribution is a limiting case.

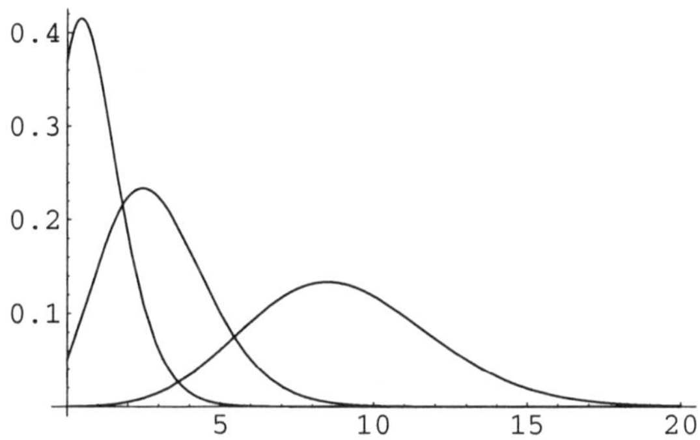

Fig. B.7 Poisson distribution at different wavelengths.

The probability is

$$\mathcal{P}(n|N) = \frac{N!}{n!\,(N-n)!}\left(\frac{m}{N}\right)^n\left(1-\frac{m}{n}\right)^{N-n}$$
$$\simeq \frac{N^n}{n!}\left(\frac{m}{N}\right)^n\left(1-\frac{m}{N}\right)^N, \tag{B.63}$$

in which $m = Np$, $\binom{N}{n}$ is a binomial coefficient, p the true probability, and hence

$$\mathcal{P}(n|m) = e^{-m}\frac{m^n}{n!}. \tag{B.64}$$

(4) Continuous distribution: It is a distribution, which has a continuous distribution function, such as a polynomial or exponential function, for example the normal distribution, the gamma distribution, and the exponential distribution. A continuous random variable, N, assigns a probability density function, $\mathcal{P}_N(n)$, to every outcome, n. The continuous probability functions are referred to as probability density function,

while discrete probability functions are referred to as probability mass function, $\mathcal{P}(n_i)$, in which $i = 1, 2, \cdots$. A cumulative density function of N is defined as,

$$\mathcal{P}'_N(n) = \mathcal{P}(N \leq n). \tag{B.65}$$

The probability density function of N,

$$\mathcal{P}_N(n) = \frac{d\mathcal{P}'_N(n)}{dn}. \tag{B.66}$$

Mathematically such a function satisfies the properties, namely: (i) $\mathcal{P}_N(n) \geq 0$; it is non-negative for all real n, and (ii) the integral of the probability density function is one, i.e.,

$$\int_{-\infty}^{\infty} \mathcal{P}_N(n)dn = 1. \tag{B.67}$$

Since continuous probability functions are defined for an infinite number of points over a continuous interval, the probability at a single point is zero. Probabilities are measured over intervals. The property that the integral must equal one is equivalent to the property for discrete distributions that the sum of all the probabilities should equal one. The mean of a continuous random variable is:

$$E[N] = \int_{-\infty}^{\infty} n\mathcal{P}_N(n)dn, \tag{B.68}$$

with variance,

$$Var(N) = \int_{-\infty}^{\infty} (n - \langle N \rangle)^2 \mathcal{P}_N(n)dn. \tag{B.69}$$

(5) Gaussian distribution: Many random variables are assumed to be Gaussian distributed. Such a distribution, also called normal distribution, is a continuous function which approximates the exact binomial distribution of events and is given by,

$$\mathcal{P}_N(n) = \frac{1}{\sqrt{2\pi}\langle\sigma\rangle} e^{-\frac{(n - \langle N \rangle)^2}{2\langle\sigma\rangle^2}}. \tag{B.70}$$

B.3.2 *Parameter estimation*

Parameter estimation is a discipline that provides tools for the efficient use of data in modeling of phenomena. The estimators are mathematical formulations to extract an estimate, $\hat{\theta}$, of the parameter, θ from measurements x and prior information about the value of the parameter. An ideal estimator has the properties such as: (i) it should depend on the measurement x, but not on the parameter, θ, (ii) it must be unbiased, and (iii) it should be consistent.

(1) Maximum likelihood (ML) estimation: It provides a consistent approach to parameter estimation problems. The ML estimator provides the estimate that maximizes the likelihood function. It commences with a likelihood function of the sample data. The maximum likelihood estimate for a parameter, θ is denoted by $\hat{\theta}$. Considering that N measurements of x are taken, i.e., $x_1, x_2, \cdots, x_N$. The likelihood function, $\prod_{i=1}^{N} f(x_i|\theta)$ is the conditional probability density function of finding those measurements for a given value of the parameter, θ, the estimate of which depends on the form of $f(x|\theta)$. Mathematically, the ML estimator is given by,

$$\hat{\theta} = \theta_{max} \left\{ \prod_{i=1}^{N} f(x_i|\theta) \right\}. \tag{B.71}$$

It is a non-linear solution and should be solved using a non-linear maximization algorithm. The notable drawback of this method is that it can be biased for small samples and it can be sensitive to the choice starting values.

For a Gaussian distribution with a variance, $\langle\sigma\rangle^2$ and a mean of θ,

$$\begin{aligned} f(x_i|\theta, \langle\sigma\rangle) &= \prod_{1}^{N} \frac{1}{\langle\sigma\rangle \sqrt{2\pi}} e^{-(x_i - \theta)^2/2\langle\sigma\rangle^2} \\ &= \frac{(2\pi)^{-N/2}}{\langle\sigma\rangle^N} e^{-\sum_{1}^{N}(x_i - \theta)^2/2\langle\sigma\rangle^2}. \end{aligned} \tag{B.72}$$

and the log-likelihood function,

$$\log f = -\frac{N}{2} \log\left(2\pi \langle\sigma\rangle^2\right) - \frac{\sum_1^N (x_i - \theta)^2}{2\langle\sigma\rangle^2}. \tag{B.73}$$

The derivatives with respect to θ,

$$\frac{\partial(\log f)}{\partial\theta} = \frac{\sum_1^N (x_i - \theta)}{\langle\sigma\rangle^2} = 0, \tag{B.74}$$

provides the ML estimate,

$$\hat{\theta} = \frac{\sum_1^N x_i}{N}, \tag{B.75}$$

which is known as centroid estimator and is obtained by taking the centre-of-mass of the measurements.

(2) Maximum *a posteriori* (MAP) estimator: The *posterior* probability comes from Bayesian approaches,

$$\mathcal{P}(B|A) = \frac{\mathcal{P}(A|B)\mathcal{P}(B)}{\mathcal{P}(A)},$$

in which $\mathcal{P}(B)$ is the probability of image B, $\mathcal{P}(AB) = \mathcal{P}(A|B)\mathcal{P}(B)$ (follows product rule), A and B the outcomes of random experiments, and $\mathcal{P}(B|A)$ the probability of B given that A has occurred and for imaging, $\mathcal{P}(A|B)$ is the likelihood of the data given B, $\mathcal{P}(A)$ is a constant which normalizes $\mathcal{P}(B|A)$ to a sum of unity, and provides the probability of the data.

The MAP estimation can be used to obtain a point estimate of an unobserved quantity on the basis of empirical data. It provides the most likely value of θ from the observed data and prior knowledge of the distribution of θ, $f(\theta)$:

$$\begin{aligned}\hat{\theta} &= \theta_{max}\left\{\prod_{i=1}^N f(\theta|x_i)\right\} \\ &= \theta_{max}\left\{\frac{f(\theta)}{\prod_{i=1}^N f(x_i)}\prod_{i=1}^N f(x_i|\theta)\right\} \\ &= \theta_{max}\left\{f(\theta)\prod_{i=1}^N f(x_i|\theta)\right\}\end{aligned} \tag{B.76}$$

The expression $f(\theta)\prod_{i=1}^N f(x_i|\theta)$ is known as *a posteriori* distribution.

B.3.3 *Central-limit theorem*

If the convolved functions possess a few simple properties, at the limit of an infinitely increasing number of convolution the result tends to be Gaussian. Let $X_i, i = 1, 2, \cdots, N$ be a sequence of random variables, satisfying,

- the random variables are statistically independent and
- the random variables have the same probability distribution with mean μ and variance $\langle\sigma\rangle^2$.

Considering the following random variable,

$$U_N = \sum_{i=1}^{N} X_i. \tag{B.77}$$

According to the central-limit theorem, in the limit as N tends to infinity, the probability distribution of U_N approaches that of a Gaussian random variable with mean $N\mu$ and variance $N\langle\sigma\rangle^2$. The implications of such a theorem are:

- it explains the common occurrence of Gaussian distributed random variables in nature and
- with N measurements from a population of mean, μ, and variance, $\langle\sigma\rangle^2$, the sample means are approximately Gaussian distributed with a mean of μ and a variance of $\langle\sigma\rangle^2/N$.

B.3.4 *Random fields*

A random process is defined as an ensemble of functions together with a probability rule that assigns a probability to a given observation of one of these functions. In turbulence theory, the structure function is used (Tatarski, 1961), i.e., instead of the stationary random function $f(t)$, the difference $F_\tau(t) = f(t+\tau) - f(t)$ is considered. Using the identity,

$$(a-b)(c-d) = \frac{1}{2}[(a-d)^2 + (b-c)^2 - (a-c)^2 - (b-d)^2],$$

one may represent the correlation (or coherence) function of the increments. In a random field, let $f(\vec{r})$ be a random function of three variables, for which the autocorrelation function is defined as,

$$\mathcal{B}_f(\vec{r}_1, \vec{r}_2) = \langle[f(\vec{r}_1) - \langle f(\vec{r}_1)\rangle]\rangle \langle[f(\vec{r}_2) - \langle f(\vec{r}_2)\rangle]\rangle. \tag{B.78}$$

where $\langle\,\rangle$ denotes the ensemble average.

The average value of a function can be a constant or change with time; a random function $f(t)$ is known to be constant if $\langle f(t)\rangle =$ constant. Similarly, a random field is called homogeneous, when $f(\vec{r}) =$ constant, and the autocorrelation function is independent of the translation of $\vec{r}_1$ and $\vec{r}_2$ by equal amount in the same direction, i.e.,

$$\mathcal{B}_f(\vec{r}_1, \vec{r}_2) = \mathcal{B}_f(\vec{r}_1 - \vec{r}_2). \tag{B.79}$$

The autocorrelation function is a function of the separation, $(\vec{r}_1 - \vec{r}_2)$. The homogeneous random field is called isotropic if $\mathcal{B}_f(\vec{r})$ depends only on $\vec{r} = |\vec{r}|$. This field can be represented in the form of three-dimensional (3-D) stochastic Fourier-Stieltjes integral,

$$f(\vec{r}) = \int_{-\infty}^{\infty} e^{i\vec{\kappa}\cdot\vec{r}} d\psi(\vec{\kappa}), \tag{B.80}$$

where $\vec{\kappa}$ is the wave vector and the amplitude $d\psi(\vec{\kappa})$ satisfy the relation,

$$d\psi(\vec{\kappa}_1)d\psi^*(\vec{\kappa}) = \delta(\vec{\kappa}_1 - \vec{\kappa}_2)\Phi_f(\vec{\kappa}_1)d\vec{\kappa}_1 d\vec{\kappa}_2, \tag{B.81}$$

with $\Phi_f(\kappa)(\geq 0)$ as the spectral density, and therefore one gets,

$$\mathcal{B}_f(\vec{r}_1 - \vec{r}_2) = \int_{-\infty}^{\infty} e^{i\omega(\vec{r}_1 - \vec{r}_2)} \widehat{\mathcal{B}}_f(\omega) d\omega. \tag{B.82}$$

The functions, $\mathcal{B}_f(\vec{r})$ and $\widehat{\mathcal{B}}_f(\vec{\kappa})$, are the Fourier transforms of each other. Thus, the Fourier transform of a correlation function, $\mathcal{B}_f(\vec{r})$, must be non-negative and the non-random function, $\widehat{\mathcal{B}}_f(\vec{\kappa})$ is known as the spectral density of the stationary random function $f(t)$. When dealing with atmospheric turbulence, random processes with infinite covariances are encountered. In order to avoid such an anomaly, the structure function, $\mathcal{D}_f(\vec{r})$ is introduced,

$$\begin{aligned} \mathcal{D}_f(\vec{r}) &= \langle [f(\vec{r} + \vec{\rho}) - f(\vec{r})]^2 \rangle \\ &= 2[\mathcal{B}_f(\vec{0}) - \mathcal{B}_f(\vec{r})]. \end{aligned} \tag{B.83}$$

The structure function has small values for the small separation distances and times of interest.

Appendix C

Bispectrum and phase values using triple-correlation algorithm

The algorithm based on the triple-correlation method to estimate the phase of the object's Fourier transform of an image of size 4 $\times$ 4 pixels is given below. The bispectrum values for a 4 $\times$ 4 array for the lower half (and extreme left in the upper half) of the Fourier plane are entered. The remaining values are determined using the Hermitian symmetry property. The phase values are estimated as well. Again these phase values are only for the lower half (and extreme left in the upper half) of the Fourier plane. By using the Hermitian symmetry, the phase values at the upper half plane are also determined. The bispectrum and phase values are:

$$b((-1,0),(0,0)) = I(-1,0)I(0,0)I^*(-1,0)$$
$$\psi(-1,0) = \psi(-1,0) + \psi(0,0) - \psi_b((-1,0),(0,0))$$

$$b((1,0),(0,0)) = I(1,0)I(0,0)I^*(1,0)$$
$$\psi(1,0) = \psi(1,0) + \psi(0,0) - \psi_b((1,0),(0,0))$$

$$b((-1,0),(-1,0)) = I(-1,0)I(-1,0)I^*(-2,0)$$
$$\psi(-2,0) = \psi(-1,0) + \psi(-1,0) - \psi_b((-1,0),(-1,0))$$

$$b((0,0),(0,-1)) = I(0,0)I(0,-1)I^*(0,-1)$$
$$\psi(0,-1) = \psi(0,0) + \psi(0,-1) - \psi_b((0,0),(0,-1))$$

$$b((0,-1),(0,-1)) = I(0,-1)I(0,-1)I^*(0,-2)$$
$$\psi(0,-2) = \psi(0,-1) + \psi(0,-1) - \psi_b((0,-1),(0,-1))$$

$$b((0,-1),(-1,0)) = I(0,-1)I(-1,0)I^*(-1,-1)$$
$$\psi(-1,-1) = \psi(0,-1) + \psi(-1,0) - \psi_b((0,-1),(-1,0))$$

$$b((0,-1),(1,0)) = I(0,-1)I(1,0)I^*(1,-1)$$
$$\psi(1,-1) = \psi(0,-1) + \psi(1,0) - \psi_b((0,-1),(1,0))$$

$$b((0,-1),(-2,0)) = I(0,-1)I(-2,0)I^*(-2,-1)$$
$$\psi(-2,-1) = \psi(0,-1) + \psi(-2,0) - \psi_b((0,-1),(-2,0))$$
$$b((-1,0),(-1,-1)) = I(-1,0)I(-1,-1)I^*(-2,-1)$$
$$\psi(-2,-1) = \psi(-1,0) + \psi(-1,-1) - \psi_b((-1,0),(-1,-1))$$

$$b((0,-1),(-1,-1)) = I(0,-1)I(-1,-1)I^*(-1,-2)$$
$$\psi(-1,-2) = \psi(0,-1) + \psi(-1,-1) - \psi_b((0,-1),(-1,-1))$$
$$\mathrm{b((0,\text{-}2),(\text{-}1,0)) = I(0,\text{-}2)\ I(\text{-}1,0)\ I^*(\text{-}1,\text{-}2)}$$
$$\psi(\text{-}1,\text{-}2) = \psi(0,\text{-}2)+\psi(\text{-}1,0)\text{-}\psi_b((0,\text{-}2),(\text{-}1,0))$$

$$b((0,-1),(1,-1)) = I(0,-1)I(1,-1)I^*(1,-2)$$
$$\psi(1,-2) = \psi(0,-1) + \psi(1,-1) - \psi_b((0,-1),(1,-1))$$
$$b((0,-2),(1,0)) = I(0,-2)I(1,0)I^*(1,-2)$$
$$\psi(1,-2) = \psi(0,-2) + \psi(1,0) - \psi_b((0,-2),(1,0))$$

$$b((0,-1),(-2,-1)) = I(0,-1)I(-2,-1)I^*(-2,-2)$$
$$\psi(-2,-2) = \psi(0,-1) + \psi(-2,-1) - \psi_b((0,-1),(-2,-1))$$
$$b((0,-2),(-2,0)) = I(0,-2)I(-2,0)I^*(-2,-2)$$
$$\psi(-2,-2) = \psi(0,-2) + \psi(-2,0) - \psi_b((0,-2),(-2,0))$$
$$b((-1,0),(-1,-2)) = I(-1,0)I(-1,-2)I^*(-2,-2)$$
$$\psi(-2,-2) = \psi(-1,0) + \psi(-1,-2) - \psi_b((-1,0),(-1,-2))$$
$$b((-1,-1),(-1,-1)) = I(-1,-1)I(-1,-1)I^*(-2,-2)$$
$$\psi(-2,-2) = \psi(-1,-1) + \psi(-1,-1) - \psi_b((-1,-1),(-1,-1))$$

$$b((0,1),(-2,0)) = I(0,1)I(-2,0)I^*(-2,1)$$
$$\psi(-2,1) = \psi(0,1) + \psi(-2,0) - \psi_b((0,1),(-2,0))$$
$$b((-1,0),(-1,1)) = I(-1,0)I(-1,1)I^*(-2,1)$$
$$\psi(-2,1) = \psi(-1,0) + \psi(-1,1) - \psi_b((-1,0),(-1,-1))$$

Bibliography

Abell G. O., 1955, Pub. Astron. Soc. Pac., **67**, 258.

Abell G. O., 1975, Galaxies and Universe, Eds. A. Sandage, M. Sandage, and J. Kristian, Chicago University Press, Chicago.

Abhayankar K. D., 1992, 'Astrophysics: Stars and Galaxies', Tata McGraw Hill Pub. Co. Ltd.

Ables J. G., 1974, Astron. Astrophys. Suppl., **15**, 383.

Acton D. S., Smithson R. C., 1992, Appl. Opt., **31**, 3161.

Aime C., 2000, J. Opt. A: Pure & Appl. Opt., **2**, 411.

Aime C., Petrov R., Martin F., Ricort G., Borgnino J., 1985, SPIE., **556**, 297.

Aime C., Ricort G., Grec G., 1975, Astron. Astrophys., **43**, 313.

Alard C., 2001, Astron. Astrophys. **379**, L44.

Alexander J. B., Andrews P. J., Catchpole R. M., Feast M. W., Lloyd Evans T., Menzies J. W., Wisse P. N. J., Wisse M., 1972. Mon. Not. R. Astron. Soc., **158**, 305.

Alladin S. M., Parthasarathy M., 1978, Mon. Not. R. Astron. Soc., **184**, 871.

Allen C. W., 1976, Astrophysical Quantities, Athlone Press, London.

Angel J. R. P., 1994, Nature, **368**, 203.

Anger H. O., 1952, Nature, **170**, 220.

Anger H. O., 1966, Trans. Instr. Soc. Am., **5**, 311.

Antonucci R. R. J., Miller J. S., 1985, Astrophys. J., **297**, 621.

Anupama G. C., Sahu D. K., Jose J., 2005, Astron. Astrophys., **429**, 667.

Appenzeller I. Mundt R., 1989, Astron. Astrophys. Rev., **1**, 291.

Aretxaga I., Mignant D. L., Melnick J., Terlevich R. J., Boyle B. J., 1998, astro-ph/9804322.

Arnulf M. A., 1936, Compt. Rend., **202**, 115.

Arsenault R., Salmon D. A., Kerr J., Rigaut F., Crampton D., Grundmann W. A., 1994, SPIE, **2201**, 883.

Asplund M., Gustafsson B., Kiselman D., Eriksson K., 1996, Astron. Astrophys. **318**, 521.

Ayers G. R., Dainty J. C., 1988, Opt. Lett., **13**, 547.

Ayers G. R., Northcott M. J., Dainty J. C., 1988, J. Opt. Soc. Am. A., **5**, 963.

Baba N., Kuwamura S., Miura N. Norimoto Y., 1994b, Astrophys. J., **431**, L111.

Baba N., Kuwamura S., Norimoto Y., 1994, Appl. Opt., **33**, 6662.

Babcock H. W, 1953, Pub. Astron. Soc. Pac., **65**, 229.

Babcock H. W., 1990, Science, **249**, 253.

Baldwin J. E., Haniff C. A., Mackay C. D., Warner P. J., 1986, Nature, **320**, 595.

Baldwin J., Tubbs R., Cox G., Mackay C., Wilson R., Andersen M., 2001, Astron. Astrophys., **368**, L1.

Balega Y., Blazit A., Bonneau D., Koechlin L., Foy R., Labeyrie A., 1982, Astron. Astrophys., **115**, 253.

Balega I. I., Balega Y. Y., Falcke, H., Osterbart R., Reinheimer T., Schöeller M., Weigelt G., 1997, Astron. Letters, **23**, 172.

Balick B., 1987, Astron. J., **94**, 671.

Banachiewicz T., 1955, Vistas in Astronomy, **1**, 200.

Barakat R., Nisenson P., 1981, J. Opt. Soc. Am., **71**, 1390.

Barletti R., Ceppatelli G., Paterno L., Righini A., Speroni N., 1976, J. Opt. Soc. Am., **66**, 1380.

Barnes J. E., Hernquist L. E., 1992, Ann. Rev. Astron. Astrophys., **30**, 705.

Barr L. D., Fox J., Poczulp G. A., Roddier C. A., 1990, SPIE, **1236**, 492.

Bates R., McDonnell M., 1986, 'Image Restoration and Reconstruction', Oxford Eng. Sc., Clarendon Press.

Bates W. J., 1947, Proc. Phys. Soc., **59**, 940.

Battaglia et al., 2005, Mon. Not. R. Astron. Soc., **364**, 433.

Beckers J. M., 1982, Opt. Acta., 29, 361.

Beckers J., 1982 Optica Acta, **29**, 361.

Beckers J. M., 1999, 'Adaptive Optics in Astronomy', ed. F. Roddier, Cambridge Univ. Press, 235.

Beckers J. M., Hege E. K., Murphy H. P., 1983, SPIE, **444**, 85.

Beckwith S., Sargent A. I., 1993, in 'Protostars and Planets III', Eds., E. H. Levy & J. I. Lunine, 521.

Bedding T. R., Minniti D., Courbin F., Sams B., 1997b, Astron. Astrophys., **326**, 936.

Bedding T. R., Robertson J. G., Marson R. G., Gillingham P. R., Frater R. H., O'Sullivan J. D., 1992, Proc. 'High Resolution Imaging Interferometry', Eds., J. M. Beckers & F. Merkle, 391.

Bedding T. R., Zijlstra A. A., Von der Lühe O., Robertson J. G., Marson R. G., Barton J. R., Carter B. S., 1997a, Mon. Not. R. Astron. Soc., **286**, 957.

Benedict G. F. et al., 2002, Astron. J., **123**, 473.

Berman L., 1935. Astrophys. J., **81**, 369.

Bertiau F. C., 1958, Astrophys. J., **128**, 533.

Bertout C., 1989, Ann. Rev. Astron. Astrophys., **27**, 351.

Bessell M. S., 1976, Pub. Astron. Soc. Pac., **88**, 557.

Bessell M. S., 2005, Ann. Rev. Astron. Astrophys., **43**, 293.

Binney J., Merrifield M., 1998, Galactic Astronomy, Princeton Series in Astrophysics, Princeton, New Jersey.

Blazit A., 1976, Thesis, University of Paris.

Blazit A., 1986, SPIE, **702**, 259.

Blazit A., Bonneau D., Koechlin L., Labeyrie A., 1977, Ap J, **214**, L79.

Blöcker T. Balega A., Hofmann K. -H., Lichtenthäler J., Osterbart R., Weigelt G., 1999, astro-ph/9906473.
Boccaletti A., 2001, Private communication.
Boccaletti A., Labeyrie A., Ragazzoni R., 1998a, astro-ph/9806144.
Boccaletti A., Moutou C., Labeyrie A., Kohler D., Vakili F., 1998b, Astron. Astrophys., **340**, 629.
Boccaletti A., Moutou C., Mouillet D., Lagrange A., Augereau J., 2001, Astron. Astrophys., **367**, 371.
Boccaletti A., Riaud P., Moutou C., Labeyrie A., 2000, Icarus, **145**, 628.
Boksenburg A., 1975, Proc., 'Image Processing Techniques in Astronomy', Eds. C. de Jager & H. Nieuwenhuizen.
Bonanos A. Z., 2006, "Eclipsing Binaries: Tools for Calibrating the Extragalactic Distance Scale", Binary Stars as Critical Tools and Tests in Contemporary Astrophysics, IAU Symposium no. 240, 240.
Bone D. J., Bachor H. A., Sandeman R. J., 1986, Appl. Opt., **25**, 1653.
Bonneau D., Foy R., 1980, Astron. Astrophys., **92**, L1.
Bonneau D., Labeyrie A., 1973, Astrophys. J., **181**, L1.
Bordovitsyn V. A., 1999, 'Synchrotron Radiation in Astrophysics', Synchrotron Radiation Theory and Its Development, ISBN 981-02-3156-3.
Born M., Wolf E., 1984, Principle of Optics, Pergamon Press.
Bouvier J., Rigaut F., Nadeau D., 1997, Astron. Astrophys., **323**, 139.
Boyle W. S. Smith G. E., 1970, Bell System Tech. J., **49**, 587.
Bracewell R., 1965, The Fourier transform and its Applications, McGraw-Hill, NY.
Brandl B. et al. 1996, Astrophys. J., **466**, 254.
Breckinridge J. B., McAlister H. A., Robinson W. A., 1979, App. Opt., **18**, 1034.
Breger M., 1979, Astrophys. J., **233**, 97.
Brosseau C., 1998, Fundamentals of Polarized light, John Wiley & Sons, INC.
Brown R. G. W., Ridley K. D., Rarity J. G., 1986, Appl. Opt., **25**, 4122.
Brown R. G. W., Ridley K. D., Rarity J. G., 1987, Appl. Opt., **26**, 2383.
Bruns D., Barnett T., Sandler D., 1997, SPIE., **2871**, 890.
Cadot O., Couder Y., Daerr A., Douady S., Tsinocber A., 1997, Phys. Rev. E1, **56**, 427.
Callados M., Vàzquez M., 1987, Astron. Astrophys., **180**, 223.
Carlson R. W., Bhattacharyya J. C., Smith B. A., Johnson T. V., Hidayat B., Smith S. A., Taylor G. E., O'Leary B., Brinkmann R. T., 1973, Science, **182**, 52.
Cassinelli J. P., Mathis J. C., Savage B. D., 1981, Science, **212**, 1497.
Chakrabarti S. K., Anandarao B. G., Pal S., Mondal S., Nandi A., Bhattacharyya A., Mandal S., Ram Sagar, Pandey J. C., Pati, A., Saha, S.K., 2005, Mon. Not. R. Astron. Soc., **362**, 957.
Chandrasekhar S., 1931, Mon. Not. R. Astron. Soc., **91**, 456.
Chapman C. R., Morrison D., Zellner B., 1975, Icarus, **25**, 104.
Chinnappan V., 2006, Ph. D thesis, Bangalore University.
Chwolson, O., 1924, Astron. Nachr., **221**, 329.
Clampin M., Crocker J., Paresce F., Rafal M., 1988, Rev. Sci. Instr., **59**, 1269.

Clampin M., Nota A., Golimowski D., Leitherer C., Ferrari A., 1993, Astrophys. J., **410**, L35.

Close L., 2003, http://athene.as.arizona.edu/ lclose/AOPRESS/.

Close L. M., Roddier F., Hora J. L., Graves J. E., Northcott M. J., Roddier C., Hoffman W. F., Doyal A., Fazio G. G., Deutsch L. K., 1997, Astrophys. J., **489**, 210.

Cognet M., 1973, Opt. Comm., **8**, 430.

Colavita M., Shao M., Staelin D. H., 1987, Appl. Opt., **26**, 4106.

Collett E., 1993, Polarized light: Fundamentals and Applications, Marcel Dekkar, Inc. N. Y.

Conan J. -M., Mugnier L. M., Fusco T., Michau V., Rousset G., 1998, Appl. Opt., **37**, 4614.

Conan R., Ziad A., Borgnino J., Martin F., Tokovinin A., 2000, SPIE, **4006**, 963.

Connors T. W., Kawata D., Gibson B. K., 2006, Mon. Not. R. Astron. Soc., **371**, 108.

Cooper D., Bui D., Bailey R., Kozlowski L., Vural K., 1993, SPIE, **1946**, 170.

Coulman C. E., 1974, Solar Phys., **34**, 491.

Coutancier J., 1940, Revue Générale de l'Électricité, **48**, 31.

Cowling T. G., 1946, Mon. Not. R. Astron. Soc., **106**, 446.

Cromwell R. H., Haemmede V. R., Woolf N. J., 1988, in Very large telecopes, their instrumention and Programs', Eds., M. -H. Ulrich & Kjär, 917.

Cuby J. -G., Baudrand J., Chevreton M., 1988, Astron. Astrophys., **203**, 203.

Currie D. Kissel K., Shaya E., Avizonis P., Dowling D., Bonnacini D., 1996, The Messenger, no. 86, 31.

Dantowitz R., Teare S., Kozubal M., 2000, Astron. J., **119**, 2455.

Denker C., 1998, Solar Phys., **81**, 108.

Denker C., de Boer C. R., Volkmer R., Kneer F., 1995, Astron. Astrophys., **296**, 567.

Diericks P., Gilmozzi R., 1999, Proc. 'Extremely Large Telescopes', Eds., T. Andersen, A. Ardeberg, and R. Gilmozzi, 43.

Drummond J., Eckart A., Hege E. K., 1988, Icarus, **73**, 1.

Dyck H. M., van Belle G. T., Thompson R. R., 1998, Astron. J., **116**, 981.

Ealey M. A., 1991, SPIE, **1543**, 2.

Ebstein S., Carleton N. P., Papaliolios C., 1989, Astrophys. J., **336**, 103.

Eddington A. E., 1909, Mon. Not. R. Astron. Soc., **69**, 178.

Eggen O. J., 1958, Mon. Not. R. Astron. Soc., **118**, 65.

Eggen O. J., 1960, Mon. Not. R. Astron. Soc., **120**, 563.

Hege E. K., Hubbard E. N., Strittmatter P. A., Worden S. P., 1981, Astrophys. J., **248**, 1.

Einstein A., 1905, Ann. der Physik, **17**, 132.

Eisberg R., Resnick R., 1974, Quantum Physics of Atoms, Molecules, Solids, Nuclei, and Particles, John Wiley & Sons, Inc.

Elster J., Geitel H., 1916, Zeitschrift Phys., **17**, 268.

Eke V., 2001, Mon. Not. R. Astron. Soc., **320**, 106.

Esposito S., Riccardi A., 2001, Astron. Astrophys., **369**, L9.

Evershed J., 1909 Mon. Not. R. Astron. Soc., **69**, 454.

Fabbiano G. et al., 2004, Astrophys. J. Lett., 605, L21.

Falcke H., Davidson K., Hofmann K. -H., Weigelt G., 1996, Astron. Astrophys., **306**, L17.

Fanaroff B. L., Riley J. M., 1974, Mon. Not. R. Astron. Soc., **167**, 31.

Feast M. W., Catchpole R. M., 1997, Mon. Not. R. Astron. Soc., **286**, L1.

Fender R., 2002, astro-ph/0109502.

Fienup J. R., 1978, Opt. Lett., **3**, 27.

Fienup J. R., 1982, Appl. Opt., **21**, 2758.

Fienup J. R., Marron J. C., Schulz T. J., Seldin J. H., 1993, Appl. Opt., **32**, 1747.

Fischer O., Stecklum B., Leinert Ch., 1998, Astron. Astrophys. **334**, 969.

Foy R., 2000, in 'Laser Guide Star Adaptive Optics', Eds. N. Ageorges and C. Dainty, 147.

Foy R., Bonneau D., Blazit A., 1985, Astron. Astrophys., **149**, L13.

Foy R., Labeyrie A., 1985, Astron. Astrophys., **152**, L29.

Francon M., 1966, Optical interferometry, Academic press, NY.

Frank J., King A. R., Raine D. J., 2002, Accretion Power in Astrophysics, Cambridge: Cambridge Univ. Press.

Fried D. L., 1965, J. Opt. Soc. Am., **55**, 1427.

Fried D. L., 1966, J. Opt. Soc. Am., **56**, 1372.

Fried D. L., 1993, in 'Adaptive Optics for Astronomy' Eds. D. M. Alloin & J. -M Mariotti, 25.

Fried D. L., Belsher J., 1994, J. Opt. Soc. Am., A, **11**, 277.

Fried D. L. Vaughn J. L., 1992, Appl. Opt., **31**.

Fugate R. Q., Fried D. L., Ameer G. A., Boeke B. R., Browne S. L., Roberts P. H., Roberti P. H., Ruane R. E., Tyler G. A., Wopat L. M., 1991, Nature, 353, 144.

Gabor D., 1948, Nature, **161**, 777.

Gallimore J. F., Baum S. A., O'Dea C. P., 1997, Nature, **388**, 852.

Gamow G., 1948, Nature **162**, 680.

Gandhi P., 2005, '21st Century Astrophysics', Eds. S. K. Saha, & V. K. Rastogi, Anita Publications, New Delhi, 90.

Gautier D., Conrath B., Flasar M., Hanel R., Kunde V., Chedin A., Scott N., 1981, J. Geophys. Res. **86**, 8713.

Gauger A., Balega Y. Y., Irrgang P., Osterbart R., Weigelt G., 1999, Astron, Astrophys., **346**, 505.

Geary J., 1995, SPIE .

Geiger H., Müller W., 1928, Zeitschrift Phys., **29**, 389.

Geiger W, 1955, Zeitschrift Phys., **140**, 608.

Gerchberg R. W., Saxton W. O., 1972, Optik, **35**, 237.

Gezari D. Y., Labeyrie A., Stachnik R., 1972, Astrophys. J., **173**, L1.

Gibson E. G., 1973, The Quiet Sun, US Govt. printing office, Washington.

Gies R. D., Mason B. D., Bagnuolo W. G. (Jr.), Haula M. E., Hartkopf W. I., McAlister H. A., Thaller M. L., McKibben W. P., Penny L. R., 1997, Astrophys. J., **475**, L49.

Gillingham P. R., 1984, 'Advanced Technology Optical Telescopes II', Eds., L. D. Barr & B. Mark, SPIE, **444**, 165.

Glindemann A., 1997, Pub. Astron. Soc. Pac, **109**, 68.

Glindemann A., Hippler S., Berkefeld T., Hackenberg W., 2000, Exp. Astron., **10**, 5.

Goldfisher L. I., 1965, J. Opt. Soc. Am, **55**, 247.

Goldstein H., 1980, Classical Mechanics, Addison-Wesley, MA.

Golimowski D. A., Nakajima T., Kulkarni S. R., Oppenheimer B. R., 1995, Astrophys. J., **444**, L101.

Gonsalves S. A., 1982, Opt. Eng., **21**, 829.

Goodman J. W., 1968, Introduction to Fourier Optics, McGraw Hill book Co. NY.

Goodman J. W., 1975, in 'Laser Speckle and Related Phenomena', Ed. J. C Dainty, Springer-Verlag, N.Y.

Goodman J. W., 1985, Statistical Optics, Wiley, NY.

Goodrich G. W., Wiley W. C., 1961, Rev. Sci. Instr., **32**, 846.

Goodrich G. W., Wiley W. C., 1962, Rev. Sci. Instr., **33**, 761.

Greenwood D. P., 1977, J. Opt. Soc. Am., **67**, 390.

Grieger F., Fleischman F., Weigelt G. P., 1988, Proc. 'High Resolution Imaging Interferometry', Ed. F. Merkle, 225.

Haas M., Leinert C., Richichi A., 1997, Astron. Astrophys., **326**, 1076.

Halliday D., Rasnick R., Walkar J., 2001, Fundamentals of Physics, John Wiley & Sons, NY.

Hamann W. -R., 1996. in 'Hydrogen-deficient stars', Eds. C. S. Jeffery, U. Heber, ASP Conf Ser. **96**, 127.

Hanbury Brown R., 1974, The Intensity Interferometry, its Applications to Astronomy, Taylor & Francis, London.

Haniff C. A., Mackay C. D., Titterington D. J., Sivia D., Baldwin J. E., Warner P. J., 1987, Nature, **328**, 694.

Hardie R. H., 1962, Astronomical Techniques, Ed. W. A. Hiltner, University of Chicago Press: Chicago, 178.

Hardy J. W., 1991, SPIE, **1542**, 229.

Hariharan P., Sen D., 1961, J. Sci. Instrum., **38**, 428.

Hartkopf W. I., Mason B. D., McAlister H. A., 1996, Astron. J., **111**, 370.

Hartkopf W. I., McAlister H. A., Franz O. G., 1989, Astron. J., **98**, 1014.

Hartkopf W. I., McAlister H. A., Mason B. D., 1997, CHARA Contrib. No. 4, 'Third Catalog of Interferometric Measurements of Binary Stars', W.I.

Hartmann J, 1900, Z. Instrum., **24**, 47.

Harvey J. W., 1972, Nature, **235**, 90.

Hawley S. A., Miller J. S., 1977, Astrophys. J., **212**, 94.

Hecht E., 1987, Optics, 333.

Heintz W. D., 1978, 'Double Stars', Reidel, Dordrecht.

Henden A. A, Kaitchuck R H., 1982, Astronomical Photometry, Van Nostrand Reinhold Co. NY.

Henry T. J., Soderblom D. R., Donahue R. A., Baliunas S. L., 1996, Astron. J., **111**, 439.

Heroux L., Hinterreger H. E., 1960, Rev. Sci. Instr., **31**, 280.

Herrmann H, Kunze C, 1969, Advances in Electron. & Electron. Phys., Academic

Press, London, **28**B, 955.
Hess S. L., 1959, Introduction to Theoretical Meteorology (Holt, New York).
Heydari M., Beuzit J. L., 1994, Astron. Astrophys., **287**, L17.
Hiltner W. A., 1962, Astronomical Techniques, University of Chicago press.
Hillwig T. C., Gies D. R., Huang W., McSwain M. V., Stark M. A., van der Meer A., Kaper L., 2004, Astroph. J., **615**, 422.
Hoeflich P., 2005, '21st Century Astrophysics', Eds. S. K. Saha, & V. K. Rastogi, Anita Publications, New Delhi, 57.
Hofmann K. -H., Seggewiss W., Weigelt G., 1995, Astron. Astrophys. **300**, 403.
Hofmann K. -H., Weigelt G., 1993, Astron. Astrophys., **278**, 328.
Holst G. C., 1996, CCD Arrays, Cameras, and Displays, SPIE Opt. Eng. Press, Washington, USA.
Holst G., DeBoer J., Teves M., Veenemans C., 1934, Physica, **1**, 297.
Hubble E. P., 1929, Proc. Nat. Acad. Sci. (Wash.), **15**, 168.
Hubble E. P., 1936, The Realm of the Nebulae, Yale University Press.
Hufnagel R. E., 1974, in Proc. 'Optical Propagaion through Turbulence', Opt. Soc. Am., Washington, D. C. WAI 1.
Hull A W, 1918, Proc. Inst. Radio Electron. Eng. Austr., **6**, 5.
Hutchings J., Morris S., Crampton D., 2001, Astron. J., **121**, 80.
IAU statement, February 28, 2003.
Icko I., 1986, 'Binary Star Evolution and Type I Supernovae', Cosmogonical Processes, 155.
Iijima T., 1998, Mon. Not. R. Astron. Soc., **297**, 347.
Ingerson T. E., Kearney R. J., Coulter R. L., 1983, Appl. Opt., **22**, 2013.
Iredale P., Hinder G., Smout D., 1969, Advances in Electron. & Electron. Phys., Academic Press, London, **28**B, 965.
Ishimaru A., 1978, 'Wave Propagation and Scattering in Random Media', Academic Press, N. Y.
Iye M., Nishihara E., Hayano Y., Okada T., Takato N., 1992, Pub. Astron. Soc. Pac., **104**, 760.
Iye M., Noguchi T., Torti Y., Mikami Y., Ando H., 1991, Pub. Astron. Soc. Pac., **103**, 712.
Jaynes E. T., 1982, Proc. IEEE, **70**, 939.
Jeffery C. S., 1996. in 'Hydrogen-deficient stars', Eds. C. S. Jeffery, U. Heber, ASP Conf Ser. **96**, 152.
Jennison R. C., 1958, Mon. Not. R. Astron. Soc., **118**, 276.
Jennison R. C., Das Gupta M. K., 1953, Nature, **172**, 996.
Jerram P., Pool P., Bell R., Burt D., Bowring S., Spencer S., Hazelwood M., Moody L., Carlett N., Heyes P., 2001, Marconi Appl. Techn.
Johnson H. L., 1966, Ann. Rev. Astron. Astrophys., **4**, 201.
Johnson H. L., Morgan W. W., 1953, Astrophys. J., **117**, 313.
Jones R. C., 1941, J. Opt. Soc. A., **31**, 488.
Jones R., Wykes C., 1983, Holographic & Speckle Interferometry, Cambridge Univ. Press, Cambridge.
Julian H. K., 1999, 'Active Galactic Nuclei' Princeton University Press.
Kallistratova M. A., Timanovskiy D. F., 1971, Izv. Akad. Nauk. S S S R., Atmos.

Ocean Phys., **7**, 46.
Karachentsev I. D., Kashibadze O. G., 2006, Astrophysics **49**, 3.
Karttunen H., Kröger P., Oja H., Poutanen M., Donner K. J., 2000, Fundamental Astronomy, Springer.
Karovska M., Nisenson P., 1992, Proc. 'High Resolution Imaging Interferometry', Eds., J. M. Beckers & F. Merkle, 141.
Karovska M., Nisenson P., Noyes R., 1986, Astrophys. J., **308**, 260.
Karovska M., Nisenson P., Papaliolios C., Boyle R. P., 1991, Astrophys. J., **374**, L51.
Keller C. U., Johannesson A., 1995, Astron. Astrophys. Suppl. Ser., **110**, 565.
Keller C. U., Von der Lühe O., 1992, Proc. 'High Resolution Imaging Interferometry', Eds., J. M. Beckers & F. Merkle, 453.
Kilkenny D., Whittet D. C. B., 1984. Mon. Not. R. Astron. Soc., **208**, 25.
Klein M. V., T. E. Furtak, 1986, Optics, John Wiley & Sons, N. Y.
Klückers V. A., Edmunds G., Morris R. H., Wooder N., 1997, Mon. Not. R. Astron. Soc., **284**, 711.
Knapp G. R., Morris M., 1985, Astrophys. J., **292**, 640.
Knox K.T., Thomson, B.J., 1974 Astrophys. J **193**, L45.
Kocinsli J., 2002, Int. J. Theoretical Phys., **41**, No. 2.
Kolmogorov A., 1941a, in 'Turbulence', Eds., S. K. Friedlander & L. Topper, 1961, Wiley-Interscience, N. Y., 151.
Kolmogorov A., 1941b, in 'Turbulence', Eds., S. K. Friedlander & L. Topper, 1961, Wiley-Interscience, N. Y., 156.
Kolmogorov A., 1941c, in 'Turbulence', Eds., S. K. Friedlander & L. Topper, 1961, Wiley-Interscience, N. Y., 159.
Kopal Z., 1959, 'Close Binary Systems', Vol. 5, The International Astrophys. Series, Chapman & Hall Ltd.
Korff D., 1973, J. Opt. Soc. Am., **63**, 971.
Kormendy J., Richstone D., 1995, Ann. Rev. Astron. Astrophys., **33**, 581.
Koutchmy S. et al., 1994, Astron. Astrophys., **281**, 249.
Koutchmy S., Zirker J. B., Steinolfson R. S., Zhugzda J. D., 1991, in Solar Interior and Atmosphere, Eds. A. N. Cox, W. C. Livingston, & M. S. Matthews, 1044.
Krasinsky G. A., Pitjeva E. V., Vasilyev M. V., Yagudina E. I., 2002, Icarus, **158**, 98.
Krolik J. H., 1999, Active Galactic Nuclei, Prin'ceton University Press.
Kunde V. G., 2004, Science **305**, 1582.
Kuwamura S., Baba N., Miura N., Noguchi M., Norimoto Y., Isobe S, 1992, Proc. 'High Resolution Imaging Interferometry', Eds., J. M. Beckers & F. Merkle, 461.
Kwok S., 1993, Ann. Rev. Astron. Astrophys., **31**, 63.
Kwok S., 2000, The Origin and Evolution of Planetary Nebulae, Cambridge University Press, Cambridge.
Labeyrie A., 1970, Astron. Astrophys., **6**, 85.
Labeyrie A., 1974, Nouv. Rev. Optique, **5**, 141.
Labeyrie A., 1975, Astrophys. J., **196**, L71.

Labeyrie A., 1985, 15th. Advanced Course, Swiss Society of Astrophys. and Astron., Eds. A. Benz, M. Huber & M. Mayor, 170.
Labeyrie A., 1995, Astron. Astrophys., **298**, 544.
Labeyrie A., 2000, Private communication.
Labeyrie A., 2005, '21st Century Astrophysics', Eds. S. K. Saha, & V. K. Rastogi, Anita Publications, New Delhi, 228.
Labeyrie A., Koechlin L., Bonneau D., Blazit A., Foy R., 1977, Astrophys. J., **218**, L75.
Lai O., Rouan D., Rigaut F., Arsenault R., Gendron E., 1998, Astron. Astrophys., **334**, 783.
Lamers Henny J. G. L. M., Cassinelli J. P., 1999, Introduction to Stellar Winds, Cambridge.
Lallemand A., 1936, C. R. Acad. Sci. Paris, **203**, 243.
Lallemand A., Duchesne M., 1951, C. R. Acad. Sci. Paris, **233**, 305.
Lancelot J. P., 2006, Private communication.
Land E. H., 1951, J. Opt. Soc. A., **41**, 957.
Lane R. G., Bates R. H. T., 1987, J. Opt. Soc. Am. A, **4**, 180.
Lang N. D., Kohn W., 1971, Phys. Rev. B, **3**, 1215.
Lawrence R. S., Ochs G. R., Clifford S. F., 1970, J. Opt. Soc. Am., **60**, 826.
Ledoux C., Théodore B., Petitjean P., Bremer M. N., Lewis G. F., Ibata R. A., Irwin M. J., Totten E. J., 1998, Astron. Astrophys., **339**, L77.
Lee J., Bigelow B., Walker D., Doel A., Bingham R., 2000, Pub. Astron. Soc. Pac, **112**, 97.
Leendertz J. A., 1970, J. Phys. E: Sci. Instru., **3**, 214.
Leinert C., Richichi A., Haas M., 1997, Astron. Astrophys., **318**, 472.
Liang J., Williams D. R., Miller D. T., 1997, J. Opt. Soc. Am. A, **14**, 2884.
Liu Y. C., Lohmann A. W., 1973, Opt. Comm., **8**, 372.
Lloyd-Hart M., 2000, Pub. Astron. Soc. Pac, **112**, 264.
Locher G. L., 1932, Phys. Rev., **42**, 525.
Lohmann A.W. Weigelt G P, Wirnitzer B, 1983, Appl. Opt., **22**, 4028.
Lopez B., 1991, 'Last Mission at La Silla, April 19 – May 8, on the Measure of the Wave-front Evolution Velocity', E S O Internal Report.
Love G., Andrews N, Birch P. et al., 1995, Appl. Opt., **34**, 6058.
Love G. D., Gourlay J., 1996, Opt. Lett., **21**, 1496.
Lucy L., 1974, Astron. J., **79**, 745.
Lynds C., Worden S., Harvey J., 1976, Astrophys. J, **207**, 174.
MacMahon P. A., 1909, Mon. Not. R. Astron. Soc., **69**, 126.
Magain P., Courbin F., Sohy S., 1998, Astrophys. J., **494**, 472.
Mahajan V. N., 1998, Optical Imaging and aberrations, Part II, SPIE Press, Washington, USA.
Mahajan V. N., 2000, J. Opt. Soc. Am, **17**, 2216.
Manley B., Guest A. J., Holmshaw R., 1969, Advances in Electronics and Electron Physics, Academic Press, London, **28**A, 471.
Mariotti J. -M., 1988, Proc. NATO-ASI workshop, Eds., D. M. Alloin & J. -M. Mariotti, 3.
Márquez I., Petitjean P., Théodore B., Bremer M., Monnet G., Beuzit J., 2001,

Astron. Astrophys., **371**, 97.

Marscher, A.P., et al., 2002, Nature, **417**, 625.

Marshall H., Miller B., Davis D., Perlman E., Wise M., Canizares C., Harris D., 2002, Astrophys. J., **564**, 683.

Masciadri E., Vernin J., Bougeault P., 1999, Astron. Astrophys. Suppl., **137**, 203.

Mason B. D., 1995, Pub. Astron. Soc. Pac., **107**, 299.

Mason B. D., 1996, Astron. J., **112**, 2260.

Mason B. D., Martin C., Hartkopf W. I., Barry D. J., Germain M. E., Douglass G. G., Worley C. H., Wycoff G. L., Brummelaar t. T., Franz O. G., 1999, Astron. J., **117**, 1890.

Maxted P. F. L., Napiwotzki R., Dobbie P. D., Burleigh M. R., 2006, Nature, **442**, 543.

McAlister H. A., 1985, Ann. Rev. Astron. Astrophys., **23**, 59.

McAlister H. A., Mason B. D., Hartkopf W. I., Shara M. M., 1993, Astron. J., **106**, 1639.

McCaughrean M. J., O'dell C. R., 1996, Astron. J., **111**, 1977.

Meixner M., 2000, astro-ph/0002373.

Mendel L., Wolf E., 1995, 'Optical Coherence and Quantum Optics', Cambridge University Press, Cambridge.

Men'shchikov A., Henning T., 1997, Astron. Astrophys., **318**, 879.

Monet D. G., 1988, Ann. Rev. Astron. Astrophys., **26**, 413.

Monnier J., Tuthill P., Lopez B., Cruzalébes P., Danchi W., Haniff C., 1999, Astrophys. J, **512**, 351.

Morel S., Saha S. K., 2005, '21st Century Astrophysics', Eds. S. K. Saha, & V. K. Rastogi, Anita Publications, New Delhi, 237.

Mouillet D., Larwood J., Papaloizou J., Lagrange A., 1997, Mon. Not. R. Astron. Soc., **292**, 896.

Nakajima T., Golimowski D., 1995, Astron. J., **109**, 1181.

Nakajima T., Kulkarni S. R., Gorham P. W., Ghez A. M., Neugebauer G., Oke J. B., Prince T. A., Readhead A. C. S., 1989, Astron. J., **97**, 1510.

Nather R. E., Evans D. S., 1970, Astron. J., **75**, 575.

Navier C. L. M. H., 1823, Mém. Acad. Roy. Sci., **6**, 389.

Nelkin M., 2000, Am. J. Phys., **68**, 310.

Nisenson P., 1988, Proc. NATO-ASI workshop, Eds., D. M. Alloin & J. -M. Mariotti, 157.

Nisenson P., 1992, Proc. 'High Resolution Imaging Interferometry', Eds., J. M. Beckers & F. Merkle, 299.

Nisenson P., Papaliolios C., 1999, Astrophys. J., **518**, L29.

Nisenson P., Papaliolios C., Karovska M., Noyes R., 1987, Astrophys. J, **320**, L15.

Nisenson P., Standley C., Gay D., 1990, Proc. 'HST Image Processing', Baltimore, Md.

Noll R. J., 1976, J. Opt. Soc. Am., **66**, 207.

Northcott M. J., Ayers G. R., Dainty J. C., 1988, J. Opt. Soc. Am. A, **5**, 986.

Nota A., Leitherer C., Clampin M., Greenfield P., Golimowski D. A., 1992, Astron. J., **398**, 621.

Obukhov A. M., 1941, Dokl. Akad. Nauk. SSSR., **32**, 22.

Osterbart R., Balega Y. Y., Weigelt G., Langer N., 1996, Proc. 'Planetary Nabulae', Eds., H. J. Habing & G. L. M. Lamers, 362.

Osterbart R., Langer N., Weigelt G., 1997, Astron. Astrophys., **325**, 609.

Osterbrock D. E., 1989, Astrophysics of gaseous nebulae and active galactic nuclei, University Science Books.

Papaliolios C., Mertz L., 1982, SPIE, **331**, 360.

Papoulis A., 1968, Systems and Transforms with Applications in Optics, McGraw-Hill, N. Y.

Parenti R., Sasiela R. J., 1994, J. Opt. Soc. Am. A., **11**, 288.

Pasachoff J. M., 2006, Black hole (html). MSN Encarta.

Paxman R., Schulz T., Fienup J., 1992, J. Opt. Soc. Am., **9**, 1072.

Peacock T., Verhoeve P., Rando N., van Dordrecht A., Taylor B. G., Erd C., Perryman M. A. C., Venn R., Howlett J., Goldie D. J., Lumley J., Wallis M., 1996, Nature, **381**, 135.

Pedrotti F. L., Pedrotti L. S., 1987, Introduction to Optics, Prentice Hall Inc., New Jersey.

Penzias A. A., Wilson R. W., 1965, Astrophys. J. **142**, 419.

Perryman M. A. C. et al., 1995, The Hipparcos and Tycho Catalogues, Noordwijk, ESA.

Perryman M. A. C., Foden C. L., Peacock A., 1993, Nucl. Instr. Meth. Phys. Res. A., **325**, 319.

Peter H., Gudiksen B. V., Nordlund A., 2006, Astrophys. J., **638**, 1086.

Peterson B. M., 1993, Proc. Astron. Soc. Pacific., **105**, 247.

Petr M. G., Du Foresto V., Beckwith S. V. W., Richichi A., McCaughrean M. J., 1998, Astrophys. J., **500**, 825.

Petrov R., Cuevas S. 1991, Proc. 'High Resolution Imaging Interferometry', Eds., J. M. Beckers & F. Merkle, 413.

Pickering E. C., 1910, Harvard Coll. Obs. Circ., **155**, 1.

Pirola V., 1973, Astron. J., **27**, 382.

Planck M., 1901, Ann. d. Physik, **4**, 553.

Pogson N. R., 1857, Mon. Not. R. Astron. Soc., **17**, 12.

Pottasch S. R., 1984, Planetary Nebulae, D. Reidel, Dordrecht.

Poynting J. H., 1883, Phil. Trans., **174**, 343.

Pratt T., 1947, J. Sci. Instr., **24**, 312.

Prialnik D., 2001. 'Novae', Encyclopaedia Astron. Astrophys., 1846.

Prieur J., Oblak E., Lampens P., Kurpinska-Winiarska M., Aristidi E., Koechlin L., Ruymaekers G., 2001, Astron. Astrophys., **367**, 865.

Priest E. R., 1982, Solar Magneto-hydrodynamics, D. Reidel publishing Co. Holland.

Primmerman C. A., Murphy D. V., Page D. A., Zollars B. G., Barclays H. T., 1991, Nature, 353, 141.

Pustylnik I., 1995, Baltic Astron. **4**, 64.

Ragazzoni R., 1996, J. Mod. Opt. **43**, 289.

Ragazzoni R., Marchetti E., Valente G., 2000, Nature, **403**, 54.

Racine R., 1984, in 'Very Large Telescopes, their Instrumentation and Programs',

Eds., M. -H. Ulrich & Kjär, 235.

Ragland S. et al., 2006, Astrophys. J., **652**, 650.

Puetter R., and A. Yahil, 1999, astro-ph/9901063.

Racine R., Salmon D., Cowley D., Sovka J., 1991, Pub. Astron. Soc. Pac., **103**, 1020.

Rando N., Peacock T., Favata, Perryman M. A. C., 2000, Exp. Astron., **10**, 499.

Rao N. K., et al. 2004, Asian J. Phys., **13**, 367.

Rees W. G., 1990, Physical Principles of Remote Sensing, Cambridge University Press.

Rees M. J., 2002, Lighthouses of the Universe: The Most Luminous Celestial Objects and Their Use for Cosmology, Proceedings of the MPA/ESO/, 345.

Richardson W. H., 1972, J. Opt. Soc. Am., **62**, 55.

Richichi A., 1988, Proc. NATO-ASI workshop, Eds., D. M. Alloin & J. -M. Mariotti, 415.

Rigaut F., Salmon D., Arsenault R., Thomas J., Lai O., Rouan D., Véran J. P., Gigan P., Crampton D., Fletcher J. M., Stilburn J., Boyer C., Jagourel P., 1998, Pub. Astron. Soc. Pac., **110**, 152.

Robbe S., Sorrentei B., Cassaing F., Rabbia Y., Rousset G., 1997, Astron. Astrophys. Suppl. **125**, 367.

Robertson J. G., Bedding T. R., Aerts C., Waelkens C., Marson R. G., Barton J. R., 1999, Mon. Not. R. Astron. Soc., **302**, 245.

Robinson C. R., Baliunas S. L., Bopp B. W., Dempsey R. C., 1984, Bull. Am. Astron. Soc., **20**, 954.

Roddier C., Roddier F., 1983, Astrophys. J, **270**, L23.

Roddier C., Roddier F., 1988, Proc. NATO-ASI workshop, Eds., D. M. Alloin & J. -M. Mariotti, 221.

Roddier C., Roddier F., Northcott M. J., Graves J. E., Jim K., 1996, Astrophys. J., **463**, 326.

Roddier F., 1981, Progress in optics, **XIX**, 281.

Roddier F., 1994, SPIE, **1487**, 123.

Roddier F., 1999, 'Adaptive Optics in Astronomy', Ed., F. Roddier, Cambridge Univ. Press.

Roddier F. J., Graves J. E., McKenna D., Northcott M. J., 1991, SPIE, **1524**, 248.

Roddier F., Roddier C., Graves J. E., Northcott M. J., 1995, Astrophys. J., **443**, 249.

Roddier F., Roddier C., Roddier N., 1988, SPIE, **976**, 203.

Roggemann M. C., Welsh B. M., Fugate R. Q., 1997, Rev. Mod. Phys., **69**, 437.

Rouan D., Field D., Lemaire J. -L., Lai O., de Foresto G. P., Falgarone E., Deltorn J. -M., 1997, Mon. Not. R. Astron. Soc., **284**, 395.

Rouan D., Rigaut F., Alloin D., Doyon R., Lai O., Crampton D., Gendron E., Arsenault R., 1998, astro-ph/9807053.

Rouaux E., Richard J. -C., Piaget C., 1985, Advances in Electronics and Electron Physics, Academic Press, London, **64**A, 71.

Rousset G., 1999, 'Adaptive Optics in Astronomy', Ed. F. Roddier, Cambridge

Univ. Press, 91.
Rutherford E., Geiger H., 1908, Proc. Roy. Soc. London A, **81**, 141.
Ruze J., 1966, Proc. IEEE, **54**, 633.
Ryan S. G., Wood P. G., 1995, Pub. ASA., **12**, 89.
Ryden B., 2003, Lecture notes.
Saha S. K., 1999a, Ind. J. Phys., **73B**, 552.
Saha S. K., 1999b, Bull. Astron. Soc. Ind., **27**, 443.
Saha S. K., 2002, Rev. Mod. Phys., **74**, 551.
Saha S. K., Chinnappan V., 1999, Bull. Astron. Soc. Ind., **27**, 327.
Saha S. K., Chinnappan V., 2002, Bull. Astron. Soc. Ind., **30**, 819.
Saha et al. 2007, Obtaining binary star orbits from speckle and other interferometric data (in preparation).
Saha S. K., Jayarajan A. P., Rangarajan K. E., Chatterjee S., 1988, Proc. 'High Resolution Imaging Interferometry', Ed. F. Merkle, 661.
Saha S. K., Jayarajan A. P., Sudheendra G., Umesh Chandra A., 1997a, Bull. Astron. Soc. Ind., **25**, 379.
Saha S. K., Maitra D., 2001, Ind. J. Phys., **75B**, 391.
Saha S. K., Nagabhushana B. S., Ananth A. V., Venkatakrishnan P., 1997b, Kod. Obs. Bull., **13**, 91.
Saha S. K., Rajamohan R., Vivekananda Rao P., Som Sunder G., Swaminathan R., Lokanadham B., 1997c, Bull. Astron. Soc. Ind., **25**, 563.
Saha S. K., Sridharan R., Sankarasubramanian K., 1999b, 'Speckle image reconstruction of binary stars', Presented at the ASI conference.
Saha S. K., Sudheendra G., Umesh Chandra A., Chinnappan V., 1999a, Exp. Astr., **9**, 39.
Saha S. K., Venkatakrishnan P., 1997, Bull. Astron. Soc. Ind., **25**, 329.
Saha S. K., Venkatakrishnan P., Jayarajan A. P., Jayavel N., 1987, Curr. Sci., **56**, 985.
Saha S. K., Yeswanth L., 2004, Asian J. Phys., **13**, 227.
Sahai R., Trauger J.T., 1998, Astron. J., **116**, 1357.
Sahu D. K., Anupama G. C., Srividya S., Munner S., 2006, Mon. Not. R. Astron. Soc., **372**, 1315.
Sams B. J., Schuster K., Brandl B., 1996, Astrophys. J., **459**, 491.
Sandage A., Gustav A. T., 1968, Astrophys. J., **151**, 531.
Schertl D., Balega Y. Y., Preibisch Th., Weigelt G., 2003, Astron. Astrophys., **402**, 267.
Schertl D., Hofmann K. -H., Seggewiss W., Weigelt G., 1996, Astron. Astrophys. **302**, 327.
Schmidt M., 1963, Nature **197**, 1040.
Schmidt M. R., Zacs L., Mikolajewska J., Hinkle K., 2006, Astron. Astrophys., **446**, 603.
Schöller M., Brandner W., Lehmann T., Weigelt G., Zinnecker H., 1996, Astron. astrophys., **315**, 445.
Seldin J., Paxman R., 1994, SPIE., **2302**, 268.
Seldin J., Paxman R., Keller C., 1996, SPIE., **2804**, 166.
Serbowski K., 1947, Planets, Stars, and Nabulae Studied with Photopolarimetry,

Ed., T. Gehrels, Tuscon, University of Arizona Press, 135.
Shack R. V., Hopkins G. W., 1977, SPIE, 'Clever Optics', **126**, 139.
Shakura N. I., Sunyaev R. A., 1973, Astron. Astrophys., **24**, 337.
Shannon C. J., 1949, Proc. IRE, **37**, 10.
Shelton J. C., Baliunas S. L., 1993, SPIE., **1920**, 371.
Shields G. A., 1999, astro-ph/9903401.
Sicardy B., Roddier F., Roddier C., Perozzi E., Graves J. E., Guyon O., Northcott M. J., 1999, Nature, **400**, 731.
Siegmund O. H. W., Clossier S., Thornton J., Lemen J., Harper R., Mason I. M., Culhane J. L., 1983, IEEE Trans. Nucl. Sci., NS-30(1), 503.
Simon M., Close L. M., Beck T. L., 1999, Astron. J., **117**, 1375.
Sinclair A. G., Kasevich M. A., 1997, Rev. Sci. Instr., **68**, 1657.
Smart W. M., 1947, 'Text book on Spherical Astronomy', Cambridge University Press.
Smith G. L., 1997, An Introduction to Classical Electromagnetic radiation, Cambridge University Press, UK.
Sobottka S., Williams M., 1988, IEEE Trans. Nucl. Sci., **35**, 348.
Soker N., 1998, Astrophys. J., **468**, 774.
Stassun K. G., Mathieu R. D., Valenti J. A., 2006, Nature, **440**, 311.
Stefanik R. P., Latham D. W., 1985, in Stellar Radial Velocities, Eds. A. G. D. Philip, & D. W. Latham, L. Davis Press, 213.
Steward E. G., 1983, Fourier Optics and Introduction, John Wiley & Sons, NY.
Stokes G. G., 1845, Trans. Camb. Phil. Soc., **8**, 287.
Strömgren B., 1956, Vistas in Astron., **2**, 1336.
Struve O., 1950, Stellar Evolution, Princeton University Press, Princeton, N. J.
Suomi V. E., Limaye S. S., Johnson D. R., 1991, Science **251**, 929.
Tallon M., Foy R., 1990, Astron. Astrophys., **235**, 549.
Tatarski V. I., 1961, Wave Propagation in a Turbulent Medium, Dover, NY.
Tatarski V. I., 1993, J. Opt. Soc. Am. A, **56**, 1380.
Taylor G. L., 1921, in 'Turbulence', Eds., S. K. Friedlander & L. Topper, 1961, Wiley-Interscience, New York, 1.
Taylor G. A., 1994, J. Opt. Soc. Am. A., **11**, 358.
Taylor J. H., 1966, Nature, **210**, 1105.
Tej A., Chandrasekhar T., Ashok M. N., Ragland S., Richichi A., Stecklum B., 1999, A J, **117**, 1857.
Thiebaut E., Abe L., Blazit A., Dubois J. -P., Foy R., Tallon M., Vakili F., 2003, SPIE, **4841**, 1527.
Thompson L. A., Gardner C. S., 1988, Nature, **328**, 229.
Timothy J. G., 1983, Publ. Astron. Soc. Pac., **95**, 810.
Torres G., Stefanik R. P., Latham D. W., 1997, Astrophys. J., **485**, 167.
Tremsin A. S., Pearson J. F., Lees J. E., Fraser G. W., 1996, Nucl. Instr. Meth. Phys. Res. A., **368**, 719.
Troxel S. E., Welsh B. M., Roggemann M. C., 1994, J. Opt. Soc. Am A, **11**, 2100.
Tsvang L. R., 1969, Radio Sci., **4**, 1175.
Tuthill P. G., Haniff C. A., Baldwin J. E., 1997, Mon. Not. R. Astron. Soc., **285**, 529.

Tuthill P. G., Monnier J. D., Danchi W. C., 1999, Nature, **398**, 487.
Tuthill P. G., Monnier J. D., Danchi W. C., 2001, Nature, **409**, 1012.
Tuthill P. G., Monnier J. D., Danchi W. C., and Wishnow, 2000, Pub. Astron. Soc. Pac., **116**, 2536.
Tyson R. K., 1991, Principles of Adaptive Optics, Academic Press.
Tyson R. K., 2000, 'Introduction' in Adaptive optics engineering handbook, Ed. R. K. Tyson, Dekkar, NY, 1.
Uchino K., Cross L. E., Nomura S., 1980, J. Mat. Sci., **15**, 2643.
Ulrich M. -H., 1981, Astron. Astrophys., **103**, L1.
Valley G. C. 1980, Appl. Opt. **19**, 574.
van Altena W. F., 1974, Astron. J., **86**, 217.
van Cittert P. H., 1934, Physica, **1**, 201.
Van de Hulst H. C., 1957, Light Scattering by Small Particles, John Wiley & Sons, N.Y.
van den Ancker M. E., de Winter D., Tjin A Djie H. R. E., 1997, Astron. Astrophys., **330**, 145.
van Leeuwen F., Hansen Ruiz C. S., 1997, in Hipparcos Venice, Ed. B. Battrick, 689.
Venkatakrishnan P., Saha S. K., Shevgaonkar R. K., 1989, Proc. 'Image Processing in Astronomy, Ed. T. Velusamy, 57.
Vermeulen R. C., Ogle H. D., Tran H. D., Browne I. W. A., Cohen M. H., Readhead A. C. S., Taylor G. B., Goodrich R. W., 1995, Astrophys. J., **452**, L5.
Von der Lühe O., 1984, J. Opt. Soc. Am. A, **1**, 510.
Von der Lühe O., 1989, Proc. 'High Spatial Resolution Solar Observation', Ed., O. Von der Lühe, Sunspot, New Mexico.
Von der Lühe O., Dunn R. B., 1987, Astron. Astrophys., **177**, 265.
Von der Lühe O., Zirker J. B., 1988, Proc. 'High Resolution Imaging Interferometry', Eds., F. Merkle, 77.
Voss R., Tauris T. M., 2003, Mon. Not. R. Astron. Soc., **342**, 1169.
Wehinger P. A., 2002, Private communication.
Weigelt G.P., 1977, Opt Communication, **21**, 55.
Weigelt G., Baier G., 1985, Astron. Astrophys. **150**, L18.
Weigelt G., Balega Y., Blöcker T., Fleischer A. J., Osterbart R., Winters J. M., 1998, Astron. Astrophys., **333**, L51.
Weigelt G.P., Balega Y. Y., Hofmann K. -H., Scholz M., 1996, Astron. Astrophys., **316**, L21.
Weigelt G., Balega Y., Preibisch T., Schertl D., Schöller M., Zinnecker H., 1999, astro-ph/9906233.
Weigelt G.P., Balega Y. Y., Preibisch T., Schertl D., Smith M. D., 2002, Astron. Astrophys., **381**, 905.
Weil M., Hernquist L., 1996, Astrphys. J., **460**, 101.
Weinberger A., Neugebauer, G., Matthews K., 1999, Astron. J., **117**, 2748.
Weitzel N., Haas M., Leinert Ch., 1992, Proc. 'High Resolution Imaging Interferometry', Eds., J. M. Beckers & F. Merkle, 511.
Wilken V., de Boer C. R., Denker C., Kneer F., 1997, Astron. Astrophys., **325**,

819.
Wilson O. C., Bappu M. K. V. 1957, Astrophy. J., **125**, 661.
Wilson R. W., Dhillon V. S., Haniff C. A., 1997, Mon. Not. R. Astron. Soc., **291**, 819.
Wittkowski M., Balega Y., Beckert T., Duschi W. J., Hofmann K. -H., Weigelt G., 1998b, Astron. Astrophys. **329**, L45.
Wittkowski M., Langer N., Weigelt G., 1998, Astron. Astrophys., **340**, L39.
Wizinovitch P. L., Nelson J. E., Mast T. S., Glecker A. D., 1994, SPIE., **2201**, 22.
Wolf E., 1954, Proc. Roy. Soc. A, **225**, 96.
Wolf E., 1955, Proc. Roy. Soc. A, **230**, 246.
Worden S. P., Lynds C. R., Harvey J. W., 1976, J. Opt. Soc. Am., **66**, 1243.
Wyngaard J. C., Izumi Y., Collins S. A., 1971, J. Opt. Soc. Am., **60**, 1495.
Young A. T., 1967, Astron. J., **72**, 747.
Young A. T., 1974, Astrophys. J., **189**, 587.
Young A. T., 1970, Appl. Opt., **9**, 1874.
Young A. T. Irvine W. M., 1967, Astron. J, **72**, 945.
Young T., 1802, Phil. Trans. Roy. Soc., London, XCII, **12**, 387.
Zago L., 1995, http://www.eso.org/gen-fac/pubs/astclim/lz-thesis/node4-html.
Zeidler P., Appenzeller I., Hofmann K. -H., Mauder W., Wagner S., Weigelt G., 1992, Proc. 'High Resolution Imaging Interferometry', Eds., J. M. Beckers & F. Merkle, 67.
Zernike F., 1934, Physica, **1**, 689.
Zernike F., 1938, Physica, **5**, 785.
Zienkiewicz O. C., 1967, 'The Finite Element Methods in Structural and Continuum Mechanics', McGrawhill Publication.
Zworykin V. K., 1936, L'Onde Électrique, **15**, 265.

Index